MW00845028

Silicon Photonics Design

From design and simulation through to fabrication and testing, this hands-on introduction to silicon photonics engineering equips students with everything they need to begin creating foundry-ready designs.

Acquire practical understanding and experience
In-depth discussion of real-world issues and fabrication challenges ensures that students are fully equipped for future careers in industry, designing complex integrated systems-on-chip.

Cut design time and development cost
Step-by-step tutorials, straightforward examples, and illustrative source code fragments guide students through every aspect of the design process, and provide a practical framework for developing and refining key skills.

Industry-ready expertise
Providing both guidance on how a process design kit (PDK) is constructed and how to best utilize the types of PDKs currently available, this text will enable students to understand the design process for building even very complex photonic systems-on-chip.
Accompanied by additional online resources to support students, this is the perfect learning package for senior undergraduate and graduate students studying silicon photonics design, and academic and industrial researchers involved in the development and manufacture of new silicon photonics systems.

Lukas Chrostowski is Associate Professor of Electrical and Computer Engineering at the University of British Columbia. He is the Program Director of the NSERC CREATE Silicon Electronic-Photonic Integrated Circuits (Si-EPIC) training program, has been teaching silicon photonics courses and workshops since 2008, and has been awarded the Killiam Teaching Prize (2014).

Michael Hochberg is Director of Architecture and Strategy for Coriant Advanced Technology Group, based in Manhattan, NY, where he holds a visiting appointment at Columbia University. He has held faculty positions at the University of Washington, University of Delaware, and National University of Singapore, and was Director of the OpSIS foundry-access service. He has co-founded several startups, including Simulant and Luxtera and received a Presidential Early Career Award in Science and Engineering (2009).

"Photonics technology has created some of the most stunning achievements in human history, but the challenges of implementing even simple systems have made it the domain of a few specialized laboratories.

Silicon photonics enables the design of photonic systems in a much more streamlined manner, and the resulting designs can be fabricated by highly evolved silicon manufacturing facilities.

This book provides a complete guide, from physical principles of device operation through fabrication and testing, using real system examples. It gives non-specialists access to what may be the most important next step in information technology."

Carver Mead, California Institute of Technology

"The book covers everything one would need to design, lay out, simulate, and fabricate an actual silicon chip for processing, detecting, and modulating light signals. The book's focus on the practical side of chip implementation means that it is quite different, and frankly more useful, for chip designers than other photonics books. I highly recommend *Silicon Photonics Design* for both experienced designers and those wishing to get up-to-speed quickly in the nascent field of silicon photonics chip design."

R. Jacob Baker, University of Nevada

"*Silicon Photonics Design* is an essential text for anyone with an interest in the application of silicon-based optical circuits, either in a commercial or academic research environment. The authors have captured all of the essential elements of silicon photonics while ensuring the text remains accessible. The inclusion of so many worked examples mixed with detailed fundamental physical descriptions is an approach that must be applauded."

A. P. Knights, McMaster University

Silicon Photonics Design

LUKAS CHROSTOWSKI
University of British Columbia

MICHAEL HOCHBERG
Coriant Advanced Technology Group

CAMBRIDGE
UNIVERSITY PRESS

University Printing House, Cambridge CB2 8BS, United Kingdom

Cambridge University Press is part of the University of Cambridge.

It furthers the University's mission by disseminating knowledge in the pursuit of education, learning and research at the highest international levels of excellence.

www.cambridge.org
Information on this title: www.cambridge.org/9781107085459

First published 2015
Reprinted 2015

Printed in the United Kingdom by Bell and Bain Ltd, Glasgow

A catalogue record for this publication is available from the British Library

Library of Congress Cataloguing in Publication data
Chrostowski, Lukas, author.
Silicon photonics design / Lukas Chrostowski, University of British Columbia, Michael Hochberg, University of Delaware.
pages cm
Includes bibliographical references and index.
ISBN 978-1-107-08545-9 (Hardback)
1. Silicon–Optical properties. 2. Photonics. 3. Microwave integrated circuits–Design and construction. I. Hochberg, Michael E., author. II. Title.
TK7871.15.S55C47 2015
621.36–dc23 2014034057

ISBN 978-1-107-08545-9 Hardback

Additional resources for this publication at www.cambridge.org/chrostowski

Contents

Contributors

Arghavan Arjmand
Lumerical Solutions, Inc., Canada

Tom Baehr-Jones
University of Delaware, USA

Robert Boeck
University of British Columbia, Canada

Chris Cone
Mentor Graphics Corporation, USA

Dan Deptuck
CMC Microsystems, Canada

Ran Ding
University of Delaware, USA

Jonas Flueckiger
University of British Columbia, Canada

Samantha Grist
University of British Columbia, Canada

Li He
University of Minnesota, USA

Nicolas A. F. Jaeger
University of British Columbia, Canada

Odile Liboiron-Ladouceur
McGill University, Canada

Charlie Lin
University of Delaware, USA

Amy Liu
Lumerical Solutions, Inc., Canada

Yang Liu
University of Delaware, USA

Dylan McGuire
Lumerical Solutions, Inc., Canada

Kyle Murray
University of British Columbia, Canada

Ari Novack
National University of Singapore, Singapore

James Pond
Lumerical Solutions, Inc., Canada

Wei Shi
University of British Columbia, Canada
Université of Laval, Canada

Matt Streshinsky
National University of Singapore, Singapore

Miguel Ángel Guillén Torres
University of British Columbia, Canada

Xu Wang
University of British Columbia, Canada
Lumerical Solutions, Inc., Canada

Yun Wang
University of British Columbia, Canada

Han Yun
University of British Columbia, Canada

Preface

The academic literature on silicon photonics is sufficiently rich that one might legitimately ask whether another book in this field is needed. Certainly all of the basic physics of waveguides, modulators, lasers, and photodetectors is covered in great detail in a series of landmark texts, from Yariv and Yeh [1] to Sze and Ng [2] to Siegman [3] and Snyder and Love [4]. More specifically integrated photonics theory is covered comprehensively in texts by Hunsberger [5], Coldren *et al.* [6], Kaminow *et al.* [7], etc. Several excellent volumes have come out in recent years describing the state of the field in silicon photonics, and discussing design considerations for a variety of devices [8–15].

So what are we aiming to add to this body of literature? Our aim is not to replicate any of the existing texts' approach, but instead to provide a practical, examples-driven introduction to the practice of designing practical devices and systems. Our (admittedly ambitious) goal for this text is to do something similar to what Mead and Conway did with their landmark text on VLSI [16]: to treat the minimal possible level of device physics, and to focus primarily on the practical design considerations associated with using state-of-the-art silicon photonic foundry processes to build real, useful systems-on-chip.

In order to do this, we focus on a series of tutorials, using the tools that are in use in our own labs. That doesn't mean that these tools are perfect, or that they are necessarily the best tools for any given application: they are just what we have used. Wherever there are alternative approaches, we highlight them and provide some context for why we choose to do things in a certain way. This is obviously an area where errors of omission are very easy to make: we welcome feedback and input.

The vendors of the commercial software we use provide in-kind access for educational institutions. For example, Lumerical Solutions software is available via the Commitment to University Education (CUE) program [17], which provides access to students in undergraduate and graduate classes. Similarly, Mentor Graphics has a higher education program [18] that provides software for classroom instruction and university research. The software has been available at the silicon photonics instructional workshops we have offered.

We also provide a cursory literature review in each chapter, as well as some exercises.

Silicon photonics – training programs

This book was developed in the context of training programs in silicon photonics led by the authors, specifically in the NSERC Si-EPIC Program (Canada) [19] and the OpSIS Workshops (United States) [20].

There are and have been several training opportunities in silicon photonics around the world, including the following.

- CMC Microsystems – University of British Columbia Silicon Nanophotonics Course (Canada) [21, 22], 2007–
- OpSIS Workshops (United States), 2011–2014

  OpSIS offered five-day intensive training workshops that have trained over 100 researchers and students in the design of silicon photonic systems.
- ePIXfab Europractice (Europe) [23]
- JSPS International Schooling on Si Photonics (Japan) [24], 2011
- Silicon Photonics Summer School (St. Andrews, UK) [25], 2011
- Summer School on Silicon Photonics (Peking University, China) [26], 2011–
- NSERC Silicon Electronic Photonics Integrated Circuits (Si-EPIC) Program (Canada) [19], 2012–

  This program offers four annual workshop/courses each of which includes a design-fabrication-test cycle. The workshops are on the topics of: (1) Passive silicon photonics, (2) Active silicon photonics, (3) CMOS electronics for photonics and (4) Systems Integration and Packaging
- plat4M Summer School Silicon Photonics (Ghent University, Ghent, Belgium) [27], 2014

Hochberg's acknowledgements

I'd like to thank Lukas for all the work that he's put into this volume over the past few years. Lukas is a truly gifted educator, and I'm constantly impressed by his ability to communicate complex ideas to students, both in a classroom and in writing. The overwhelming majority of the work that went into this book was his, with help from a number of the students in both his and my groups.

I'd like to thank both my and Lukas' graduate students, and Dr. Tom Baehr-Jones. Individuals contributing to the book are acknowledged below.

I thank Gernot Pomrenke for his support of the OpSIS program over the past several years, and Mario Paniccia and Justin Rattner of Intel for their help in getting the program started. The people who have made OpSIS possible are too many to mention. In particular, Juan Rey, of Mentor Graphics, Klaus Engenhardt and Stan Kaveckis of Tektronix, Lukas Chrostowski of UBC, Andy Pomerene, Stewart Ocheltree and Steve Danziger of BAE Systems, Thierry Pinguet, Marek Tlalka and Chris Bergey of Luxtera and Andy Lim Eu-Jin, Jason Liow Tsung-Yang, and Patrick Lo Guo-Qiang of IME have all been immensely helpful. Lastly, I would like to thank Carver Mead for his time and for productive discussions.

Chrostowski's acknowledgements

I would like to thank Michael and Tom Baehr-Jones for their vision and pioneering efforts in silicon photonics over the past 15 years. In particular, their leadership at establishing a multi-project wafer foundry service for silicon photonics is greatly appreciated by me, my students, and colleagues throughout the world. I have enjoyed learning about silicon photonics design from both Michael and his group, and I appreciate Michael's willingness and efforts at educating silicon photonics designers and supporting the development of instructional workshops.

I am grateful for my colleague, Professor Nicolas Jaeger, with whom I have worked closely on both research and educational initiatives such as the silicon photonics workshops and the SiEPIC program. He has given me tremendous technical insight into guided-wave optics, microwave design, and high-speed testing, just to name a few topics.

I thank the numerous students and colleagues who contributed to this book, including those in Michael's group, students at UBC, and students across Canada and around the world with whom I have interacted via collaborations and at silicon photonics workshops. Numerous topics described in this book are a result of questions asked by participants at workshops and the interesting discussions that ensued. I also thank the readers of this book who have provided feedback over the past two years, particularly Robert Boeck and Megan Chrostowski. I thank colleagues for insightful discussions and collaborations that led to topics discussed in this book, including: James Pond, Dylan McGuire, Jackson Klein, Todd Kleckner, and Amy Liu at Lumerical Solutions; Chris Cone, John Ferguson, Angela Wong, and Kostas Adam at Mentor Graphics; Professors Shahriar Mirabbasi and Sudip Shekhar, and Han Yun at the University of British Columbia; Professors David Plant, Odile Liboiron-Ladouceur, and Lawrence Chen at McGill University; Professor Andrew Knights and Edgar Huante-Cerón at McMaster University; Professors Sophie Larochelle and Wei Shi at Laval University; Professor Dan Ratner and Dr. Richard Bojko at the University of Washington; Professor Jose Azana and Dr. Maurizio Burla at INRS; and Professors Joyce Poon and Mo Mojahedi, and Jan Nikas Caspers at the University of Toronto. I thank the foundries and services that have provided access to silicon photonic fabrication from which I have benefited, including CMC Microsystems, Imec, IME, OpSIS, BAE, and the University of Washington. I thank NSERC for funding our research, and in particular for funding the Silicon Electronic Photonics Integrated Circuits (Si-EPIC) CREATE research training Program.

Finally, I thank my wife, children and parents for their love and support.

Contributions

We acknowledge the direct contributions to the content in this book: Ari Novack – photodetector theory and experimental data (Chapter 7); Wei Shi – ring resonator model (Section 4.4, pn-junction model (Section 6.2) and ring modulator model (6.3);

Yun Wang and Li He – fibre grating couplers (Section 5.2); Dylan McGuire – model development for the modulator and detector (Section 6.2.4, 7.5); Amy Liu – scripts for the modulator (Section 6.2.4); Arghavan Arjmand – scripts for the detector (Section 7.5); Jonas Flueckiger – photonic circuit circuit simulations (Chapter 9) and automated testing (Section 12.2); Miguel Guillén Ángel Torres – directional coupler FDTD S-parameters (Section 9.4); Robert Boeck – fabrication corner analysis (Section 11.1.2) and directional couplers (Section 4.1); Dan Deptuck and Odile Liboiron-Ladouceur – design for test and check-list (Section 12.3); Yang Liu and Ran Ding – WDM transmitter (Section 13.1); Matt Streshinsky – Mach–Zehnder modulators; Kyle Murray – parasitic coupling (Section 4.1.7); Chris Cone – process design kit; Samantha Grist – SEM images; Han Yun – test setup and diagrams; and Nicolas Jaeger and Dan Deptuck – co-development of the CMC-UBC Silicon Nanophotonics Fabrication workshops and Si-EPIC workshops which formed the basis for this text.

References

[1] Amnon Yariv and Pochi Yeh. *Photonics: Optical Electronics in Modern Communications (The Oxford Series in Electrical and Computer Engineering).* Oxford University Press, Inc., 2006 (cit. on p. xv).

[2] S. M. Sze and K. K. Ng. *Physics of Semiconductor Devices.* Wiley-Interscience, 2006 (cit. on p. xv).

[3] A. E. Siegman. *Lasers University Science Books.* Mill Valley, CA, 1986 (cit. on p. xv).

[4] A. W. Snyder and J. D. Love. *Optical Waveguide Theory.* Vol. 190. Springer, 1983 (cit. on p. xv).

[5] R. G. Hunsperger. *Integrated Optics: Theory and Technology.* Advanced texts in physics. Springer, 2009. ISBN: 9780387897745 (cit. on p. xv).

[6] L. A. Coldren, S. W. Corzine, and M. L. Mashanovitch. *Diode Lasers and Photonic Integrated Circuits.* Wiley Series in Microwave and Optical Engineering. John Wiley & Sons, 2012. ISBN: 9781118148181 (cit. on p. xv).

[7] I. P. Kaminow, T. Li, and A. E. Willner. *Optical Fiber Telecommunications V A: Components and Subsystems.* Optics and Photonics. Academic Press, 2008. ISBN: 9780123741714 (cit. on p. xv).

[8] G. T. Reed and A. P. Knights. *Silicon Photonics: an Introduction.* John Wiley & Sons, 2004. ISBN: 9780470870341 (cit. on p. xv).

[9] L. Pavesi and D. J. Lockwood. *Silicon Photonics.* Vol. 1. Springer, 2004 (cit. on p. xv).

[10] G. T. Reed and A. P. Knights. *Silicon Photonics.* Wiley Online Library, 2008 (cit. on p. xv).

[11] Lorenzo Pavesi and Gérard Guillot. *Optical Interconnects: The Silicon Approach.* 978-3-540-28910-4. Springer Berlin / Heidelberg, 2006 (cit. on p. xv).

[12] D. J. Lockwood and L. Pavesi. *Silicon Photonics II: Components and Integration.* Topics in Applied Physics vol. 2. Springer, 2010. ISBN: 9783642105050 (cit. on p. xv).

[13] B. Jalali and S. Fathpour. *Silicon Photonics for Telecommunications and Biomedicine.* Taylor & Francis Group, 2011. ISBN: 9781439806371 (cit. on p. xv).

[14] M. J. Deen and P. K. Basu. *Silicon Photonics: Fundamentals and Devices.* Vol. 43. Wiley, 2012 (cit. on p. xv).

[15] Laurent Vivien and Lorenzo Pavesi. *Handbook of Silicon Photonics.* CRC Press, 2013 (cit. on p. xv).

[16] C. Mead and L. Conway. *Introduction to VLSI Systems.* Addison-Wesley series in computer science. Addison-Wesley, 1980. ISBN: 9780201043587 (cit. on p. xv).

[17] *Commitment to University Education – Lumerical Solutions.* [Accessed 2014/04/14]. URL: https://www.lumerical.com/company/initiatives/cue.html (cit. on p. xv).

[18] *Higher Education Program – Mentor Graphics.* [Accessed 2014/04/14]. URL: http://www.mentor.com/company/higher_ed/ (cit. on p. xv).

[19] *NSERC CREATE Silicon Electronic Photonic Integrated Circuits (Si-EPIC) program.* [Accessed 2014/04/14]. URL: http://www.siepic.ubc.ca (cit. on p. xvi).

[20] Tom Baehr-Jones, Ran Ding, Ali Ayazi, *et al.* "A 25 Gb/s Silicon Photonics Platform". *arXiv:1203.0767v1 (2012)* (cit. on p. xvi).

[21] *EECE 584 – Silicon Nanophotonics Fabrication – Microsystems and Nanotechnology Group (MiNa) UBC.* [Accessed 2014/04/14]. URL: http://www.mina.ubc.ca/eece584 (cit. on p. xvi).

[22] Lukas Chrostowski, Nicolas Rouger, Dan Deptuck, and Nicolas A. F. Jaeger. "Silicon Nanophotonics Fabrication: an Innovative Graduate Course". (invited). Doha, Qatar: (invited), 2010. DOI: 10.1109/ICTEL.2010.5478599 (cit. on p. xvi).

[23] Amit Khanna, Youssef Drissi, Pieter Dumon, *et al.* "ePIX-fab: the silicon photonics platform". *SPIE Microtechnologies.* International Society for Optics and Photonics. 2013, 87670H–87670H (cit. on p. xvi).

[24] *JSPS International Schooling on Si Photonics.* [Accessed 2014/04/14]. URL: https://sites.google.com/site/coretocore2011 (cit. on p. xvi).

[25] *Silicon Photonics Summer School.* [Accessed 2014/04/14]. URL: http://www.st-andrews.ac.uk/microphotonics/spschool/index.php (cit. on p. xvi).

[26] *2014 Peking silicon photonics technology and application Summer School.* [Accessed 2014/04/14]. URL: http://spm.pku.edu.cn/summerschool.html (cit. on p. xvi).

[27] *plat4m Summer School Silicon Photonics 2014.* [Accessed 2014/04/14]. URL: http://plat4m-fp7.eu/spsummerschool (cit. on p. xvi).

Part I

Introduction

1 Fabless silicon photonics

1.1 Introduction

We are on the cusp of revolutionary changes in communication and microsystems technology through the marriage of photonics and electronics on a single platform. By marrying large-scale photonic integration with large-scale electronic integration, wholly new types of systems-on-chip will emerge over the next few years.

Electronic-photonic circuits will play a ubiquitous role globally, impacting such areas as high-speed communications for mobile devices (smartphones, tablets), optical communications within computers and within data centres, sensor systems, and medical applications. In particular, we can expect the earliest impacts to emerge in telecommunications, data centers and high-performance computing, with the technology eventually migrating into higher-volume, shorter-reach consumer applications.

In the emerging field of electronics in the 1970s, Lynn Conway at Xerox PARC and Professor Carver Mead at Caltech developed an electronics design methodology, wrote a textbook, taught students how to design electronic integrated circuits, and had their designs fabricated by Intel and HP as multi-project wafers, where several different designs were shared in a single manufacturing run [1]. These efforts led to the foundation of an organization named MOSIS in 1981 that introduced cost sharing of fabrication runs with public access. The inexpensive design-build-test cycle enabled by MOSIS trained, and continues to train, thousands of designers who are responsible for the ubiquity of electronics we see today. MOSIS got started based on commercial processes that were already in production, and opened them up to the design community for prototyping and research purposes.

One of the keys to the long-term success of the microelectronics community, and in particular of the CMOS community, has been this type of access. By making these volume production processes publicly available for research and development at modest cost, anyone with a very modest level of funding is able to do cutting edge, creative work *in a process that can instantly go into large-scale production.* Training student engineers to use the production tools and processes, and then letting them loose to build cutting-edge circuits which can, with modest funding, be translated into fabless IC start-ups, has been the source of countless successful companies. It is hard to over-emphasize the difference between this and the situation in photonics (and most engineering fields), where getting from research into production involves huge barriers.

Silicon photonics is currently at the same early stage of expansion as electronics was in the 1970s, but with a major advantage for chip fabrication: existing silicon foundries

Figure 1.1 Photograph of 8" SOI wafer with various photonic components and circuits. With permission [10].

that produce highly controlled wafers (Figure 1.1) for microelectronics already exist. The microfabrication infrastructure to do silicon photonics already exists, in the microelectronics industry. Several companies are manufacturing silicon photonic chips, e.g., Luxtera's chips are already used in some high-performance computer clusters [2]. We are presently in an important transition in that academics, students, and industry worldwide now have access to active silicon photonics fabrication, e.g., via the multi-project wafer services offered by ePIXfab [3], IME [4–6], CMC Microsystems [9], and others. However, none of the processes that are currently available to the wider community are production-qualified; they are all prototyping and R&D processes, which support only very limited volumes. The inability to leverage pre-existing commercial processes was a significant impediment to the success of OpSIS, which recently shut down, since commercial users were rightly unwilling to rely on non-production-proven processes for product development. Funding from the research users and funders was not sufficient to keep the effort going, and would not allow the development of processes suitable for commercial use, which was the logical next step for the program.

We are fortunate that silicon allows us to perform all of the key optical functions at a reasonably competitive performance level, as shown in Figure 1.2, with the exception of a laser. There is a lot of recent and very interesting work going on related to growing quantum dots and germanium for lasers monolithically integrated in silicon [7]. And with the bonding technologies that have been inherited from the electronics industry, it is possible to bond lasers at relatively low cost, either through front-end integration similar to the approaches of Intel and Aurrion, or through die-scale bonding of finished laser chips [8]. These approaches are still in development, but it's clear that there are several practical approaches to making cheap light sources at various levels of integration with the silicon platform. It remains to be seen which approaches will be the most successful.

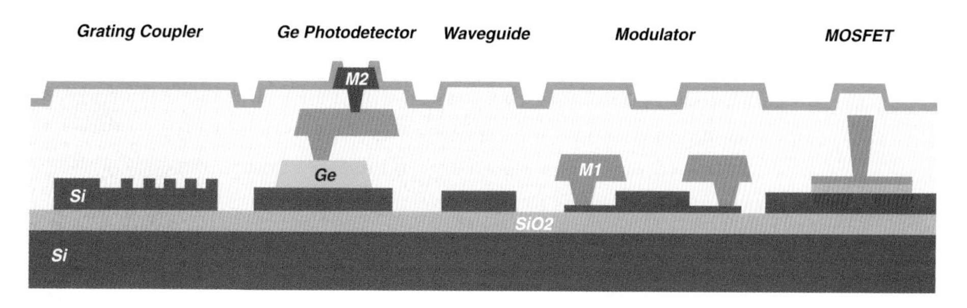

Figure 1.2 Typical process stack representing a silicon photonics platform with grating couplers, germanium photodetectors, waveguides, modulators and MOSFETs, on a silicon-on-insulator wafer. Note that most of the advanced work being done in silicon photonics today does not utilize monolithic integration with transistors, but instead leverages 3d or 2.5d integration. With permission [10].

1.2 Silicon photonics: the next fabless semiconductor industry[1]

The same foundries and processes that were developed to build transistors are being re-purposed to build chips that can generate, detect, modulate, and otherwise manipulate light. This is somewhat counter-intuitive, since the electronics industry spends billions of dollars to develop tools, processes and facilities that lend themselves to building the very best transistors, without any thought about how to make these processes compatible with photonics (with the exception of the processes designed to make devices such as CMOS and CCD camera chips). How are we so lucky that these capabilities can be directly reused for photonics?

In reality, they can't be directly reused. Every attempt to directly integrate photonic functionality into CMOS or bipolar silicon wafers, without making any process changes, has yielded poorly-performing devices. Electronics processes are designed for making electronics; it stands to reason they cannot be used for competitive photonics products. And even if they could, it wouldn't make economic sense; silicon photonics chips require relatively primitive processing (90 nm type capabilities) compared to advanced microelectronics chips (16 nm). Using the tools for truly advanced microelectronics to try to build photonics is a mistake, and would be impossible to justify from a performance or economic standpoint.

There is no reason to expect that the integration flows used to build electronic circuits would be in any way compatible with making components that manipulate light. But, over the last decade, it has emerged that silicon is actually a fantastic material system for building photonic devices, as well as electronic ones. And, even more surprisingly, the silicon photonics community has developed process flows that permit the re-use of CMOS fabrication infrastructure to build complex photonic circuits, where information is transferred seamlessly from the electronic to the optical domain and back again.

[1] A version of this chapter has been published in the *IEEE Journal of Solid-State Circuits* [10]: Michael Hochberg, Nicholas C. Harris, Ran Ding, Yi Zhang, Ari Novack, Choice Xuan, Tom Baehr-Jones. "Silicon photonics: The next fabless semiconductor industry", *IEEE Journal of Solid-State Circuits*, Vol. 5, No. 1, pp. 48–58, March 2013.

While the fully-integrated processes used to make transistors are not reused, modular process steps can be rearranged and reused, with distinct process flows being developed to build silicon photonics. This is not a trivial endeavor, but several organizations have shown that it is possible.

What has emerged is a vibrant community of companies and academics using the materials and techniques that have been developed over the past 50 years in the silicon microelectronics industry, and repurposing them to build photonic devices and circuits. What is particularly compelling about this work is that many of the efforts don't just make use of the same kinds of equipment in separate facilities, but actually use the exact same tools and facilities where CMOS transistors are routinely built. The constraints associated with working in such facilities are significant: materials that are not proven to be compatible with the CMOS processes are banned, and both processes and circuits have to be designed in such a way that processing them will not harm or contaminate the tools. The cost of mask sets and process development in the more advanced CMOS-compatible manufacturing facilities can also be very high [11] in advanced processes. But if the billion-dollar scale investments that go into building modern CMOS facilities can be directly leveraged to build silicon photonic systems on chip, it means that there is an immediate and a rapid path to commercialization and large-scale production is available.

1.2.1 Historical context – Photonics

Up to the present day, there has been very little in the way of opportunities for fabless photonics companies.

One of the key problems in photonics, historically, is that processes have been highly specialized for the particular application, utilizing different materials. With individual devices separately packaged and connected together by fibres, it is not unusual to see communications systems which incorporate chips made in half a dozen different material systems: RF CMOS or bipolar processes for the high-bandwidth electronics (e.g., serializers and deserializers), FPGA's or highly scaled CMOS for the digital parts (e.g., control circuits), diffused waveguides on glass for optical multiplexers (e.g., arrayed waveguide gratings) and passives, lithium niobate for modulators, indium phosphide for lasers, germanium for photodetectors, and MEMS-based switches, for instance. Each of these devices is made in a process that is fundamentally and irreconcilably incompatible with the ones used to make the other components. Each material system is chosen to provide ultimate performance for a single type of device. This means, in most cases, that the photonic components are produced in specialty fabrication facilities, in very low volumes. This results in high-cost components, since very few photonic devices are truly high-volume on the scale of the electronics industry. The only things that come close are VCSELs (which are made on a wafer scale, but are used as discrete devices), and components for PON networks (again, wafer-scale fabrication of directly modulated lasers, but they are used as singulated discrete components).

While discrete photonic devices can be interfaced to one another with standard optical fibres and connectors, a large fraction of the final device cost and yield loss

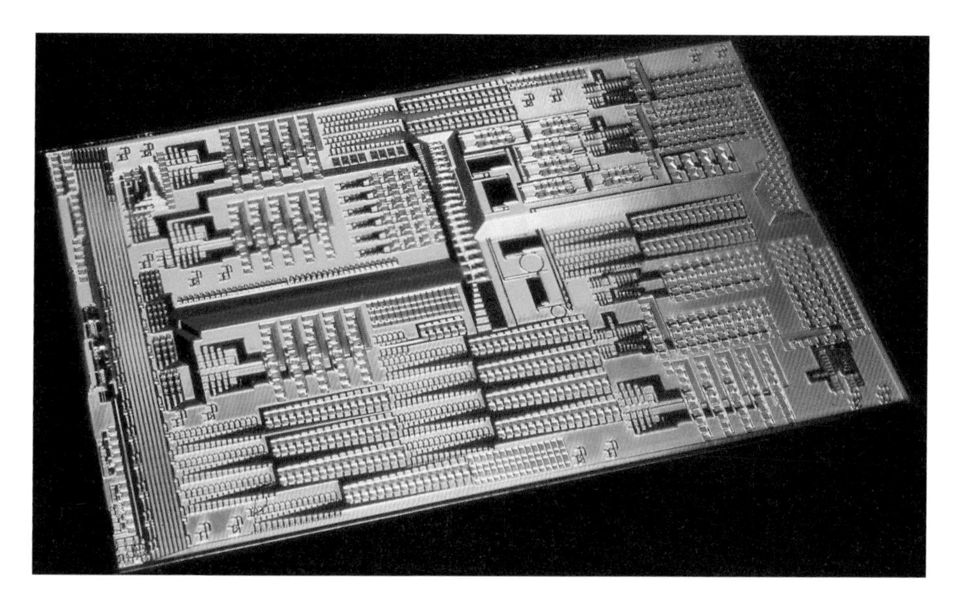

Figure 1.3 Photograph of SOI photonic chip fabricated at IME A*STAR.

emerges from the photonic packaging processes, which generally require 5- and 6-axis alignments with submicron accuracy, and from the packages themselves, which are often hermetically sealed and are sometimes quite literally gold-coated. Again, this contributes to the high cost of photonic components and systems.

The great promise of silicon photonics lies in integrating multiple functions into a single package, and manufacturing most or all of them using the same fabrication facilities that are used to build advanced microelectronics, as part of a single chip or chip stack (see Figures 1.3-1.4). Doing so will radically drive down the cost of moving data through fibres, and will create the opportunity for a variety of fundamentally new applications of photonics, where high-complexity systems can be built at very modest costs.

1.3 Applications

There are a number of applications that are emerging for complex silicon photonic systems, the most common being data communication. This includes high-bandwidth digital communications for short-reach applications, complex modulation schemes and coherent communications for long-reach applications, and so on.

Beyond data communications, there are a huge number of new applications being explored in both the commercial and academic worlds for this technology. These include: nano-optomechanics and condensed matter physics [12], biosensing [13, 14], nonlinear optics [15], LIDAR systems [16], optical gyroscopes [17,18], radio frequency integrated optoelectronics [19,20], integrated radio transceivers [21], coherent communications [22], novel light sources [23,24], laser noise reduction [25], gas sensors [26], very long wavelength integrated photonics [27], high-speed and microwave signal

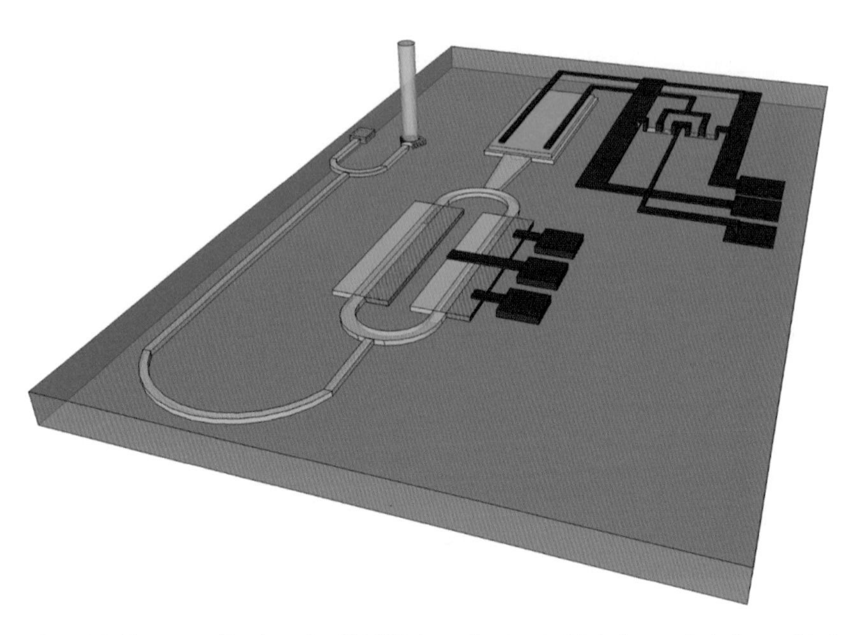

Figure 1.4 Conceptualization of a CMOS/photonic circuit. Light is coupled via on-chip laser or vertical fibre into a grating coupler. The light is then modulated, transduced by a photodetector, and inverted via a CMOS inverting circuit. Silicon photonic-electronic circuits can support systems with hundreds or thousands of such components today. Ref. [10]

processing [20], and many more. Areas of particular promise include biosensing, imaging, LIDAR, inertial sensing, hybrid photonic-RFICs, and signal processing.

1.3.1 Data communication

There are several categories of photonic devices where the silicon photonics components are competitive with best-in-class devices, discussed below. But what we regard as perhaps the most transformative work being done in the optical communications field is concerned with creating integrated platforms with modulators, detectors, waveguides and other components on the same chip, all talking to one another. In some cases, transistors are also included in these platforms, allowing amplifiers, serialization, and feedback to all be integrated onto the same chip. Because of the expense of developing such processes, this effort has largely been led by commercial players, mainly aimed at applications in point-to-point data communications. And because of the expense of developing transistor manufacturing processes, the emerging consensus in the field is that integrating electronics through bonding, either at the wafer or a the die level, makes the most sense for the foreseeable future from both a performance and cost perspective.

There is a lot of obvious value in being able to make chips that can perform computations with electronics, and transmit data optically. The vast majority of the early applications of silicon photonics are in digital data communications. This is driven by the fundamental differences in physics between electrons, which are Fermions, and photons, which are Bosons. Electrons are great for computation, because two of them cannot be in the same place at the same time. That means that they interact very strongly with each other. As a result, it is possible to build massively nonlinear switching devices – transistors – using electronics.

Photons have a different set of properties: many photons can all be in the same place at the same time, and except under very special circumstances they don't interact with each other. That's why it is possible [28] to transmit literally terabits of data per second through a single optical fibre: this is not done by creating a single terabit-bandwidth stream of data. Instead, typical high-bandwidth fibre optic transmission systems today make use of a variety of techniques to take a large number of streams of electrical data, at 10, 28 and 40 Gbits/second speeds, and multiplex them together onto single fibres. One such technique, commonly used in telecommunications, is wavelength-division multiplexing [29], where each colour of light is modulated separately and the various colours are all combined onto a single fibre. Other techniques, where coherence is exploited in order to encode information into phase, amplitude and polarization separately, have become common in long-haul and metro telecommunications as well [30]. In fact, these different approaches can be combined, with each wavelength representing a separate stream of data, and also being modulated in phase, amplitude and polarization, in order to transmit many bits per symbol. Thus it is possible to build systems that transmit terabits of data through a single fibre, without any electronics operating faster than 28 or 40 Gbits/second.

These techniques are critical, because fibre optic data transmission over long distances is limited by the expense of laying the fibres: Obtaining right-of-way and physically laying the conduits is expensive on land, and frightfully expensive across oceans [31]. In such systems, even if it requires millions of dollars of equipment at the endpoints to make efficient use of existing fibres, that is economically sensible, since the cost of laying more fibre is prohibitive. What we've seen over the past 50 years is that, because of the very low losses of fibres and the availability of erbium-doped fibre amplifiers [32] (which amplify all of the different optical wavelengths in a given window, around 1.5 μm) fibre optics first became the dominant technology for transmitting data over long distances, between cities. Over time, it came to dominate metro links, and is now dominating the relatively short distances within the data center. In many parts of the world, fiber-to-the-home is the dominant access paradigm, though this has not proven to be true in the United States, where it competes with DSL and other technologies. With the constant demand for increasing bandwidth [33], demand for ever-more-efficient ways of pushing data through fibres has grown steadily.

The broad trend in the data communications market is that as distances get shorter, the price per part drops precipitously, while the volumes go up. Unsurprisingly, silicon photonics commercialization efforts have focused a lot of effort onto the higher-volume, shorter-reach applications, aiming at data centres and high performance computing. Future applications will include board-to-board, short reach connections on the scale of USB, and perhaps eventually CPU core-to-core communications [34], though the case for on-chip core-to-core applications remains quite speculative.

Though not yet on the scale of the CMOS industry, silicon photonics is beginning to be a significant industry in its own right. The first commercial products integrating chip-scale electronics and photonics have recently hit the market [35], and Intel has announced its intention to standardize a format for optical data communication for personal computing [REF-INTEL]. Luxtera announced the sale of their millionth silicon photonic data channel [37], and they are now selling a 100 Gbits/second-class optically active cable (4x28G) [38], fabricated using the Freescale CMOS foundry in Texas [39]. Numerous startups and established companies (Kotura, Luxtera, Oracle, Genalyte, Lightwire/Cisco, APIC, Skorpios, TeraXion, and many others) are actively developing silicon photonic products. Many of the top American defense companies have programs in this area. Moreover, many of the major semiconductor players now have active programs in silicon photonics. Intel, Samsung, IBM, ST and many others have publicly announced activity in this area. Although there have been some over-blown predictions by market research firms [40] claiming that the field will be generating $2B/year in revenue by 2015, it does seem likely that revenues around $1B will be achieved before 2020, as predicted by the authors [11].

1.4 Technical challenges and the state of the art

1.4.1 Waveguides and passive components

There are a wide variety of waveguide geometries that have been developed in silicon-compatible systems; nearly any transparent material with a higher refractive index than glass can be deposited on top of an oxidized silicon substrate and turned into a waveguide. For purposes of CMOS process compatibility, however, the community has converged on a few classes of geometries. The most common are high-confinement waveguides made out of the active device layer of an SOI wafer, etched either fully to the bottom oxide layer or partially etched with a timed process stop [41,42] (Figure 1.2 and 3.4). It took several years of work to reduce the losses of these sub-micron waveguides to acceptable levels, since the strong interaction of the optical fields with the sidewalls can lead to substantial losses, driven by roughness [43]. Propagation loss can be reduced either by process optimization to smooth the sidewall [43] or by waveguide geometry optimization to reduce modal field strength at the sidewall [44]. Typical losses for high-confinement guides are in the 2 dB/cm range today for cutting-edge processes [45]. Low loss multimode straight waveguides in combination with tight single mode waveguide bends turn out to be an optimal choice for routing, achieving 0.026 dB/cm [46]. Other key passive components such as grating couplers [47] (Figures 1.2, 1.5, and 1.6), distributed Bragg gratings [48], waveguide crossings [49], and arrayed waveguide gratings (AWG) [50] have all been demonstrated, in each case with very low losses. More recently, CMOS compatible waveguides that can be formed into the dielectric back-end process have become available, made out of silicon nitride. With dedicated processing, the losses of these waveguides are extraordinarily low (< 0.1 dB/m), though the compatibility of such processes with front-end active devices is an open issue, given the requirement for high temperature growth [51]. It should be noted that considerable work has been done on low-confinement silicon waveguides [52,53].

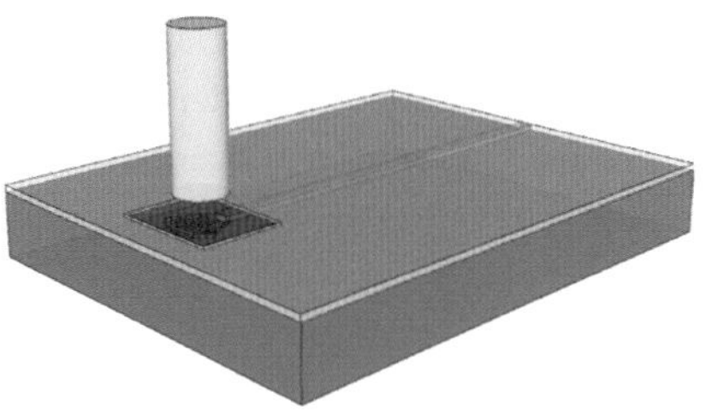

(a) Vertical coupling of optical fibre to on-chip grating coupler

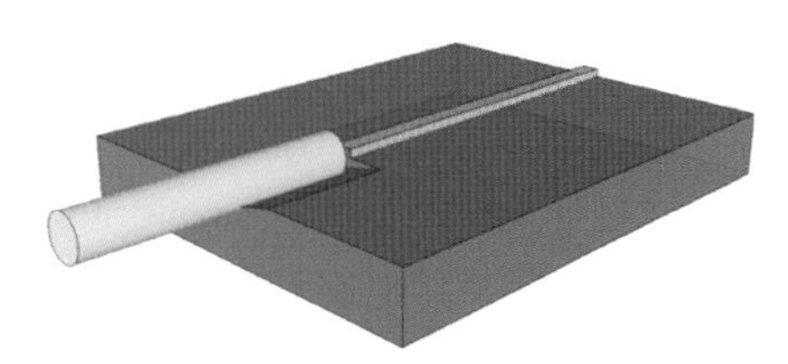

(b) V-groove coupling of optical fibre to on-chip waveguide

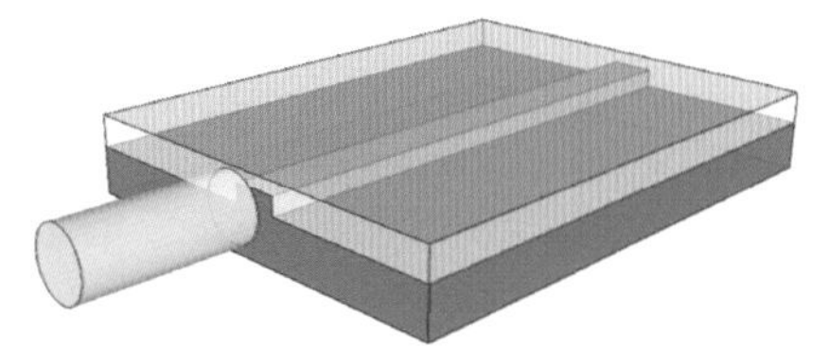

(c) Edge-coupled optical fibre to on-chip waveguide

Figure 1.5 Light coupling techniques. With permission [10].

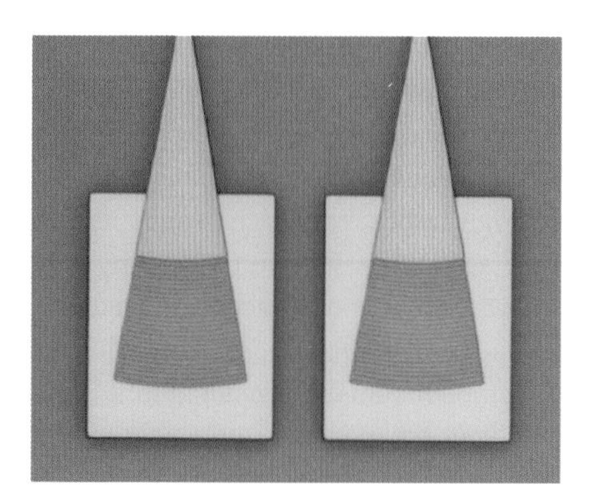

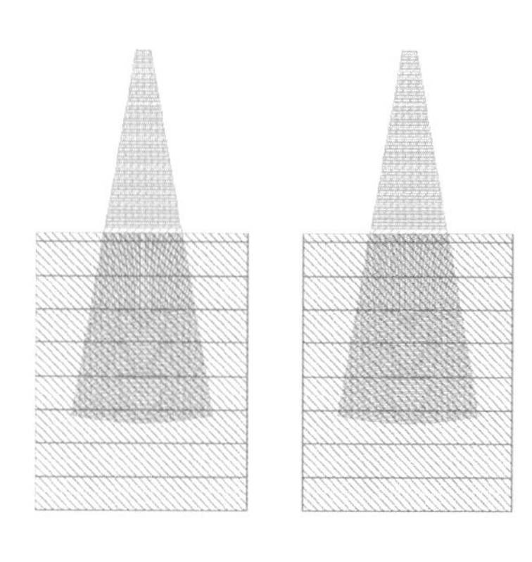

Figure 1.6 Optical micrograph and layout of an optical grating coupler fabricated at IME A*STAR. With permission [10].

There is considerable difficulty in creating compatible high-speed modulators and detectors in these platforms, making them less promising for large-scale integration.

One of the challenges with silicon photonics is coupling light between the chip and optical fibres, and achieving this is in a packaging method that is cost effective [54]. This is typically done using edge or grating couplers, as in Figure 1.5, and described in Chapter 5. Both approaches have demonstrated below 1 dB loss per interface [54–56]. Dealing with polarization is also a challenge, since silicon photonic waveguides are, by default, highly birefringent, namely the optical propagation constants in the waveguides are different for the two different polarizations. The common approach is to build circuits using a single polarization, and to duplicate them when both polarizations are required. This is termed polarization diversity, and takes advantage of polarization-splitting grating couplers [57], polarization splitters [58], splitter-rotators [59], or other related components. Other approaches (ref kotura?) have been explored which rely on using square waveguides to eliminate birefringence, but these approaches impose significant design constraints.

1.4.2 Modulators

Modulation in silicon is most commonly achieved by the plasma dispersion effect [60], where free carrier density changes can induce changes in the refractive index and modulate the light. Several different mechanisms of manipulating free-carrier densities have been implemented in monolithic devices [61]. Among them, carrier-depletion mode devices, usually based on a reverse-biased pn junction, are widely used to achieve high-speed operation.

Since the first GHz silicon modulator was demonstrated by the Intel group [62], the modulators' performance metrics have been significantly improved. Mach-Zehnder interferometer (MZI) structures [63] are typically used for amplitude modulation. With traveling-wave designs, data rates up to 50 Gbit/s have been experimentally demonstrated [64,65]. At 20 Gbit/s, a record low power consumption of 200 fJ/bit has recently been reported [66]. An example MZI modulator is shown in Figure 1.7a.

Resonant structures can be used to dramatically reduce the device footprint and further reduce power consumption, although at the cost of significantly narrower operating wavelength window and high thermal sensitivity. First introduced by Xu et al. [42] in 2005, high-speed ring modulators have been demonstrated to operate up to 40 Gbit/s [67] and recently were reported to achieve 7 fJ/bit at 25 Gb/s with thermal tuning capabilities [68]. An example of a ring modulator is shown in Figure 1.7b, and described in Chapter 6. Recent development includes breaking the cavity photon lifetime limit by coupling modulation [69], and building WDM transmitters using ring modulators, as described in Section 13.1.

In addition to pure silicon solutions, other materials can be integrated on to the silicon platform, for example by bonding III/V material [70], epitaxially growing germanium [71] or cladding graphene [72] to build efficient electro-absorption modulators. Chemically engineered active electro-optic polymers have also been introduced to the silicon platform in slot waveguides [73] and photonic crystals [74] to build efficient phase shifters. The broader field of hybrid integration of new materials into silicon, either through post-processing or through various kinds of encapsulation approaches to make the new materials acceptable to a CMOS foundry, is emerging as an active subfield in silicon photonics. These approaches all tend to involve challenging fabrication, and are likely to be limited to specialized applications with very specific high-end requirements. The possible exception to this is in direct integration of light sources, though the jury is out on the various technological approaches to this problem. It remains unclear whether there is a technological or economic imperative driving the integration of the light source onto the same chip as the modulators, detectors, and other components.

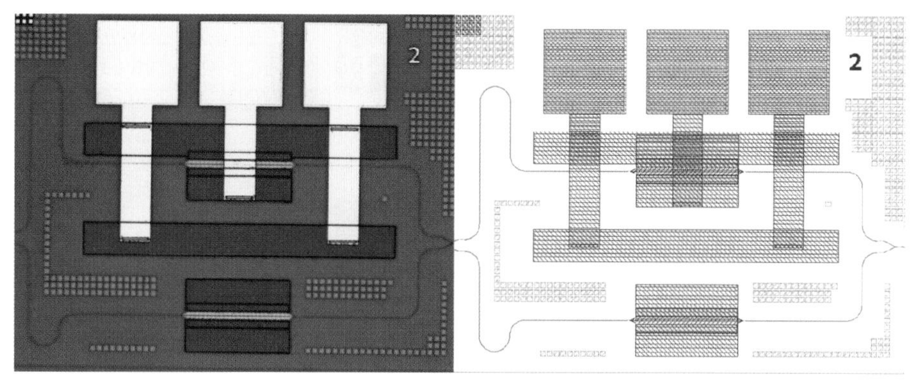

(a) Mach-Zehnder modulator

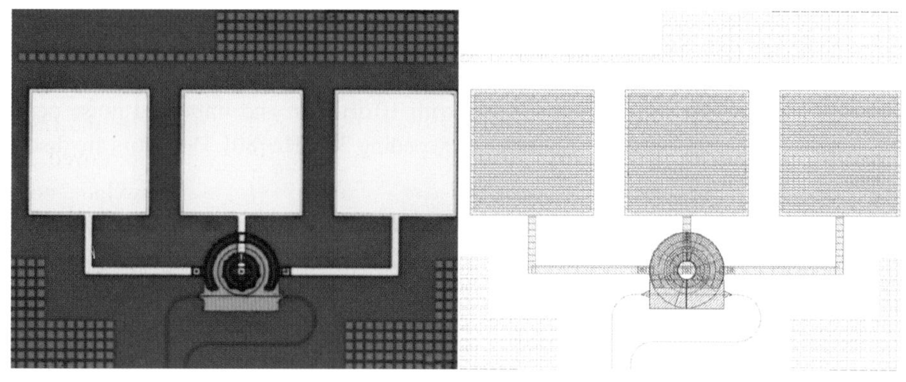

(b) Tuneable ring resonator modulator

Figure 1.7 Optical micrograph of devices fabricated at IME A*STAR as well as their layouts. With permission [10].

1.4.3 Photodetectors

At the operating wavelengths of a silicon photonic chip, materials with a narrower bandgap than silicon need to be integrated to function as the detection (absorption) medium. Germanium can be grown epitaxially and germanium absorbs light at the telecommunication wavelengths [75]. This is necessary for compatibility with standard infrastructure, but not necessarily required for short-reach applications where both ends of the link can be defined, without adhering to interoperability standards. Bonded III/V materials are also used for photodetection [76]. These materials are integrated close to, or directly connected to, a silicon waveguide, so that the guided light can be evanescently coupled or butt-coupled into the photodetector and the photodetector can have a small cross-section to reduce device capacitance and improve speed [77]. The state-of-the-art for germanium photodetectors, in a photodiode configuration, is a 120 GHz bandwidth with 0.8 A/W responsivity [77]. A responsivity of 1.05 A/W was demonstrated at 20 GHz for the wavelength 1550 nm [78], which amounts to 84% quantum efficiency. An example germanium detector is shown in Figure 1.8. A very low photodetector capacitance of 2.4 fF has been achieved in a photoconductive device with

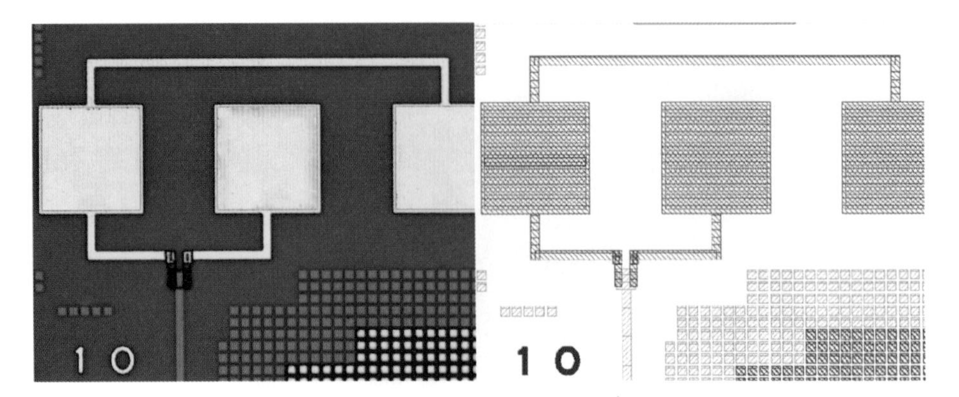

Figure 1.8 Optical micrograph and layout of a photodetector fabricated at IME A*STAR. With permission [10].

an estimated quantum efficiency of 90% operating with a bandwidth of 40 GHz [79]. Low bias avalanche detection with 10 dB gain and improved noise performance was achieved by IBM with a speed exceeding 30 GHz [80]. Detectors are discussed in detail in Chapter 7.

1.4.4 Light sources

One of the main challenges of the silicon photonics platform is the lack of an on-chip light source. The current generation of silicon photonic chips in production all couple light from an external laser. While edge and grating couplers have both seen improvements in coupling efficiency, the lack of an on-chip source limits the potential applications of these chips.

A number of techniques have been proposed to address the light source issue; these are discussed briefly here, and in Chapter 8. Hybrid Silicon lasers have been demonstrated using both bonding [81,82] and epitaxial growth [24] to transfer III-V materials to the silicon wafer (Figure 1.9a). However these techniques are hindered by the incompatibility of III-V materials with the standard CMOS process along with the high cost and low yield of bonding, and the small size of the available III/V wafers. Germanium has been proposed as a CMOS compatible gain medium though its photo-emission efficiency is hindered by its indirect band gap. The small (134 meV) difference between the indirect and direct band gaps can be overcome by a combination of strain and heavy n-type doping [83] and just recently, the first electrically driven laser using germanium gain medium on silicon was demonstrated [23].

All of the products currently in the market use more pedestrian approaches (e.g., Figure 1.9b–Figure 1.9c). These include both off-chip light sources, connected to the silicon chip by fibres, and lasers integrated within the same package as the silicon photonic chip. Technologies for this kind of integration have been inherited from the MEMS community, including micro-packaging techniques, and are both low-cost and very mature. These types of co-packaging approaches are already proven in medium scale production at Luxtera, for example.

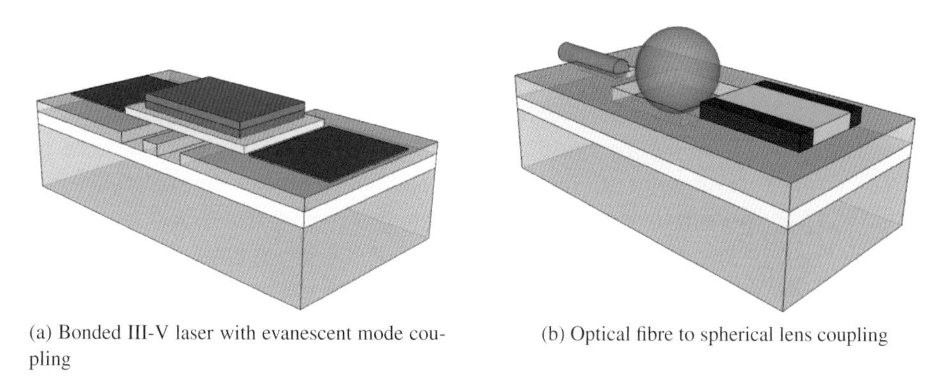

(a) Bonded III-V laser with evanescent mode coupling

(b) Optical fibre to spherical lens coupling

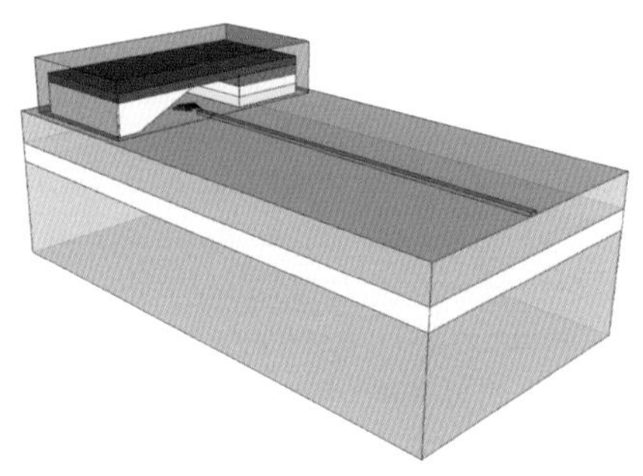

(c) Surface-mounted laser package with vertical reflector aligned to on-chip grating coupler

Figure 1.9 Laser integration techniques. With permission [10].

1.4.5 Approaches to photonic–electronic integration

Bringing electronics physically close to photonics can significantly improve system performance, reduce cost and open up tremendous new system design opportunities. Integration approaches generally fall into two categories: monolithic and multi-chip.

Monolithic integration

As the pioneer in monolithic electronics-photonics integration, Luxtera developed a CMOS-photonics foundry process based on a 130 nm CMOS process at Freescale and brought the first silicon photonics product to the market [35], and they recently teamed with ST Microelectronics on a development project aimed at bonded integration (see below) [84]. In their current process, the electronics and photonic components are fabricated together, in a single process flow. This required considerable process development effort, with additional layers being inserted in the flow in such a way as to allow the existing CMOS transistors to continue working, while adding germanium for photodetectors and silicon etches to support the fabrication of waveguides. Another process, from IHP, is integrating bipolar transistors with silicon photonics; this is a promising approach from a technical perspective. It remains to be seen whether cutting-edge bipolar transistors can be integrated with silicon photonic without performance compromises; if so, this could be a very powerful technology. Thus far, the integration efforts have centred around non-leading-edge bipolar transistors.

Another approach to front-end integration is to use unmodified advanced electronics process nodes, such as a 45 nm SOI CMOS process, and to attempt to integrate photonic devices within the constraints of the design rules [85]. This is an attractive approach at first blush, but the constraints imposed by the inability to modify the process mean that wavelengths are highly constrained to being near the silicon absorption edge by the low-percentage silicon germanium. Fast photodetectors and modulators have yet to be proven in these technologies, which are as yet in the early stages of development. We do not expect that this approach will yield competitive commercial products in the foreseeable future.

Multi-chip integration

On the other hand, multi-chip integration is being investigated and demonstrated, in which electronics and photonics circuitries come from different processes and are electrically bonded to each other (1.10). These can be stacked in different orders depending on the applications, e.g., the photonic chip on top with the electronics on the bottom, or vice-versa. Proposed and demonstrated approaches include using wire-bonding [86], flip-chip bump bonding (providing reduced parasitic capacitance and higher density than wire-bonding) [87], low-capacitance copper pillar interconnects, and through-silicon vias (TSV) [88] that offer even higher density and lower parasitics. The benefits of these approaches are significant: as the CMOS world continues to improve with Moore's law, it will be possible to rapidly adapt each successive generation of CMOS or Bi-CMOS onto an inexpensive silicon photonics wafer, getting the benefits of best-in-class processes for both the photonics and the electronics. Because the photonics generally does not require the fabrication of very small structures (~100 nm critical dimensions are typical) and devices are large compared to transistors, economic

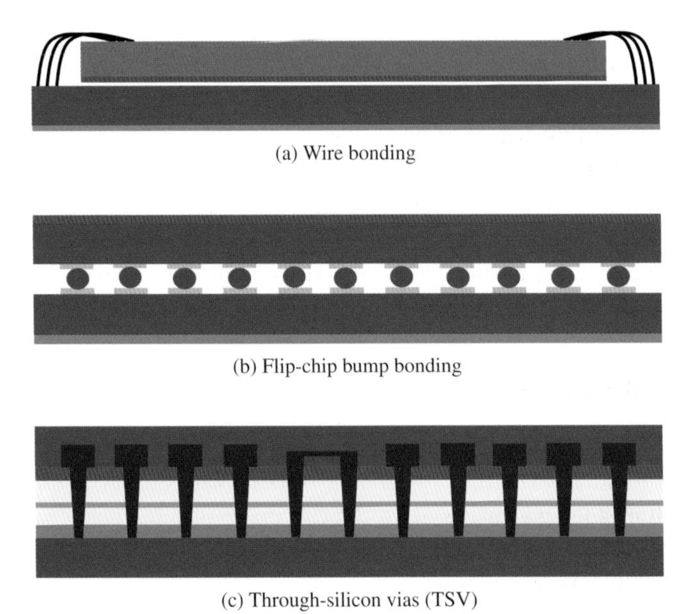

Figure 1.10 Electronic/Photonic chip-to-chip bonding schemes. With permission [10].

considerations will tend to drive the photonic devices to be made in a separate process from any advanced electronics that are needed in a final product. We expect these types of bonding processes to figure prominently in the next few years of silicon photonic research and product development.

1.5 Opportunities

We describe five fundamentally different kinds of opportunities in silicon photonics:

1.5.1 Device engineering

The first opportunity, *device engineering*, is the one that has driven perhaps 90% of the work in the field to date – particularly in terms of academic publications. Device engineering centres around answering a simple question: what are the best photonic devices that can be built in silicon?

We are now at the point where silicon Mach-Zehnder modulators are competitive with lithium niobate for many applications, and the silicon modulators continue to improve by significant factors with every passing year. By the same token, germanium waveguide-coupled photodiodes are competitive, on technical merits, with other uncooled photodetectors in the near infrared. Some of the world's lowest-loss waveguides are in silicon-nitride, which is fundamentally compatible with large scale photonic integration. Low-loss fibre couplers, various kinds of high-performance passive optics and even efficient lasers have been demonstrated within the silicon system, with co-packaging of III/V materials. More recently germanium lasers have been demonstrated, but thus far their efficiency limits their competitiveness. Much of the work on device engineering has centered around efforts at universities, because they have significant advantages in that they can close the simulation-fabrication-test loop very quickly for prototyping.

It is in the nature of device-level research to generate results that are not fully compatible with existing integrated processes. Those designing the latest and greatest new transistor may not initially be concerned about whether it fits into an existing process flow. The same is true in photonics: there will naturally be a pipeline from new devices being proven in specialty processes, to these new devices making their way into integrated platforms over time.

1.5.2 Photonic system engineering

The second major opportunity is in *photonic system engineering*. Once there is a supporting infrastructure that provides stable processes with a variety of photonic devices already included as accessible library elements, the question becomes: "What can we do with these library elements to build something useful?" Obviously one answer is to simply take single devices and package them as discrete components, but this approach does not really leverage the core advantage of silicon processing, which is to enable a rapid scaling of complexity.

In the electronics world, circuit designers do not need to be experts on the physics and fabrication of the transistor: instead, they rely on phenomenological models (termed "compact" models in the electronics industry) implemented in an environment like SPICE or VERILOG-A to simulate complex systems. Since the device models are guaranteed by the foundry, looking 'inside' of the device models is only required if a design is pushing the ultimate limits and needs something unusual and special. This kind of work is done in only a very small fraction of the electronic chip designs. In general, the device physicists work for the foundries, and develop the PDKs; very few foundry users will ever run the kind of TCAD simulations required to simulate the innards of the transistors.

A major near-term goal for silicon photonics foundry service providers is to make this same kind of infrastructure available for silicon photonics. Typical silicon photonic foundry service providers offer design kits with fixed designs for leading-edge photonics devices like modulators and detectors. The hope is that users will take these devices and either use them as starting points to build their own high-performance devices, or to use copies of these devices as part of complex systems containing numerous components (Figure 1.11). Photonic circuit modelling is discussed in Chapter 9.

One particularly rich area for innovation in this area is in photonics-electronic integration. Once transistors and photonics components are available in the same platform, with low-capacitance interconnects between them, it becomes possible to build cutting-edge devices that blend the functionality of what were formerly discrete devices. Luxtera, a company co-founded by one of the authors, has shown a number of results in this direction.

A transition from devices to systems

One of the most exciting aspects of the OpSIS-IME multi-project wafer shuttle runs was that many of the users, when presented with a library of functioning device designs,

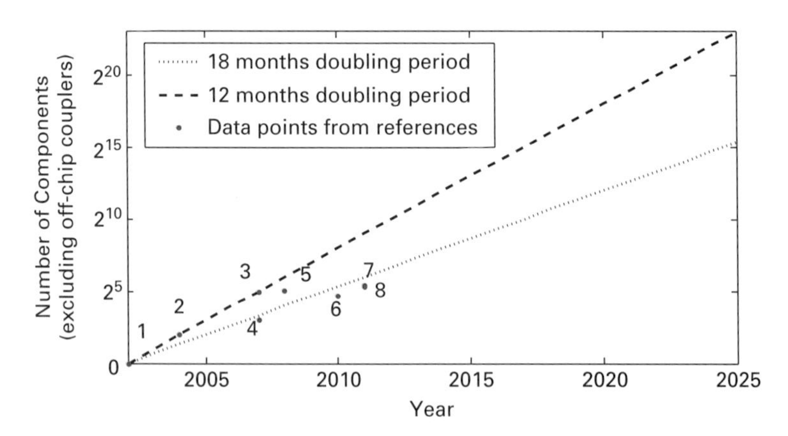

Figure 1.11 Plot of component number data points extracted from various publications on silicon photonics system design. (1) Luxtera: first silicon photonic system, (2) Intel: first CMOS modulator, (3) Kotura: VOA array, (4) Luxtera: 40Gb/s photonic chip, (5) Intel: 200 Gb/s PIC, (6) Intel: 4-channel WDM link with hybrid lasers, (7) Alcatel-Lucent Bell Lab: 10 x 25 Gb/s chip and (8) UC Davis group: fast-reconfigurable filter. With permission [10].

chose to focus their efforts on building integrated systems rather than on modifying the library elements. This was a sea-change in the field, because it meant that the basic devices were "good enough" to stop iterating on them at a device level, and shift focus toward building application-specific photonic integrated circuits (AS-PICs). This represented a very exciting moment for the field, as we moved from focusing most of the effort on device innovation into a period of very rapid growth, driven by the desire to build complex chip-scale systems.

Device innovations will continue to be important, but we can expect that over the next few years, the size of the community of people designing systems will grow much more rapidly than the community of device designers. The ecosystem now exists to build AS-PICS and to have a reasonable chance of them working in the first tape-out, at least to the point of developing prototypes. This is a new capability, and one that we expect will spark the rapid growth of a fabless photonics design community. The manufacturing transition remains riskier, and a major focus of efforts over the coming years will need to be on creating a smooth transition directly from prototype to production, by opening production processes for use as a prototyping vehicles.

1.5.3 Tools and support infrastructure

A third area of opportunity is in *developing tools and support infrastructure*. The silicon photonics design ecosystem is still very much in the development phase. Device-level simulation is now a very mature field, but higher-level simulation is just beginning to emerge. Wafer-scale automation of testing, design rule checking, layout-versus-schematic checkers, and design-for-test tools either do not exist or are extremely immature. Design-of-experiment for test structures to easily extract critical second order parameters like nonlinearity, power handling, coherence length, SFDR, etc. is an open problem in the literature.

Electronic–photonic co-design

One of the greatest innovations in the microelectronics world was the development of process design kits, PDKs, that allow design to be almost completely separated from fabrication, and which abstract the interaction between the two. There is a great deal of active work in the silicon photonics world in developing dedicated tools and design flows that can ensure first-silicon success, but these efforts are relatively immature. Luxtera was the first to develop an advanced design kit, living in a Cadence environment, which includes layout-versus-schematic and design rule checking for both the photonic and electronic components, plus statistical corner models for their standard components. Dedicated simulation tools for integrated photonics at the TCAD level have been developed by a number of companies [89–92], and existing tools for thermal and RF simulation [93] are routinely reused for simulating these aspects of photonic devices. In terms of photonic system simulation, tools include OptiSPICE by Optiwave [94], VPItransmissionMaker and Lumerical INTERCONNECT [95]. The design tools are described in more detail in Chapter 2.

One of the key advantages of electronic-photonic integration is the ability to integrate thermal feedback and management into photonic chips at the device level. Doing so is likely to dramatically simplify functional design, since thermal changes of a few degrees can radically alter the performance of photonic devices, particularly if they make use of on-chip resonant effects.

DFM and yield management

One significant area for innovation will be in developing design-for-manufacturing rules and methods, and in designing the right structures and procedures for yield management. What are the key structures to test to ensure photonic yield? What are the right ways to implement optical proximity correction in these systems? How do you map simple and inexpensive tests onto photonic system-on-chip performance? How does one pick the right mask technology for a given product, and predict the effects on performance? There is very sparse information in the public literature to address these questions, which will be critical issues over the coming years. We know empirically that even without doing any work in yield control, we routinely build wafers containing devices that operate within relatively tight tolerances (couple of dB in the electrooptic S-parameters) [6]. These numbers are quite good by optical device variation standards, but lot-to-lot data does not exist in the public domain.

1.5.4 Basic science

The fourth opportunity is in the area of *basic science*. There has been a lot of really exciting basic work in opto-mechanics and quantum optics, for instance, enabled by the high field concentrations and low losses of the silicon system. Work on novel materials like polymers, vanadium oxide, graphene, quantum dots, and many others, is ongoing, leveraging silicon photonics for various purposes. As these processes become more available over the coming years, we can expect photonics-enabled silicon to become a platform for exploring many areas of basic materials science and physics, much like VLSI electronics has proven to be an enabling technology across the sciences and engineering. A considerable amount of science in quantum optics [96] and low-temperature physics is ongoing [97]. Furthermore, silicon photonic circuits are being used as testbeds for exciting network modeling work, because of the ability to rapidly prototype new kinds of switches and transceivers [34].

1.5.5 Process standardization and a history of MPW services

A key innovation in the microelectronics industry has been the development of multi-project wafer (MPW) services, which make advanced processes available to the wider community. Making production processes available to the research and development community has been a key part of the fabless ecosystem in microelectronics. MOSIS has been an essential organization in making these processes available and accessible over the past several decades [98].

As is the case in the microelectronics world, a common commercial, production-ready foundry platform provides the following advantages:

(1) Resource Efficiency – eliminates the need for in-house cleanrooms and costly process development.

(2) Design Re-Use – the electronics industry benefits by reusing building blocks termed Intellectual Property (IP) cores rather than designing systems (such as processors) from the transistor up for each new design (over 1 billion transistors in some cases). Similarly, the same concept could allow our community to develop libraries of components. This will facilitate systems-level research and development.

(3) Resource Accessibility – makes the fabrication process and the libraries available to the vast community of photonics and electronics designers.

(4) Transferability for commercialization – allows industry to use the components and systems developed by the community to design new products that will be fabricated by a common foundry.

(5) Encourages collaborative design – facilitated by common goals, language and fabrication processes.

ePIXfab and Europractice

The first publicly available silicon photonics multi-project wafer foundry services were offered by imec (formerly the Interuniversity Microelectronics Centre; in Leuven, Belgium) and CEA-Leti (Grenoble, France) via ePIXfab beginning in 2006 [3,99]. Numerous academic and industrial users worldwide have benefited from access to silicon photonics fabrication. Initially, the fabrication process was a passive silicon photonic process using 193 nm deep-UV lithography. In 2012, CEA-Leti began offering a fabrication process with heaters, to enable thermo-optic silicon photonic devices. ePIXfab then offered two further fabrication services: one offered modulators fabricated at imec, and the other offered fabrication of detectors at CEA-Leti. In 2013, ePIXfab announced a full platform (passives, modulators, detectors) with fabrication at both imec and CEA-Leti [3]. ePIXfab also offers fabrication at IHP GmbH (Innovations for High Performance Microelectronics; in Frankfurt, Germany), VTT (Finland), and packaging services at Tyndall National Institute, University College (Cork, Ireland). Silicon photonics technologies are also available via the Europractice IC Service [100].

IME

The Institute of Microelectronics (IME) is a member of the Agency for Science, Technology and Research (A*STAR), in Singapore. IME conducts research and development in silicon photonics, and has made its fabrication process available to worldwide academic and industrial customers [4]. IME has a wide range of fabrication capabilities and services which allows their customers to tailor the process based on their needs. It has been used, for example, by OpSIS and CMC Microsystems, for MPW runs.

OpSIS

Optoelectronic Systems In Silicon (OpSIS) [6], which was led by one of the co-authors and by Thomas Baehr-Jones, was a non-profit Silicon Photonics foundry access and state-of-the-art PDK design service operated by the University of Delaware in the United States. It offered the first public MPW run to include passives, modulators, and detectors on a single platform, with the fabrication performed by IME in Singapore, in 2012.

OpSIS provided high-performance silicon photonics cells and design tools, including an advanced Process Design Kit (PDK) with some of the world's highest performance active, passive and thermal photonic devices. OpSIS also provided an eco-system of service and equipment providers (design, simulation, fabrication, and packaging) for the users, simplifying the development and testing of silicon photonic device and system prototypes. OpSIS' organizational mission was to prove that photonic MPW runs could support a large community of users; at peak OpSIS had a couple of hundred active users. It offered six MPW runs, ending in 2014, when OpSIS' 5-year core funding expired and the organization shut down.

CMC Microsystems

CMC Microsystems is Canadian government funded centre for the creation and application of micro- and nano-system knowledge by providing a national infrastructure for excellence in research and a path to commercialization [101]. CMC Microsystems supports access to international foundries for the manufacturing of silicon photonics and CMOS electronics, access to packaging technologies, and access to design software. CMC supports the Canadian industry and academic institutions in their design cycles including by offering training such as workshops offered by the Si-EPIC program.

Other organizations

There are several other organizations around the world involved in silicon photonics fabrication, including offering multi-project wafers and/or providing access to fabrication. Numerous universities, organizations, companies, and institutes, are involved in silicon photonics research and have contributed immensely to the growing field of silicon photonics. While the above list is not exhaustive, it identifies that there is an international critical mass of foundries, service providers, software vendors, designers, companies, universities, etc. This growing list provides both confidence and several sources for manufacturing services to the designer seeking to develop silicon photonic systems.

References

[1] Conway, L., "Reminiscences of the VLSI Revolution: How a Series of Failures Triggered a Paradigm Shift in Digital Design," *Solid-State Circuits Magazine, IEEE*, vol.4, **4**, pp. 8–31, Dec. 2012, DOI: 10.1109/MSSC.2012.2215752 (cit. on p. 3).

[2] A. Mekis, S. Gloeckner, G. Masini, *et al.* "A grating-coupler-enabled CMOS photonics platform". *IEEE Journal of Selected Topics in Quantum Electronics* **17**.3 (2011), pp. 597–608. DOI: 10.1109/JSTQE.2010.2086049 (cit. on p. 4).

[3] Amit Khanna, Youssef Drissi, Pieter Dumon, *et al.* "ePIX-fab: the silicon photonics platform". *SPIE Microtechnologies*. International Society for Optics and Photonics (2013), 87670H–87670H (cit. on pp. 4, 21).

[4] *Agency for Science, Technology and Research (A*STAR) Institute of Microelectronics (IME).* [Accessed 2014/07/21]. URL: http://www.a-star.edu.sg/ime/ (cit. on pp. 4, 21).

[5] M. Hochberg and T. Baehr-Jones. "Towards fabless silicon photonics". *Nature Photonics* **4**.8 (2010), pp. 492–494 (cit. on p. 4).

[6] Tom Baehr-Jones, Ran Ding, Ali Ayazi, *et al.* "A 25 Gb/s silicon photonics platform". *arXiv:1203.0767v1* (2012) (cit. on pp. 4, 20, 21).
[7] Liu, Alan Y., Chong Zhang, Justin Norman, *et al.* "High performance continuous wave 1.3 μm quantum dot lasers on silicon." *Applied Physics Letters* **104**.4 (2014): 041104. (cit. on p. 4).
[8] Peter De Dobbelaere, Ali Ayazi, Yuemeng Chi, *et al.* "Packaging of Silicon Photonics Systems". Optical Fiber Communication Conference. Optical Society of America. 2014, W3I.2 http://dx.doi.org/10.1364/OFC.2014.W3I.2 (cit. on p. 4).
[9] *CMC Microsystems - Fab: IME Silicon Photonics General-Purpose Fabrication Process.* [Accessed 2014/07/21]. URL: https://www.cmc.ca/en/WhatWeOffer/Products/CMC-00200-03001.aspx (cit. on p. 4).
[10] M. Hochberg, N.C. Harris, R. Ding, *et al.* "Silicon photonics: the next fabless semiconductor industry". *IEEE Solid-State Circuits Magazine* **5**.1 (2013), pp. 48–58. DOI: 10.1109/MSSC.2012.2232791 (cit. on pp. 4, 5, 8, 11, 13, 14, 15, 16, 18).
[11] T. Baehr-Jones, T. Pinguet, P. L. Guo-Qiang, *et al.* "Myths and rumours of silicon photonics". *Nature Photonics* **6**.4 (2012), pp. 206–208 (cit. on pp. 6, 10).
[12] M. Li, W. H. P. Pernice, C. Xiong, T. Baehr-Jones, M. Hochberg, and H. X. Tang. "Harnessing optical forces in integrated photonic circuits". *Nature* 456.27 (2008), pp. 480–484 (cit. on p. 7).
[13] J. Hu, X. Sun, A. Agarwal, and L. Kimerling. "Design guidelines for optical resonator biochemical sensors". *Journal Optics Society America B* **26** (2009), pp. 1032–1041 (cit. on p. 7).
[14] Muzammil Iqbal, Martin A Gleeson, Bradley Spaugh, *et al.* "Label-free biosensor arrays based on silicon ring resonators and high-speed optical scanning instrumentation". *IEEE Journal of Selected Topics in Quantum Electronics* **16**.3 (2010), pp. 654–661 (cit. on p. 7).
[15] M. Foster, A. Turner, M. Lipson, and A. Gaeta. "Nonlinear optics in photonic nanowires". *Optics Express* **16**.2 (2008), pp. 1300–1320 (cit. on p. 7).
[16] J. K. Doylend, M. J. R. Heck, J. T. Bovington, *et al.* "Two-dimensional free-space beam steering with an optical phased array on silicon-on-insulator". *Optics Express* **19**.22 (2011), pp. 21 595–21 604 (cit. on p. 7).
[17] A. A. Trusov, I. P. Prikhodko, S. A. Zotov, A. R. Schofield, and A. M. Shkel. "Ultra-high Q silicon gyroscopes with interchangeable rate and whole angle modes of operation". *Proc. IEEE Sensors* **2010** (2010), pp. 864–867 (cit. on p. 7).
[18] M. Guilln-Torres, E. Cretu, N.A.F. Jaeger, and L. Chrostowski. "Ring resonator optical gyroscopes – parameter optimization and robustness analysis". *Journal of Lightwave Technology* **30**.12 (2012), pp. 1802–1817 (cit. on p. 7).
[19] J. Capmany and D. Novak. "Microwave photonics combines two words". *Nature Photonics* **1**.6 (2007), pp. 319–330 (cit. on p. 7).
[20] Maurizio Burla, Luis Romero Cortés, Ming Li, *et al.* "Integrated waveguide Bragg gratings for microwave photonics signal processing". *Optics Express* **21**.21 (2013), pp. 25 120–25 147. DOI: 10.1364/OE.21.025120 (cit. on pp. 7, 8).
[21] M. Ko, J. Youn, M. Lee, *et al.* "Silicon photonics-wireless interface IC for 60-GHz wireless link". *IEEE Photonics Technology Letters* **24**.13 (2012), pp. 1112–1114 (cit. on p. 7).
[22] C. R. Doerr, L. L. Buhl, Y. Baeyens, *et al.* "Packaged monolithic silicon 112-Gb/s coherent receiver". *IEEE Photonics Technology Letters* **23**.12 (2011), pp. 762–764 (cit. on p. 7).
[23] R. Camacho-Aguilera, Y. Cai, N. Patel, *et al.* "An electrically pumped germanium laser". *Optics Express* **20** (2012), pp. 11 316–11 320(cit. on pp. 7, 14).

[24] H. Y. Liu, T. Wang, Q. Jiang, *et al.* "Long-wavelength InAs/GaAs quantum-dot laser diode monolithically grown on Ge substrate". *Nature Photonics* **5**.7 (2011), pp. 416–419 (cit. on pp. 7, 14).

[25] Firooz Aflatouni and Hossein Hashemi. "Wideband tunable laser phase noise reduction using single sideband modulation in an electro-optical feed-forward scheme". *Optics Letters* **37**.2 (2012), pp. 196–198 (cit. on p. 7).

[26] R. B. Wehrspohn, S. L. Schweizer, T. Geppert, *et al.* "Chapter 12. Application of Photonic Crystals for Gas Detection and Sensing". *Advances in Design, Fabrication, and Characterization*, K. Busch, S. Lalkes, R. B. Wehrspohn, and H. Fall (eds.), in *Photonic Crystals*: Wiley-VCH Verlag GmbH, 2006 (cit. on p. 7).

[27] R. Soref. "Mid-infrared photonics in silicon and germanium". *Nature Photonics* **4**.8 (2010), pp. 495–497 (cit. on p. 7).

[28] John Senior. *Optical Fiber Communications: Principles and Practice.* Prentice Hall, 2008 (cit. on p. 9).

[29] C. A. Brackett. "Dense wavelength division multiplexing networks: principles and applications". *IEEE Journal on Selected Areas in Communications* **8**.6 (1990), pp. 948–964 (cit. on p. 9).

[30] G. Li. "Recent advances in coherent optical communication". *Advances in Optics and Photonics* **1** (2009), pp. 279–307 (cit. on p. 9).

[31] Neal Stephenson, "Mother Earth Mother Board", Wired, Issue **4**.12, December 1996. http://archive.wired.com/wired/archive/4.12/ffglass.html (cit. on p. 9).

[32] E. Desurvire. *Erbium Doped Fiber Amplifiers: Principles and Applications.* Wiley-Interscience, 1994 (cit. on p. 9).

[33] *Nielsen's Law of Internet Bandwidth.* [Accessed 2014/04/14]. URL: http://www.useit.com/alertbox/980405.html (cit. on p. 9).

[34] A. Shacham, K. Bergman, and L. P. Carloni. "Photonic networks-on-chip for future generations of chip multiprocessors". *IEEE Transactions on Computers* **57**.9 (2008), pp. 1246–1260 (cit. on pp. 9, 20).

[35] *Luxtera Introduces Industrys First 40G Optical Active Cable, Worlds First CMOS Photonics Product.* [Accessed 2014/04/14]. URL: http://www.luxtera.com/2007081341/luxtera-introduces-industry-s-first-40g-optical-active-cable-world-s-first-cmos-photonics-product.html (cit. on pp. 10, 15).

[36] *Intel Silicon Photonics Research*, [Accessed 2014/04/14]. URL: http://www.intel.com/content/www/us/en/research/intel-labs-ces-2010-keynote-light-peak-future-io-video.html

[37] *Luxtera Ships One-Millionth Silicon CMOS Photonics Enabled 10Gbit Channel.* [Accessed 2014/04/14]. URL: http://www.luxtera.com/20120221252/luxtera-ships-one-millionth-silicon-cmos-photonics-enabled-10gbit-channel.html (cit. on p. 10).

[38] *Luxtera Delivers Worlds First Single Chip 100Gbps Integrated Opto-Electronic Transceiver.* [Accessed 2014/04/14]. URL: http://www.luxtera.com/20111108239/luxtera-delivers-world's-first-single-chip-100gbps-integrated-opto-electronic-transceiver.html (cit. on p. 10).

[39] *Luxtera Announces Production Status of Worlds First Commercial Silicon CMOS Photonics Fabrication Process.* [Accessed 2014/04/14]. URL: http://www.luxtera.com/20090603183/luxtera-announces-production-status-of-worlds-1st-commercial-silicon-cmos-photonics-fabrication-process.html (cit. on p. 10).

[40] *Silicon Photonics Market by Product & Applications 2020.* [Accessed 2014/04/14]. URL: http://www.marketsandmarkets.com/Market-Reports/silicon-photonics-116.html (cit. on p. 10).

[41] P. Dumon, W. Bogaerts, V. Wiaux, *et al.* "Low-loss SOI photonic wires and ring resonators fabricated with deep UV lithography". *IEEE Photonics Technology Letters* **16** (2004), pp. 1328–1330 (cit. on p. 10).
[42] Q. Xu, B. Schmidt, S. Pradhan, and M. Lipson. "Micrometre-scale silicon electro-optic modulator". *Nature* **435** (2005), pp. 325–327 (cit. on pp. 10, 12).
[43] K. Lee, D. Lim, L. Kimerling, J. Shin, and F. Cerrina. "Fabrication of ultralow-loss Si/SiO2 waveguides by roughness reduction". *Optics Letters* **26** (2001), pp. 1888–1890 (cit. on p. 10).
[44] F. Grillot, L. Vivien, S. Laval, D. Pascal, and E. Cassan. "Size influence on the propagation loss induced by sidewall roughness in ultrasmall SOI waveguides". *IEEE Photonics Technology Letters* **16**.7 (2004), pp. 1661–1663 (cit. on p. 10).
[45] *ePIXfab – The silicon photonics platform – IMEC Standard Passives.* [Accessed 2014/04/14]. URL: http://www.epixfab.eu/technologies/49-imecpassive-general (cit. on p. 10).
[46] Guoliang Li, Jin Yao, Hiren Thacker, *et al.* "Ultralow-loss, high-density SOI optical waveguide routing for macrochip interconnects". *Optics Express* **20**.11 (May 2012), pp. 12035–12039. DOI: 10.1364/OE.20.012035 (cit. on p. 10).
[47] N. Na, H. Frish, I. W. Hsieh, *et al.* "Efficient broadband silicon-on-insulator grating coupler with low backre-flection". *Optics Letters* **36**.11 (2011), pp. 2101–2103 (cit. on p. 10).
[48] Xu Wang, Wei Shi, Han Yun, *et al.* "Narrow-band waveguide Bragg gratings on SOI wafers with CMOS-compatible fabrication process". *Optics Express* **20**.14 (2012), pp. 15 547–15 558. DOI: 10.1364/OE.20.015547 (cit. on p. 10).
[49] W. Bogaerts, P. Dumon, D. Thourhout, and R. Baets. "Low-loss, low-crosstalk crossings for silicon-on-insulator nanophotonic waveguides". *Optics Letters* **32** (2007), pp. 2801–2803 (cit. on p. 10).
[50] X. Fu and D. Dai. "Ultra-small Si-nanowire-based 400 GHz-spacing 15 X 15 arrayed-waveguide grating router with microbends". *Electronics Letters* **47**.4 (2011), pp. 266–268 (cit. on p. 10).
[51] D. Dai, Z. Wang, J. Bauters, *et al.* "Low-loss Si3N4 arrayed-waveguide grating (de)multiplexer using nano-core optical waveguides". *Optics Express* **19** (2011), pp. 14 130–14 136 (cit. on p. 10).
[52] *VTT Si Photonics Technology*, [Accessed 2014/12/17] URL: http://www.epixfab.eu/technologies/vttsip (cit. on p. 10).
[53] G.T. Reed and A.P. Knights. Silicon photonics. Wiley Online Library, 2008 (cit. on p. 10).
[54] A. Mekis, S. Abdalla, D. Foltz, *et al.* "A CMOS photonics platform for high-speed optical interconnects". *Photonics Conference (IPC).* IEEE. 2012, pp. 356–357 (cit. on p. 11).
[55] R. Takei, M. Suzuki, E. Omoda, *et al.* "Silicon knife-edge taper waveguide for ultralow-loss spot-size converter fabricated by photolithography". *Applied Physics Letters* **102**.10 (2013), p. 101108 (cit. on p. 11).
[56] Wissem Sfar Zaoui, Andreas Kunze, Wolfgang Vogel, *et al.* "Bridging the gap between optical fibers and silicon photonic integrated circuits". *Opt. Express* **22**.2 (2014), pp. 1277–1286. DOI: 10.1364/OE.22.001277 (cit. on p. 11).
[57] Dirk Taillaert, Harold Chong, Peter I. Borel, *et al.* "A compact two-dimensional grating coupler used as a polarization splitter". *Photonics Technology Letters, IEEE* **15**.9 (2003), pp. 1249–1251 (cit. on p. 12).
[58] M. R. Watts, H. A. Haus, and E. P. Ippen. "Integrated mode-evolution-based polarization splitter". *Optics Letters* **30**.9 (2005), pp. 967–969 (cit. on p. 12).

[59] Daoxin Dai and John E Bowers. "Novel concept for ultracompact polarization splitter-rotator based on silicon nanowires". *Optics Express* **19**.11 (2011), pp. 10 940–10 949 (cit. on p. 12).
[60] R. Soref and B. Bennett. "Electrooptical effects in silicon". *IEEE Journal of Quantum Electronics* **23**.1 (1987), pp. 123–129 (cit. on p. 12).
[61] G. T. Reed, G. Mashanovich, F. Y. Gardes, and D. J. Thomson. "Silicon optical modulators". *Nature Photonics* **4**.8 (2010), pp. 518–526 (cit. on p. 12).
[62] A. Liu, R. Jones, L. Liao, *et al.* "A high-speed silicon optical modulator based on a metal-oxide-semiconductor capacitor". *Nature* **427** (2004), pp. 615–618 (cit. on p. 12).
[63] G. V. Treyz. "Silicon Mach-Zehnder waveguide interferometers operating at 1.3 m". *Electronics Letters* **27** (1991), pp. 118–120 (cit. on p. 12).
[64] P. Dong, L. Chen, and Y. Chen. "High-speed low-voltage single-drive push-pull silicon Mach-Zehnder modulators". *Optics Express* **20** (2012), pp. 6163–6169 (cit. on p. 12).
[65] D. J. Thomson, F. Y. Gardes, J. M. Fedeli, *et al.* "50-Gb/s silicon optical modulator". *IEEE Photonics Technology Letters* **24**.4 (2012), pp. 234–236 (cit. on p. 12).
[66] T. Baehr-Jones, R. Ding, Y. Liu, *et al.* "Ultralow drive voltage silicon traveling-wave modulator". *Optics Express* **20**.11 (2012), pp. 12 014–12 020 (cit. on p. 12).
[67] Y. Hu, X. Xiao, H. Xu, *et al.* "High-speed silicon modulator based on cascaded microring resonators". *Optics Express* **20**.14 (2012), pp. 15 079–15 085 (cit. on p. 12).
[68] G. Li, X. Zheng, J. Yao, *et al.* "25Gb/s 1V-driving CMOS ring modulator with integrated thermal tuning". *Optics Express* **19** (2011), pp. 20 435–20 443 (cit. on p. 12).
[69] W. D. Sacher, W. M. J. Green, S. Assefa, *et al.* "Breaking the cavity linewidth limit of resonant optical modulators". *arXiv preprint arXiv:1206.5337* (2012) (cit. on p. 12).
[70] Y. Tang, H. Chen, S. Jain, *et al.* "50 Gb/s hybrid silicon traveling-wave electroabsorption modulator". *Optics Express* **19** (2011), pp. 5811–5816 (cit. on p. 12).
[71] Y. Kuo, Y. Lee, Y. Ge, *et al.* "Strong quantum-confined Stark effect in germanium quantum-well structures on silicon". *Nature* **437** (2005), pp. 1334–1336 (cit. on p. 12).
[72] M. Liu, X. Yin, E. Ulin-Avila, *et al.* "A graphene-based broadband optical modulator". *Nature* **474**.7349 (2011), pp. 64–67 (cit. on p. 12).
[73] R. Ding, T. Baehr-Jones, Y. Liu, *et al.* "Demonstration of a low $V\pi L$ modulator with GHz bandwidth based on electro-optic polymer-clad silicon slot waveguides". *Optics Express* **18** (2010), pp. 15618–15623 (cit. on p. 12).
[74] J. Brosi, C. Koos, L. Andreani, *et al.* "High-speed low-voltage electro-optic modulator with a polymer-infiltrated silicon photonic crystal waveguide". *Optics Express* **16** (2008), pp. 4177–4191 (cit. on p. 12).
[75] J. Wang and S. Lee. "Ge-photodetectors for Si-based optoelectronic integration". *Sensors (Basel Switzerland)* **11**.1 (2011), pp. 696–718 (cit. on p. 13).
[76] H. Park, A. Fang, R. Jones, *et al.* "A hybrid AlGaInAs-silicon evanescent waveguide photodetector". *Optics Express* **15** (2007), pp. 6044–6052 (cit. on p. 13).
[77] L. Vivien, A. Polzer, D. Marris-Morini, *et al.* "Zero-bias 40Gbit/s germanium waveguide photodetector on silicon". *Optics Express* **20** (2012), pp. 1096–1101. DOI: 10.1364/OE.20.001096 (cit. on p. 13).
[78] S. Liao, N. Feng, D. Feng, *et al.* "36 GHz submicron silicon waveguide germanium photodetector". *Optics Express* **19** (2011), pp. 10 967–10 972 (cit. on p. 13).
[79] L. Chen and M. Lipson. "Ultra-low capacitance and high speed germanium photodetectors on silicon". *Optics Express* **17** (2009), pp. 7901–7906 (cit. on p. 14).
[80] S. Assefa, F. Xia, and Y. A. Vlasov. "Reinventing germanium avalanche photodetector for nanophotonic on-chip optical interconnects". *Nature* **464**.7285 (2010), pp. 80–84 (cit. on p. 14).

[81] A. Fang, H. Park, O. Cohen, *et al.* "Electrically pumped hybrid AlGaInAs-silicon evanescent laser". *Optics Express* **14** (2006), pp. 9203–9210 (cit. on p. 14).
[82] B. Ben Bakir, A. Descos, N. Olivier, *et al.* "Electrically driven hybrid Si/III–V Fabry–Perot lasers based on adiabatic mode transformers". *Optics Express* **19** (2011), pp. 10 317–10 325 (cit. on p. 14).
[83] J. Liu, X. Sun, D. Pan, *et al.* "Tensile-strained, n-type Ge as a gain medium for monolithic laser integration on Si". *Optics Express* **15** (2007), pp. 11272–11277 (cit. on p. 14).
[84] *Luxtera and STMicroelectronics to Enable High-Volume Silicon Photonics Solutions.* [Accessed 2014/04/14]. URL: http://web.archive.org/web/20140415052434/http://www.st.com/web/en/press/en/t3279 (cit. on p. 15).
[85] J. Orcutt, B. Moss, C. Sun, *et al.* "Open foundry platform for high-performance electronic-photonic integration". *Optics Express* **20** (2012), pp. 12 222–12 232 (cit. on p. 16).
[86] B. Lee, C. Schow, A. Rylyakov, *et al.* "Demonstration of a digital CMOS driver codesigned and integrated with a broadband silicon photonic switch". *Journal of Lightwave Technology* **29** (2011), pp. 1136–1142 (cit. on p. 16).
[87] H. D. Thacker, Y. Luo, J. Shi, *et al* "Flip-chip integrated silicon photonic bridge chips for sub-picojoule per bit optical links". In *IEEE Electronic Components and Technology Conference* (2010), pp. 240–246 (cit. on p. 16).
[88] N. Sillon, A. Astier, H. Boutry, *et al.* "Enabling technologies for 3D integration: From packaging miniaturization to advanced stacked ICs". *IEDM Tech.* (2008), pp. 595–598 (cit. on p. 16).
[89] *Lumerical Solutions Inc. – Innovative Photonic Design Tools.* [Accessed 2014/04/14]. URL: http://www.lumerical.com/ (cit. on p. 19).
[90] *Optiwave.* URL: http://www.optiwave.com/ (cit. on p. 19).
[91] *Photon Design.* [Accessed 2014/04/14]. URL: http://www.photond.com/ (cit. on p. 19).
[92] *RSoft Products - Synopsys Optical Solutions.* [Accessed 2014/04/14]. URL: http://optics.synopsys.com/rsoft/ (cit. on p. 19).
[93] *ANSYS HFSS.* [Accessed 2014/04/14]. URL: http://www.ansys.com/Products/Simulation+Technology/Electronics/Signal+Integrity/ANSYS+HFSS (cit. on p. 19).
[94] *OptiSPICE Archives – Optiwave.* [Accessed 2014/04/14]. URL: http://optiwave.com/category/products/system-and-amplifier-design/optispice/# (cit. on p. 19).
[95] *Lumerical INTERCONNECT – Photonic Integrated Circuit Design Tool.* [Accessed 2014/04/14]. URL: http://www.lumerical.com/tcad-products/interconnect/ (cit. on p. 19).
[96] Nicholas C. Harris, Davide Grassani, Angelica Simbula, et al. An integrated source of spectrally filtered correlated photons for large scale quantum photonic systems, arXiv:1409.8215 [quant-ph] (cit. on p. 20).
[97] M.K. Akhlaghi, E. Schelew and J.F. Young, "Waveguide Integrated Superconducting Single Photon Detectors Implemented as Coherent Perfect Absorbers" arXiv:1409.1962 [physics.ins-det], (5 Sep 2014). (cit. on p. 20).
[98] C. Tomovich. "MOSIS – A gateway to silicon". *IEEE Circuits and Devices Magazine* **4**.2 (1988), pp. 22–23 (cit. on p. 20).
[99] *ePIXfab – The silicon photonics platform – MPW Technologies.* [Accessed 2014/04/14]. URL: http://www.epixfab.eu/technologies (cit. on p. 21).
[100] *Europractice Silicon Photonics Technologies.* [Accessed 2014/07/21]. URL: http://www.europractice-ic.com/SiPhotonics_technology.php (cit. on p. 21).
[101] *CMC Microsystems.* [Accessed 2014/04/14]. URL: http://www.cmc.ca (cit. on p. 22).

2 Modelling and design approaches

In this chapter, we present an overview of the simulation and design tools useful for silicon photonics component and circuit design.

The design methodology for silicon photonic systems is illustrated in Figure 2.1. The order of the material presented in this book follows this illustration from top down. The design of passive silicon photonic components is considered in Part II, Chapters 3–5, while active component design is considered in Part III, Chapters 6–7. Model synthesis is described throughout these sections, and in more detail in Chapter 9, in which we describe optical circuit modelling techniques. Circuit modelling initially is focused on predicting the system behaviour in the presence of external stimulus, namely electrical and optical signals. Once a circuit is designed, the designer uses the schematic to lay out the components in a physical mask layout using a variety of design aids, as described in Chapter 10. This is followed by verification, including manufacturing design rule checking (DRC, DFM), layout versus schematic checking (LVS), test considerations, lithography simulation, and parasitic extraction. The results of the verification are fed back into the circuit simulations to predict the system response including effects due to the physical implementation (e.g. lithography effects, fabrication non-uniformity, temperature, waveguide lengths, and component placement). In this step, the circuit simulation takes into account not only the external stimulus but also the fabrication process (Chapter 11) and environmental variations. The design methodology was presented in Reference [1].

2.1 Optical waveguide mode solver

An eigenmode solver (or mode solver) solves for optical modes in a cross-section of an arbitrary waveguide geometry (or a 3D geometry) at a particular frequency. A waveguide mode is a transverse field distribution that propagates along the waveguide without changing shape; the solution is time-invariant. An example mode profile is shown in Figure 2.2a.

Eigenmode solvers determine time-harmonic solutions to Maxwell's equations in the frequency domain. Since they provide a solution for a single optical frequency, numerous simulations are required to obtain wavelength sweeps needed to study waveguide dispersion. There are numerous approaches to solving this problem, including the Finite Element Method (FEM), Finite Difference (FD) algorithm, and various approximations

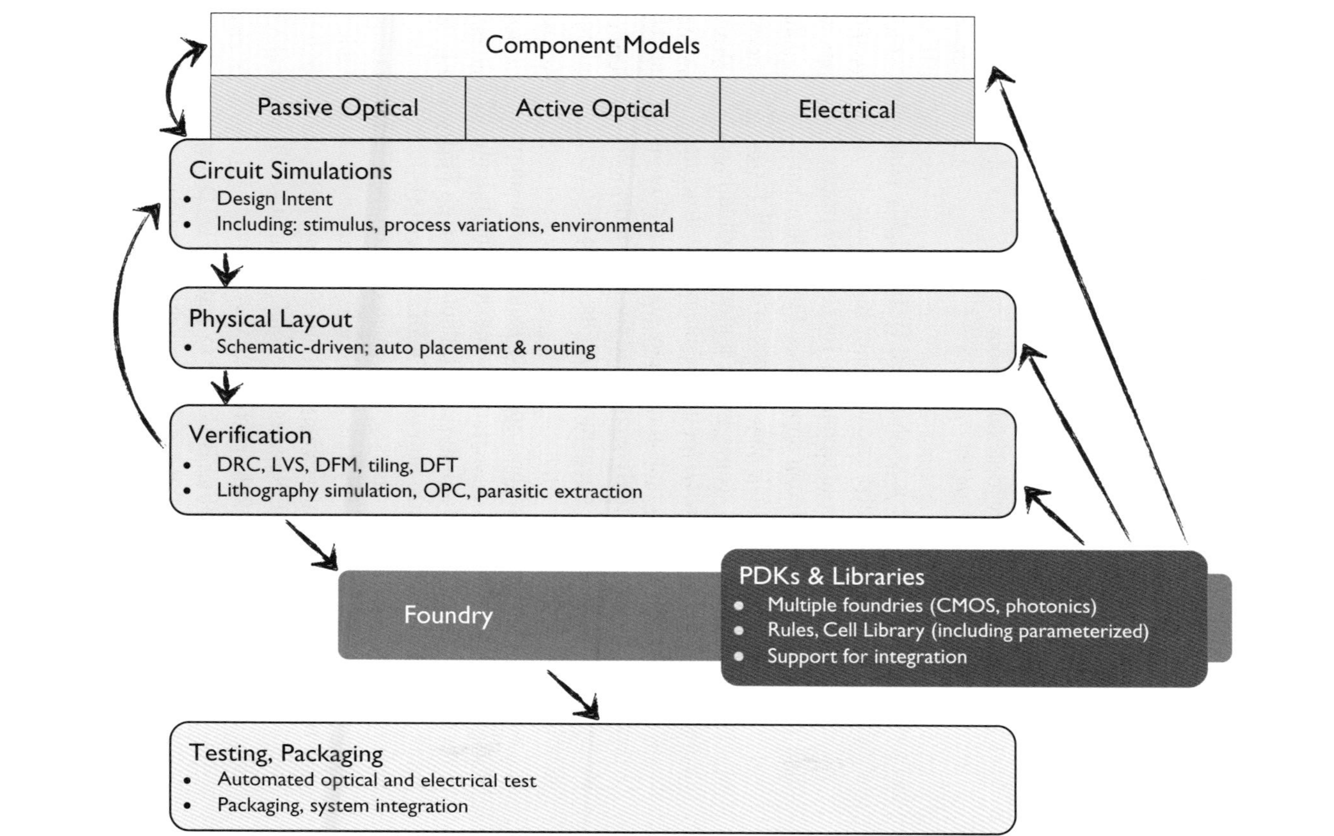

Figure 2.1 Graphical overview of a silicon photonic system design workflow, starting from the simulation of individual devices, the circuit simulations, verification, fabrication, testing, and packaging. The flow is supported by foundry-provided design rules and component libraries.

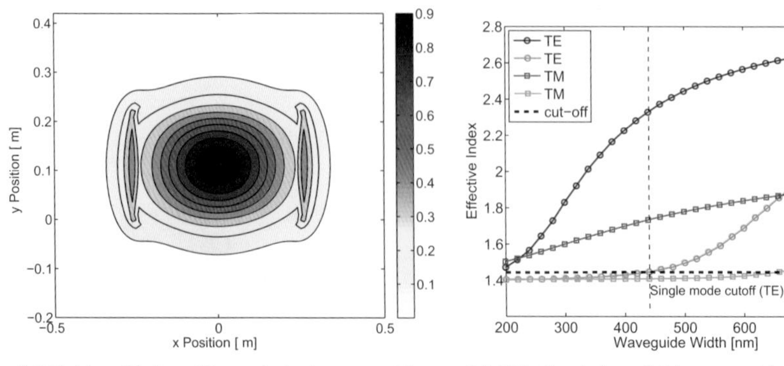

(a) Field profile in a silicon photonic waveguide.

(b) Effective index of this waveguide, simulated for numerous waveguide widths.

Figure 2.2 Example results from mode calculations.

including the Effective Index Method (EIM). For highly confined waveguides found in silicon photonic integrated circuits, with high index of refraction contrasts, it is important to use fully vectorial techniques for accurate results. This is available in both finite element and finite different approaches. Finite element solvers have the advantage that the mesh is unstructured, thus particularly useful for studying non-rectangular structures such as rounded shapes where a more flexible meshing is desirable. Finite element modelling is also useful for three-dimensional structures, such as for finding the resonant mode profile of an optical resonator such as a photonic crystal. The finite difference technique is particularly well suited for high-index contrast structures, and has the benefit that the mesh is compatible with the finite difference time domain (FDTD) technique described below.

Numerical tools that offer this capability include open-source implementations such as WGMODES in MATLAB [2, 3], and commercial tools including Lumerical MODE Solutions [4], COMSOL Multiphysics [5], Photon Design FIMMWAVE [6], Synopsys RSoft FemSIM [7], PhoeniX Software FieldDesigner [8], JCMwave [9], and others. Eigenmode calculations are also useful for determining the band structures in periodic structures such as photonic crystals and Bragg gratings, and are available in packages such as MIT Photonic-Bands (MPB) [10] and Synopsys RSoft BandSOLVE [7]. This book uses Lumerical MODE Solutions for mode calculations. Lumerical MODE uses a finite difference algorithm with sparse matrix techniques.

The procedure for modelling is similar to that of FDTD, described in Section 2.2.1. Briefly, the simulation steps are: define waveguide structure including the waveguide and the cladding; specify the materials; choose a mesh and accuracy; choose boundary conditions; and specify the wavelength, or range of wavelengths. Maxwell's equations are then formulated into a matrix eigenvalue problem and solved to obtain the effective index and mode profiles of the waveguide modes [11]. The simulations can be repeated for varied geometries, e.g. sweeping the width of the waveguide as shown in Figure 2.2b.

Mode calculations are used throughout this book for calculating: waveguide propagation parameters (Section 3.2); radiation in waveguide bends (Section 3.3.2); coupling coefficients in directional couplers (estimate) (Section 4.1.1); reflection coefficients for Bragg gratings (Section 4.5.2); mode overlap for edge coupling efficiency calculations and optimization (Section 5.3.1); and for calculating the effective index change and phase shift in pn-junction phase modulators, e.g. for ring and travelling wave modulators.

This book also uses "exact" analytic solutions of the 1D problem – the slab waveguide – to calculate its effective index in order to design fibre grating couplers (Section 5.2). The effective index method is used for approximate mode calculations, such as for the design for pn-junction phase shifters (Section 6.2.2). These are implemented both in MATLAB and within the mask layout tool, Mentor Graphics Pyxis, to create parameterized designs.

2.2 Wave propagation

The design of photonic devices usually requires an understanding of how light will propagate in the structure. For uniform devices like waveguides, the mode solvers described above will suffice since the mode does not change as it propagates down the waveguide. However, most other devices include a variation of the structure as the light propagates, which can lead to numerous reflections, interference, scattering and radiation, interaction between multiple modes, and changes in the mode profile. Numerous techniques exist for solving wave propagation. Many techniques were developed for specific limited problems, while other approaches are more general. The most general and rigorous time-domain approach is the finite difference time domain (FDTD) method, and is the typical work-horse in silicon photonics design and extensively used in this book.

2.2.1 3D FDTD

The FDTD technique is a numerical method for solving the three-dimensional (3D) Maxwell equations [12–15]. This technique is particularly useful for analyzing the interaction of light with complicated structures employing sub-wavelength-scale features. FDTD is an "exact" numerical calculation of Maxwell's equations, where the accuracy converges to the exact solution as the spatial discretization of the volume is reduced, i.e. smaller mesh size. As the name suggests, FDTD operates in the time domain. FDTD simulates the propagation of a pulse of light (tens to hundreds of femtoseconds long), which contains a broad spectrum of wavelength components. The system's response to this short pulse is related to the transmission spectrum via the Fourier transform. Thus, a single simulation provides the response of the optical system for a wide range of wavelengths at once. It is analogous to the impulse response, $h(t)$, which completely characterizes a linear, time-invariant (LTI) system such as an electrical filter. FDTD can model materials that are dispersive and nonlinear, and can even be extended to include electronic interactions such as those found in semiconductor lasers and optical amplifiers [16].

The disadvantage of the 3D FDTD technique is that it is computationally intensive, since the simulation time-step is sub-femtosecond. However, the algorithm scales well to multiple processor and compute cluster simulations, thus the simulation volume and accuracy can be improved with additional computing resources [17, 18], including graphic processing units (GPUs) [19].

This book uses the FDTD tool Lumerical FDTD Solutions [17]. This tool efficiently and accurately models the device response over wide optical bandwidths, including taking into account material dispersion. The software is able to generate the scattering parameters (S-Parameters) for components (Section 9.4.2), which is useful for system modelling. There are several alternative FDTD programs available, including Synopsys RSoft FullWAVE [7], Photon Design OmniSim [6], MEEP [18, 20], PhoeniX Software OptoDesigner [8], Acceleware FDTD [19], etc.

FDTD calculations are used throughout this book for calculating: waveguide bends (Section 3.3.1); coupling coefficients in directional couplers (Section 4.1.4); Bragg gratings with the inclusion of lithography smoothing effects (Section 4.5.2); fibre grating couplers (Section 5.2); edge couplers (Section 5.3.1); and for calculating the optical field profile and photo-current generation rates in detectors (Section 7.5). Examples of other devices that are suitable for FDTD design include Y-branch splitters, waveguide crossings, and polarization splitter rotators.

FDTD modelling procedure

The general procedure for performing simulations using FDTD is as follows.

(1) The optical materials are defined, ensuring they are appropriate for the intended simulation (e.g. material dispersion is properly accounted for). Typical materials in silicon photonics include silicon, silicon oxide, germanium, and metals. Accurate and consistent material models are important for comparing designs, particularly when models are created and compared by a team of designers. The materials can be defined via a graphic user interface (GUI), or via a script as in Listing 3.1.

(2) The structure is drawn, either via a GUI as in Figure 2.3a, or generated by a script as in the example in Listing 3.16. This includes defining the geometries for the silicon substrate, oxide cladding, silicon waveguides, and other materials such as germanium for detectors, and metal electrodes. The structure geometry can be parameterized, which is useful for design optimization.

(3) The simulation volume is defined, as illustrated in Figure 2.3b, and in the Listing 3.17. The simulation volume needs to be smaller than the structure to be simulated, i.e. the waveguides are extended past the simulation boundaries. Effectively, the simulation considers a fraction of the total device extent. Several simulation parameters need to be specified. (a) Mesh: the mesh can be defined by the number of mesh points per wavelength in the material. Advanced FDTD algorithms provide very accurate results for meshes of 14–18 points per wavelength, which in silicon is a mesh of 20–25 nm (note that classic FDTD algorithms typically required meshes of 1–10 nm for accurate results), and shown in Figure 2.4b. The choice of mesh has an enormous impact on simulation time,

(a) Materials and geometries are defined. Here a silicon waveguide surrounded by silicon dioxide cladding is drawn.

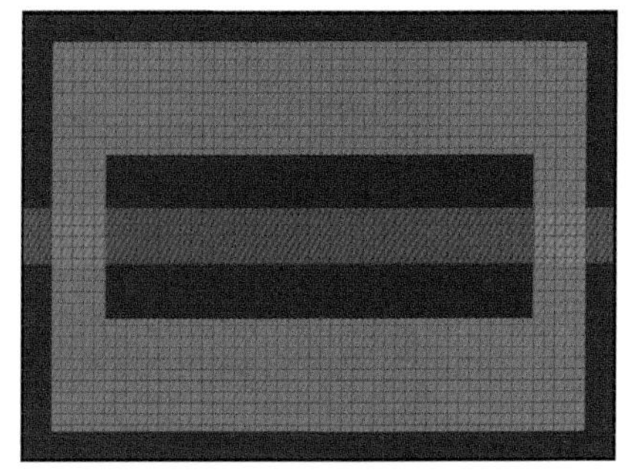

(b) Simulation volume, mesh, and boundary conditions are specified. This illustration shows the PML boundary conditions, which are approximately 500 nm wide and 1000 nm tall.

Figure 2.3 Steps for creating an FDTD simulation. This illustrative example is for a simple optical waveguide.

since the simulation time is proportional to $1/\Delta x^4$, where Δx is the mesh size. (b) The boundary conditions are defined. Perfect Matching Layers (PML) are the most general ones, which are used to absorb all the light leaving the simulation. However, these PMLs are implemented by numerous (e.g. 16) layers, thus adding to the computation time. In many cases, periodicity or symmetry can simplify the problem. In other cases, where there is an expectation that no light will be incident on a boundary, metallic boundaries (perfect reflectors) can be used. These are useful for rapid simulations and during debugging. (c) The simulation time needs to be defined. For structures where the light is expected to pass through once, such as directional couplers, grating couplers, Y-branches, etc., the simulation time is typically estimated by $t = L/v_g$, where L is the propagation length, and $v_g = c/n_g$ is the group velocity and is determined by the waveguide's group index, n_g, found via mode calculations (see Section 3.2.9). However, FDTD solvers can terminate automatically once most of the light has exited the simulation, hence it is typical to specify a much larger time and let the simulation terminate automatically.

(4) The optical sources are added. The most useful source in silicon photonics is a mode source, whereby the light is injected into a waveguide mode, as illustrated in Figure 2.4a, and in the Listing 3.17. This is achieved by calculating the modes for the waveguide, choosing the appropriate mode (e.g. fundamental TE mode), and launching this field into the waveguide. The centre wavelength (or time-domain parameters) need to be specified, e.g. centre wavelength of 1550 nm, with an optical bandwidth of 200 nm. Given that FDTD is a time-domain approach, it is generally desirable to choose a short pulse length, which translates to a broad optical spectrum. Finally, the source needs to be placed within the simulation volume, away from the boundary (e.g., 1–2 mesh points away).

(5) Monitors are added. Monitors are used to measure the optical field quantities, both E and H, at the chosen locations. Since FDTD simulations are in the

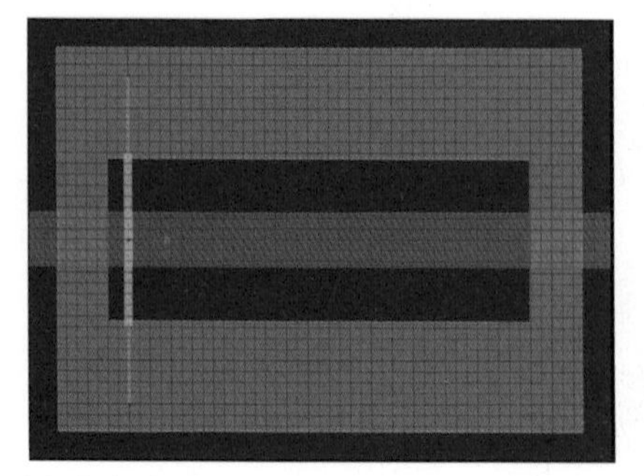

(a) Optical source is added (left side). Here a mode source is launching light into the silicon waveguide.

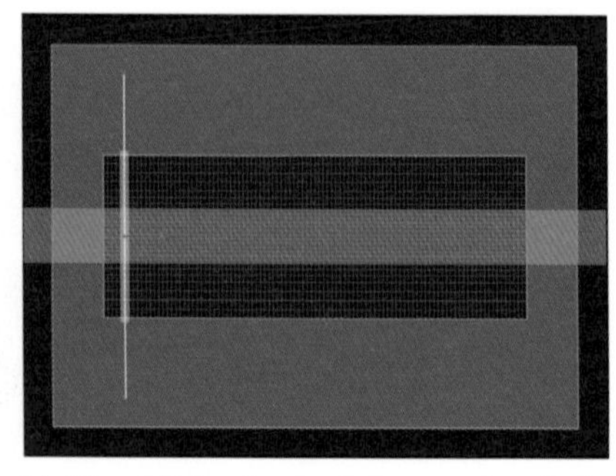

(b) After defining the simulation wavelength, the mesh is recalculated. A non-uniform mesh is used in this simulation.

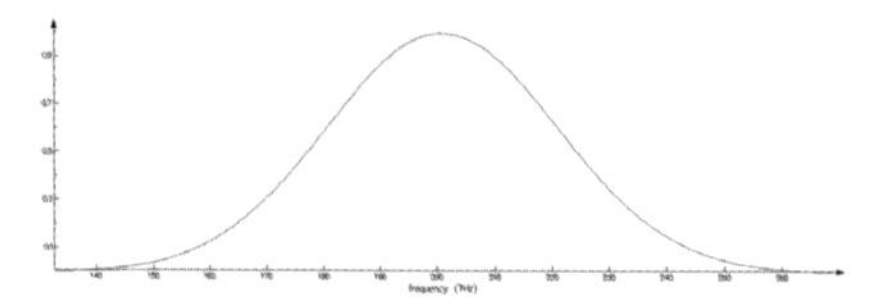

(c) Source wavelength/frequency settings. Here, a 200 THz centre frequency is chosen, with a 40 THz bandwidth.

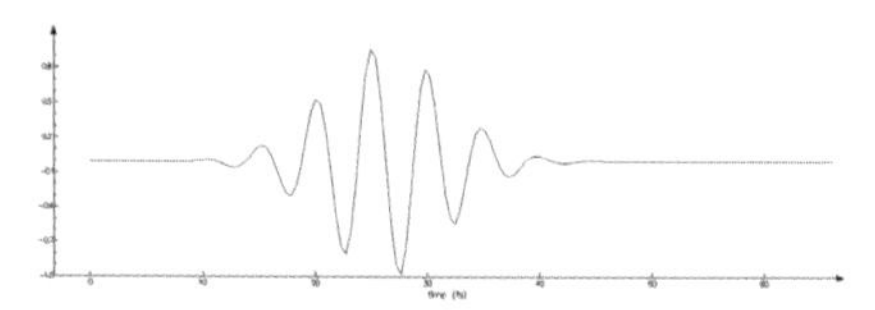

(d) The source in the time-domain consists of several optical cycles at the centre frequency. Here, the pulse duration is 30 fs.

Figure 2.4 Steps for creating an FDTD simulation, continued from Figure 2.3.

time domain, the fundamental monitor is the time-domain monitor. However, it is useful to calculate the frequency response of the device, and to normalize this response relative to the launched power spectral density. The result is a frequency-domain field profile monitor. This monitor can plot the field profile, or generate the optical transmission spectrum of the device. It is important to note that the field profile data are wavelength dependent, that is a single FDTD simulation generates a set of field profiles for a range of wavelengths. This is useful to study wavelength-dependent (directional couplers) and interference (e.g., interferometers, resonators) phenomena. Typical locations of interest include a single point, a line, a plane, or the entire 3D volume. As shown in Figure 2.5, a 2D monitor is used to measure the field profile at the output of the waveguide. Field profile monitors are also useful to generate a top-view profile of a device, for example, to help understand where the light is scattering. Movies can also be generated.

(6) Convergence testing is performed. In all numerical methods, the validity of the simulation needs to be ascertained. FDTD and mode calculations are accurate in the limit of small mesh size, but only to the degree that the simulation boundaries do not interfere with the results. Thus, simulation results need to be tested to ensure the results are stable, by decreasing the mesh size and increasing the simulation volume, etc. This enables the designer to understand and limit the numerical error in the simulation. This is described in more detail in Section 3.2.4 in the context of both mode and FDTD calculations.

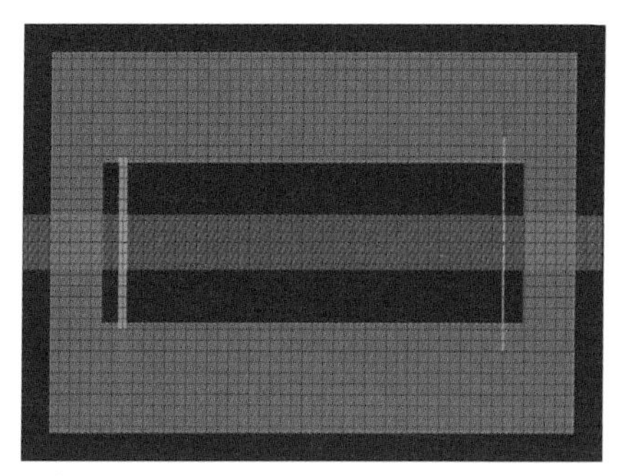

A field profile monitor is added at the waveguide output (right side).

Figure 2.5 Steps for creating an FDTD simulation, continued from Figure 2.4.

(7) Analyses are performed. With the knowledge of the modes of the waveguide, and the field profile at the output, it is possible to decompose the output field into eigenmodes. This is useful to study the excitation of higher-order modes by components. For example, the waveguide bend can excite higher-order modes, and it is of interest to determine how much power is transmitted into the fundamental mode. This is implemented in the Listing 3.17, using a mode expansion monitor. This is also useful for generating scattering parameters for the device (described in Section 9.3.2), either for single-mode, or generalized to include numerous modes and polarizations. Scattering parameters can be used to link multiple FDTD simulations [21] or for circuit simulations (Section 9.5). Other analyses include determining the quality factor of an optical resonator, by using the quality factor definition:

$$Q = 2\pi \frac{\text{Energy stored}}{\text{Energy dissipated per optical cycle}}. \tag{2.1}$$

This is accomplished by curve fitting the time-domain monitor data, in which the slope is related to the rate of energy decay, and Q.

(8) Parameter sweeping is typically performed when designing devices. Numerous simulations can be performed for various geometric parameters, wavelengths, polarizations, etc. Optimization algorithms, e.g. genetic algorithm, can also be used to find the global optimum design, provided that the designer can express the performance as a figure of merit used during the optimization.

Listings 3.16 and 3.17, and other examples throughout the book, serve the purpose of explicitly stating the simulation configuration that was used to obtain the presented results.

2.2.2 2D FDTD

The speed of FDTD simulations can be dramatically improved by reducing the problem to two dimensions. The best problems tackled with 2D FDTD are those that have one dimension in which the structure is invariant in one dimension; for example, fibre grating couplers with straight gratings can be approximated with gratings that are infinitely long. This approach can also be used for focusing grating couplers. In the case of grating couplers, full bi-directional simulations can be done in seconds in 2D FDTD, as opposed to minutes or hours with 3D FDTD (Section 5.2).

2.2.3 Additional propagation methods

For those cases where 3D FDTD is prohibitive in terms of computer resources and long simulation times are required, alternative methods, albeit approximations, can be considered.

2D FDTD with Effective Index Method

Also known as 2.5D FDTD, this approach is well suited for planar photonic integrated circuits. This method can be used for ridge/rib waveguide-based systems to more complex geometries such as photonic crystals. Since it uses the 2D FDTD algorithm, it allows for planar, omni-directional, propagation without any assumptions about an optical axis, which allows for structures like ring resonators to be efficiently modelled. The benefit of this approach is that it can quickly model devices on the scale of hundreds of microns. It is implemented in Lumerical MODE Solutions [4,22]. In this approach, a 3D structure is converted into a 2D FDTD simulation using the Effective Index Method [23–27], where the effective indices are found using the variational [26] or reciprocity [27] methods. The model is based on collapsing a 3D geometry into a 2D set of effective indices that can be solved with 2D FDTD; the resulting 2D geometry is the top-view of the photonic integrated circuit. The main assumption of this method is that there is little coupling between different slab modes. For many devices, such as SOI based slab waveguide structures, that are single-mode in each TE and TM polarization, this is an excellent assumption; specifically this applies to silicon thicknesses of less than 240 nm at a wavelength of 1550 nm (see Figure 3.10).

The calculation steps involve the following.

(1) Identification of the vertical slab modes of the core waveguide structure, over a desired range of wavelengths.
(2) Meshing of the structure and collapse of the third dimension by calculation of the corresponding effective 2D indices (taking into account the vertical slab mode profile). The generated effective materials are also dispersive, where the dispersion comes both from the original material properties (material dispersion) and the slab waveguide geometry (waveguide dispersion). These new materials are then fitted and used in the 2D FDTD simulation.
(3) 2D FDTD simulation where the original structure is replaced with the new effective materials.

2.5D versus 3D accuracy

Since 3D FDTD rigorously solves the vectorial Maxwell equations, its result is considered to be reliable. The results from the 2.5D method can be very close to that of the 3D FDTD method under some circumstances. In the case of 500 × 220 nm strip waveguides, we can determine the effective and group indices with the two approaches. The error in group index is only 2.7%; see Section 3.2.5. Using these waveguides in a ring resonator, the error introduced in the free spectral range (FSR) by this approach is approximately the same as the error in group index.

The advantage of this method, as with other approximations, is the fast simulation time. Coupled with a genetic optimization algorithm, this technique can be used to optimize a geometry by performing a large number of simulations. This technique was used, for example, to design a low-loss waveguide crossing [28].

This method, however, does not consider coupling between the TE and TM modes, hence cannot be used to model devices such as polarization rotators.

Another limitation of the technique is for cases where radiation will be present, such as in small-radius bent waveguide in Figure 3.26. Since this technique works only when the light maintains approximately the same vertical profile as the original slab modes, it will not be correct here since any radiation modes will not have the same vertical field profile. Thus, this approach typically overestimates the radiation loss of bent waveguides.

Beam Propagation Method (BPM)

BPM is an approximate solution that was popular prior to the advancements in both FDTD software and in computer speeds, and has allowed for the "exact" solution of Maxwell's equations. It was developed for slowly varying structures, approximating for paraxial (small angle) forward-only propagation for structures with small index of refraction contrasts, typically as a scalar solution. The technique has been expanded to allow for wide-angle propagation; forward and backward propagation including reflections (Synopsys RSoft BeamPROP [7]); and vectorial. This technique is useful for designing "sub-circuit" components such as arrayed waveguide gratings and Mach–Zehnder interferometers. It is available in PhoeniX Software OptoDesigner [8], Synopsys RSoft BeamPROP [7], OptiBPM by Optiwave [29], etc.

Eigenmode Expansion Method (EME)

EME considers the propagation of light by decomposing the local field into the modes at that position (known as "supermodes"). While travelling in a uniform medium (e.g. a section of waveguide, a directional coupler, the rectangle in an MMI), each mode is propagated individually simply by multiplying by the complex propagation constant. To connect to the next section of the device, scattering parameters are used. The use of S-parameters is inherently bi-directional hence includes both forward and backwards propagation. The technique is accurate for an infinite number of modes. When considering a large number of modes, this allows for propagation of modes with large angles, and with arbitrary accuracy. This technique is well suited for such structures as MMI couplers, tapers, directional couplers, gratings, etc. It is implemented in CAMFR, an open-source project from Ghent University [30], FIMMPROP by Photon Design [6], Synopsys RSoft ModePROP [7], etc. It can also be implemented in Lumerical MODE.

Coupled Mode Theory (CMT)

Coupled Mode Theory (CMT) is a widely used method for solving optical structures such as couplers, gratings, etc. This technique is based on perturbation, namely that the system is assumed to have a set of modes that are perturbed. For example, in a directional coupler, the mode is calculated for each waveguide in isolation. When the waveguides are brought together, it is assumed that the original modes are slightly perturbed and that there is a coupling between them. Similarly for Bragg gratings, it is assumed that there are forward- and backward-travelling modes for the straight waveguide; the gratings introduce small perturbations that couple the forward- and backward-going modes.

Transfer Matrix Method (TMM)

TMM is a very simple and fast technique for solving propagation in devices with a varying index of refraction profile, whereby matrices are used to describe the transmission and reflection through sections of material and at inferfaces. This approach provides an exact answer for one-dimensional structures such as thin-film reflectors, Distributed Bragg Reflectors in Vertical Cavity Lasers (VCSELs), etc. The use of this approach in silicon photonics requires an approximation of the structure as being one-dimensional, such as a Bragg grating where the structure is assumed to have a varying index of refraction in the direction of propagation, but where the transverse geometry does not change. TMM is available in commercial tools such as Synopsys RSoft GratingMOD [7] and Optiwave OptiGrating [29]. It is easily implemented in MATLAB, and used in this book to model Bragg gratings (Section 4.5.2).

2.2.4 Passive optical components

The wave propagation techniques described above (FDTD, EME, etc.) are suitable for a wide variety of passive optical devices, as well as active devices described in the next section. For passive devices, some generalizations can be made. Specifically, a passive device or system that is linear and time-invariant will fundamentally be reciprocal unless magnetic fields and magnetic materials are present (e.g. optical isolator). The implication of reciprocity is that two ports of an optical component can be switched, and the response should be identical. This is useful in modelling to verify the model – namely two simulations (e.g. forwards versus backwards simulation) should give identical results. For example, the ideal Y-branch (see Section 4.2) always results in a 50% insertion loss, regardless of whether the input is on the left (single port) or the right (one of two ports). One needs to be careful when using reciprocity to always consider the same mode at the input and output ports. Namely, reciprocity does not hold when simply measuring the total power going through a waveguide if more than one mode is present, as is the case of the Y-branch operated in the reverse direction; rather, modal decomposition must be performed to measure the power in each mode of the waveguides. Geometric symmetric, when present, is also useful in optical modelling to either reduce the simulation volume, or again to verify the simulations. Considerations for constructing compact models of passive optical components are described further in Section 9.4.4, namely in the context of generating S-parameters that satisfy reciprocity and passivity.

2.3 Optoelectronic models

Active silicon photonic devices, such as modulators, detectors, and lasers, require modelling of both optics and electro-optics. The electro-optics modelling is often referred to as Technology Computer Aided Design (TCAD). For the case of modulators and detectors, the modelling can be independently partitioned into the optical solution (i.e. mode profile, or field propagation using FDTD), and the electronic solution (e.g. electron and hole carrier density as a function of applied voltage). Thus, the simulation for a modulator consists of the electronic simulation, the results of which are fed into and perturb the optical solution. In the detector, the reverse is done. First the field is propagated in the detector. The field intensity, and more specifically the light absorbed, is used to generate a photo-current discretized throughout the volume of the detector. This spatially distributed current is used in the electronic simulations of the material, to simulate effects such as recombination, and to determine the voltage or current at the output of the detector.

TCAD electronic material simulations use physics-based electrical models for semiconductors. The tools self-consistently solve the system of equations that describe the electrostatic potential – the Poisson equation – and the spatial dependence of the free charges (electrons and holes) – the drift-diffusion equation. Numerous physical models can be included such as carrier recombination (Shockley–Read–Hall, Auger, stimulated and spontaneous emission), carrier transport across hetero-junctions (e.g. between silicon and germanium), electronic band structures, defects, traps, and interface states, mobility, temperature dependence, etc. In addition, the fabrication process can be taken into account to obtain realistic doping and stress profiles.

These types of simulations are available in Synopsis Sentaurus Process and Device [31], Silvaco TCAD tools such as Victory Process and Atlas [32], Crosslight CSUPREM and PICS3D [33], and Lumerical DEVICE [34].

Semiconductor lasers are modelled and designed using both optical and electro-optic models, solved self-consistently. Namely both the optical field profile and spatial carrier densities are interconnected, time varying and bias current dependent. The design of lasers in three dimensions can be accomplished using Crosslight PICS3D [33].

This book focuses on the design of modulators and detectors. It uses simple analytic models implemented in MATLAB to gain insight into the design of the silicon photonic modulator (Section 6.2.1). Lumerical DEVICE [34] is used for more accurate pn-junction phase modulator design, and for detector modelling.

2.4 Microwave modelling

The design of silicon photonics for high-frequency applications may require electromagnetic modelling of metal traces, particularly for large structures such as coplanar microwave waveguides used in travelling wave modulators, on-chip inductors, or on-chip capacitors. Several tools are available including SONNET [35] for planar circuits, ANSYS HFSS for a full 3D simulation [36], and Agilent Advanced Design System (ADS) [37]. HFSS and ADS are well suited for both RFIC design, as well as RF packaging.

2.5 Thermal modelling

Silicon photonic circuits are very temperature sensitive, due to the thermo-optic effect in silicon (Section 3.1.1). This is both advantageous for making thermal tuners and phase shifters (Section 6.5.2), but also problematic since the chip temperature is not likely to be uniform in a system, particularly when integrated with electronics and numerous thermal tuners. There are thus two needs for thermal modelling. The first is for device design, whereby a designer would like to optimize the efficiency of a thermal tuner. There are numerous commercial thermal modelling tools, including ANSYS [38] and COMSOL [5]. This book uses the MATLAB Partial Differential Toolbox [39] to solve the steady-state heat equation. Results from the thermal model can be used to construct a compact model of a device that includes a thermal tuner, such as a Mach–Zehnder or ring modulator. The second is to model the temperature distribution at the circuit level. This is useful to understand thermal cross-talk and for optimization of the system design. Temperature can be included as part of the photonic circuit modelling tools, described in Section 2.6.

2.6 Photonic circuit modelling

For designing silicon photonic circuits containing numerous components, there are several approaches and tools available. The general idea is to build simple models for components. The simulations thus focus on the functionality and performance of the entire circuit. Many methods are available for implementing compact models for simulating devices and sub-systems that either have analytic solutions (e.g. 1D structures such as thin-film filters), or for systems that are physically too large to be effectively handled by numerical methods. In the latter, phenomenological models, such as parameterized waveguides, can be used to simplify the simulation of larger circuits. Desired simulations include frequency-domain response of the system (optical filter characterization) and time-domain simulations (transient, eye diagram, bit error rate). The reader may be interested in journal papers that describe optical circuit modelling methods [21, 40, 41].

For optical circuit modelling, the choices include: building simple circuit models in a programming language, e.g. in MATLAB [42]; using open-source solutions such as the Python-based Caphe from University of Ghent [40, 43]; implementing optical models in an EDA environment, such as the Luxtera approach using VerilogA [44]; and using one of numerous commercial tools for simulating the steady-state response such as the Advanced Simulator for Photonic Integrated Circuits (ASPIC) [45] included as part of the PhoeniX Software Design Flow Automation [8], using tools for time-domain modelling, such as Photon Design PICWave [6] and Optiwave OptiSystem [29], and using tools that simulate both the time-domain and frequency-domain circuit response, including Synopsys RSoft OptiSim and ModeSYS [7], VPIsystems VPItransmissionMaker and VPIcomponentMaker Photonic Circuits [46], Lumerical INTERCONNECT [47], Caphe [40, 43], and others.

For the propagation of light in optical fibres, the models typically take into account both linear and nonlinear pulse propagation. This is necessary to understand dispersion, four-wave mixing, etc. This is typically accomplished by solving the Schroedinger equation, and is available in tools including the open source SSPROP [48], and Synopsys RSoft OptiSim [7]. For nonlinear silicon photonics, this approach may also be suitable for pulse propagation, or FDTD simulations can be used for the design of nonlinear optical components.

For silicon photonic systems, there will often be an electronic chip. In many cases, these can be designed separately, which allows the designer to choose the best tools for each chip. However, some systems will require co-simulation of both optics and electronics, e.g. optoelectronic oscillators. This leads to the choice of starting with a well established electronics tool (e.g. Cadence) and adding the optical functionality within the constraints of the electronic modelling approach (e.g., VHDL, Spice, VerilogA), or to start with a tool focused on optical simulations, and add on electronic modelling capabilities. The electronic-tool approach was implemented in Cadence by Luxtera using VerilogA [44]. The photonic-tool approach is available in tools such as Synopsys RSoft OptiSim [7] and Optiwave OptiSPICE [29,49]. Finally, in the future there may be a third solution, namely advanced EDA tools coupled with advanced photonic design tools, whereby the EDA tool couples to a photonic simulation engine for electronic–photonic co-simulation.

The optical circuit modelling tool used in this book is Lumerical INTERCONNECT [47]. This is a photonic integrated circuit design software package that allows for the design, simulation, and analysis of integrated circuits, silicon photonics, and optical interconnects containing such devices as Mach–Zehnder modulators, coupled ring-resonators, and arrayed waveguide gratings. INTERCONNECT includes both time- and frequency-domain simulators. In the time domain, the simulator calculates each element in order to generate time-domain waveform samples and propagate them bi-directionally. Very close coupling between components can be simulated allowing, for example, the analysis of optical resonators. Frequency-domain simulation is performed using scattering data analysis to calculate the overall circuit response. It is done by solving a sparse matrix that represents the circuit as connected scattering matrices, each one of them representing the frequency response of an individual element. The individual elements can be built using experimental data, analytic or phenomenological models, or numerically calculated models (e.g. by FDTD or mode solver).

2.7 Physical layout

There are numerous physical mask layout tools available for electronics, MEMS, photonics, and other applications. There are several open source options: KLayout [50] is primarily focused on polygon-based graphical editing, while Python [51] or MATLAB [52] toolboxes enable polygon-based layout using scripts. IPKISS [53] offers scripting in Python to create parameterized layouts of components and circuits. There are numerous commercial tools for creating layouts including Cadence Virtuso Layout

Suite [54], Mentor Graphics Pyxis [55], Synopsys IC Compiler [56], Tanner LEdit [57], LayoutEditor [58], PhoeniX Software MaskEngineer [8], WieWeb CleWin [59], and Design Workshop Technologies DW-2000 [60]. As described in Chapter 10, the mask layout aspect of the design flow also includes verification, including design rule checking, lithography simulation, layout versus schematic, etc. Several of the tools listed here offer a comprehensive solution.

The design flow presented in this book, particularly in Section 10.1, uses Mentor Graphics Pyxis for schematic capture and physical layout, Eldo for electronic modelling and design [61], together with several other Mentor Graphics tools including Calibre for physical verification, design for manufacturing and 3D-IC integration [62].

2.8 Software tools integration

Clearly there are numerous software packages that are available and necessary for silicon photonic–electronic integrated-circuit designers. This poses a challenge both for individual designers and for organizations. For the designer, one challenge is the requirement to learn and understand numerous tools. For organizations, this leads to an increased expense and an operation challenge with administering the tools and the data flow. The electronics industry has also been facing this problem for numerous years. This has been met in part by the following strategies. (1) Interoperability and data exchange between tools from different vendors allow designers to customize and optimize the design flow for the specific application; this has been aided by standards organizations. (2) Use of a common operating system simplifies the communication between tools. In the EDA world, this has been dominated by Linux operating systems. For silicon photonic designs, which will ultimately require the co-design of both electronics and photonics components, it is thus desirable to choose tools that function on this platform. Of course numerous vendors also offer their tools on multiple platforms (i.e. OSX and Windows), which can be useful for designers focusing on a specific task (e.g. passive component design), but ultimately it is desirable to have all tools on the same platform (i.e. Linux).

Full silicon photonic design flows have been, and are being, developed by several groups, including Luxtera, IPKISS [53], and PhoeniX. The Luxtera flow runs in Linux within the Cadence environment. IPKISS runs on all three platforms with the software implemented in Python. IPKISS also offers a unique design flow – namely it is available via a web browser. Finally, PhoeniX Software runs on the Windows platform.

The designers who contributed to this book have been motivated to minimize the number of tools used and identify an efficient work flow. Although the list of tools is still large, where possible, one tool is used for as many functions as possible. For example, FDTD, the most general of passive modelling tools, is used for most photonic device design, at the expense of computation time. For mask layout, the same tool is used for schematic capture, layout, design rule checking, layout versus schematic checking, etc. Given that one goal is the electronic–photonic integration, we have chosen to use a set of Linux-based electronic design automation tools from a major vendor, Mentor

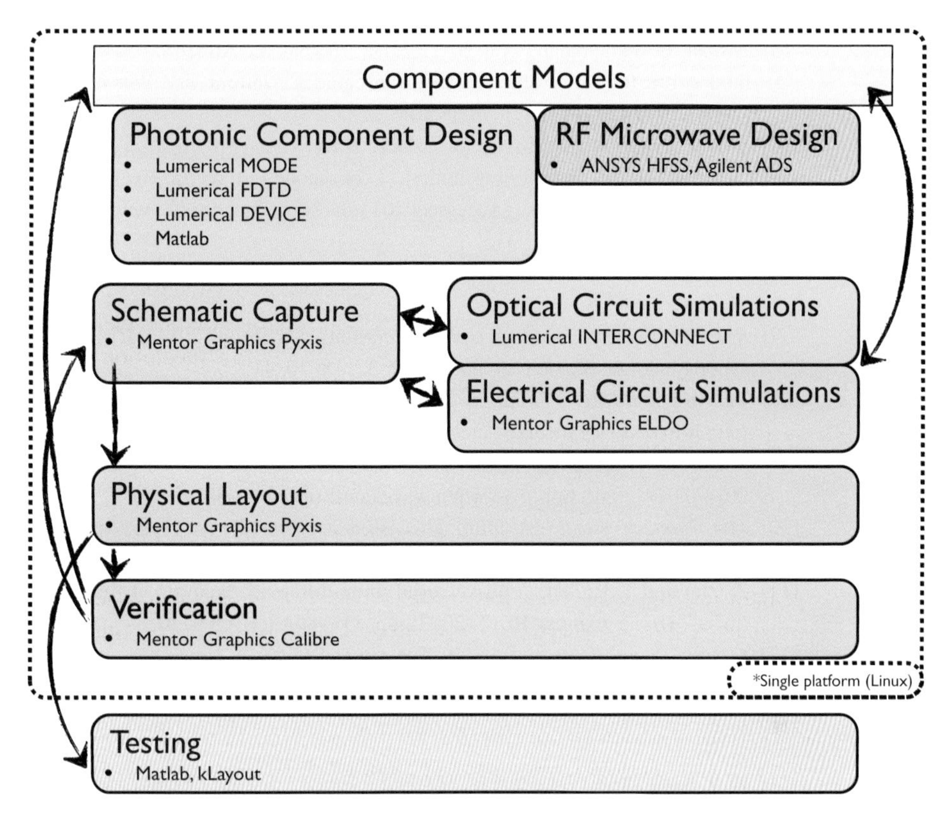

Figure 2.6 Graphical overview of the simulation tools used in the design workflow presented in this book.

Graphics, together with compatible optical circuit and component modelling software, from Lumerical Solutions, as shown in Figure 2.6. Having all tools on the same computer and operating system simplifies the system and software administration (e.g., software updates, revision control), simplifies the design flow with all files being easily accessible by all tools, enables collaborative design with multiple users, and provides a simple method for collaboration between individuals at multiple locations using remote desktops (e.g., VNC, NX). Access to a remote server has been also offered by the OpSIS foundry service, to provide designers with a complete design environment for them to complete their designs. The server has included the design environment (software) and the process design kit.

References

[1] Lukas Chrostowski, Jonas Flueckiger, Charlie Lin, *et al.* "Design methodologies for silicon photonic integrated circuits". *Proc. SPIE, Smart Photonic and Optoelectronic Integrated Circuits XVI* 8989 (2014), pp. 8989–9015 (cit. on p. 28).

[2] *WGMODES – Photonics Research Laboratory.* [Accessed 2014/04/14]. URL: http://www.photonics.umd.edu/software/wgmodes/ (cit. on p. 30).

[3] Arman B. Fallahkhair, Kai S. Li, and Thomas E. Murphy. "Vector finite difference modesolver for anisotropic dielectric waveguides". *Journal of Lightwave Technology* **26**.11 (2008), pp. 1423–1431 (cit. on p. 30).

[4] *MODE Solutions – Waveguide Mode Solver and Propagation Simulator.* [Accessed 2014/04/14]. URL: http://www.lumerical.com/mode (cit. on pp. 30, 36).

[5] *COMSOL Multiphysics.* [Accessed 2014/04/14]. URL: http://www.comsol.com (cit. on pp. 30, 40).

[6] *Photon Design.* [Accessed 2014/04/14]. URL: http://www.photond.com/ (cit. on pp. 30, 32, 37, 40).

[7] *RSoft Products – Synopsys Optical Solutions.* [Accessed 2014/04/14]. URL: http://optics.synopsys.com/rsoft/ (cit. on pp. 30, 32, 37, 38, 40, 41).

[8] *PhoeniX Software – Solutions for micro and nano technologies.* [Accessed 2014/04/14]. URL: http://www.phoenixbv.com/ (cit. on pp. 30, 32, 37, 40, 42).

[9] *JCMwave – Complete Finite Element Technology for Optical Simulations.* [Accessed 2014/04/14]. URL: http://www.jcmwave.com/ (cit. on p. 30).

[10] *MIT Photonic Bands – AbInitio.* [Accessed 2014/04/14]. URL: http://ab-initio.mit.edu/wiki/index.php/MIT_Photonic_Bands (cit. on p. 30).

[11] Z. Zhu and T. Brown. "Full-vectorial finite-difference analysis of microstructured optical fibers". *Optics Express* **10**.17 (2002), pp. 853–864 (cit. on p. 30).

[12] *Computational Electrodynamics: The Finite-Difference Time-Domain Method.* Third. Vol. ISBN: 1580538320. Artech House; 3rd edition, 2005 (cit. on p. 31).

[13] D. M. Sullivan. *Electromagnetic Simulation Using the FDTD method.* IEEE, 2000 (cit. on p. 31).

[14] K. S. Kunz and R. J. Luebbers. *The Finite Difference Time Domain Method for Electromagnetics.* CRC, 1993 (cit. on p. 31).

[15] S. D. Gedney. "Introduction to the Finite-difference Time-domain (FDTD) method for electromagnetics". *Synthesis Lectures on Computational Electromagnetics* **6**.1 (2011), pp. 1–250 (cit. on p. 31).

[16] Shih-Hui Chang and Allen Taflove. "Finite-difference time-domain model of lasing action in a four-level two-electron atomic system". *Optics Express* **12**.16 (2004), pp. 3827–3833 (cit. on p. 31).

[17] *FDTD Solutions – Lumerical's Nanophotonic FDTD Simulation Software.* [Accessed 2014/04/14]. URL: http://www.lumerical.com/fdtd (cit. on p. 32).

[18] *Meep – AbInitio.* [Accessed 2014/04/14]. URL: http://ab-initio.mit.edu/wiki/index.php/Meep (cit. on p. 32).

[19] *FDTD Solvers – Acceleware Ltd.* [Accessed 2014/04/14]. URL: http://www.acceleware.com/fdtd-solvers (cit. on p. 32).

[20] Ardavan F. Oskooi, David Roundy, Mihai Ibanescu, *et al.* "MEEP: A flexible free-software package for electromagnetic simulations by the FDTD method". *Computer Physics Communications* **181**.3 (2010), pp. 687–702 (cit. on p. 32).

[21] Tsugumichi Shibata and Tatsuo Itoh. "Generalized-scattering-matrix modeling of waveguide circuits using FDTD field simulations". *IEEE Transactions on Microwave Theory and Techniques* **46**.11 (1998), pp. 1742–1751 (cit. on pp. 35, 40).

[22] *Lumerical's 2.5D FDTD Propagation method.* [Accessed 2014/04/14]. URL: https://www.lumerical.com/solutions/innovation/2.5d_fdtd_propagation_method.html (cit. on p. 36).

[23] R. M. Knox and P. P. Toulios. "Integrated circuits for the millimeter through optical frequency range". *Proceedings of the Symposium on Submillimeter Waves.* Vol. 20. Brooklyn, NY. 1970, pp. 497–515 (cit. on p. 36).

[24] J. Buus. "The effective index method and its application to semiconductor lasers". *IEEE Journal of Quantum Electronics* **18**.7 (1982), pp. 1083–1089 (cit. on p. 36).

[25] G. B. Hocker and W. K. Burns. "Mode dispersion in diffused channel waveguides by the effective index method". *Applied Optics* **16**.1 (1977), pp. 113–118 (cit. on p. 36).

[26] M. Hammer and O. V. Ivanova. "Effective index approximations of photonic crystal slabs: a 2-to-1-D assessment". *Optical and Quantum Electronics* **41**.4 (2009), pp. 267–283 (cit. on p. 36).

[27] A. W. Snyder and J. D. Love. *Optical Waveguide Theory*. Vol. 190. Springer, 1983 (cit. on p. 36).

[28] Yangjin Ma, Yi Zhang, Shuyu Yang, *et al.* "Ultralow loss single layer submicron silicon waveguide crossing for SOI optical interconnect". *Optics Express* **21**.24 (2013), pp. 29 374–29 382. DOI: 10.1364/OE.21.029374 (cit. on p. 37).

[29] *Optiwave*. URL: http://www.optiwave.com/ (cit. on pp. 37, 38, 40, 41).

[30] *CAMFR Home Page*. [Accessed 2014/04/14]. URL: http://camfr.sourceforge.net/ (cit. on p. 37).

[31] *Synopsys TCAD*. [Accessed 2014/04/14]. URL: http://www.synopsys.com/Tools/TCAD (cit. on p. 39).

[32] *Silvaco TCAD*. [Accessed 2014/04/14]. URL: http://www.silvaco.com/products/tcad.html (cit. on p. 39).

[33] *Semiconductor TCAD Numerical Modeling and Simulation – Crosslight Software*. [Accessed 2014/04/14]. URL: http://crosslight.com/products/pics3d.shtml (cit. on p. 39).

[34] *Lumerical DEVICE – Optoelectronic TCAD Device Simulation Software*. [Accessed 2014/04/14]. URL: http://www.lumerical.com/tcad-products/device/ (cit. on p. 39).

[35] *EM Analysis and Simulation – High Frequency Electromagnetic Software Solutions – Sonnet Software*. [Accessed 2014/04/14]. URL: http://www.sonnetsoftware.com (cit. on p. 39).

[36] *ANSYS HFSS*. [Accessed 2014/04/14]. URL: http://www.ansys.com/Products/Simulation+Technology/Electronics/Signal+Integrity/ANSYS+HFSS (cit. on p. 39).

[37] *Advanced Design System (ADS) – Agilent*. [Accessed 2014/04/14]. URL: http://www.home.agilent.com/en/pc-1297113/advanced-design-system-ads (cit. on p. 39).

[38] *ANSYS Multiphysics*. [Accessed 2014/04/14]. URL: http://www.ansys.com/Products/Simulation+Technology/Systems+&+Multiphysics/Multiphysics+Enabled+Products/ANSYS+Multiphysics (cit. on p. 40).

[39] *PDE – Partial Differential Equation Toolbox – MATLAB*. [Accessed 2014/04/14]. URL: http://www.mathworks.com/products/pde/ (cit. on p. 40).

[40] Martin Fiers, Thomas Van Vaerenbergh, Ken Caluwaerts, *et al.* "Time-domain and frequency-domain modeling of nonlinear optical components at the circuit-level using a node-based approach". *Journal of the Optical Society of America B* **29**.5 (2012), pp. 896–900. DOI: 10.1364/JOSAB.29.000896 (cit. on p. 40).

[41] Daniele Melati, Francesco Morichetti, Antonio Canciamilla, *et al.* "Validation of the building-block-based approach for the design of photonic integrated circuits". *Journal of Lightwave Technology* **30**.23 (2012), pp. 3610–3616 (cit. on p. 40).

[42] Marek S. Wartak. *Computational Photonics: An Introduction with MATLAB*. Cambridge University Press, 2013 (cit. on p. 40).

[43] *Caphe – analysis of optical circuits in frequency and time domain*. [Accessed 2014/04/14]. URL: http://www.intec.ugent.be/caphe/ (cit. on p. 40).

[44] Thierry Pinguet, Steffen Gloeckner, Gianlorenzo Masini, and Attila Mekis. "CMOS photonics: a platform for optoelectronics integration". In *Silicon Photonics II*. David J.

Lockwood and Lorenzo Pavesi (eds.). Vol. 119. Topics in Applied Physics. Springer Berlin Heidelberg, 2011, pp. 187–216. ISBN: 978-3-642-10505-0. DOI: 10.1007/978-3-642-10506-7_8 (cit. on pp. 40, 41).

[45] *Aspic Design: Home Page.* [Accessed 2014/04/14]. URL: http://www.aspicdesign.com/ (cit. on p. 40).

[46] *VPIphotonics: Simulation Software and Design Services.* [Accessed 2014/04/14]. URL: http://www.vpiphotonics.com (cit. on p. 40).

[47] *Lumerical INTERCONNECT – Photonic Integrated Circuit Design Tool.* [Accessed 2014/04/14]. URL: http://www.lumerical.com/tcad-products/interconnect/ (cit. on pp. 40, 41).

[48] *SSPROP – Photonics Research Laboratory.* [Accessed 2014/04/14]. URL: http://www.photonics.umd.edu/software/ssprop/ (cit. on p. 41).

[49] Pavan Gunupudi, Tom Smy, Jackson Klein, and Z Jan Jakubczyk. "Self-consistent simulation of opto-electronic circuits using a modified nodal analysis formulation". *IEEE Transactions on Advanced Packaging* **33**.4 (2010), pp. 979–993 (cit. on p. 41).

[50] *KLayout Layout Viewer and Editor.* [Accessed 2014/04/14]. URL: http://www.klayout.de (cit. on p. 41).

[51] *GDSII for Python – Gdspy 0.6 documentation.* [Accessed 2014/04/14]. URL: http://gdspy.sourceforge.net (cit. on p. 41).

[52] *GDS II Toolbox – Ulf's Cyber Attic.* [Accessed 2014/04/14]. URL: https://sites.google.com/site/ulfgri/numerical/gdsii-toolbox (cit. on p. 41).

[53] *IPKISS.* [Accessed 2014/04/14]. URL: http://www.ipkiss.org (cit. on pp. 41, 42).

[54] *Cadence Virtuoso Layout Suite.* [Accessed 2014/04/14]. URL: http://www.cadence.com/products/cic/layoutsuite (cit. on p. 42).

[55] *Custom IC Design – Mentor Graphics.* [Accessed 2014/04/14]. URL: http://www.mentor.com/products/ic_nanometer_design/custom-ic-design (cit. on p. 42).

[56] *Synopsys Physical Verification.* [Accessed 2014/04/14]. URL: http://www.synopsys.com/Tools/Implementation/PhysicalImplementation (cit. on p. 42).

[57] *Industry-leading Productivity for Analog, Mixed Signal and MEMS Layout from Tanner EDA.* [Accessed 2014/04/14]. URL: http://www.tannereda.com/products/l-edit-pro (cit. on p. 42).

[58] *LayoutEditor.* [Accessed 2014/04/14]. URL: http://www.layouteditor.net (cit. on p. 42).

[59] *WieWeb software – Layout Software.* [Accessed 2014/04/14]. URL: http://www.wieweb.com/ns6/index.html (cit. on p. 42).

[60] *Design Workshop Technologies.* [Accessed 2014/04/14]. URL: http://www.designw.com (cit. on p. 42).

[61] *Eldo Classic – Foundry Certified SPICE Accurate Circuit Simulation – Mentor Graphics.* [Accessed 2014/04/14]. URL: http://www.mentor.com/products/ic_nanometer_design/analog-mixed-signal-verification/eldo/ (cit. on p. 42).

[62] *Physical Verification – Mentor Graphics.* [Accessed 2014/04/14]. URL: http://www.mentor.com/products/ic_nanometer_design/verification-signoff/physical-verification/ (cit. on p. 42).

Part II

Passive components

3 Optical materials and waveguides

This chapter describes the substrates used in silicon photonics and the optical properties of the materials. Two typical optical waveguides are presented – strip and rib waveguides. First, we calculate the optical modes of the substrate (slab modes). Then, we consider the modes in a confined waveguide. Methods of ensuring that numerical accuracy has been achieved are presented. We also discuss optical propagation losses and the losses in 90° bends.

3.1 Silicon-on-insulator

The wafers commonly used for silicon photonics are termed "silicon-on-insulator". These are commonly used in the electronics industry for high-performance circuits. The typical 200 mm (8") wafer consists of a 725 μm silicon substrate, 2 μm of oxide (buried oxide, or BOX), and 220 nm of crystalline silicon, as shown in Figure 3.1. It is in the top crystalline silicon layer that waveguides and devices are defined. Hence, the material properties of this silicon are important for designing optical (and optoelectronic) devices.

The examples in this book are based on a top silicon thickness of 220 nm, which has been used since before 2003 (e.g. Reference [1]). The thickness is considered in Section 3.2.4. The 220 nm thickness has become a standard used in particular by multi-project wafer foundries and foundry service providers (e.g. imec, LETI, IME, see also Section 1.5.5). However, it should be noted that other thicknesses are also in use (e.g. Luxtera, Kotura, Skorpios). It should be noted that the optimum thickness is application dependent and that 220 nm may not be the optimal choice [2].

The thickness variation of the top silicon is also an important parameter, and can have a variation range of +/− 5 nm (see References [3, 4] and Section 11.1 for discussions on non-uniformity). The material is typically intrinsic, with a light doping density of 1×10^{15} cm^{-3}.

3.1.1 Silicon

In this section, we describe silicon's refractive index wavelength and temperature dependence. The effect of free carriers is considered in Section 6.1.1.

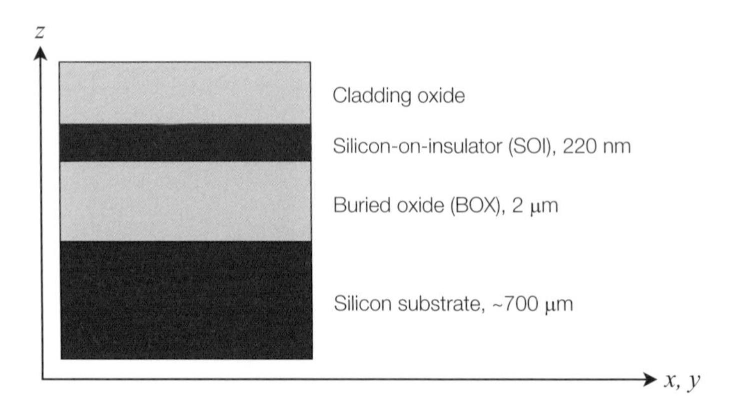

Figure 3.1 Cross-sectional view of silicon-on-insulator (SOI) wafer.

Silicon – wavelength dependence

We are interested in designing devices for a variety of wavelengths, hence the refractive index wavelength dependence of silicon and silicon dioxide should be taken into account in order to correctly describe the dispersion effects. The simplest model for silicon's wavelength dependence is a first-order dependence of -7.6×10^{-5} nm^{-1} [5]. For a more complete description, the Sellmeier equation is typically used to describe the index of refraction of materials:

$$n^2(\lambda) = \epsilon + \frac{A}{\lambda^2} + \frac{B\lambda_1^2}{\lambda^2 - \lambda_1^2}. \tag{3.1}$$

However, this model cannot be used directly for FDTD simulations. Instead, a Lorentz model [6] can be used:

$$n^2(\lambda) = \epsilon + \frac{\epsilon_{\text{Lorentz}}\omega_0^2}{\omega_0^2 - 2i\delta_0 2\pi c/\lambda - \left(\frac{2\pi c}{\lambda}\right)^2}. \tag{3.2}$$

The advantage of using this model is that the same material model for silicon in both eigenmode (Section 2.1) and FDTD calculations (Section 2.2.1) can be used to ensure consistent simulations. Appropriate coefficients are chosen to match the silicon data (from Palik's handbook [5]) over a wavelength range of 1.15 to 1.8 μm: $\epsilon = 7.9874$, $\epsilon_{\text{Lorentz}} = 3.6880$, $\omega_0 = 3.9328 \times 10^{15}$, $\delta_0 = 0$. This model satisfies Kramers–Kronig relations hence is compatible with FDTD modelling. It is also lossless for $\delta_0 \to 0$. A plot of the experimental data and Equation (3.2) is shown in Figure 3.2. Implementation of the material in a Lumerical script is provided in Listing 3.1.

Silicon – temperature dependence

The modification of the refractive index in silicon is due to the changes in the distribution functions of carriers and phonons, and the temperature-induced shrinkage of the bandgap [7]. As we will see, a slight modification of the temperature will shift the transmission spectrum of photonic devices such as ring resonators. This will be useful for thermally tuning devices.

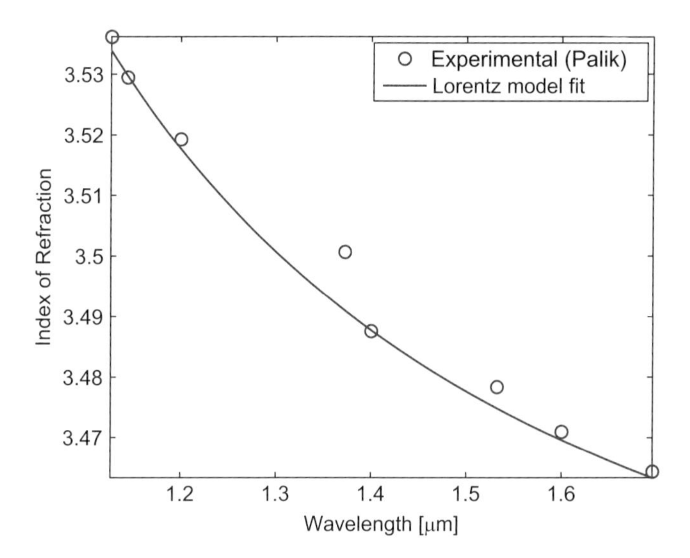

Figure 3.2 Index of refraction of silicon at room temperature, $T = 300$ K. Data fit using Lorentz model.

The temperature dependence can be approximated as $\beta = \frac{1}{n}\frac{dn}{dT}$, which for silicon is $5.2 \times 10^{-5}\ \mathrm{K}^{-1}$ [8–10]. It has also been measured to be $\frac{dn}{dT} = 1.87 \times 10^{-4}\ \mathrm{K}^{-1}$ at 1500 nm [11].

3.1.2 Silicon dioxide

Silicon dioxide, SiO_2 (also termed glass, or silica), has a nearly constant index of refraction around 1.444 at 1550 nm (material dispersion is about 6× lower than in silicon, i.e. $-1.2 \times 10^{-5}\ \mathrm{nm}^{-1}$). Silicon dioxide's temperature dependence is also 6.3× lower than silicon [10]. In addition, most of the light is confined in the silicon, hence the oxide dispersion and temperature dependence does not play a significant role in the performance of silicon photonic circuits. The inclusion of the oxide dispersion may be important for waveguides where light is present outside the waveguide, e.g. in slot waveguides, or in thin waveguides. In such a case, the model presented in Figure 3.3 can be used. Implementation of the material in a Lumerical script is provided in Listing 3.1.

However, for FDTD simulations, it is often preferable to choose a simpler model (fewer convergence issues, faster simulation time), specifically the constant index model ($n = 1.444$ at 1550 nm). The error introduced by using a constant index versus the dispersive one, for the group index of a 500 × 220 nm strip waveguide, is about 0.1%. Hence, we generally use the constant index model in FDTD and MODE simulations.

3.2 Waveguides

There are several types of waveguides used in silicon photonics. The strip waveguide (also known as channel, photonic wire, or ridge waveguide), Figure 3.4(left), is typically

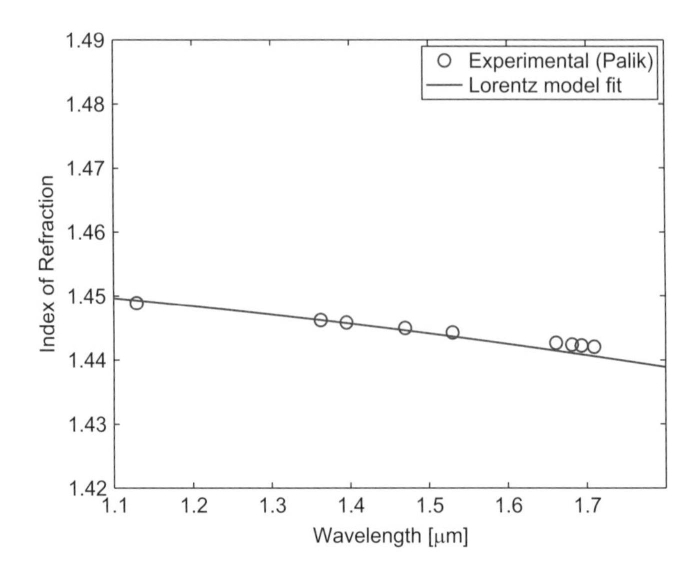

Figure 3.3 Index of refraction of silicon dioxide (SiO_2) at room temperature, T = 300 K. Data fit using Lorentz model.

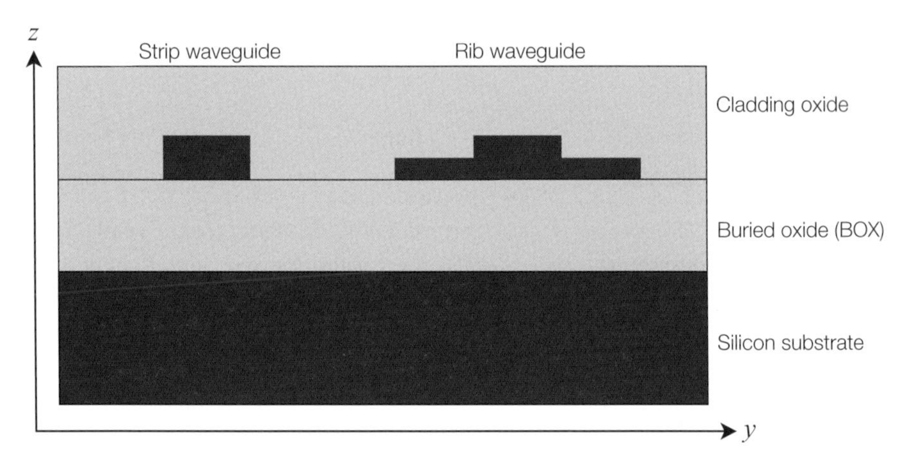

Figure 3.4 Common waveguides in silicon photonics. (Left) Strip waveguide, also known as channel waveguides, photonic wires, or ridge waveguides. (Right) Rib waveguide, also known as ridge waveguide or strip-loaded ridge waveguide.

used for routing as it offers tight bend radii. The rib waveguide (also known as strip-loaded ridge, or ridge waveguide), Figure 3.4(right), is used for electo-optic devices such as modulators, since it allows for electrical connections to be made to the waveguide. Both waveguides can be made to have a loss of less than 3 dB/cm.

Oxide cladding is used to protect the devices and to permit the fabrication of metal interconnects above the waveguides. Waveguides can also be fabricated without the cladding, for example, for evanescent field sensors for lab-on-chip applications [12–14].

3.2.1 Waveguide design

Waveguide design typically has the following procedure.

- First, one-dimensional (1D) calculations are performed to determine the slab waveguide modes, as supported by the wafer (with appropriate coating, e.g. oxide) as in Figure 3.1. This can be done analytically (Section 3.2.2) or numerically (Section 3.2.3). The silicon thickness is chosen based on requirements, e.g. supports only a single TE and TM waveguide mode. The thickness is typically constrained to what is available by the foundry, e.g. SOI thickness of 220 nm, or etched silicon with a 90 nm thickness.
- For the given thickness, find a suitable waveguide width, again, to meet requirements, e.g. supports only a single TE and TM waveguide mode. This can be done using the effective index method (EIM, Section 3.2.5), or a fully vectorial 2D method (Section 3.2.7).
- Additional consideration should be given to waveguide bend loss, substrate leakage, etc.

The simulations described next, in Sections 3.2.2–3.2.10, are focused on waveguide properties including the effective index, mode profile, and the group index in Section 3.2.9.

3.2.2 1D slab waveguide – analytic method

The MATLAB code in Listing 3.2 finds the effective index of the slab waveguide for the TE and TM modes. For the example we are considering in this section, the command is invoked using:

```
[n_te, n_tm] = wg_1D_analytic (1.55e-6, 0.22e-6, 1.444, 3.473, 1.444)
```

The effective index is found to be 2.845 for the TE mode, and 2.051 for the TM mode. Note that this example solves the waveguide at a single wavelength. To perform wavelength sweeps, the material dispersion should be included. Mode profiles can also be calculated, using MATLAB Script 3.3.

This technique is also used for the design of fibre grating couplers, in Section 5.2.

3.2.3 Numerical modelling of waveguides

In this section, we model the waveguide using a numerical eigenmode solver. First, the waveguide geometry is drawn. The code in Listing 3.4 creates the structure for the wafer and the waveguides, in Lumerical MODE Solutions, as in Figures 3.4 and 3.5. It is then solved as a slab waveguide (1D), then using the effective index method, and finally the fully vectorial 2D solution for the waveguide cross-section (2D). We use the material models defined in Listing 3.1.

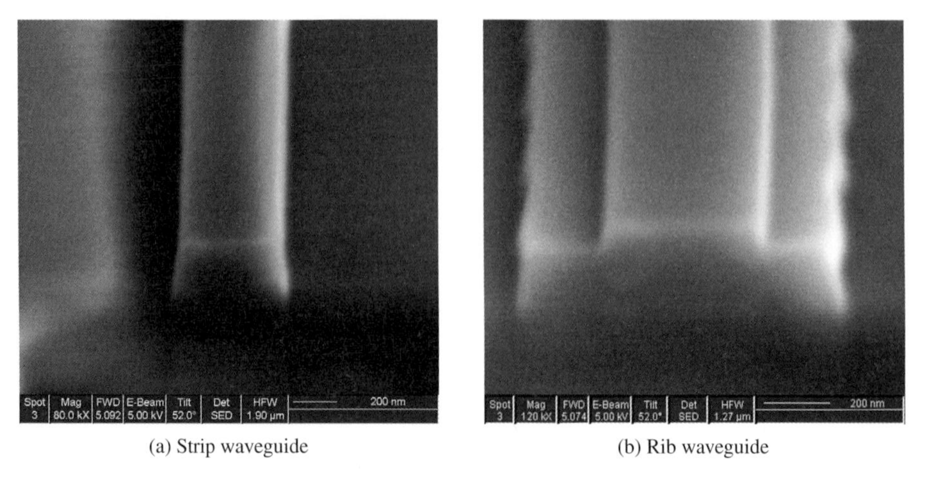

(a) Strip waveguide (b) Rib waveguide

Figure 3.5 SEM cross-sectional images of strip and rib waveguides. Note: the rib waveguide has intentional corrugations used to fabricate Bragg gratings (see Section 4.5). Reference [15].

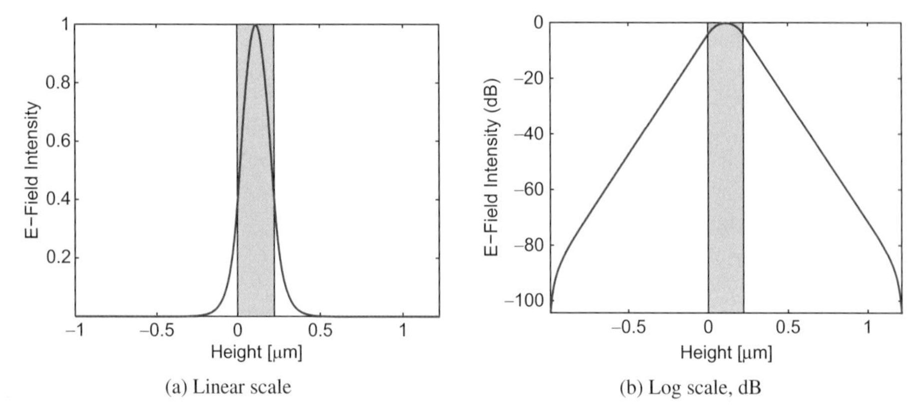

(a) Linear scale (b) Log scale, dB

Figure 3.6 The fundamental mode profile of the slab waveguide (TE). The effective index of this mode is 2.845. Calculated using the eigenmode solver. The waveguide dimension is indicated by the grey area.

3.2.4 1D slab – numerical

First, we simulate the slab waveguide modes, of the structure in Figure 3.1. The first step is configuring a 1D eigenmode solver in the cross-section of the waveguide. This is done in Script 3.5. Next, the calculation is performed in Listing 3.6 and modes profiles are plotted.

The first two modes of the slab waveguide are shown in Figure 3.6 (TE polarization) and Figure 3.7 (TM polarization). This geometry supports only two modes at 1550 nm. The log-scale plots are useful to ensure that the field profile has decayed sufficiently so that the simulation boundaries do not affect the results. The figures also show that the field profiles decay at different rates, namely the TE mode has a stronger confinement. In general, higher effective index modes are more tightly confined to the waveguide

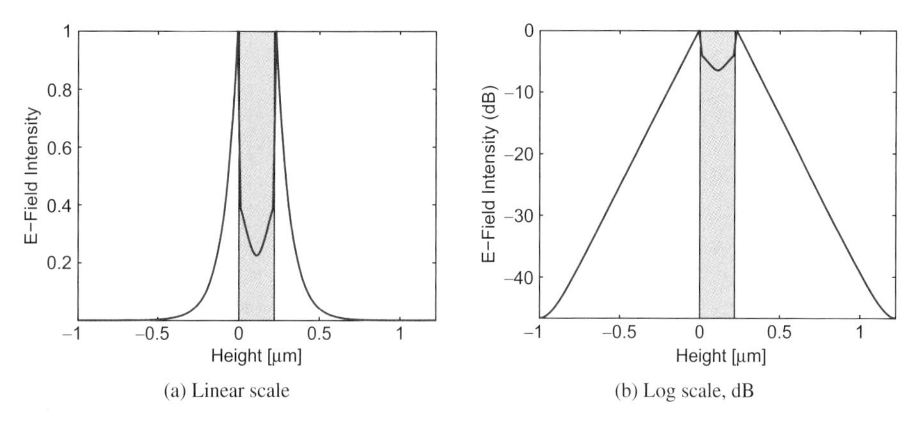

(a) Linear scale

(b) Log scale, dB

Figure 3.7 The mode profile of the 2nd mode in the slab waveguide (TM). The effective index of this mode is 2.054. Calculated using the eigenmode solver. The waveguide dimension is indicated by the grey area.

and have more rapidly decaying evanescent tails. The decay of the field profile in the cladding can be approximated by:

$$E \propto e^{-\frac{2\pi d}{\lambda}\sqrt{n_{\mathrm{eff}}^2 - n_c^2}}, \tag{3.3}$$

where E is the field amplitude at a distance, d, from the core–cladding interface, n_{eff} is the effective index of the mode, and n_c is the refractive index of the cladding.

Convergence tests

When doing numerical simulations, it is critical to do convergence tests. This ensures that the simulation is free from numerical artifacts. In the following example, we will ensure that the simulation boundaries do not interfere with the simulation results. It is easiest to perform convergence tests by ensuring that only one parameter in the simulation is changing; for example, the mesh points inside the waveguides should remain at the same positions during a parameter sweep for the simulation span, as discussed next.

Script 3.7 calculates the effective index versus the simulation span, with results shown in Figure 3.8a. The TE mode is calculated by selecting mode #1, and the TM mode is calculated by selecting mode #2. For the TE mode, simulation spans of larger than 1300 nm converge to an effective index variation of less than 0.0001. This suggests that 550 nm above and below the waveguide is sufficient for accurate simulations of TE modes. We will use this information for optimizing the 3D FDTD simulations. We can compare this span to the results in Figure 3.6 to determine that the E-field intensity should decay down to 10^{-6} of its maximum value to ensure that the boundary does not perturb the mode. Because the TM field profile has a larger tail (this can be inferred from the lower effective index), the simulation converges at larger spans, hence this mode requires a larger simulation volume for accurate modelling. In this case, the TM mode requires spans of 2000 nm for the same precision. Figure 3.8b plots the absolute error in terms of the difference in effective index (compared to the 2000 nm result). The

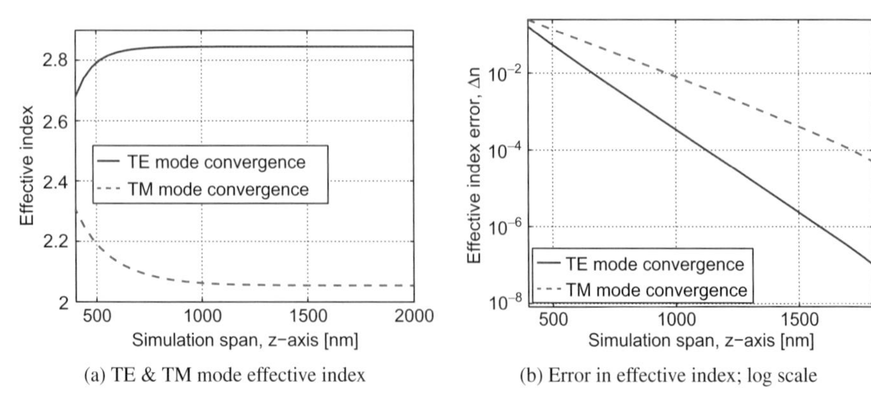

(a) TE & TM mode effective index (b) Error in effective index; log scale

Figure 3.8 Convergence tests for slab mode calculations (varying the simulation size).

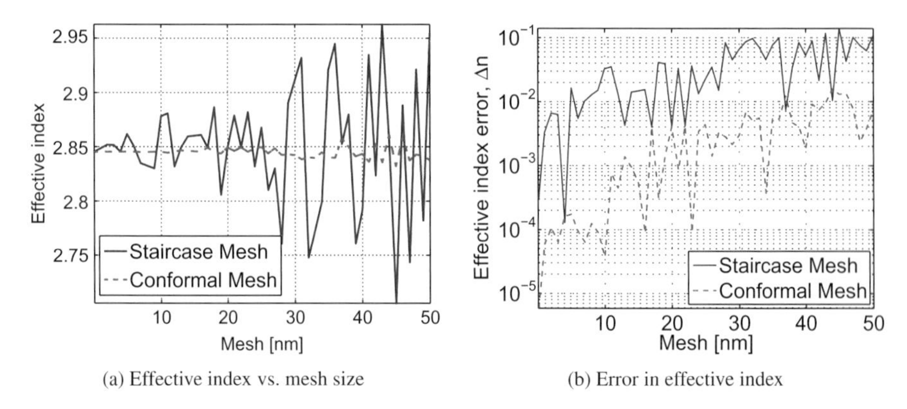

(a) Effective index vs. mesh size (b) Error in effective index

Figure 3.9 Convergence tests for slab mode calculations (varying the mesh grid), for the TE mode.

error is exponentially reduced for increasing simulation size, where a $\Delta n = 10^{-6}$ error is found for a span of 1600 nm.

Next, we study the impact of the mesh grid on the precision. In this case, we set the simulation span to be large, 2 µm in this case, and plot the results for the TE mode. Conventional algorithms use a staircase mesh, which introduces significant error unless the mesh is reduced down to 1 nm or smaller (which is typically not practical as it results in long simulation times), as shown in Figure 3.9a(staircase) and Figure 3.9b(staircase). For a typical mesh size of 10 nm, the staircase mesh introduces an error of $\delta n_{\text{eff}} = 0.05$. To improve the accuracy, mesh overrides can be used to increase the number of mesh grid points in areas of interest. It is also advisable to have a grid point lie at each material interface.

Advanced algorithms, such as the conformal mesh, improve the accuracy significantly. In fact, a 20 nm conformal mesh has a similar accuracy as the 1 nm staircase mesh. Using the conformal mesh, from Figure 3.9b(conformal), we find that a mesh of 10 nm results in an error of approximately $\Delta n = 10^{-4}$, and a mesh of 20 nm results in an error of approximately $\Delta n = 10^{-3}$. Even for large meshes such as 40–50 nm, the error is typically less than $\Delta n = 10^{-2}$.

These simulations are implemented in Script 3.8. The meshing algorithm is selected using the "mesh refinement" parameter.

Parameter sweep – slab thickness

Next, we wish to study the effective index of the slab versus the slab thickness. Script 3.9 performs the simulation sweep with results shown in Figure 3.10. The effective index as well as the polarization fraction (TE vs. TM) is returned. Note that, for a 1D waveguide structure, the mode is either a pure TE or a pure TM mode. This is in contrast to the 2D waveguides, where pure TE or TM modes do not exist. The TE/TM polarization fraction is used to label the modes in Figure 3.10. The dotted line is the index of refraction of the oxide; modes with index below this value are not guided.

As can be seen, at a wavelength of 1550 nm, the slab operates in a single TE and TM mode for thicknesses below 240 nm (indicated by the vertical line, which shows the onset of guiding for the third mode). This explains why 220 nm is commonly chosen for silicon-on-insulator substrates for a wavelength of 1550 nm. These graphs are useful in designing waveguides with different thicknesses, wavelengths, and polarization.

3.2.5 Effective Index Method

To calculate the effective index of a 2D waveguide cross-section, as in Figure 3.4, the Effective Index Method can be used (see Section 2.2.3). Although fully vectorial 2D solutions are more accurate (Section 3.2.7), this method provides important insights into how waveguides operate. Also, this method is computationally very fast and easy to implement, hence is used for designing modulators in Section 6.2.2, and for 2.5D FDTD simulations (Section 2.2.3).

To calculate the TE-like modes, we first find the 1D slab out-of-plane TE mode (this defines that it is predominantly a TE mode), and then we find the 1D in-plane TM mode using the effective index from the 1D slab. The change of polarization is necessary since we are solving a 2D problem using two 1D simulations; hence we are rotating the frame of reference with respect to the primary field component (e.g. in-plane E-field for the TE-like mode). For TM-like modes, the reverse procedure is used: first the TM mode for the slab, followed by the TE mode.

As an example, to find the TE-like mode, first, the slab effective index is found, for the out-of-plane cross-section, as per Listing 3.6. The procedure is repeated, except this time in the in-plane direction. The slab effective index is used as the input into this second step. The final result is the effective index of the 2D waveguide. The script in Listing 3.10 accomplishes this, where we have used the slab TE mode effective index (2.845) as the input, with the aim of calculating the TE mode effective index of the strip waveguide. The result is a waveguide effective index of 2.489 with a field profile as shown in Figure 3.11.

Using the mode profiles in Figure 3.6a and Figure 3.11a, we can construct the 2D mode profile. The inherent assumption in the Effective Index Method is that the fields

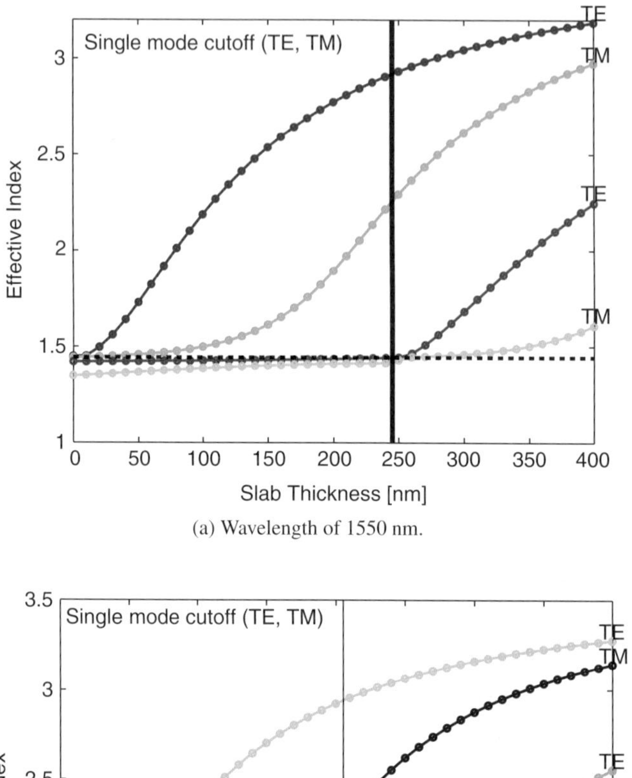

(a) Wavelength of 1550 nm.

(b) Wavelength of 1310 nm.

Figure 3.10 Simulation of the effective index of the slab waveguide modes versus the thickness of silicon. Waveguide is cladded with silicon dioxide. Only the modes above the dotted lines are guided.

are separable, similar to the method of "separation of variables" for solving differential equations. Here, we write the 2D field profile as

$$E(z, y) = E(z) \cdot E(y),$$

where $E(z)$ and $E(y)$ refer to the fields in Figure 3.6a and Figure 3.11a, respectively. The resulting mode profile $E(z, y)$ is plotted in Figure 3.12a. We compare the mode profile determined by the Effective Index Method with the "exact" 2D finite element

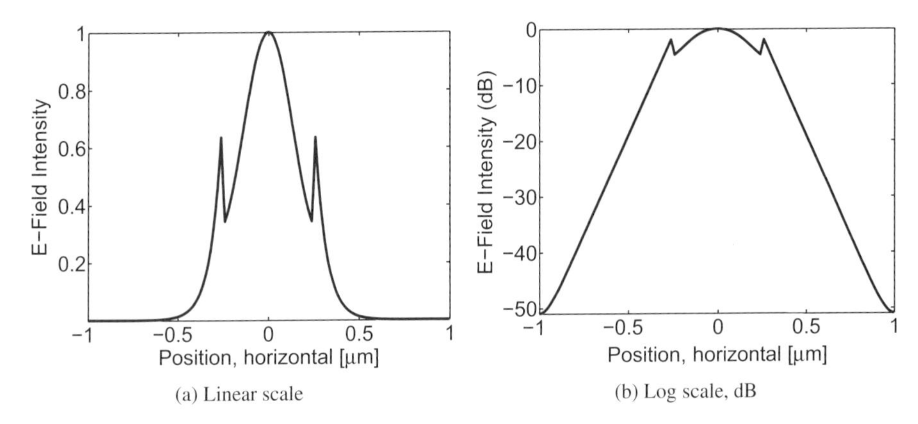

(a) Linear scale

(b) Log scale, dB

Figure 3.11 The fundamental mode profile of the waveguide (TE), using Effective Index Method (EIM). The effective index of this mode is 2.489.

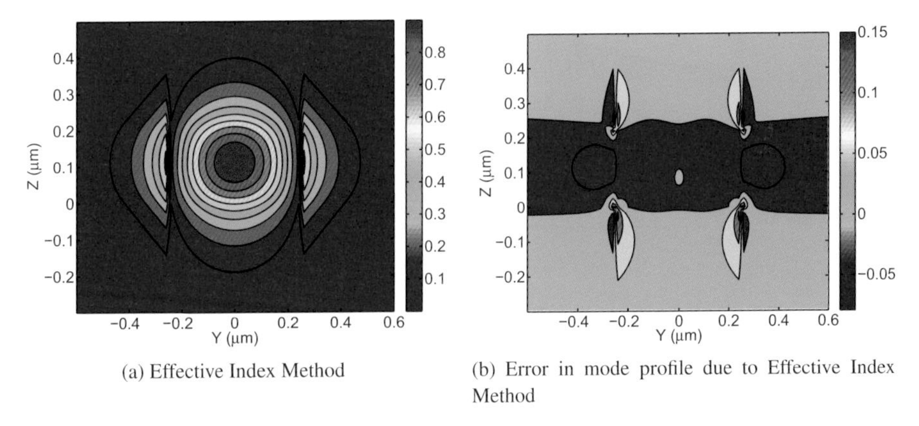

(a) Effective Index Method

(b) Error in mode profile due to Effective Index Method

Figure 3.12 Comparison of the Effective Index Method reconstructed field profile versus the 2D fully vectorial calculation for a strip waveguide.

calculations in Section 3.2.7, by taking the difference of the mode profiles, with results shown in Figure 3.12b. The error in effective and group index are $\Delta n_{\mathrm{eff}} \sim 1.2\%$ and $\Delta n_g \sim 2.7\%$ for the strip waveguide considered.

This result is useful in that we can use the Effective Index Method in combination with 2D FDTD modelling (2.5D FDTD, see Section 2.2.3), and anticipate an error in the group index of only several percent. This translates into a several percent error in the free-spectral range in resonators, but with dramatically reduced computation time as compared with full 3D simulations, which is useful when performing numerous simulations for optimization.

3.2.6 Effective Index Method – analytic

Similarly, we can use the analytic method to find the 1D field profiles (MATLAB code 3.3) and construct the 2D field profile using the Effective Index Method. This

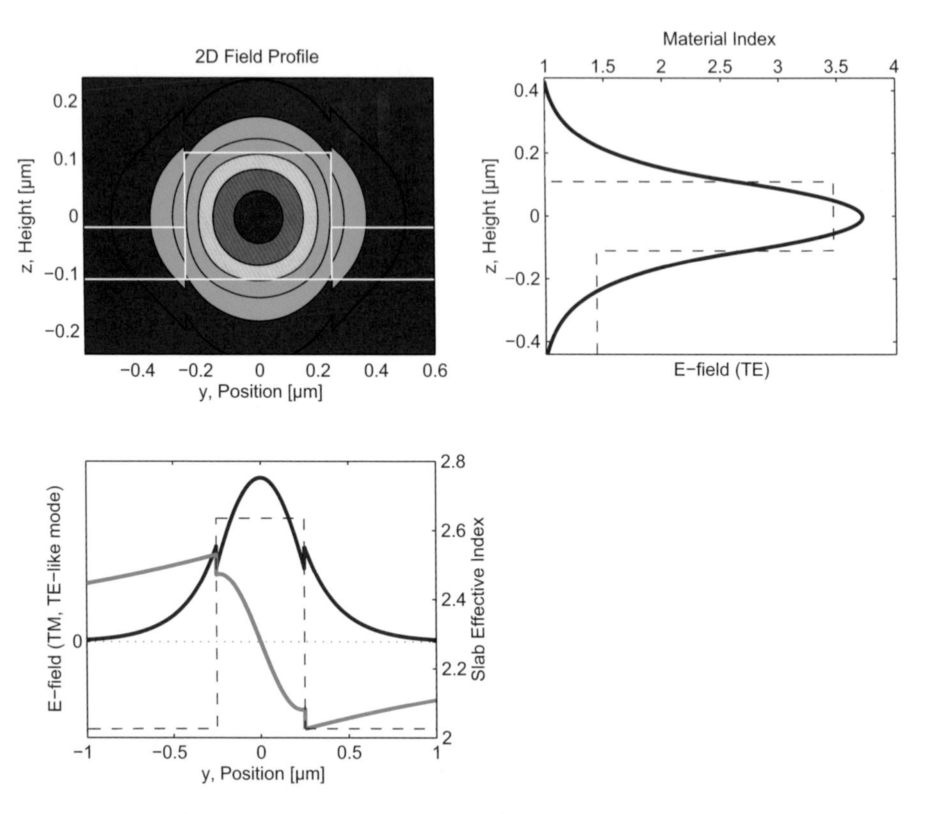

Figure 3.13 Analytic calculation of 2D waveguide field profile for a rib waveguide using the Effective Index Method, in MATLAB. (Top left) Rib waveguide geometry, with the resulting 2D field profile calculated by the Effective Index Method for the fundamental mode. (Top right) Material index of refraction for the air–silicon-oxide silicon-on-insulator wafer (shown in the dashed line), and the calculated 1D field profile in the vertical cross-section of the 220 nm waveguide; this procedure is also repeated for the slab regions (90 nm). (Bottom left) Slab effective indices calculated from step 1 (shown in the dashed line); 1D field profiles in the horizontal cross-section of the waveguide for the two guided TE-like modes.

implemented in the MATLAB code 3.11, with the results for a rib waveguide shown in Figure 3.13. This is used for modelling pn-junction modulators in Section 6.2.2.

3.2.7 Waveguide mode profiles – 2D calculations

Next, we wish to perform accurate calculations for the mode profile for the 2D cross-section of the waveguide. In this section, we draw the waveguide, define the simulation parameters, and solve for the mode solution using the fully vectorial method described in Section 2.1. We wish to find the mode profiles for quasi-TE and quasi-TM modes, and observe the electric field intensity, magnetic field intensity, and the energy density, as shown in Figures 3.14–3.17. The three field components in the x, y, and z directions are shown in Figure 3.15; from the figure, it is evident that the mode is not strictly a TE

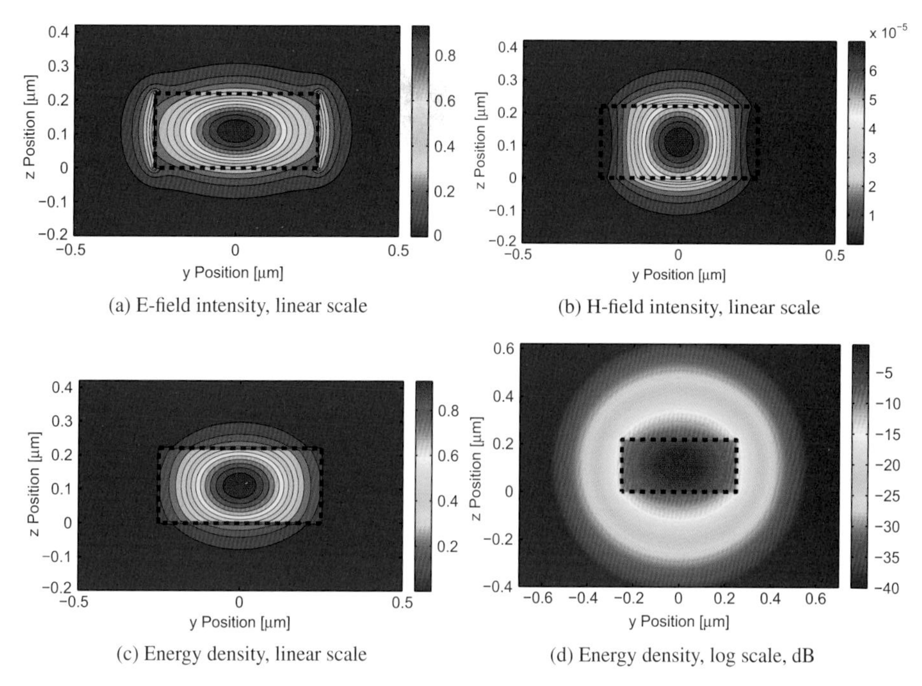

(a) E-field intensity, linear scale

(b) H-field intensity, linear scale

(c) Energy density, linear scale

(d) Energy density, log scale, dB

Figure 3.14 TE (first mode) mode profile of a 500 × 220 nm strip waveguide at 1550 nm, mode effective index is $n_{\text{eff}} = 2.443$.

mode (where only the E_y component would be present), but rather is a three-dimensional vectorial field profile.

We continue the Lumerical script by using the materials (Script 3.1) and waveguide (Script 3.4) previously described. First, we define the simulation parameters and add a 2D mode solver in the cross-section of the waveguide, in Listing 3.12. Next, in Script 3.13, we calculate the mode profiles and plot the E-field and H-field intensities, as well as the energy densities. The energy density plot requires the waveguide dispersion, hence the script calculates the mode effective index at two slightly detuned frequencies, see Reference [16].

Figures 3.14–3.17 show the results of the first three modes supported by the waveguide. Figure 3.14 shows the fundamental mode profile, in this case, with a TE-like polarization. The field and energy is strongly confined inside the waveguide, although about 10% of the field is in the cladding. This is the mode used for most silicon photonics devices for the 220 nm thickness SOI wafer.

Figure 3.16 shows the TM-like mode. This mode is well-enough confined to be useful for TM-like mode devices. However, some researchers have used a slightly thicker silicon, e.g. 260 nm [17, 18], for applications specifically targeting TM-like mode operation.

Finally, Figure 3.17 shows the TE-like mode. The effective index of the mode is close to that of the oxide, and there is a significant E-field on the edges of the waveguide. This will introduce a significant scattering loss, hence this mode will experience a high

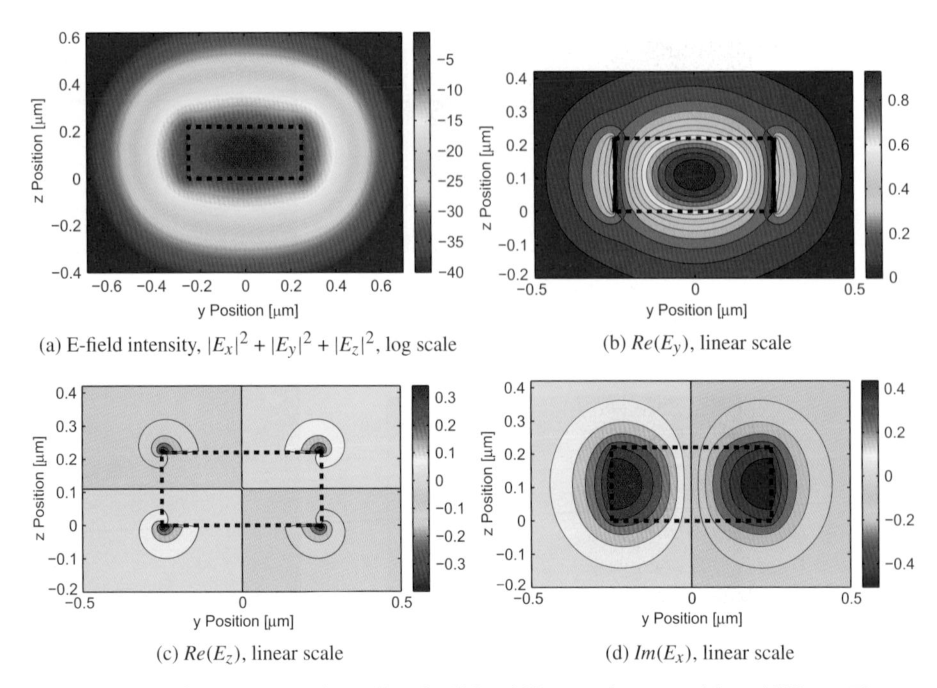

Figure 3.15 TE (first mode) mode profile of a 500 × 220 nm strip waveguide at 1550 nm. The three field components in the x, y, and z directions are shown.

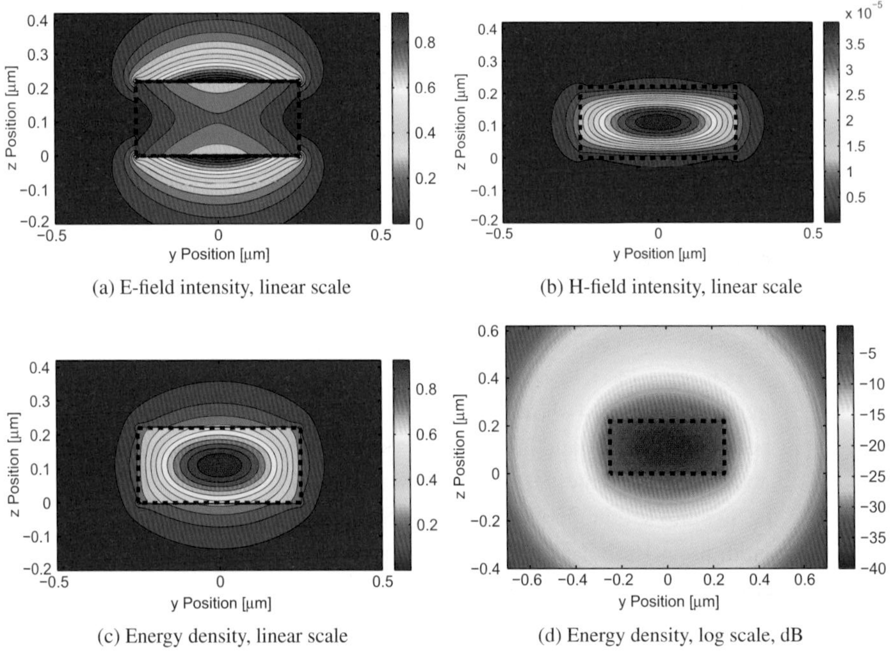

Figure 3.16 TM (second mode) mode profile of a 500 × 220 nm strip waveguide at 1550 nm, mode effective index is $n_{\text{eff}} = 1.771$.

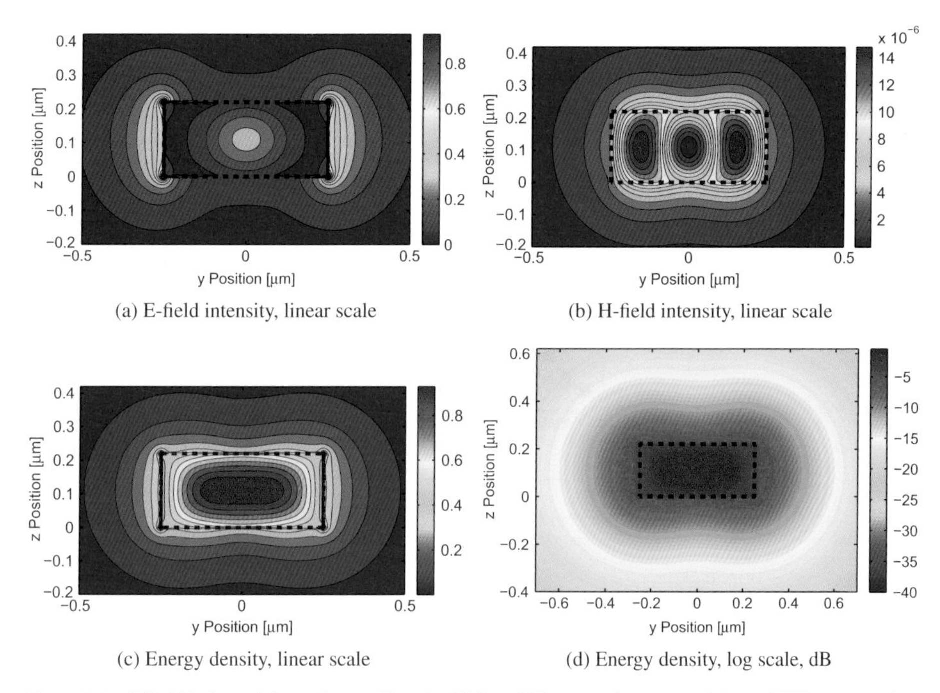

(a) E-field intensity, linear scale

(b) H-field intensity, linear scale

(c) Energy density, linear scale

(d) Energy density, log scale, dB

Figure 3.17 TE (third mode) mode profile of a 500 × 220 nm strip waveguide at 1550 nm, mode effective index is $n_{\text{eff}} = 1.493$. This mode is barely guided and the effective index is close to the oxide index. It will experience high optical loss due to strong side-wall scattering.

level of loss. Thus, this waveguide geometry can be considered to effectively operate as a single TE-like mode waveguide. If more mode selectivity is required, a narrower waveguide can be chosen, e.g. 440 nm, as will be found in the next section.

3.2.8 Waveguide width – effective index

Next, we vary the width of the waveguide (from 200 nm to 800 nm), in Listing 3.14, using fully vectorial 2D mode calculations. This allows us to determine the single-mode condition for the waveguide. The effective index of all modes is saved in a matrix, n_{eff}, and plotted.

The results are shown in Figure 3.18a for 1550 nm, and Figure 3.18b for 1310 nm.

For 1550 nm, only the modes above the dotted lines are guided. To obtain a single TE-like mode polarization at 1550 nm, a strip waveguide with a height of 220 nm, and width of 440 nm is required. In this case, the waveguide supports one TE-like and one TM-like mode. For wider waveguides, a second TE-like mode is present, and above 660 nm, a second TM-like mode appears. Note that at approximately 680 nm, there is a mode crossing: namely below 680 nm it is the TM-like mode that is the second supported mode of the waveguide, whereas above 680 nm, the TE-like mode becomes the second one and the TM-like mode becomes the third-order mode. The polarization of the mode is also recorded; note that these waveguides do not operate in a pure TE

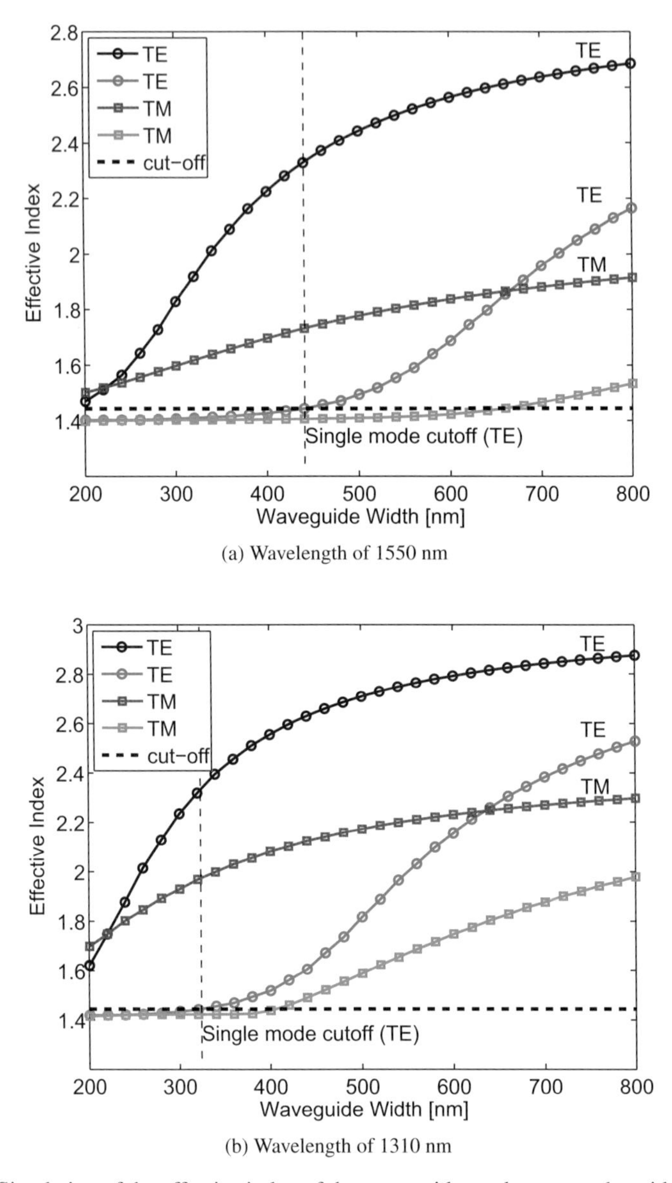

(a) Wavelength of 1550 nm

(b) Wavelength of 1310 nm

Figure 3.18 Simulation of the effective index of the waveguide modes versus the width of a strip waveguide, for a silicon thickness of 220 nm.

or TM polarization, hence the polarization is described as a polarization fraction, with 1 representing a pure TE mode. The polarization fraction is plotted in Figure 3.19. It is seen that the fundamental TE-mode polarization fraction is typically between 0.95 and 1 (nearly pure TE mode), whereas the other polarizations are not pure polarizations. This effect can be used design polarization converters, whereby efficient mixing occurs when the TE and TM polarizations have similar effective indices – e.g. at 680 nm, the first TM and second TE modes cross, as shown in Figure 3.18a.

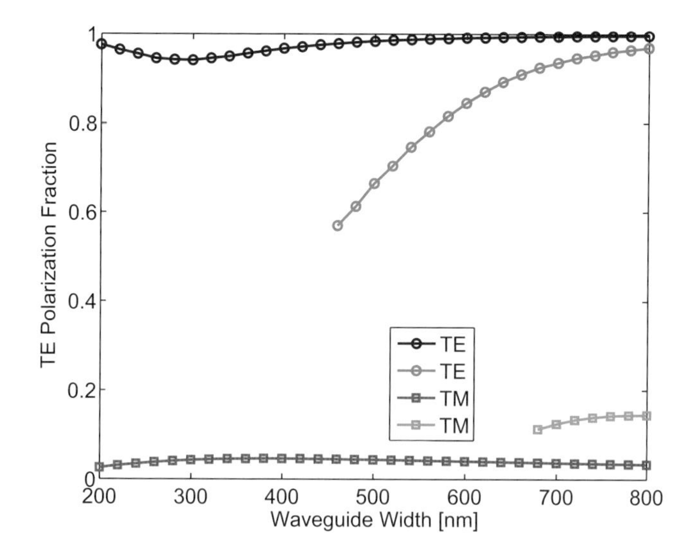

Figure 3.19 Simulation of the mode polarization fraction (relative to a pure TE mode) versus the width of the strip waveguide, for a silicon thickness of 220 nm.

For 1310 nm, the 220 nm thickness is not perfectly single-mode hence there are some weakly guided higher-order slab modes present. Hence, perfectly single-mode operation is not possible. This was already known from the slab waveguide data in Figure 3.10b, where it was found that a thickness of approximately 205 nm is required for single TE-mode operation. Hence, the four modes considered are guided. Despite this, the fundamental TE-like mode is much more confined than the other modes, hence it is still possible to design devices operating at 1310 nm using this waveguide geometry.

We also examine a rib waveguide with a 90 nm slab thickness. The results are shown in Figure 3.20 for a wavelength of 1550 nm. This waveguide does not support any TM modes. This is the type of waveguide used for electro-optic devices such as pn-junction modulators, described in Section 6.2.2.

3.2.9 Wavelength dependence

The wavelength dependence of the waveguide's effective and group index is simulated via Script 3.15. The script performs a sweep of the wavelength. The results for the effective index are shown in Figure 3.21a. The effective index of the waveguide is used to describe the phase velocity of the light, v_p. However, it is the group index that determines the propagation speed of a pulse, namely the group velocity, v_g:

$$v_p(\lambda) = \frac{c}{n_{\text{eff}}}, \quad v_g(\lambda) = \frac{c}{n_{\text{g}}}. \tag{3.4}$$

The group index is an important parameter in photonic integrated circuit design since it is the group index that determines the mode spacing (free-spectral range) in resonators

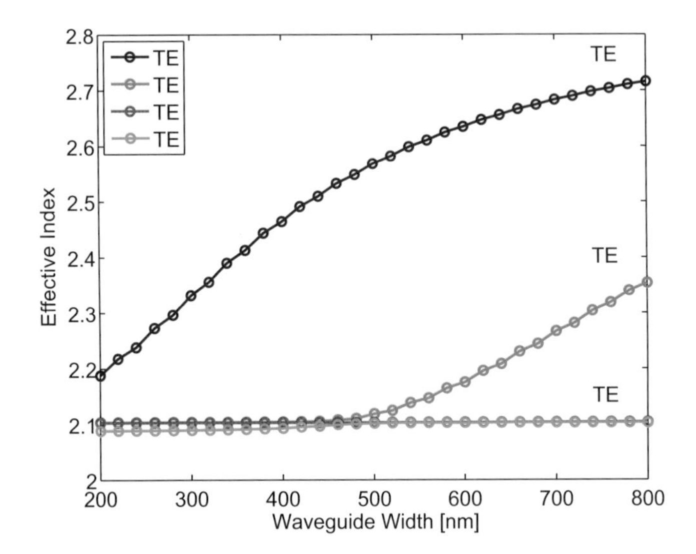

Figure 3.20 Simulation of the effective index of the waveguide modes versus the width of a rib waveguide, for a silicon thickness of 220 nm and slab thickness of 90 nm.

and interferometers. The group index can be related to the effective index:

$$n_g(\lambda) = n_{\text{eff}}(\lambda) - \lambda \frac{dn_{\text{eff}}}{d\lambda}. \tag{3.5}$$

The group index is plotted in Figure 3.21b. The simulated group index is compared with experimental results extracted from a ring resonator. Good agreement is obtained when: (a) a small enough mesh is used to ensure accuracy, in this case 10 nm; (b) both material and waveguide dispersion are taken into account. It is important to include both the material dispersion and waveguide dispersion to correctly predict the group index. Group velocity dispersion, important in understanding the spreading of optical pulses travelling down a waveguide, can also be determined from these simulations by taking the next-order derivative. This is described by the dispersion parameter:

$$\text{D}(\lambda) = \frac{d\frac{n_g}{c}}{d\lambda} = -\frac{\lambda}{c}\frac{d^2 n_{\text{eff}}}{d\lambda^2}. \tag{3.6}$$

Simulations are also performed to illustrate the effect of the waveguide width on the wavelength dependence of the waveguides. The fundamental TE-like mode is considered. The effective index and group index versus wavelength for the 220×550 nm strip waveguide are plotted in Figure 3.22. Similarly, Figure 3.23 is for the rib waveguide with a 90 nm slab.

3.2.10 Compact models for waveguides

For device (e.g. ring resonators) and system design, it is often preferable to describe the performance of the waveguides using compact models consisting of phenomenological

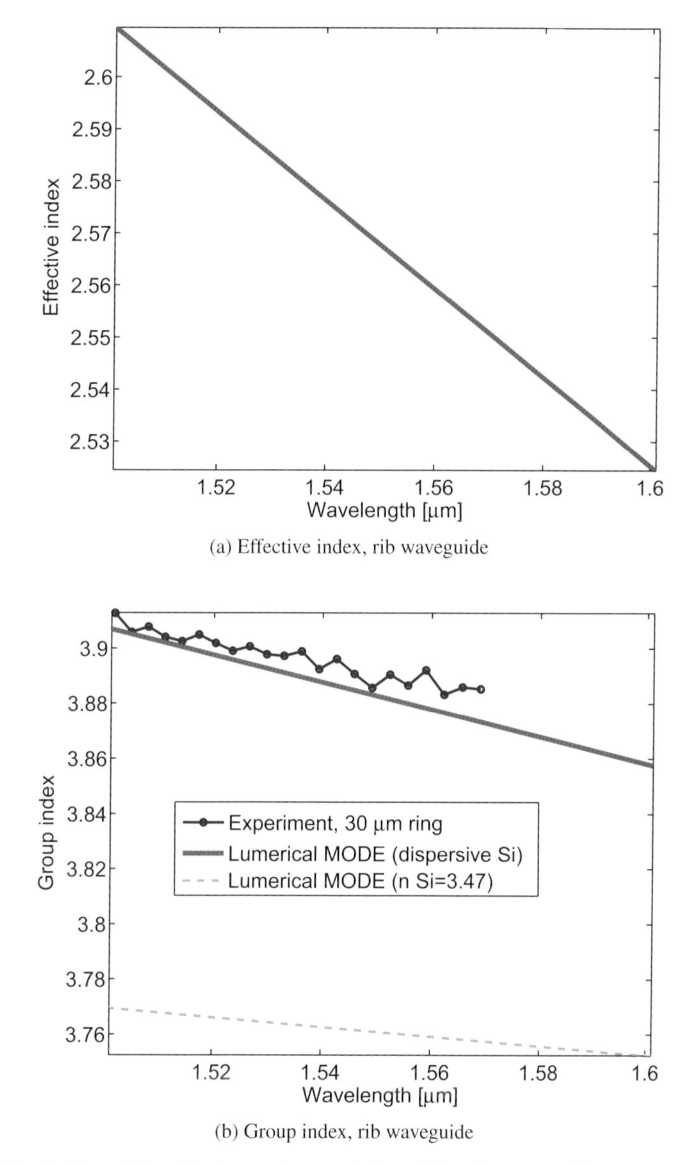

(a) Effective index, rib waveguide

(b) Group index, rib waveguide

Figure 3.21 Simulation of the effective and group index of the rib waveguide versus wavelength, for a silicon thickness of 220 nm with a slab height of 90 nm. Comparison with experimental results (see Section 4.4) for the same geometry are included.

parameters and fit functions. If there is no obvious physical dependency on the parameters, the results can be fitted using Taylor expansion approximations.

For the waveguide in Figure 3.21, the wavelength-dependent effective index can be simply approximated as:

$$n_{\text{eff}}(\lambda) = 2.57 - 0.85(\lambda[\mu\text{m}] - 1.55). \quad (3.7)$$

The waveguide effective index can be calculated for various wavelengths and temperatures, $n_{\text{eff}}(\lambda, T_i)$, and fitted using a second-order Taylor expansion for the wavelength

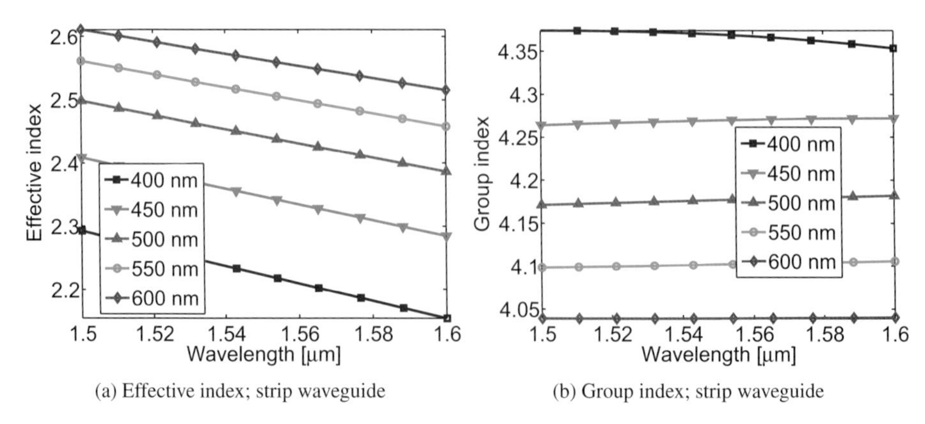

(a) Effective index; strip waveguide (b) Group index; strip waveguide

Figure 3.22 Simulation of the effective and group index of the strip waveguide versus wavelength, for a silicon thickness of 220 nm, for various waveguide widths.

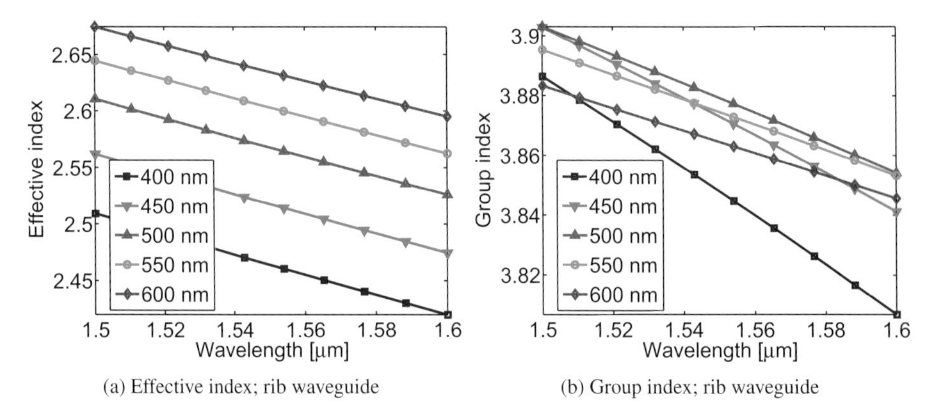

(a) Effective index; rib waveguide (b) Group index; rib waveguide

Figure 3.23 Simulation of the effective and group index of the rib waveguide versus wavelength, for a silicon thickness of 220 nm with a slab height of 90 nm, for various waveguide widths.

λ. Another second-order Taylor expansion for the temperature dependence can then be used. This results in a higher-order polynomial description [19] as follows:

$$n_{\text{eff}}(\lambda,T) = N_0(T) + N_1(T)\left(\frac{\lambda-\lambda_0}{\sigma_\lambda}\right) + N_2(T)\left(\frac{\lambda-\lambda_0}{\sigma_\lambda}\right)^2$$
$$\begin{cases} N_0(T) &= n_0 + n_1\left(\frac{T-T_0}{\sigma_T}\right) + n_2\left(\frac{T-T_0}{\sigma_T}\right)^2 \\ N_1(T) &= n_3 + n_4\left(\frac{T-T_0}{\sigma_T}\right) + n_5\left(\frac{T-T_0}{\sigma_T}\right)^2 \\ N_2(T) &= n_6 + n_7\left(\frac{T-T_0}{\sigma_T}\right) + n_8\left(\frac{T-T_0}{\sigma_T}\right)^2 \end{cases} . \tag{3.8}$$

The dependence of numerous waveguide parameters can also be included, such as: wavelength, temperature, thickness of the silicon, width of the waveguide, ridge waveguide slab thickness, carrier density, and optical nonlinearity. The designer needs to make a choice regarding the appropriate complexity for the given problem.

A compact model for waveguides that considers the wavelength dependence and the waveguide width is used in the context of synthesizing an optical grating based on a desired spectral response. The design necessarily includes a variation of the width of the waveguide, hence the compact model needs to account for this. This approach is presented in Section 4.5.2.

3.2.11 Waveguide loss

Losses in waveguides originate from several contributions.

- Absorption due to metal in proximity. For example, a 500 × 220 nm strip waveguide with metal above was measured to have an excess loss of 1.8 ± 0.2 dB/cm for metal that is 600 nm above the waveguide.
- Sidewall scattering loss is typically 2–3 dB/cm for 500 × 220 nm waveguides. Atomic force microscope measurements can be done to measure the sidewall roughness. For example, in [20], a 2.8 nm rms roughness was measured. Knowing the roughness, the loss in waveguides can be simulated [21–24].
- Material loss is typically negligible for passive structures. It becomes significant for doped silicon, as described in Section 6.1.1.
- Surface-state absorption can also contribute to the propagation loss if the waveguides are not properly passivated. Indeed, un-passivated waveguides have been used to make detectors [25, 26].
- Sidewall roughness also introduces reflections along the waveguide and phase perturbations that are wavelength dependent [27].

Waveguide losses can be reduced by using wider waveguides, though care has to be taken in single-mode applications since these waveguides are multi-mode. Hence, they require gradual tapers to convert from single-mode to the wide multi-mode waveguide. For example, in Reference [28], the authors report 0.27 dB/cm losses in 2 μm wide, 250 nm high, rib waveguides. Such waveguides are well suited for long-distance traces, e.g. global routing on-chip.

The loss of a waveguide is typically expressed in dB/cm, but can converted to an absorption coefficient, in [m^{-1}], or in terms of the imaginary coefficient, k, in the complex index of refraction, $n + ik$,

$$\alpha[\text{m}^{-1}] = \frac{\alpha[\text{dB/m}]}{10\log_{10}(e)} = \frac{\alpha[\text{dB/m}]}{4.34} \tag{3.9a}$$

$$k = \frac{\lambda \cdot \alpha[\text{dB/m}]}{4\pi \cdot 4.34}. \tag{3.9b}$$

3.3 Bent waveguides

A requirement for silicon photonics is to bend waveguides, e.g. for signal routing, and ring/racetrack resonators. Thus, we must understand how much optical loss is introduced by the bend. In this section, we model the bend losses using 3D FDTD.

We determine the relative contributions of radiation loss versus mode mismatch loss using an eigenmode solver. Finally, we compare with experimental data.

There are several mechanisms for loss in bends.

(1) Scattering losses and substrate leakage: for the short distances considered in bends, these losses are small. As we will find, these losses are small for bend radii $< 10\,\mu m$.

(2) Radiative loss: in highly confined waveguides, especially for TE modes in strip waveguides, radiation is typically small. It does contribute in rib waveguides for bend radii $< 5\,\mu m$ and for TM-polarized modes.

(3) Mode mismatch loss: this is the largest source of loss in bends. It is due to the imperfect mode overlap between the straight and the bent waveguides. This leads to scattering at the abrupt radius transition regions (start and end of fixed-radius bends). There are several methods of decreasing mode-mismatch loss: (a) to laterally offset the straight waveguide relative to the bent waveguide in order to obtain a better mode overlap [29]; (b) to vary the curvature continuously, rather than abruptly; for example a 90° bend with an effective radius of $20\,\mu m$, where curvature changes from zero to $1/15\,\mu m^{-1}$ then back to zero [30]. This has the added benefit that the bend does not excite higher order modes in the waveguide. See Figures 10.6, 10.7, and 10.8, for more details.

There have been several reports of losses in bends. For 500×220 nm strip waveguides, IBM measured losses per 90° bend to be approximately 0.09 dB for a bending radius of $1\,\mu m$ and 0.02 dB for a $2\,\mu m$ bend [29]. Similarly, 500×220 nm strip waveguides fabricated at imec, Belgium, demonstrated losses of 0.1 dB per 90° bend for a bending radius of $1\,\mu m$ and 0.01 dB per 90° bend for a bending radius of $5\,\mu m$ [31]. It was concluded that the bend loss is decreased to the point where it equals the waveguide propagation loss at a radius of approximately $10\,\mu m$ [31].

The experimental results presented in this section are based on waveguides fabricated at IME, Singapore, via the OpSIS foundry service. Optical measurements are taken on numerous test structures consisting of two bends (with radius ranging from $0.5\,\mu m$ to $50\,\mu m$, both strip and rib waveguides), where the optical input–output is accessed via fibre grating couplers. The experimental results are shown for both strip (Figures 3.24) and rib (Figures 3.25) waveguides. The experimental data have an uncertainty of approximately 0.2 dB (hence data for larger bends are not reliable). Note: the radius is defined from the centre of the waveguide.

3.3.1 3D FDTD bend simulations

The bend loss can be simulated by 3D FDTD. The scripts in Listings 3.16 and 3.17 draw the input, output, and bent waveguides, define the simulation volume and parameters, add an optical mode source in the waveguide, and add optical power monitors.

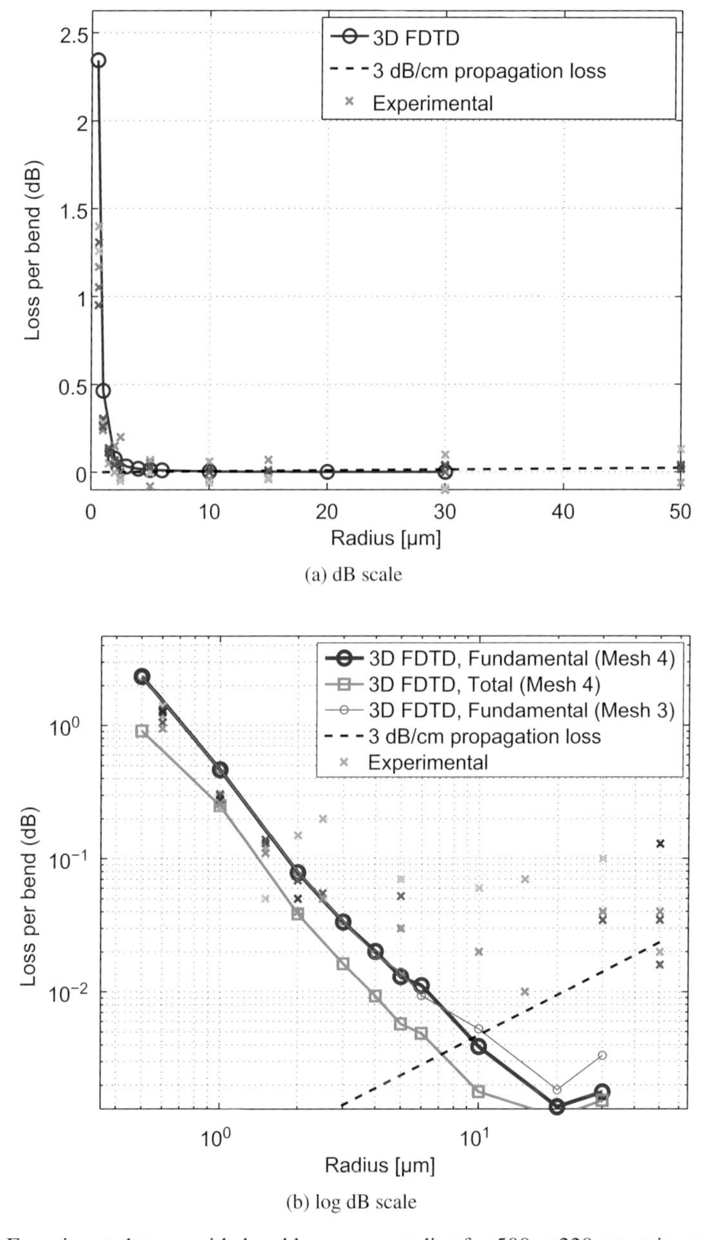

(a) dB scale

(b) log dB scale

Figure 3.24 Experimental waveguide bend loss versus radius for 500 × 220 nm strip waveguides. Data from OpSIS-IME [32] and 3D FDTD simulation data.

Included is a mode expansion power monitor to determine the power transmitted into the fundamental mode only. The script optionally creates a movie of simulation.

Next, several FDTD simulations are performed, with varied bend radius, in Listing 3.18. A mesh accuracy of 3 and 4 is employed. The optical transmission, defined as

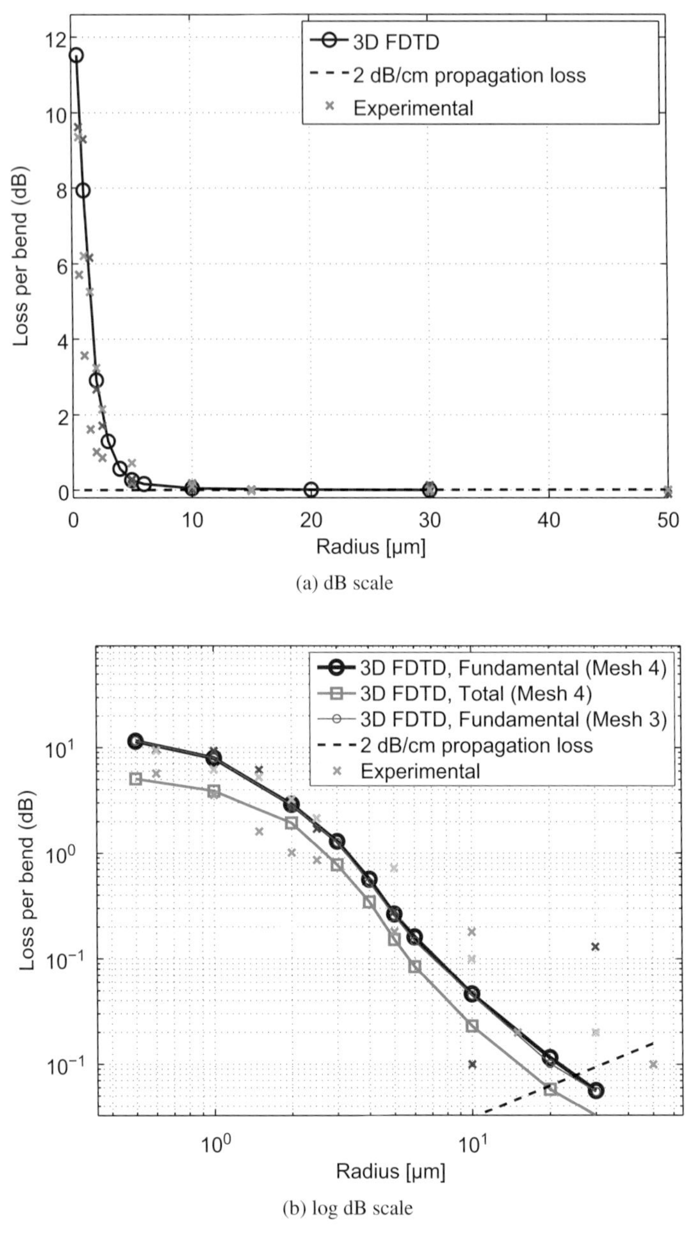

Figure 3.25 Experimental waveguide bend loss versus radius for 500 × 220 nm rib waveguides with a 90 nm slab. Data from OpSIS-IME [32] and 3D FDTD simulation data.

the ratio of the output versus the input, is calculated and plotted versus the bend radius. The transmission can be measured by two methods.

(1) Total transmission: this is a measurement of all the optical power in the output waveguide, hence it includes the power that is not guided in the fundamental

mode. This measurement underestimates the true optical losses of the device. It is included in Figures 3.24 and 3.25 as "3D FDTD, Total".

(2) Fundamental mode transmission: this is a measurement of the optical power in the fundamental mode of the output waveguide. The rationale is that the higher-order modes will scatter away as they propagate down the waveguide, or will be filtered by other mode-selective components (e.g. fibre grating couplers, directional couplers). This measurement is accomplished by performing mode overlap calculations between the FDTD-calculated field and the waveguide mode profile, as described in Reference [33]. It is included in Figures 3.24 and 3.25 as "3D FDTD, Fundamental".

The results for the strip waveguides are plotted together with the experimental results in Figure 3.24. An excellent agreement between the model and the experiments is obtained for the small radius bends (0.5–1 μm), where the losses are large, as shown in Figure 3.24b. For larger radii, the experimental uncertainty is much larger than the predicted losses. On this graph, a line is included for the optical scattering losses, assuming 3 dB/cm. Using the intersection of this propagation loss line, and the 3D FDTD results, the simulations predict that the lowest losses for this conventional 90° bend will be obtained for a 10 μm bend radius.

Similarly, the results for rib waveguides are plotted together with the experimental results in Figure 3.25. Good agreement between the model and the experiments is obtained for radii up to 4 μm, where the losses are large, as shown in Figure 3.25b. For larger radii, the experimental uncertainty is larger than the predicted losses. On this graph, a line is included for the optical scattering losses, assuming 2 dB/cm. The simulations predict that the lowest losses for this conventional 90° bend will be obtained for a 25 μm bend radius.

It should be noted that the inclusion of the dB/cm propagation loss, and the values chosen, are based on the straight waveguide propagation loss. Bent waveguides are known to have higher propagation losses [34]. Hence, the optimal bend radius is likely smaller than stated above.

Next, we are interested in visualizing where the energy is being lost in the bend. This can be done by creating a movie, as in previous script, or by observing the time-integrated E-field intensity, as defined by:

$$|E|^2 = |E_x|^2 + |E_y|^2 + |E_z|^2. \tag{3.10}$$

The script in Listing 3.19 adds a power monitor in the cross-section of the waveguide. The results for the 1 μm bend radius strip waveguide are shown in Figure 3.26, which illustrates where the energy is being lost in the bend.

3.3.2 Eigenmode bend simulations

To understand the loss mechanism, another tool is at our disposal – the waveguide mode eigensolver (Section 2.1) can be used to calculate the field profile of a bent waveguide.

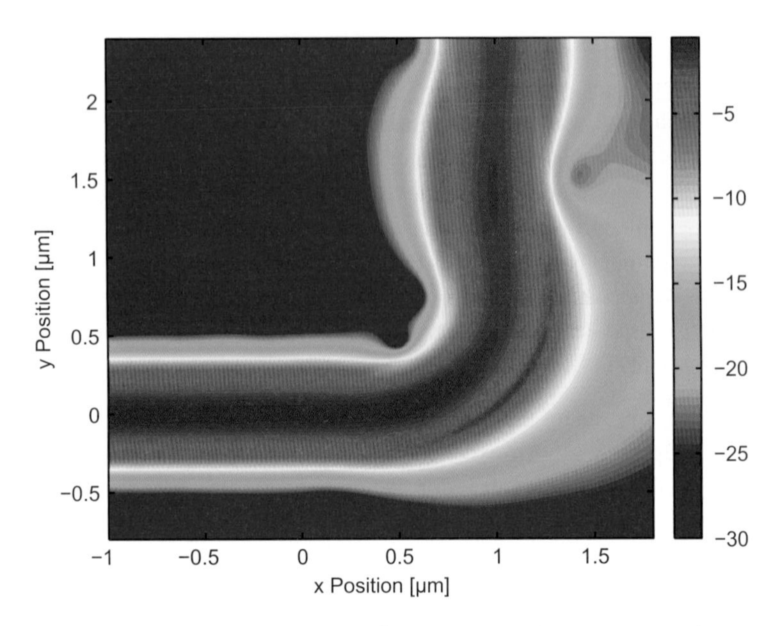

Figure 3.26 Time-integrated field profile of a 90° bend with a radius of 1 μm. Mesh accuracy = 3.

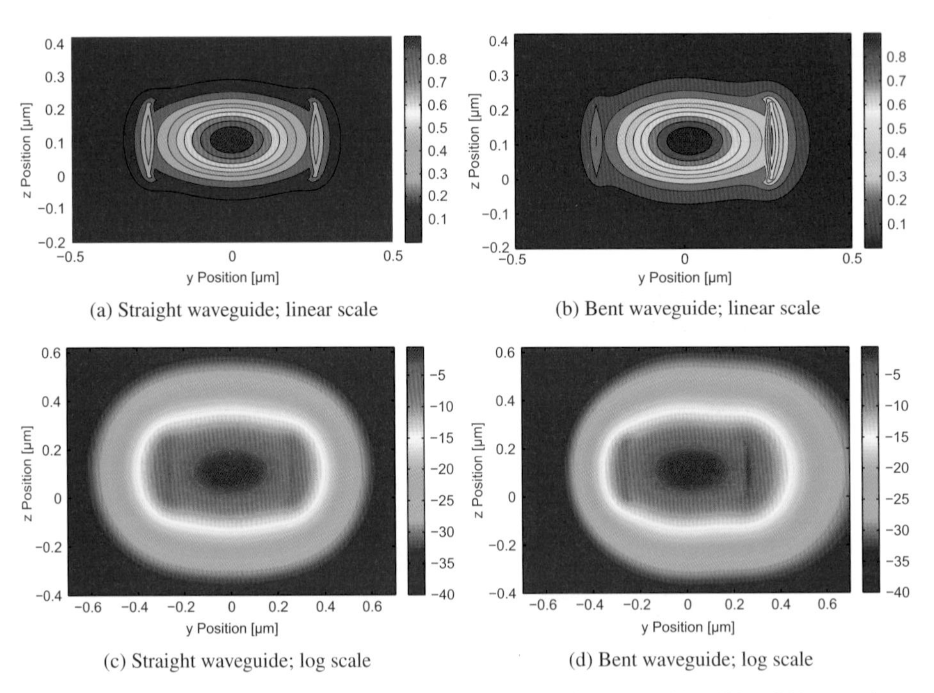

(a) Straight waveguide; linear scale

(b) Bent waveguide; linear scale

(c) Straight waveguide; log scale

(d) Bent waveguide; log scale

Figure 3.27 E-field mode profile (linear, log scale) of a straight waveguide, 500 × 220 nm strip (left), and the same waveguide bent at a radius of 2 μm (right).

The first portion of Script 3.20 calculates the field profiles for the straight waveguide, as well as for various bent waveguide radii. The field profiles are shown in Figure 3.27. The difference in the field profiles of the straight and bent waveguides is clearly visible and leads to mode-mismatch loss.

The second portion of Script 3.20 calculates mode overlap integrals to determine the power coupled from the straight waveguide to the bent waveguides. Further details on the approach can be found in Reference [35]. The mode calculations for bent waveguides with radii smaller than 3 μm are numerically challenging, hence only results above 3 μm are found.

The results of the mode power coupling calculations are plotted in Figure 3.28a for the strip waveguide and Figure 3.28b for the rib waveguide. For the strip waveguide, negligible radiation loss was found, and the simulations predict that the losses originate from the mode mismatch. However, the predicted losses from the mode calculations are higher than experiments and FDTD simulations. For the rib waveguides, the mode-mismatch losses still dominate over radiation losses; however, radiation losses do contribute significantly for radii below 4 μm. Better agreement between FDTD, experiments, and mode calculations is found in the case of the rib waveguides.

In summary, both the 3D FDTD and eigenmode solver techniques are most suitable for evaluating the loss of bends in waveguides, and the simulations agree well with experimental results.

3.4 Problems

3.1 Find the group and phase velocity of light for the fundamental TE and TM modes in the following waveguides:

- slab waveguide, 220 nm thick, at 1550 nm;
- slot waveguide, in a 500 nm wide strip, with a 150 nm fully etched gap in the middle.

3.2 What is the single-mode condition for the rib waveguide TE mode, for a slab thickness of 90 nm? What is the maximum width for the waveguide before it supports more than one mode? What is the minimum width for the silicon before it does not support a mode?

3.3 Design a single-mode (TE/TM) waveguide for 1310 nm operation, for a silicon thickness of 150 nm, and 220 nm.

3.4 Explain intuitively why the effective index of a strip waveguide typically decreases with wavelength.

3.5 Explain intuitively why the effective index of a strip waveguide typically increases as the waveguide width is increased.

3.6 Explain intuitively why the group index of a strip waveguide typically decreases as the waveguide width is increased (e.g. from 400 nm to 600 nm).

3.7 Design a waveguide with zero group velocity dispersion at 1550 nm.

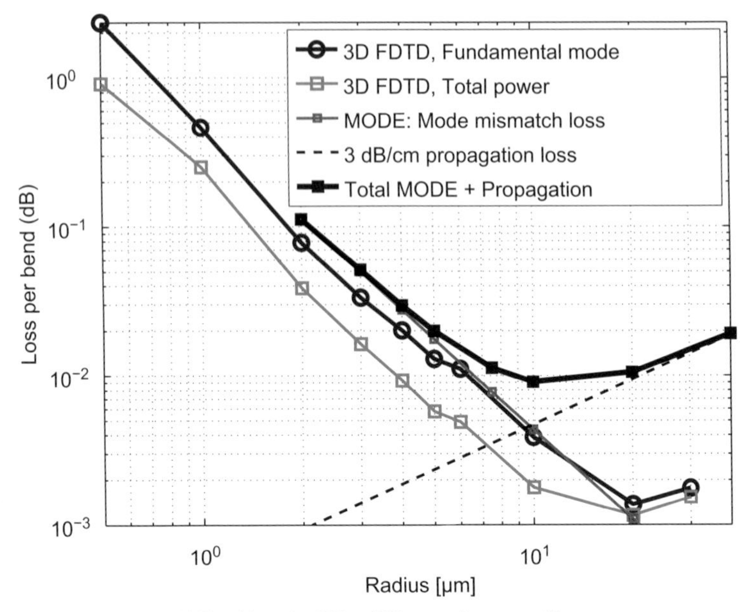

(a) Bend loss for 500 × 220 nm strip waveguides.

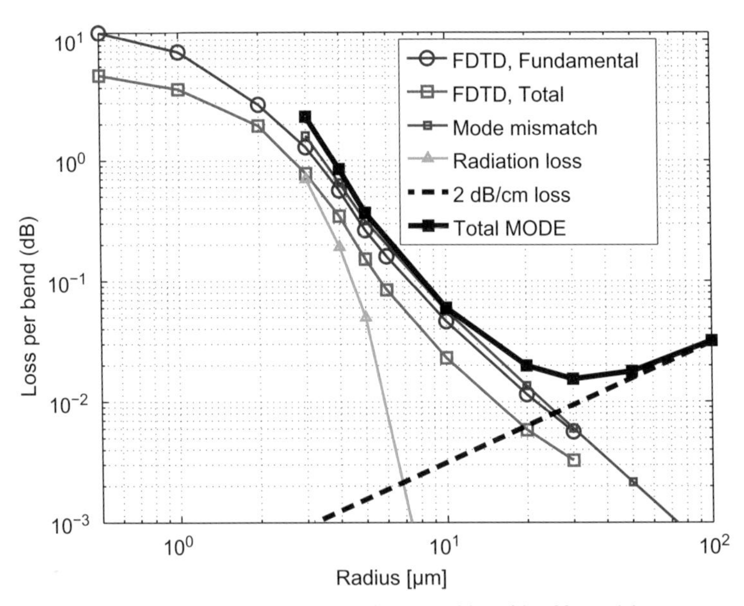

(b) Bend loss for 500 × 220 nm rib waveguides with a 90 nm slab.

Figure 3.28 Numerical waveguide bend loss versus bend radius. Comparison of 3D FDTD results with MODE calculations. MODE simulations can be used to identify the contribution mechanisms: radiation losses are negligible and are dominated by mode-mismatch losses between the straight and bent waveguides. For very large radius bends, the propagation losses (scattering, substrate leakage, etc.) become dominant.

3.5 Code listings

Listing 3.1 Material definitions for Lumerical MODE and FDTD Solutions; materials.lsf

```
# materials.lsf - creates a dispersive material model in Lumerical.
matname = "Si (Silicon) - Dispersive & Lossless";
newmaterial = addmaterial("Lorentz");
setmaterial(newmaterial,"name",matname);
setmaterial(matname,"Permittivity",7.98737492);
setmaterial(matname,"Lorentz Linewidth",1e8);
setmaterial(matname,"Lorentz Resonance",3.93282466e+15);
setmaterial(matname,"Lorentz Permittivity",3.68799143);
setmaterial(matname,"color",[0.85, 0, 0, 1]); # red

matname = "Air (1)";
if (1) { newmaterial = addmaterial("Dielectric"); }
setmaterial(newmaterial,"name",matname);
setmaterial(matname,"Refractive Index",1);
setmaterial(matname,"color",[0.85, 0.85, 0, 1]);

matname = "SiO2 (Glass) - Dispersive & Lossless";
newmaterial = addmaterial("Lorentz");
setmaterial(newmaterial,"name",matname);
setmaterial(matname,"Permittivity",2.119881);
setmaterial(matname,"Lorentz Linewidth",1e10);
setmaterial(matname,"Lorentz Resonance",3.309238e+13);
setmaterial(matname,"Lorentz Permittivity", 49.43721);
setmaterial(matname,"color",[0.5, 0.5, 0.5, 1]); # grey

matname = "SiO2 (Glass) - Const";
newmaterial = addmaterial("Dielectric");
setmaterial(newmaterial,"name",matname);
setmaterial(matname,"Permittivity",1.444^2);
setmaterial(matname,"color",[0.5, 0.5, 0.5, 1]); # grey
```

Listing 3.2 Analytic calculation of 1D waveguide mode parameters and effective index, in MATLAB; see Section 3.2 in *Photonics: Optical Electronics in Modern Communication* by A. Yariv and P. Yeh, 6th edition, Equations 3.2-1 to 3.2-5 [36]; wg_1D_analytic.m

```
% wg_1D_analytic.m - Analytic solution of waveguide
% by Lumerical Solutions, http://www.lumerical.com/mode_online_help/slab_wg.m
% modified by Lukas Chrostowski, 2012
% See Yariv Photonics book, Chapter 3
% finds the TE and TM effective indices of a 3-layer waveguide

% usage:
%  - get effective indices for supported modes:
%  [nTE, nTM] = wg_1D_analytic2 (1.55e-6, 0.22e-6, 1.444, 3.47, 1.444)
%  - TEparam,TMparam: h, q, p parameters of the mode.

function [nTE,nTM,TEparam,TMparam]=wg_1D_analytic (lambda, t, n1, n2, n3)
k0 = 2*pi/lambda;
b0 = linspace( max([n1 n3])*k0, n2*k0, 1000); %k0*n3 < b < k0*n2
b0 = b0(1:end-1);
te0=TE_eq(b0,k0,n1,n2,n3,t);
tm0=TM_eq(b0,k0,n1,n2,n3,t);

%TE
intervals=(te0>=0)-(te0<0);
izeros=find(diff(intervals)<0);
X0=[b0(izeros); b0(izeros+1)]';
[nzeros,scrap]=size(X0);
for i=1:nzeros
    nTE(i)=fzero(@(x) TE_eq(x,k0,n1,n2,n3,t),X0(i,:))/k0;
    [TEparam(i,1),TEparam(i,2),TEparam(i,3),TEparam(i,4)]= TE_eq(nTE(i)*k0,k0,n1,n2,n3,t);
end
nTE=nTE(end:-1:1);
TEparam=TEparam(end:-1:1,:);
```

```
%TM
intervals=(tm0>=0)-(tm0<0);
izeros=find(diff(intervals)<0);
X0=[b0(izeros); b0(izeros+1)]';
[nzeros,scrap]=size(X0);
for i=1:nzeros
   nTM(i)=fzero(@(x) TM_eq(x,k0,n1,n2,n3,t),X0(i,:))/k0;
   [TMparam(i,1),TMparam(i,2),TMparam(i,3),TMparam(i,4)]= TM_eq(nTM(i)*k0,k0,n1,n2,n3,t);
end
if nzeros>0
   nTM=nTM(end:-1:1);
   TMparam=TMparam(end:-1:1,:);
else
   nTM=[];
end

function [te0,h0,q0,p0]=TE_eq(b0,k0,n1,n2,n3,t)
h0 = sqrt( (n2*k0)^2 - b0.^2 );
q0 = sqrt( b0.^2 - (n1*k0)^2 );
p0 = sqrt( b0.^2 - (n3*k0)^2 );
%the objective is to find zeroes of te0 and tm0
te0 = tan( h0*t ) - (p0+q0)./h0./(1-p0.*q0./h0.^2);

function [tm0,h0,q0,p0]=TM_eq(b0,k0,n1,n2,n3,t)
h0 = sqrt( (n2*k0)^2 - b0.^2 );
q0 = sqrt( b0.^2 - (n1*k0)^2 );
p0 = sqrt( b0.^2 - (n3*k0)^2 );
pbar0 = (n2/n3)^2*p0;
qbar0 = (n2/n1)^2*q0;
tm0 = tan( h0*t ) - h0.*(pbar0+qbar0)./(h0.^2-pbar0.*qbar0);
```

Listing 3.3 Analytic calculation of 1D waveguide field profile, in MATLAB; see Section 3.2 in *Photonics: Optical Electronics in Modern Communication* by A. Yariv and P. Yeh, 6th edition, Equations 3.2-2 to 3.2-7 [36]; wg_1D_mode_profile.m

```
% wg_1D_mode_profile.m - Calculate the 1D mode profile of a waveguide
% by Lukas Chrostowski, 2012
% See Yariv Photonics book, Chapter 3.2
% - function returns mode profiles for TE and TM modes (E, H components)
% usage, e.g.:
%  [x, TE_E, TE_H, TM_E, TM_H] = wg_1D_mode_profile (1.55e-6, 0.22e-6, 1.444, 3.47, 1.444,
     100, 4)
% plot (x, TE_E);

function [x, TE_E, TE_H, TM_E, TM_H]= wg_1D_mode_profile (lambda, t, n1, n2, n3, pts, M)
[nTE,nTM,TEparam,TMparam]= wg_1D_analytic(lambda,t,n1,n2,n3);
x1=linspace( -M*t, -t/2, pts); x2=linspace( -t/2, t/2, pts);
x3=linspace( t/2, M*t, pts); x=[x1 x2 x3];
nx=[n1*ones(pts,1); n2*ones(pts,1); n3*ones(pts,1)]';
mu0=4*pi*1e-7; epsilon0=8.85e-12; eta=sqrt(mu0/epsilon0); c=3e8; % constants
for i=1:length(nTE)
   h=TEparam(i,2);q=TEparam(i,3); p=TEparam(i,4);
   beta = 2*pi*nTE(i)/lambda;
   C=2*h*sqrt ( 2*pi*c/lambda*mu0 / (beta * (t+1/q+1/p)*(h^2+q^2) ) ); % normalize to 1W
   % n1, n2, n3 regions
   TE_E(i,:)=C*[exp(q*(x1+t/2)), (cos(h*(x2+t/2))+q/h*sin(h*(x2+t/2))),
        (cos(h*t)+q/h*sin(h*t)).*exp(-p*(x3-t/2))];
end
TE_H=TE_E'.*(nx'*ones(1,length(nTE)))/eta;

for i=1:length(nTM)
   h=TMparam(i,2); q=TMparam(i,3);
   p=TMparam(i,4); qb=n2^2/n1^2*q;pb=n2^2/n3^2*p;
   beta = 2*pi*nTM(i)/lambda;
   temp=(qb^2+h^2)/qb^2 * (t/n2^2 + (q^2+h^2)/(qb^2+h^2)/n1^2/q + (
        p^2+h^2)/(p^2+h^2)/n3^2/p) ;
```

```
    C=2*sqrt ( 2*pi*c/lambda*epsilon0 / (beta * temp )); % normalize to 1W
    TM_H(i,:)=C*[h/qb*exp(q*(x1+t/2)), (h/qb*cos(h*(x2+t/2))+sin(h*(x2+t/2))),
         (h/qb*cos(h*t)+sin(h*t)).*exp(-p*(x3-t/2))];
end
TM_E=TM_H'./(nx'*ones(1,length(nTM)))*eta;
```

Listing 3.4 Draw the 2D waveguide structure, in Lumerical MODE Solutions; wg_2D_draw.lsf

```
# wg_2D_draw.lsf - draw the waveguide geometry in Lumerical MODE
newmode; newmode; redrawoff;

# define wafer and waveguide structure
thick_Clad = 2.0e-6;
thick_Si  = 0.22e-6;
thick_BOX = 2.0e-6;
thick_Slab = 0;              # for strip waveguides
# thick_Slab = 0.13e-6;      # for rib waveguides
width_ridge = 0.5e-6;        # width of the waveguide

# define materials
material_Clad = "SiO2 (Glass) - Const";
material_BOX = "SiO2 (Glass) - Const";
material_Si  = "Si (Silicon) - Dispersive & Lossless";
materials;   # run script to add materials

# define simulation region
width_margin = 2.0e-6; # space to include on the side of the waveguide
height_margin = 1.0e-6; # space to include above and below the waveguide

# calculate simulation volume
# propagation in the x-axis direction; z-axis is wafer-normal
Xmin = -2e-6; Xmax = 2e-6; # length of the waveguide
Zmin = -height_margin; Zmax = thick_Si + height_margin;
Y_span = 2*width_margin + width_ridge; Ymin = -Y_span/2; Ymax = -Ymin;

# draw cladding
addrect; set("name","Clad"); set("material", material_Clad);
set("y", 0);            set("y span", Y_span+1e-6);
set("z min", 0);        set("z max", thick_Clad);
set("x min", Xmin);     set("x max", Xmax);
set("override mesh order from material database",1);
set("mesh order",3); # similar to "send to back", put the cladding as a background.
set("alpha", 0.05);

# draw buried oxide
addrect; set("name", "BOX"); set("material", material_BOX);
set("x min", Xmin);     set("x max", Xmax);
set("z min", -thick_BOX); set("z max", 0);
set("y", 0);            set("y span", Y_span+1e-6);
set("alpha", 0.05);

# draw silicon wafer
addrect; set("name", "Wafer"); set("material", material_Si);
set("x min", Xmin);     set("x max", Xmax);
set("z max", -thick_BOX); set("z min", -thick_BOX-2e-6);
set("y", 0);            set("y span", Y_span+1e-6);
set("alpha", 0.1);

# draw waveguide
addrect; set("name", "waveguide"); set("material",material_Si);
set("y", 0);      set("y span", width_ridge);
set("z min", 0);  set("z max", thick_Si);
set("x min", Xmin); set("x max", Xmax);

# draw slab for rib waveguides
addrect; set("name", "slab"); set("material",material_Si);
if (thick_Slab==0) {
 set("y min", 0);   set("y max", 0);
```

```
} else {
 set("y", 0);        set("y span", Y_span+1e-6);
}
set("z min", 0);     set("z max", thick_Slab);
set("x min", Xmin); set("x max", Xmax);
set("alpha", 0.2);
```

Listing 3.5 One-dimensional slab waveguide simulation parameters, in Lumerical MODE Solutions; wg_1D_slab.lsf

```
# wg_1D_slab.lsf - setup the Lumerical MODE 1D simulation

wg_2D_draw;              # draw the waveguide

wavelength=1.55e-6;
meshsize = 10e-9; # mesh size

# add 1D mode solver (waveguide cross-section)
addfde; set("solver type","1D Z:X prop");
set("x", 0); set("y", 0);
set("z max", Zmax); set("z min", Zmin);
set("wavelength", wavelength);
set("define z mesh by","maximum mesh step");
set("dz", meshsize);
modes=2; # modes to output
set("number of trial modes",modes);
```

Listing 3.6 One-dimensional slab waveguide simulation of the mode profile, in Lumerical MODE Solutions; wg_1D_slab_mode.lsf

```
# wg_1D_slab_mode.lsf - calculate mode profiles in Lumerical MODE

wg_1D_slab;     # Draw waveguides and setup the simulation

n=findmodes;    # calculate the modes
for (m=1:modes) {
  ? neff=getdata ("FDE::data::mode"+num2str(m),"neff"); # display effective index
  z=getdata("FDE::data::mode1","z");
  E3=pinch(getelectric("FDE::data::mode"+num2str(m)));
  plot(z,E3); # plot the mode profile
}
```

Listing 3.7 One-dimensional slab waveguide simulation convergence test (for the simulation span), in Lumerical MODE Solutions; wg_1D_slab_convergence_z.lsf

```
# wg_1D_slab_convergence_z.lsf - perform convergence test, Lumerical MODE

wg_1D_slab;     # Draw waveguides and setup the simulation

zspan_list=[.4:0.2:2]*1e-6;  # sweep for the simulation region
neff=matrix(length(zspan_list),2); # initialize empty matrix
for (x=1:length(zspan_list)) {
  switchtolayout;
  select("MODE");
  set("z span", zspan_list(x));
  n=findmodes;
  neff(x,1)=getdata ("MODE::data::mode1","neff");
  neff(x,2)=getdata ("MODE::data::mode2","neff");
}
plot(zspan_list,real(neff)); legend('Mode 1','Mode 2');
matlabsave ('out/wg_slab_convergence_z');
```

Listing 3.8 One-dimensional slab waveguide simulation convergence test (for the simulation mesh), in Lumerical MODE Solutions; wg_1D_slab_convergence_mesh.lsf

```
# wg_1D_slab_convergence_mesh.lsf - perform convergence test, Lumerical MODE

wg_1D_slab;     # Draw waveguides and setup the simulation

mesh_list=[0.01; 0.1; 1:1:50]*1e-9;      # vary the mesh size
refine_list=[1,3];                 # 1=Staircase, 3=Conformal mesh
neff=matrix(length(mesh_list),length(refine_list));
for (y=1:length(refine_list)) {
  switchtolayout; select ("MODE"); set ("z span",2e-6);
  set ("mesh refinement", refine_list(y));
  for (x=1:length(mesh_list)) {
    switchtolayout; select("MODE"); set("dz", mesh_list(x));
    n=findmodes;
    neff(x,y)=getdata ("MODE::data::mode1","neff");
  }
}
plot(mesh_list,real(neff)); legend("Staircase mesh","Conformal mesh");
matlabsave ('out/wg_slab_convergence_mesh');
```

Listing 3.9 One-dimensional slab waveguide simulation, for slab thickness parameter sweep, in Lumerical MODE Solutions; wg_1D_slab_neff_sweep.lsf

```
# wg_1D_slab_neff_sweep.lsf - perform sweep mode calculations on the slab

wg_1D_slab;     # Draw waveguides and setup the simulation

thick_Si_list = [0:.01:.4]*1e-6; # sweep waveguide thickness
mode_list=[1:4];

neff_slab = matrix (length(thick_Si_list), length(mode_list) );
TE_pol = matrix (length(thick_Si_list), length(mode_list) );

select("MODE");
set("number of trial modes",max(mode_list)+2);

for(kk=1:length(thick_Si_list))
{
  switchtolayout;
  setnamed('waveguide','z max', thick_Si_list(kk));
  n=findmodes;
  for (m=1:length(mode_list))
  {
    neff_slab (kk,m) =abs( getdata ("MODE::data::mode"+num2str(m),"neff") );
    TE_pol(kk,m) = getdata("MODE::data::mode"+num2str(m),"TE polarization fraction");
    if ( TE_pol(kk,m) > 0.5 )
      {    pol = "TE"; } else  {  pol = "TM"; }
  }
}
plot ( thick_Si_list, neff_slab);
# matlabsave ("wg_mode_neff_sweep_slab"); # save the data for plotting in Matlab.
```

Listing 3.10 Waveguide Effective Index Method; wg_EIM.lsf

```
# wg_EIM.lsf - setup the Lumerical MODE 1D simulation for Effective index method

wg_2D_draw;              # draw the waveguide

material_Si  = "<Object defined dielectric>";
select("waveguide");
set("material",material_Si);
set("index", 2.845); # effective index taken from the TE slab mode
```

```
wavelength=1.55e-6;
meshsize = 10e-9; # mesh size

# add 1D mode solver (horizontal waveguide cross-section)
addfde; set("solver type","1D Y:X prop");
set("x", 0); set("z", 0.1e-6);
set("y max", 1e-6); set("y min", -1e-6);
set("wavelength", wavelength);
set("define y mesh by","maximum mesh step");
set("dy", meshsize);
modes=2; # modes to output
set("number of trial modes",modes);

n=findmodes;   # calculate the modes
for (m=1:modes) {
  ? neff=getdata ("FDE::data::mode"+num2str(m),"neff"); # display effective index
  y=getdata("FDE::data::mode1","y");
  E3=pinch(getelectric("FDE::data::mode"+num2str(m)));
  plot(y,E3); # plot the mode profile
  matlabsave("wg_EIM"+num2str(m));
}
```

Listing 3.11 Analytic calculation of 2D waveguide field profile using the Effective Index Method from Listing 3.2, in MATLAB; wg_EIM_profile.m

```
% wg_EIM_profile.m - Effective Index Method - mode profile
% Lukas Chrostowski, 2012
% usage, e.g.:
%  wg_EIM_profile (1.55e-6, 0.22e-6, 0.5e-6, 90e-9, 3.47, 1, 1.44, 100, 2)

function wg_EIM_profile (lambda, t, w, t_slab, n_core, n_clad, n_oxide, pts, M)

% find TE (TM) modes of slab waveguide (waveguide core and slab portions):
[nTE,nTM]=wg_1D_analytic (lambda, t, n_oxide, n_core, n_clad);
if t_slab>0
   [nTE_slab,nTM_slab]=wg_1D_analytic (lambda, t_slab, n_oxide, n_core, n_clad);
else
   nTE_slab=n_clad; nTM_slab=n_clad;
end
[xslab,TE_Eslab,TE_Hslab,TM_Eslab,TM_Hslab]=wg_1D_mode_profile (lambda, t, n_oxide,
     n_core, n_clad, pts, M);

figure(1); clf; subplot (2,2,2); Fontsize=9;
plot(TE_Eslab/max(max(TE_Eslab)),xslab*1e9,'LineWidth',2);hold all;
ylabel('Height [nm]','FontSize',Fontsize);
xlabel('E-field (TE)','FontSize',Fontsize);
set(gca,'FontSize',Fontsize,'XTick',[]);
axis tight; a=axis; axis ([a(1)*1.1, a(2)*1.1, a(3), a(4)]);
Ax1 = gca; Ax2 = axes('Position',get(Ax1,'Position'));
get(Ax1,'Position');
nx=[n_oxide*ones(pts,1); n_core*ones(pts,1); n_clad*ones(pts,1)]';
plot (nx, xslab*1e9, 'LineWidth',0.5,'LineStyle','--','parent',Ax2);
a2=axis; axis ([a2(1), a2(2), a(3), a(4)]);
set(Ax2,'Color','none','XAxisLocation','top', 'YTick',[],'TickDir','in');
set(gca,'YAxisLocation','right'); box off;
xlabel('Material Index','FontSize',Fontsize);
set(gca,'FontSize',Fontsize);

% TE-like modes of the etched waveguide (for fundamental slab mode)
%  solve for the "TM" modes:
[nTE,nTM]=wg_1D_analytic (lambda, w, nTE_slab(1), nTE(1), nTE_slab(1));
neff_TEwg=nTM;
[xwg,TE_E_TEwg,TE_H_TEwg,TM_E_TEwg,TM_H_TEwg]=wg_1D_mode_profile (lambda, w, nTE_slab(1),
     nTE(1), nTE_slab(1), pts, M);
```

```
subplot (2,2,3);
plot (xwg*1e9, TM_E_TEwg/max(max(TM_E_TEwg)), 'LineWidth',2,'LineStyle','-');
xlabel('Position [nm]','FontSize',Fontsize);
ylabel('E-field (TM, TE-like mode)','FontSize',Fontsize);
set(gca,'FontSize',Fontsize,'YTick',[]);
axis tight; a=axis; axis ([a(1), a(2), a(3)*1.1, a(4)*1.1]);
Ax1 = gca; Ax2 = axes('Position',get(Ax1,'Position'));
nx=[nTE_slab(1)*ones(pts,1); nTE(1)*ones(pts,1); nTE_slab(1)*ones(pts,1)]';
plot (xwg*1e9, nx, 'LineWidth',0.5,'LineStyle','--','parent',Ax2);
set(Ax2,'Color','none','YAxisLocation','right'); box off;
a2=axis; axis ([a(1), a(2), a2(3), a2(4)]);
ylabel('Slab Effective Index','FontSize',Fontsize);
set(gca,'FontSize',Fontsize);

% Plot the product of the two fields
subplot (2,2,1); Exy=TM_E_TEwg(:,1)*(TE_Eslab(1,:));
contourf(xwg*1e9,xslab*1e9,abs(Exy)/max(max(Exy))')
xlabel ('X (nm)','FontSize',Fontsize);
ylabel ('Y (nm)','FontSize',Fontsize);
set (gca, 'FontSize',Fontsize);
A=axis; axis([A(1)+0.4, A(2)-0.4, A(3)+.2, A(4)-0.2]);
title('Effective Index Method');
% Draw the waveguide:
rectangle ('Position',[-w/2,-t/2,w,t]*1e9, 'LineWidth',1, 'EdgeColor','white')
if t_slab>0
   rectangle ('Position',[-M*w,-t/2,(M-0.5)*w, t_slab]*1e9, 'LineWidth',1,
        'EdgeColor','white')
   rectangle ('Position',[w/2,-t/2,(M-0.5),t_slab]*1e9, 'LineWidth',1,
        'EdgeColor','white')
end

function draw_WG_vertical(M)
pP=get(gca,'Position');pPw=pP(3);
pPc=pP(3)/2+pP(1); pP2=pPw/4/M;
annotation ('line',[pPc-pP2,pPc-pP2], [pP(2),pP(4)+pP(2)],'LineStyle','--');
annotation ('line',[pPc+pP2,pPc+pP2], [pP(2),pP(4)+pP(2)],'LineStyle','--');
axis tight; a=axis; axis ([a(1), a(2), a(3)*1.1, a(4)*1.1]);

function draw_WG_horiz(M)
pP=get(gca,'Position');pPw=pP(4);
pPc=pP(4)/2+pP(2); pP2=pPw/4/M;
annotation ('line',[pP(1),pP(3)+pP(1)], [pPc-pP2,pPc-pP2],'LineStyle','--');
annotation ('line',[pP(1),pP(3)+pP(1)], [pPc+pP2,pPc+pP2],'LineStyle','--');
axis tight; a=axis; axis ([a(1)*1.1, a(2)*1.1, a(3), a(4)]);
```

Listing 3.12 Waveguide 2D eigenmode calculation setup; wg_2D.lsf

```
# wg_2D.lsf - set up the mode profile simulation solver

wg_2D_draw; # run script to draw the waveguide

# define simulation parameters
wavelength   = 1.55e-6;
meshsize     = 20e-9;      # maximum mesh size
modes        = 4;          # modes to output

# add 2D mode solver (waveguide cross-section)
addfde; set("solver type", "2D X normal");
set("x", 0);
set("y", 0);         set("y span", Y_span);
set("z max", Zmax); set("z min", Zmin);
set("wavelength", wavelength); set("solver type","2D X normal");
set("define y mesh by","maximum mesh step"); set("dy", meshsize);
set("define z mesh by","maximum mesh step"); set("dz", meshsize);
set("number of trial modes",modes);
```

Listing 3.13 Waveguide 2D eigenmode profile calculation; wg_2D_mode_profile.lsf

```
# wg_2D_mode_profile.lsf - calculate the mode profiles of the waveguide

wg_2D;  # run the script to draw the waveguide and set up the simulation

# output filename
clad = substring(material_Clad, 1, (findstring(material_Clad,' '))-1);
?FILE = "out/wl" + num2str(wavelength*1e9) +"nm_" + clad + "-clad" +
       "_"+ num2str(thick_Si*1e9) + "nm-wg_" + num2str(thick_Slab*1e9)+"nm-ridge";

# find the material dispersion (using 2 frequency points), for energy density calculation
switchtolayout; set("wavelength", wavelength*(1 + .001) );
run; mesh;
f1 = getdata("FDE::data::material","f");
eps1 = pinch(getdata("FDE::data::material","index_x"))^2;
switchtolayout; set("wavelength", wavelength*(1 - .001) );
run; mesh;
f3 = getdata("FDE::data::material","f");
eps3 = pinch(getdata("FDE::data::material","index_x"))^2;
re_dwepsdw = real((f3*eps3-f1*eps1)/(f3-f1));

switchtolayout; set("wavelength", wavelength);
n=findmodes;
neff = matrix ( modes ); TE_pol = matrix (modes );
for (m=1:modes) {        # extract mode data
  neff(m) = abs( getdata ("FDE::data::mode"+num2str(m),"neff") );
  TE_pol(m) = getdata("FDE::data::mode"+num2str(m),"TE polarization fraction");
  if ( TE_pol(m) > 0.5 ) # identify the TE-like or TM-like modes
    {   pol = "TE"; }  else  {  pol = "TM"; }
  z  = getdata("FDE::data::mode"+num2str(m),"z");
       y=getdata("FDE::data::mode"+num2str(m),"y");
  E1 = pinch(getelectric("FDE::data::mode"+num2str(m)));
  H1 = pinch(getmagnetic("FDE::data::mode"+num2str(m)));
  W1 = 0.5*(re_dwepsdw*eps0*E1+mu0*H1);
  image(y,z,E1); # plot E-field intensity of mode
  setplot("title","mode" + num2str(m) + "("+pol+"): "+"neff:" + num2str(neff(m)));
  image(y,z,W1); # plot energy density of mode
  setplot("title","mode" + num2str(m) + "("+pol+"): "+"neff:" + num2str(neff(m)));
# matlabsave ( FILE + "_"+ num2str(m) );
}
```

Listing 3.14 Effective index versus waveguide width (2D simulation); wg_2D_neff_sweep_width.lsf

```
# wg_2D_neff_sweep_width.lsf - perform mode calculations on the waveguide

wg_2D;    # run the script to draw the waveguide and set up the simulation

modes=4;         # modes to output
set("number of trial modes",modes+2);

# define parameters to sweep
width_ridge_list=[.2:.02:.8]*1e-6; # sweep waveguide width

neff = matrix (length(width_ridge_list), modes );
TE_pol = matrix (length(width_ridge_list), modes );

for(ii=1:length(width_ridge_list)) {
  switchtolayout;
  setnamed("waveguide","y span", width_ridge_list(ii));
  setnamed("slab","z max", thick_Slab);
  n=findmodes;
  for (m=1:modes) { # extract mode data
    neff (ii,m) = abs( getdata ("FDE::data::mode"+num2str(m),"neff") );
    TE_pol(ii,m) = getdata("FDE::data::mode"+num2str(m),"TE polarization fraction");
  }
}
plot (width_ridge_list, neff (1:length(width_ridge_list), 1)); # plots the 1st mode.
plot (width_ridge_list, neff);
```

Listing 3.15 Effective and group index versus wavelength; wg_2D_neff_sweep_wavelength.lsf

```
# wg_2D_neff_sweep_wavelength.lsf - Calculate the wavelength dependence of waveguide's
     neff and ng

wg_2D; # draw waveguide

run; mesh;
setanalysis('wavelength',1.6e-6);
findmodes; selectmode(1); # find the fundamental mode

setanalysis("track selected mode",1);
setanalysis("number of test modes",5);
setanalysis("detailed dispersion calculation",0); # This feature is useful for
     higher-order dispersion.
setanalysis('stop wavelength',1.5e-6);
frequencysweep;  # perform sweep of wavelength and plot
f=getdata("frequencysweep","f");
neff=getdata("frequencysweep","neff");
f_vg=getdata("frequencysweep","f_vg");
ng=c/getdata("frequencysweep","vg");
plot(c/f*1e6,neff,"Wavelength (um)", "Effective Index");
plot(c/f_vg*1e6,ng,"Wavelength (um)", "Group Index");
matlabsave ('wg_2D_neff_sweep_wavelength.mat',f, neff, f_vg, ng);
```

Listing 3.16 Draw the bent waveguides; bend_draw.lsf

```
# bend_draw.lsf - Define simulation parameters, draw the bend
# input: variable "bend_radius" pre-defined

# define wafer structure
thick_Clad = 3e-6;
thick_Si = 0.22e-6;
thick_BOX = 2e-6;
thick_Slab = 0;         # for strip waveguides
#thick_Slab = 0.09e-6; # for rib waveguides
width_ridge = 0.5e-6;  # width of the waveguide

# define materials
material_Clad = "SiO2 (Glass) - Const";
material_BOX = "SiO2 (Glass) - Const";
material_Si  = "Si (Silicon) - Dispersive & Lossless";
materials;    # run script to add materials

Extra=0.5e-6;
thick_margin = 500e-9;
width_margin=2e-6;
length_input=1e-6;

Xmin = 0-width_ridge/2-width_margin;
Xmax = bend_radius+length_input;
Zmin =-thick_margin; Zmax=thick_Si+thick_margin;
Ymin = 0;
Ymax = bend_radius+width_ridge/2+width_margin+length_input/2;

addrect; set('name','Clad'); set("material", material_Clad);
set('y min', Ymin-Extra); set('y max', Ymax+Extra);
set('z min', 0);         set('z max', Zmax);
set('x min', Xmin-Extra); set('x max', Xmax+Extra);
set('alpha', 0.2);

addrect; set("name", "BOX"); set("material", material_BOX);
set('x min', Xmin-Extra); set('x max', Xmax+Extra);
set('z min', -thick_BOX); set('z max', 0);
set('y min', Ymin-Extra); set('y max', Ymax+Extra);
set('alpha', 0.3);

addrect; set("name", "slab"); set("material",material_Si);
set('y min', Ymin-Extra); set('y max', Ymax+Extra);
```

```
set('z min', 0);           set('z max', thick_Slab);
set('x min', Xmin-Extra); set('x max', Xmax+Extra);
set('alpha', 0.4);

addrect; set('name', 'input_wg'); set("material",material_Si);
set('x min', -width_ridge/2); set('x max', width_ridge/2);
set('z min', 0);              set('z max', thick_Si);
set('y min', Ymin-2e-6);      set('y max', Ymin+length_input);

addrect; set('name', 'output_wg'); set("material",material_Si);
set('y', length_input+bend_radius); set('y span', width_ridge);
set('z min', 0);           set('z max', thick_Si);
set('x min', bend_radius); set('x max', bend_radius+length_input+2e-6);

addring; set('name', 'bend'); set("material",material_Si);
set('x', bend_radius);
set('y', length_input);
set('z min', 0); set('z max', thick_Si);
set('theta start', 90); set('theta stop', 180);
set('outer radius', bend_radius+0.5*width_ridge);
set('inner radius', bend_radius-0.5*width_ridge);
```

Listing 3.17 Set up the 3D FDTD simulation for bent waveguides; bend_FDTD_setup.lsf

```
# bend_FDTD_setup.lsf - setup FDTD simulation for bend calculations

wavelength=1.55e-6;
Mode_Selection = 'fundamental TE mode';
Mesh_level=1; # Mesh of 3 is suitable for high accuracy

addfdtd;
set('x min', Xmin); set('x max', Xmax);
set('y min', Ymin); set('y max', Ymax);
set('z min', Zmin); set('z max', Zmax);
set('mesh accuracy', Mesh_level);

addmode;
set('injection axis', 'y-axis');
set('direction', 'forward');
set('y', Ymin+100e-9);
set('x', 0); set('x span', width_ridge+width_margin);
set('z min', Zmin); set('z max', Zmax);
set('set wavelength','true');
set('wavelength start', wavelength);
set('wavelength stop',wavelength);
set('mode selection', Mode_Selection);
updatesourcemode;

addpower;  # Power monitor, output
set('name', 'transmission');
set('monitor type', '2D X-normal');
set('y', length_input+bend_radius);
set('y span', width_ridge +width_margin);
set('z min', Zmin); set('z max', Zmax);
set('x', Xmax-0.5e-6);

addmodeexpansion;
set('name', 'expansion');
set('monitor type', '2D X-normal');
set('y', length_input+bend_radius);
set('y span', width_ridge +width_margin);
set('z min', Zmin); set('z max', Zmax);
set('x', Xmax-0.3e-6);
set('frequency points',10);
set('mode selection', Mode_Selection);
setexpansion('T','transmission');

addpower;  # Power monitor, input
set('name', 'input');
```

```
set('monitor type', '2D Y-normal');
set('y', Ymin+500e-9); set('x', 0);
set('x span', width_ridge+width_margin);
set('z min', Zmin); set('z max', Zmax);

if (0) {
  addmovie;
  set('name', 'movie');
  set('lockAspectRatio', 1);
  set('monitor type', '2D Z-normal');
  set('x min', Xmin); set('x max', Xmax);
  set('y min', Ymin); set('y max', Ymax);
  set('z', 0.5*thick_Si);
}
```

Listing 3.18 Run 3D FDTD simulations for bent waveguides; bend_FDTD_radius_sweep.lsf

```
# bend_FDTD_radius_sweep.lsf - 3D-FDTD script to calculate the loss in a 90 degree bend
      versus bend-radius, including mode expansion

bend_radius_sweep=[0.5,1,2,3,4,5,6,10,20,30]*1e-6; # bend radii to sweep

L=length(bend_radius_sweep); T=matrix(2,L);
for(ii=1:L)
{
  newproject; switchtolayout; redrawoff;
  selectall; delete;
  bend_radius = bend_radius_sweep(ii);
  bend_draw;       # draw the waveguides
  bend_FDTD_setup; # setup the FDTD simulations
  save('bend_radius_'+ num2str(ii));
  run;
        T_fund=getresult('expansion', 'expansion for T');
        T_forward=T_fund.getattribute('T forward');
  T(1,ii)=T_forward;
        T(2,ii)=transmission('transmission'); # total output power in WG
        T(3,ii)=transmission('input'); # output power in fundamental mode
}
plot(bend_radius_sweep, -10*log10(T(1,1:ii)/T(3,1:ii)), -10*log10(T(2,1:ii)/T(3,1:ii)));
legend ('Transmission, in fundamental mode', 'Transmission, total');
matlabsave ('bend.mat', bend_radius_sweep, T);
```

Listing 3.19 3D FDTD simulations for bent waveguides to visualize the E-field intensity; bend_FDTD_top_field.lsf

```
# bend_top_field.lsf - simulate a bend, observe the top-view field profile

bend_radius = 1e-6;  # 1 micron bend radius
bend_draw;           # Call script to draw the bent waveguide
bend_FDTD_setup;     # setup the FDTD simulations

addpower;  # Power monitor, top-view
set('name', 'top');
set('monitor type', '2D Z-normal');
set('x min', Xmin); set('x max', Xmax);
set('y min', Ymin); set('y max', Ymax);
set('z', thick_Si/2); # cross-section through the middle of the waveguide

save('./out/bend_FDTD_top_field');
run;

X=getdata('top','x'); Y=getdata('top','y');
I2=abs(getdata('top','Ex'))^2 + abs(getdata('top','Ey'))^2 + abs(getdata('top','Ez'))^2;
image(X,Y,I2);            # E-field intensity image plot
image(X,Y,10*log10(I2)); # E-field intensity image plot
```

Listing 3.20 MODE profiles for bent waveguides and calculations of mode-mismatch loss; bend_MODE.lsf

```
# bend_MODE.lsf: script to:
# 1) calculate the mode profile in a waveguide with varying bend radius
# 2) calculate mode mismatch loss with straight waveguide and radiation loss vs. radius

# Example with default parameters requires 1.2 GB ram.

radii= [0, 100, 50, 30, 20, 10, 5, 4, 3]*1e-6;
# min radius as defined in:
# http://docs.lumerical.com/en/solvers_finite_difference_eigenmode_bend.html
wg_2D_draw; # run script to draw the waveguide

# define simulation parameters
wavelength   = 1.55e-6;
# maximum mesh size; 40 gives reasonable results
meshsize     = 10e-9;
modes        = 4;         # modes to output

# add 2D mode solver (waveguide cross-section)
addfde; set("solver type", "2D X normal");
set("x", 0);
width_margin = 2e-6; # ensure it is big enough to accurately measure radiation loss via
     PMLs
height_margin = 0.5e-6;
Zmin = -height_margin; Zmax = thick_Si + height_margin;
set('z max', Zmax); set('z min', Zmin);
Y_span = 2*width_margin + width_ridge;
Ymin = -Y_span/2; Ymax = -Ymin;
set('y',0); set('y span', Y_span);
set("wavelength", wavelength); set("solver type","2D X normal");
set("y min bc","PML"); set("y max bc","PML"); # radiation loss
set("z min bc","metal"); set("z max bc","metal"); # faster
set("define y mesh by","maximum mesh step");
set("dy", meshsize);
set("define z mesh by","maximum mesh step");
set("dz", meshsize);
set("number of trial modes",modes);
cleardcard; # Clears all the global d-cards.

# solve modes in the waveguide:
n=length(radii); Neff=matrix(n); LossdB_m=matrix(n);
LossPerBend=matrix(n); power_coupling=matrix(n);
for (i=1:n) {
  if (radii(i)==0) {
    setanalysis ('bent waveguide', 0); # Cartesian
  } else {
    setanalysis ('bent waveguide', 1); # cylindrical
    setanalysis ('bend radius', radii(i));
  }
  setanalysis ('number of trial modes', 4);
  nn = findmodes;
  if (nn>0) {
    Neff(i) = getdata('FDE::data::mode1','neff');
    LossdB_m(i) = getdata('FDE::data::mode1','loss'); # per m
    LossPerBend(i) = LossdB_m(i) * 2*pi*radii(i)/4;
    copydcard( 'mode1', 'radius' + num2str(radii(i)) );

    # Perform mode-overlap calculations between the straight and bent waveguides
    if (radii(i)>0) {
      out = overlap('::radius0','::radius'+num2str(radii(i)));
      power_coupling(i)=out(2); # power coupling
    }

    # plot mode profile:
    E3=pinch(getelectric('FDE::data::mode1')); y=getdata('FDE::data::mode1','y');
         z=getdata('FDE::data::mode1','z');
    image(y,z,E3);
    exportfigure('out/bend_mode_profile_radius'+ num2str(radii(i)));
```

```
    matlabsave('out/bend_mode_profile_radius'+ num2str(radii(i)), y,z,E3);
  }
}
PropagationLoss=2 *100; # dB/cm *100 --- dB/m
LossMM=-10*log10( power_coupling(2:n)^2 ); # plot 2X couplings per 90 degree bend vs
     radius (^2 for two)
LossR=LossPerBend (2:n)-LossPerBend(1);
LossP=PropagationLoss*2*pi*radii(2:n)/4; # quarter turn
plot ( radii (2:n)*1e6, LossMM, LossR, LossP, LossMM+LossR+LossP, "Radius [micron]", "Loss
     [dB]" ,"Bend Loss", "loglog, plot points");
legend ('Mode Mismatch Loss', 'Radiation loss','2 dB/cm propagation loss', 'Total Loss');
matlabsave ('out/bend_MODE_profiles_coupling', radii, power_coupling, LossPerBend);
```

References

[1] Dirk Taillaert, Harold Chong, Peter I. Borel, *et al.* "A compact two-dimensional grating coupler used as a polarization splitter". *IEEE Photonics Technology Letters* **15**.9 (2003), pp. 1249–1251 (cit. on p. 49).

[2] Dan-Xia Xu, J. H. Schmid, G. T. Reed, *et al.* "Silicon Photonic Integration Platform – Have We Found The Sweet Spot?". *IEEE Journal of Selected Topics in Quantum Electronics* **20**.4 (2014), pp. 1–17. ISSN: 1077-260X. DOI: 10.1109/JSTQE.2014.2299634 (cit. on p. 49).

[3] W. A. Zortman, D. C. Trotter, and M. R. Watts. "Silicon photonics manufacturing". *Optics Express* **18**.23 (2010), pp. 23598–23607 (cit. on p. 49).

[4] A. V. Krishnamoorthy, Xuezhe Zheng, Guoliang Li, *et al.* "Exploiting CMOS manufacturing to reduce tuning requirements for resonant optical devices". *IEEE Photonics Journal* **3**.3 (2011), pp. 567–579. DOI: 10.1109/JPHOT.2011.2140367 (cit. on p. 49).

[5] Edward Palik. *Handbook of Optical Constants of Solids.* Elsevier, 1998 (cit. on p. 50).

[6] Kurt Oughstun and Natalie Cartwright. "On the Lorentz–Lorenz formula and the Lorentz model of dielectric dispersion". *Optics Express* **11** (2003), pp. 1541–1546 (cit. on p. 50).

[7] Lorenzo Pavesi and Gérard Guillot. *Optical Interconnects: The Silicon Approach.* 978-3-540-28910-4. Springer Berlin/Heidelberg, 2006 (cit. on p. 50).

[8] G. Cocorullo and I. Rendina. "Thermo-optical modulation at 1.5 μm in silicon etalon". *Electronics Letters* **28**.1 (1992), pp. 83–85. DOI: 10.1049/el:19920051 (cit. on p. 51).

[9] J. A. McCaulley, V. M. Donnelly, M. Vernon, and I. Taha. "Temperature dependence of the near-infrared refractive index of silicon, gallium arsenide, and indium phosphide". *Physical Review B* **49**.11 (1994), p. 7408 (cit. on p. 51).

[10] Pieter Dumon. "Ultra-compact integrated optical filters in silicon-on-insulator by means of wafer-scale technology". PhD thesis. Gent University, 2007 (cit. on p. 51).

[11] Bradley J. Frey, Douglas B. Leviton, and Timothy J. Madison. "Temperature-dependent refractive index of silicon and germanium". *Proceedings SPIE.* Vol. 6273. 2006, 62732J–62732J–10. DOI: 10.1117/12.672850 (cit. on p. 51).

[12] Muzammil Iqbal, Martin A. Gleeson, Bradley Spaugh, *et al.* "Label-free biosensor arrays based on silicon ring resonators and high-speed optical scanning instrumentation". *IEEE Journal of Selected Topics in Quantum Electronics* **16**.3 (2010), pp. 654–661 (cit. on p. 52).

[13] Lukas Chrostowski, Samantha Grist, Jonas Flueckiger, *et al.* "Silicon photonic resonator sensors and devices". *Proceedings of SPIE Volume 8236; Laser Resonators, Microresonators, and Beam Control XIV* (Jan. 2012) (cit. on p. 52).

[14] Xu Wang, Samantha Grist, Jonas Flueckiger, Nicolas A. F. Jaeger, and Lukas Chrostowski. "Silicon photonic slot waveguide Bragg gratings and resonators". *Optics Express* **21** (2013), pp. 19 029–19 039 (cit. on p. 52).

[15] Xu Wang. "Silicon photonic waveguide Bragg gratings". PhD thesis. University of British Columbia, 2013 (cit. on p. 54).

[16] *Effective Mode Area – FDTD Solutions Knowledge Base.* [Accessed 2014/04/14]. URL: http://docs.lumerical.com/en/fdtd/user_guide_effective_mode_area.html (cit. on p. 61).

[17] A. Densmore, D. X. Xu, P. Waldron, *et al.* "A silicon-on-insulator photonic wire based evanescent field sensor". *IEEE Photonics Technology Letters* **18**.23 (2006), pp. 2520–2522 (cit. on p. 61).

[18] D. X. Xu, A. Delge, J. H. Schmid, *et al.* "Selecting the polarization in silicon photonic wire components". *Proceedings of SPIE.* Vol. 8266 (2012), 82660G (cit. on p. 61).

[19] N. Rouger, L. Chrostowski, and R. Vafaei. "Temperature effects on silicon-on-insulator (SOI) racetrack resonators: a coupled analytic and 2-D finite difference approach". *Journal of Lightwave Technology* **28**.9 (2010), pp. 1380–1391. DOI: 10.1109/JLT.2010.2041528 (cit. on p. 68).

[20] K. P. Yap, J. Lapointe, B. Lamontagne, *et al.* "SOI waveguide fabrication process development using star coupler scattering loss measurements". *Proceedings Device and Process Technologies for Microelectronics, MEMS, Photonics, and Nanotechnology IV, SPIE* (2008), p. 680014 (cit. on p. 69).

[21] Dietrich Marcuse. *Theory of Dielectric Optical Waveguides.* Elsevier, 1974 (cit. on p. 69).

[22] F. P. Payne and J. P. R. Lacey. "A theoretical analysis of scattering loss from planar optical waveguides". *Optical and Quantum Electronics* **26**.10 (1994), pp. 977–986 (cit. on p. 69).

[23] Christopher G. Poulton, Christian Koos, Masafumi Fujii, *et al.* "Radiation modes and roughness loss in high index-contrast waveguides". *IEEE Journal of Selected Topics in Quantum Electronics* **12**.6 (2006), pp. 1306–1321 (cit. on p. 69).

[24] Frdric Grillot, Laurent Vivien, Suzanne Laval, and Eric Cassan. "Propagation loss in single-mode ultrasmall square silicon-on-insulator optical waveguides". *Journal of Lightwave Technology* **24**.2 (2006), p. 891 (cit. on p. 69).

[25] Tom Baehr-Jones, Michael Hochberg, and Axel Scherer. "Photodetection in silicon beyond the band edge with surface states". *Optics Express* **16**.3 (2008), pp. 1659–1668 (cit. on p. 69).

[26] Jason J. Ackert, Abdullah S. Karar, John C. Cartledge, Paul E. Jessop, and Andrew P. Knights. "Monolithic silicon waveguide photodiode utilizing surface-state absorption and operating at 10 Gb/s". *Optics Express* **22**.9 (2014), pp. 10710–10715 (cit. on p. 69).

[27] A. D. Simard, N. Ayotte, Y. Painchaud, S. Bedard, and S. LaRochelle. "Impact of sidewall roughness on integrated Bragg gratings". *Journal of Lightwave Technology* **29**.24 (2011), pp. 3693–3704 (cit. on p. 69).

[28] Po Dong, Wei Qian, Shirong Liao, *et al.* "Low loss shallow-ridge silicon waveguides". *Optics Express* **18**.14 (2010), pp. 14 474–14 479 (cit. on p. 69).

[29] Yurii Vlasov and Sharee McNab. "Losses in single-mode silicon-on-insulator strip waveguides and bends". *Optics Express* **12**.8 (2004), pp. 1622–1631. DOI: 10.1364/OPEX.12.001622 (cit. on p. 70).

[30] Guoliang Li, Jin Yao, Hiren Thacker, *et al.* "Ultralow-loss, high-density SOI optical waveguide routing for macrochip interconnects". *Optics Express* **20**.11 (May 2012), pp. 12 035–12 039. DOI: 10.1364/OE.20.012035 (cit. on p. 70).

[31] Wim Bogaerts, Pieter Dumon, *et al.* "Compact wavelength-selective functions in silicon-on-insulator photonic wires". *IEEE Journal of Selected Topics in Quantum Electronics* **12**.6 (2006) (cit. on p. 70).

[32] Tom Baehr-Jones, Ran Ding, Ali Ayazi, *et al.* "A 25 Gb/s silicon photonics platform". *arXiv:1203.0767v1* (2012) (cit. on pp. 71, 72).

[33] *Using Mode Expansion Monitors – FDTD Solutions Knowledge Base.* [Accessed 2014/04/14]. URL: http://docs.lumerical.com/en/fdtd/user_guide_using_mode_expansion_monitors.html (cit. on p. 73).

[34] R. J. Bojko, J. Li, L. He, *et al.* "Electron beam lithography writing strategies for low loss, high confinement silicon optical waveguides". *Journal of Vacuum Science & Technology B: Microelectronics and Nanometer Structures* **29**.6 (2011), 06F309–06F309 (cit. on p. 73).

[35] *Bent Waveguide Calculation – MODE Solutions Knowledge Base.* [Accessed 2014/04/14]. URL: http://docs.lumerical.com/en/mode/usr_waveguide_bend.html (cit. on p. 75).

[36] Amnon Yariv and Pochi Yeh. *Photonics: Optical Electronics in Modern Communications (The Oxford Series in Electrical and Computer Engineering).* Oxford University Press, Inc., 2006 (cit. on pp. 77, 78).

4 Fundamental building blocks

This chapter describes fundamental components for silicon photonic integrated circuits. Specifically, this chapter describes methods of splitting and combining light (directional coupler and Y-branch), the Mach–Zehnder interferometer, the ring resonator, and Bragg gratings.

4.1 Directional couplers

The directional coupler is the most common method of splitting and combining light in photonic systems, especially those in optical fibres. The directional coupler consists of two parallel waveguides, where the coupling coefficient is controlled by both the length of the coupler and the spacing between the two waveguides. In silicon photonics, directional couplers can be implemented using any type of waveguide, including rib and strip waveguides. This chapter focuses on rib waveguides with quasi-TE polarization since these are subsequently used to build ring modulators (strip waveguide directional couplers are considered in Section 4.1.6). A diagram of a directional coupler is given in Figure 4.1a.

The behaviour of a directional coupler can be found using coupled mode theory [2,3]. The fraction of the power coupled from one waveguide to the other can be expressed as:

$$\kappa^2 = \frac{P_{\text{cross}}}{P_0} = \sin^2 (C \cdot L) , \tag{4.1}$$

where P_0 is the input optical power, P_{cross} is the power coupled across the directional coupler, L is the length of the coupler, and C is the coupling coefficient. The fraction of the power remaining in the original "through" waveguide, assuming a lossless coupler ($\kappa^2 + t^2 = 1$), is:

$$t^2 = \frac{P_{\text{through}}}{P_0} = \cos^2 (C \cdot L) . \tag{4.2}$$

To obtain the coupling coefficient, we use "supermode" analysis, based on a numerical calculation of the effective indices, n_1 and n_2, of the first two eigenmodes of the coupled waveguides. These two modes, shown in Figure 4.2, are known as the symmetric and antisymmetric supermodes.

The supermode approach, and what follows next, is often called the eigenmode expansion method (see Section 2.2.3 EME). This approach is more accurate (especially

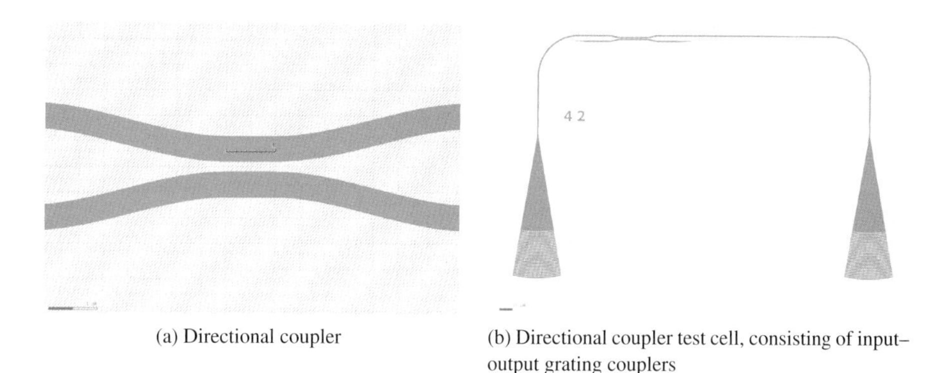

(a) Directional coupler

(b) Directional coupler test cell, consisting of input–output grating couplers

Figure 4.1 GDS layout of the directional coupler [1]; DC.gds.

for high-index contrast waveguides with strong coupling) than the traditional Coupled Mode Theory approach, where the coupling coefficient is found by perturbation methods (Section 2.2.3 CMT). From these two supermodes, the coupling coefficient is found:

$$C = \frac{\pi \Delta n}{\lambda}, \tag{4.3}$$

where Δn is the difference between the effective indices, $n_1 - n_2$.

The concept behind the coupler can be explained via the propagation of the two modes (Figure 4.2a and 4.2b) with different propagation constants:

$$\beta_1 = \frac{2\pi n_1}{\lambda} \tag{4.4a}$$

$$\beta_2 = \frac{2\pi n_2}{\lambda}. \tag{4.4b}$$

As the modes travel, the field intensity oscillates between the two waveguides, as shown in Figures 4.3 and 4.6. With the two modes in phase, the power is localized in the first waveguides. After a π phase shift difference between the modes, the power becomes localized in the second waveguide. This occurs after a distance called the cross-over length, L_x, and is found by:

$$\beta_1 L_x - \beta_2 L_x = \pi$$

$$L_x \left[\frac{2\pi n_1}{\lambda} - \frac{2\pi n_2}{\lambda} \right] = \pi$$

$$L_x = \frac{\lambda}{2\Delta n}. \tag{4.5}$$

4.1.1 Waveguide mode solver approach

The numerical simulation of the directional coupler, using the eigenmode solver, is considered next. First, we draw the waveguide geometry, in Listing 4.5. Next, we calculate the mode profiles, in Listing 4.6, with the results shown in Figure 4.2.

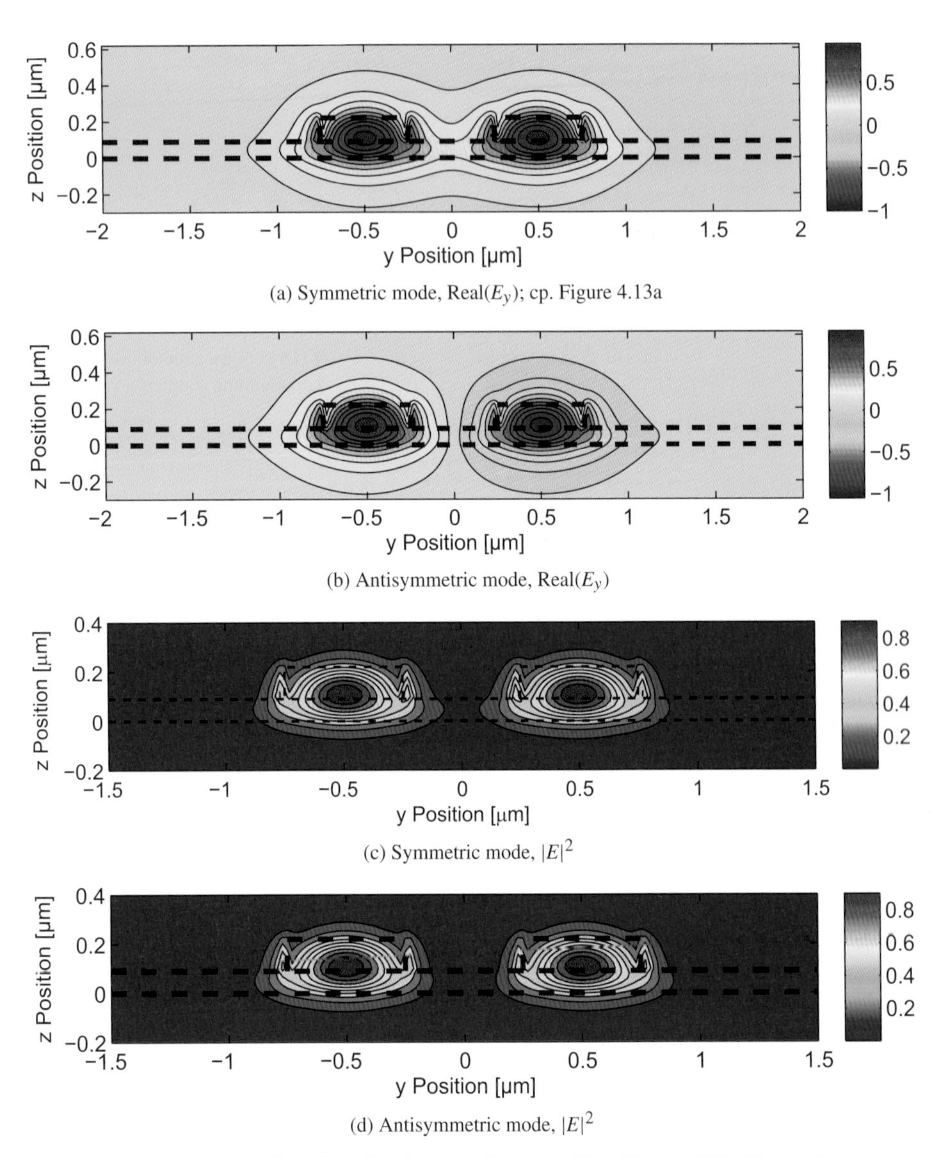

(a) Symmetric mode, Real(E_y); cp. Figure 4.13a

(b) Antisymmetric mode, Real(E_y)

(c) Symmetric mode, $|E|^2$

(d) Antisymmetric mode, $|E|^2$

Figure 4.2 Two fundamental modes of a directional coupler. Real(E_y) and E-field intensity. The asymmetric nature of the 2nd mode profile is evident in the zero-crossing in the centre of the gap. $\lambda = 1550$ nm, coupler gap $g = 200$ nm, 500 nm × 220 nm waveguides with a 90 nm slab.

Coupler-gap dependence

Next, we solve for the coupling coefficient versus the gap, in Listing 4.7. The cross-over length, L_x, is plotted in Figure 4.4. It is found that the coupling coefficient has a dependence on the distance between the waveguides, g, that follows an exponential behaviour:

$$C = B \cdot e^{-A \cdot g}, \tag{4.6}$$

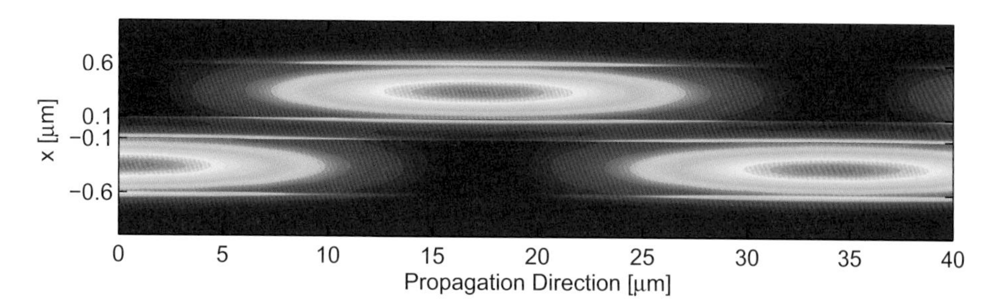

Figure 4.3 Field propagation along a directional coupler. Modes are calculated using the waveguide mode solver and the modes are propagated using the eigenmode expansion method.

where A and B are dependent on the geometry of the coupler, optical wavelength, etc.

We can curve-fit the cross-over length data from Figure 4.4, similar to Reference [4], with the result for this case (500 × 220 nm waveguide with 90 nm slab) being:

$$L_x = 10^{(0.0026084*g[nm]+0.657094)}[\mu\text{m}]. \tag{4.7}$$

For example, the coupler with gap $g = 200$ nm has a cross-over length of $L_x = 15.1$ μm.

Using the fit in Equation (4.7), we can calculate the field coupling coefficient, κ, using Equation (4.1), for any directional coupler gap or length:

$$\kappa = \left[\frac{P_{\text{Coupled}}}{P_0}\right]^{1/2} = \left|\sin\left(\frac{\pi\,\Delta n}{\lambda}\cdot L\right)\right| = \left|\sin\left(\frac{\pi}{2}\cdot\frac{z}{L_x}\right)\right|. \tag{4.8}$$

As an example, the field coupling, κ, for a 5 μm and 15 μm long coupler is plotted versus gap in Figure 4.5. The data points are taken from the cross-over length values calculated from the mode solver, whereas the solid lines are taken from Equation (4.7). The fit function is most accurate for larger gaps. For small gaps, the coupling coefficient slightly deviates from the exponential form in Equation (4.7).

In Figure 4.5a, notice that the 15 μm long coupler has an almost zero coupling for a 100 nm gap. This is because the cross-over length is 8 μm, hence the light goes through (almost) a complete cycle where (almost) all the power is returned to the original input waveguide.

Coupler-length dependence

The coupler-length dependence can be found using Equation (4.8), and is plotted in Figure 4.6. To design a coupler with a target coupling, e.g. 10%, these equations are used to identify a set of device parameters (e.g. $g = 200$ nm, $L = 3$ μm).

Wavelength dependence

Next, we solve for the wavelength dependence, in Listing 4.8. The results are plotted in Figure 4.7. Figure 4.7a shows the effective indices of the two modes, while Figure 4.7b

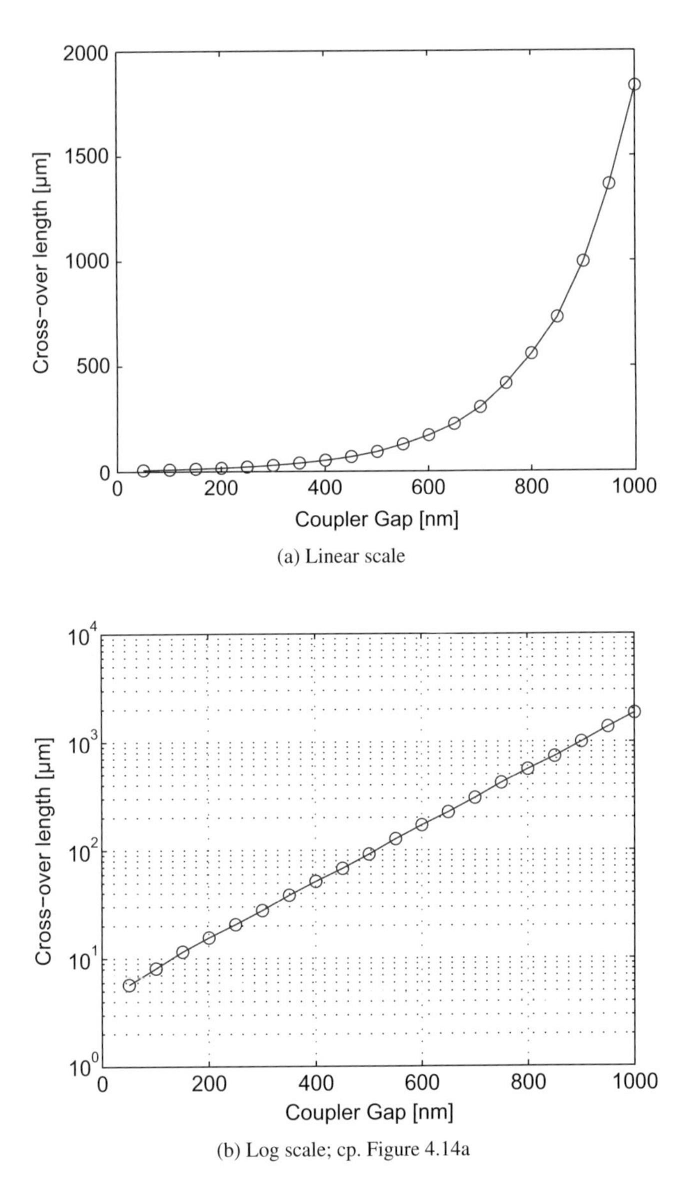

(a) Linear scale

(b) Log scale; cp. Figure 4.14a

Figure 4.4 Cross-over length, L_x, versus gap, λ = 1550 nm, 500 nm × 220 nm waveguides with a 90 nm slab, calculated using the waveguide mode solver with a 20 nm mesh size.

shows the cross-over length of the coupler, calculated using Equation (4.5). See also Figure 4.8.

4.1.2 Phase

Section 4.1.1 considered only the magnitude of the coupling coefficients, t and κ. In this section, we determine the phase relationship. The discussion begins with the field profiles of the symmetric (Figure 4.2a) and asymmetric (Figure 4.2b) modes of the

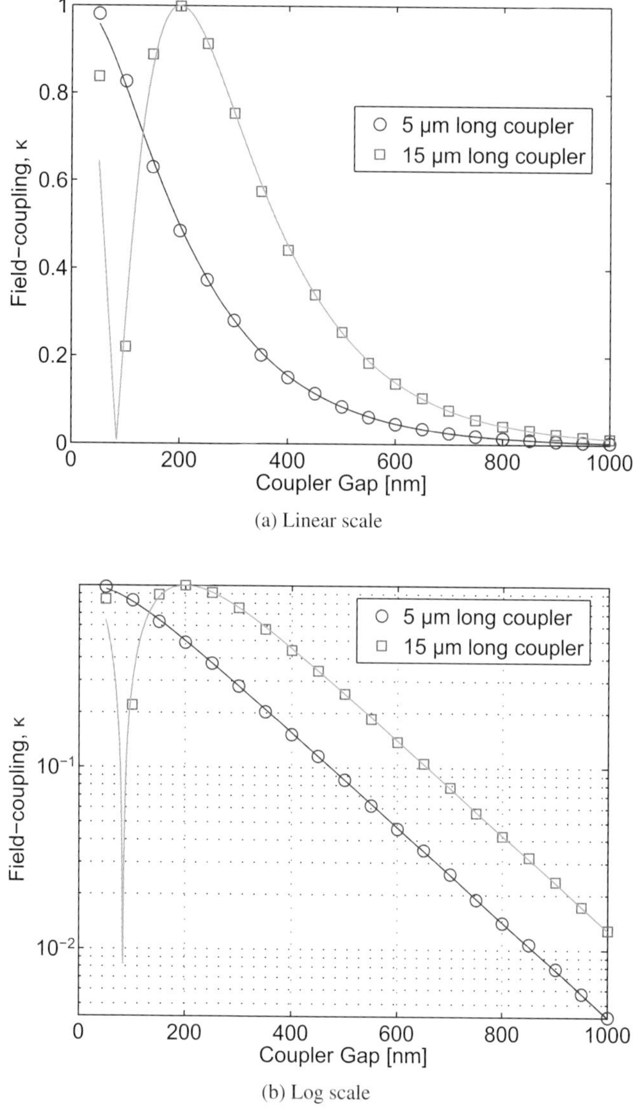

(a) Linear scale

(b) Log scale

Figure 4.5 κ versus gap, $\lambda = 1550$ nm, coupler length $L = 5\,\mu$m and $L = 15\,\mu$m, 500 nm × 220 nm waveguides with a 90 nm slab, calculated using the waveguide mode solver with a 20 nm mesh size.

directional coupler. These are illustrated as "Mode 1" and "Mode 2" in Figure 4.9, with the two waveguides labelled as "wg A" and "wg B". Based on the eigenmodes, the light present in each waveguide is found by the vector summation:

$$E_{\text{wg A}} = \frac{1}{\sqrt{2}}\left(e^{i\beta_1 L} + e^{i\beta_2 L}\right) \tag{4.9a}$$

$$E_{\text{wg B}} = \frac{1}{\sqrt{2}}\left(e^{i\beta_1 L} + e^{i\beta_2 L - i\pi}\right). \tag{4.9b}$$

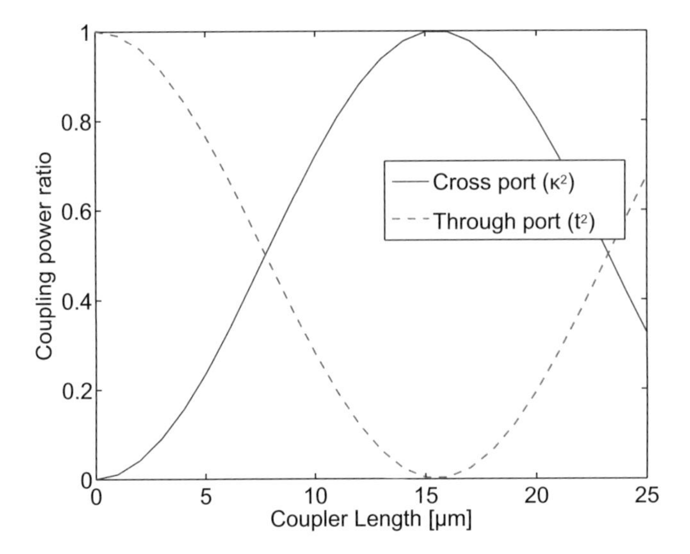

Figure 4.6 Plot of κ versus length of the ideal coupler (only parallel waveguides, no bends), λ = 1550 nm, coupler gap g = 200 nm, 500 nm × 220 nm waveguides with a 90 nm slab, calculated using the waveguide mode solver with a 20 nm mesh size.

The π phase shift is introduced in "wg B" since the asymmetric "Mode 2" has a negative field component in "wg B". This is illustrated by the vectors in Figure 4.9.

The phase of the light within each waveguide, and the phase difference, can be found:

$$\angle E_{\text{wg A}} = \frac{\beta_1 + \beta_2}{2} L \tag{4.10a}$$

$$\angle E_{\text{wg B}} = \frac{\beta_1 + \beta_2}{2} L - \frac{\pi}{2} \tag{4.10b}$$

$$\angle E_{\text{wg B}} - \angle E_{\text{wg A}} = -\frac{\pi}{2}. \tag{4.10c}$$

The 90° difference between t ("wg A") and κ ("wg B") is also illustrated in Figure 4.9.

Finally, the phase can be included in the coupling coefficients, t and κ, found in Equations (4.1) and (4.2):

$$t = |t|^2 \cdot e^{\frac{\beta_1+\beta_2}{2} L} \tag{4.11a}$$

$$\kappa = |\kappa|^2 \cdot e^{i\frac{\beta_1+\beta_2}{2} L - i\frac{\pi}{2}}. \tag{4.11b}$$

In photonic circuit modelling, the phase of the directional coupler can be included within the compact model for the directional coupler. Alternatively, as for the ring resonator model in Section 4.4.1, it can be considered separately assuming the average propagation constant of the coupler is the same as single waveguide. In this case, the coefficients become:

$$t = |t|^2 \tag{4.12a}$$

$$\kappa = |\kappa|^2 \cdot e^{-i\frac{\pi}{2}} = -i|\kappa|^2. \tag{4.12b}$$

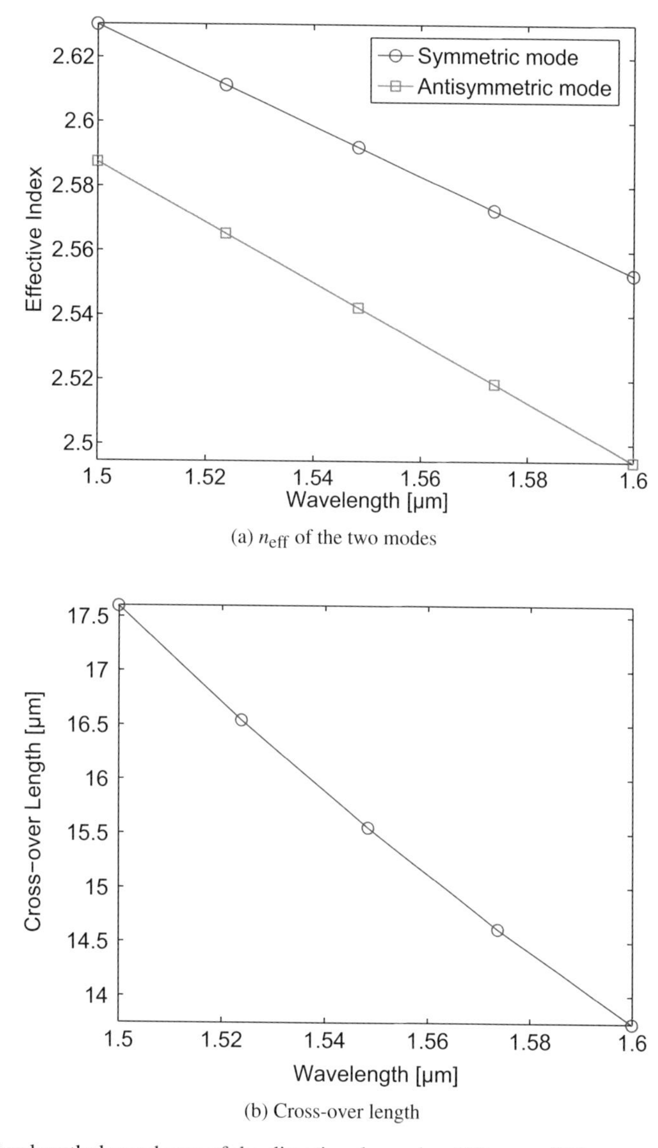

(a) n_{eff} of the two modes

(b) Cross-over length

Figure 4.7 Wavelength dependence of the directional coupler, 500 nm × 220 nm waveguides with a 90 nm slab, 200 nm gap, using the waveguide mode solver with a 20 nm mesh size.

4.1.3 Experimental data

To experimentally measure the coupling coefficient, cross-over length, and the contribution of the bends to the coupling, one approach is to fabricate numerous directional couplers with variation in the coupler length. The directional coupler in Figure 4.1a was used as the basic cell, in which the length of the coupler (in the figure, it is 1 μm long) is varied from 0 to 25 μm in steps of 0.5 μm. The optical power in the through and cross ports are measured. The measurements are conducted using an automated probe station (see Section 12.2) on 100 devices on two separate die on a wafer, with each

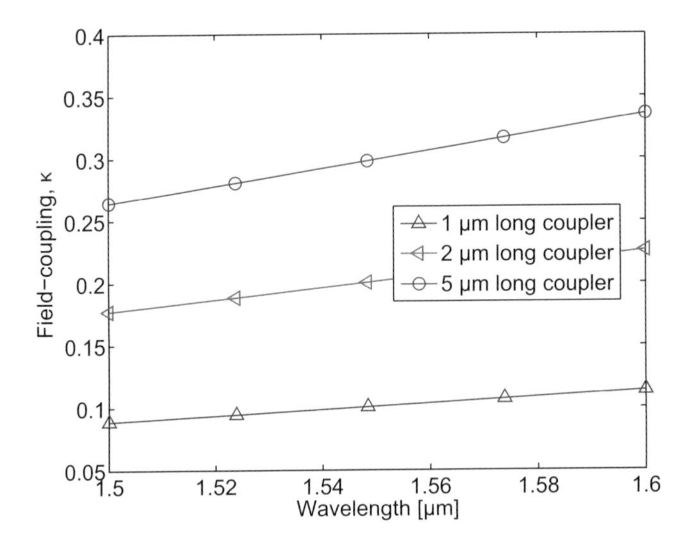

Figure 4.8 Plot of κ versus wavelength, $g = 200$ nm, for several coupler lengths, calculated using the waveguide mode solver.

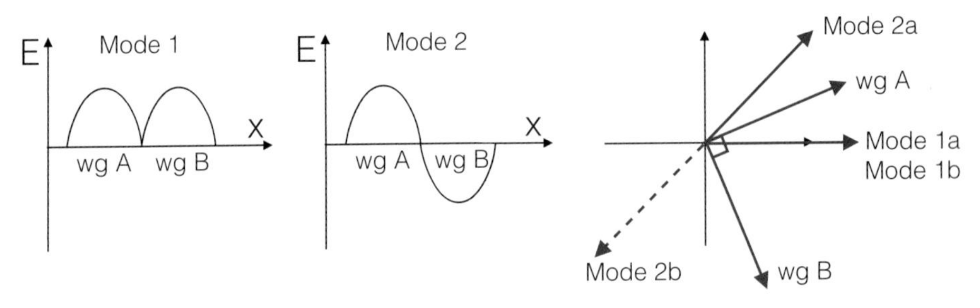

Figure 4.9 The two eigenmodes of the directional coupler, "Mode 1" and "Mode 2." Phase relationship between the light in the two waveguides in the directional coupler: (1) in terms of the components of the two eigenmodes within each of the waveguides ("Mode 1a", "Mode 1b", "Mode 2a", "Mode 2b"), and (2) as the summation ("wg A" and "wg B") in the waveguides.

measurement conducted twice to verify repeatability. Optical spectra are measured over a wavelength range of 1500–1570 nm. The centre wavelength of the grating coupler response is adjusted to be 1550 nm, by adjusting the fibre angle. The optical power traces are integrated over the wavelength range, normalized to the maximum power, and the data are plotted as shown in Figure 4.10. The data for the two ports is subsequently fit to the following equations:

$$\kappa^2 = \sin\left(\frac{\pi}{2L_x} \cdot [L + z_{bend}]\right)^2 \tag{4.13a}$$

$$t^2 = \cos\left(\frac{\pi}{2L_x} \cdot [L + z_{bend}]\right)^2. \tag{4.13b}$$

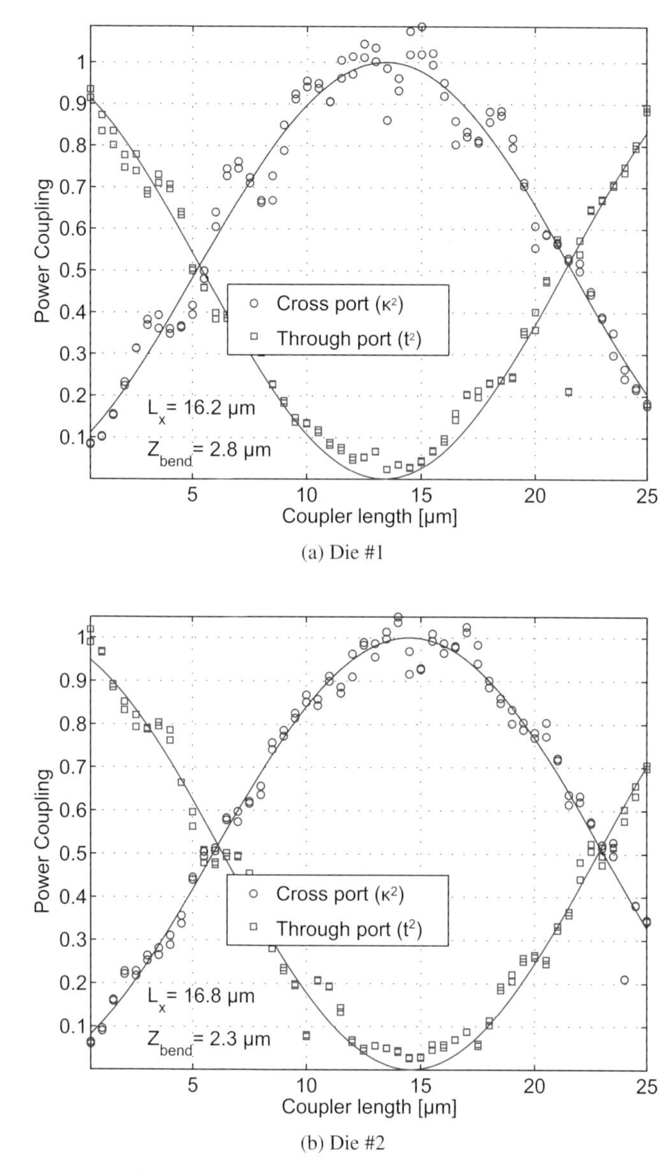

Figure 4.10 Experimental directional coupler performance, with curve fitting using Equation (4.4a,b); devices fabricated by OpSIS-IME [1].

This allows us to extract both the cross-over length, L_x, and the contribution of the bends to the coupling, via the parameter z_{bend}. Here, z_{bend} represents the effective extra coupler distance introduced by the coupling from the non-parallel waveguides (bends). We find that the cross-over length, L_x, is in the range of 16.2 μm to 16.8 μm, while z_{bend} is in the range of 2.3 μm to 2.8 μm. The cross-over length found from the mode calculations was 15.5 μm in Figure 4.7b, comparable to the experimental data.

4.1.4 FDTD modelling

The model in Section 4.1.1 only considers the two-dimensional cross-section of the coupler. Three-dimension simulations, on the other hand, are necessary to accurately model directional couplers, to correctly account for:

- optical losses due to mode mismatches or mode conversion loss [5],
- contribution of the bends to the coupling.

Prior research has identified approximations for the coupling coefficient taking into account the bend region, e.g. Eq. 3 in Ref. [5]. However, such approximations are not accurate for the high index contrast silicon photonic waveguides considered here. Thus, we perform 3D FDTD simulations on the same structure as used in the experiments, Figure 4.1a. The script in Listing 4.9 imports the mask layout file (GDS format) and sets the material properties and silicon thickness, adds the wafer (oxide cladding, substrate), and adds an FDTD simulation region. We also add a mesh override region in the space between the two waveguides in the directional coupler, in order to improve the simulation accuracy.

We set up the simulation by adding a mode source in the top-left waveguide, and two power monitors in each of the output waveguides. We record the power transmitted through the waveguides only at one wavelength (1550 nm in this case). The code is provided in Listing 4.10.

Next, we parameterize the directional coupler by splitting the design into two parts: the left side, and the right side. We keep the design centred at $x = 0$. We insert straight sections of waveguides in this middle region, and move all other parts of the simulation. The code is provided in Listing 4.10.

The results of the FDTD simulations are shown in Figure 4.11. Included is a comparison of the experimental data with the 3D FDTD simulation.

Using Equation (4.13) and the method described, we find that the simulated cross-over length is $L_x = 15.5\,\mu m$, while $z_{bend} = 2.8\,\mu m$. The results show that the experimental cross-over length is slightly longer than that simulated by both FDTD and eigenmode approach. The discrepancy is attributed to the uncertainty in fabrication: silicon thickness, waveguide width, and sidewall angle (see Section 11.1). The impacts of the bends are in good agreement with the experiment (within the large experimental uncertainty).

FDTD versus mode solver

In this section, we compare the results of 3D FDTD simulations of the directional coupler, including the effects of bends, versus the results based on mode calculations. We know that the contribution of the bends to the coupling, via the parameter z_{bend} is 2.3–2.8 μm (experimental, Figure 4.10), or 2.8 μm (3D-FDTD, Figure 4.11). As shown in Figure 4.12, the 5 μm long coupler mode simulations do not match with the 3D FDTD results, owing to the bend region effect. The effect of the bend can be nicely incorporated into the parameterized model (Equations (4.7) and (4.13)) by simply increasing the length of the coupler by 2.7 μm, as done in Figure 4.12. For this case, the results

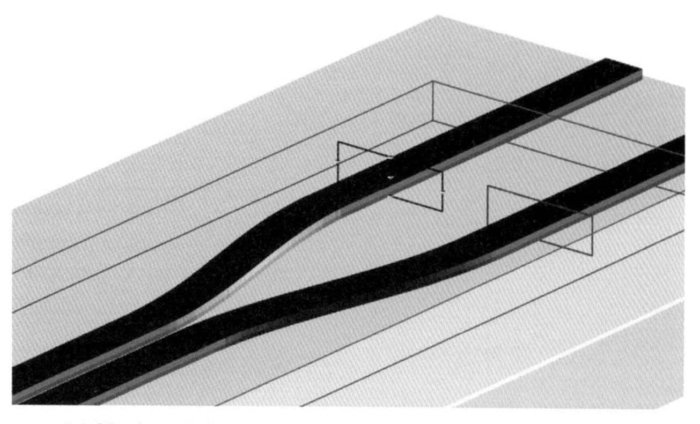

(a) 3D view of the coupler geometry, for 3D-FDTD simulations

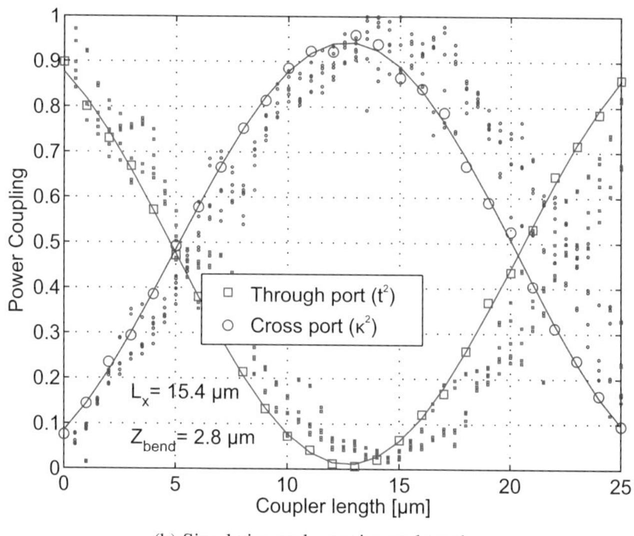

(b) Simulation and experimental results

Figure 4.11 3D FDTD simulation of directional couplers, shown in solid lines with simulation data points. Simulation parameters: mesh accuracy = 3; simulation boundary 300 nm above/below waveguides; 3 PML and 3 metal boundaries. Experimental data from three die, as per Figure 4.10, included as small data points.

of the two simulations are in agreement with each other in a range of 50 nm to 400 nm. Beyond 400 nm, the 3D FDTD data appears to saturate, most likely due to optical mode-mismatch losses from the directional coupler leading to an insertion loss. Finally, these results are in agreement with the experimental results.

4.1.5 Sensitivity to fabrication

The fabrication uncertainty, specifically lithography exposure, etch, and silicon thickness variations, led to variations in the waveguide width, coupler gaps, and thicknesses. Typically when the waveguide width shrinks, the gap increases, and vice versa. One method of analyzing the sensitivity to fabrication error is as follows.

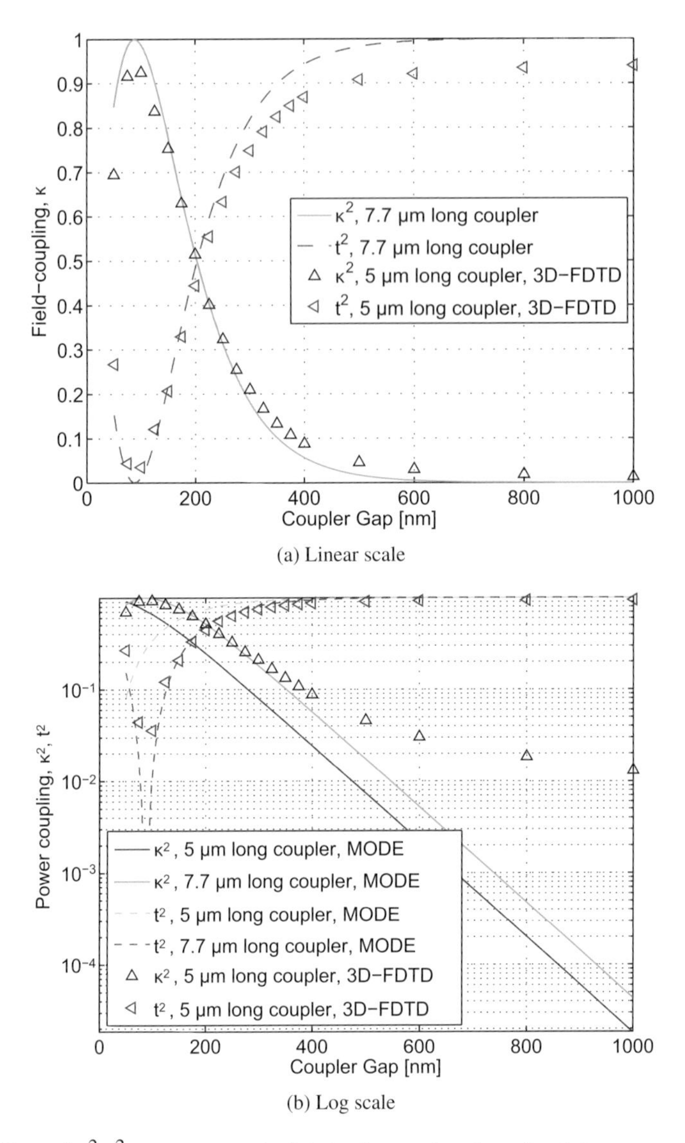

(a) Linear scale

(b) Log scale

Figure 4.12 Plots of κ^2, t^2 versus gap, calculated using mode solver simulations parameterized in Equations (4.7) and (4.13) for the ideal parallel waveguide coupler, compared with the coupler in Figure 4.1a (including bends) simulated using 3D FDTD (mesh accuracy 2; higher meshes do not significantly change results).

(1) Identify a family of designs that achieve a particular coupling, e.g. $t = 0.95$. This requires choices in the waveguide geometry, coupler length, and coupler gap.

(2) Simulate each design for variation in lithography, e.g. a variation in waveguide width of $\delta w = \pm 30$ nm. Keep in mind that the change in the waveguides' width leads to the same but opposite change in the coupler gap.

(3) Plot the results versus δw. The slope of these graphs can be interpreted as the sensitivity to fabrication.

One of the observations from such an analysis is that directional couplers are very sensitive to fabrication variations. Thus, the coupling coefficients vary from wafer to wafer.

For more discussion on fabrication errors and non-uniformity, see Section 11.1.

4.1.6 Strip waveguide directional couplers

The same analysis as presented in this section can be repeated for strip waveguides. Here, a 500×220 nm waveguide directional coupler is analyzed using the waveguide mode solution method. Results are shown in Figures 4.13–4.16. The main conclusion

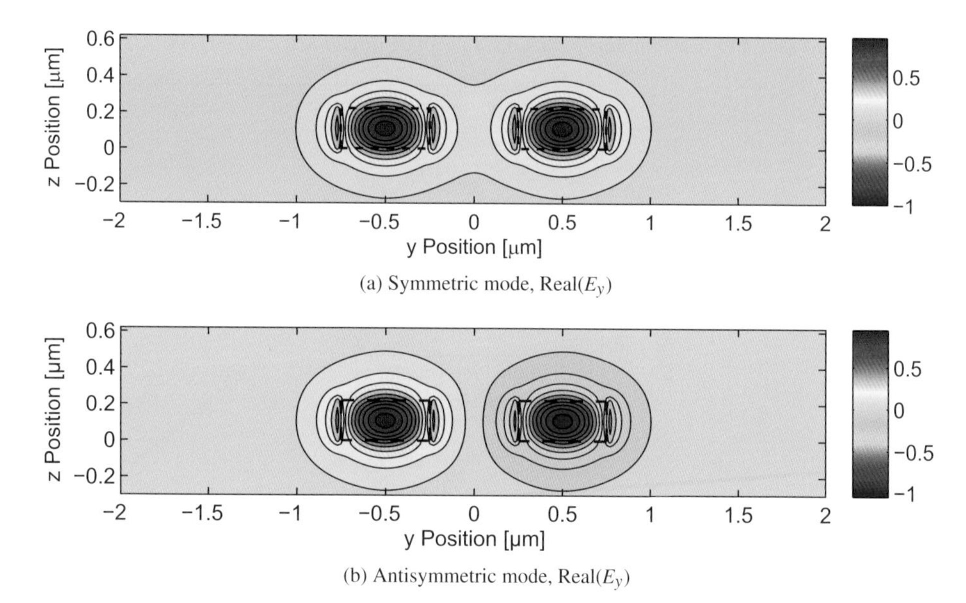

(a) Symmetric mode, Real(E_y)

(b) Antisymmetric mode, Real(E_y)

Figure 4.13 Two fundamental modes of a strip waveguide directional coupler, with waveguides being 500×220 nm; cf. Figure 4.2a.

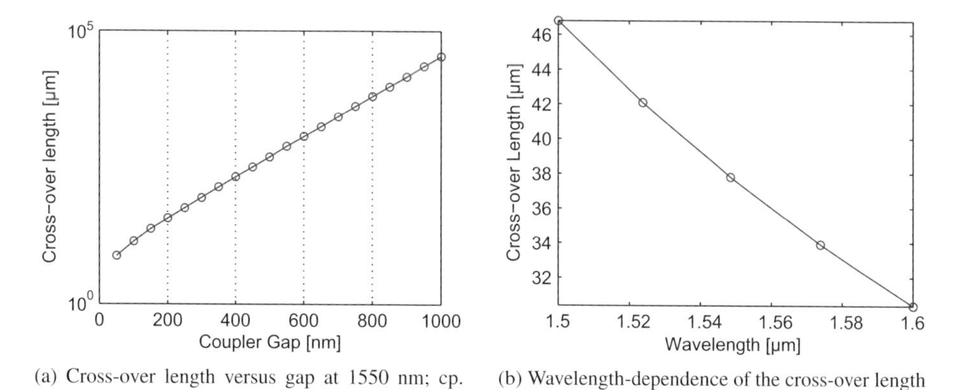

(a) Cross-over length versus gap at 1550 nm; cp. Figure 4.4b.

(b) Wavelength-dependence of the cross-over length for $g = 200$ nm; cf. Figure 4.7b

Figure 4.14 Calculations for a strip waveguide directional coupler, 500×220 nm.

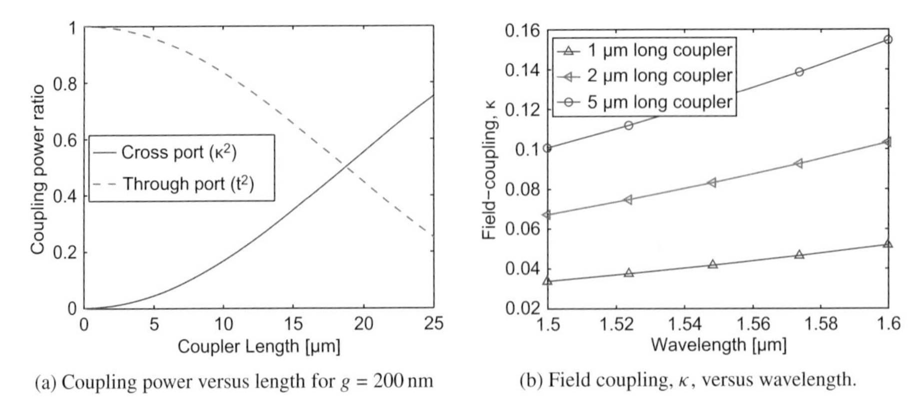

(a) Coupling power versus length for $g = 200\,\text{nm}$

(b) Field coupling, κ, versus wavelength.

Figure 4.15 Calculations for a strip waveguide directional coupler, 500 × 220 nm.

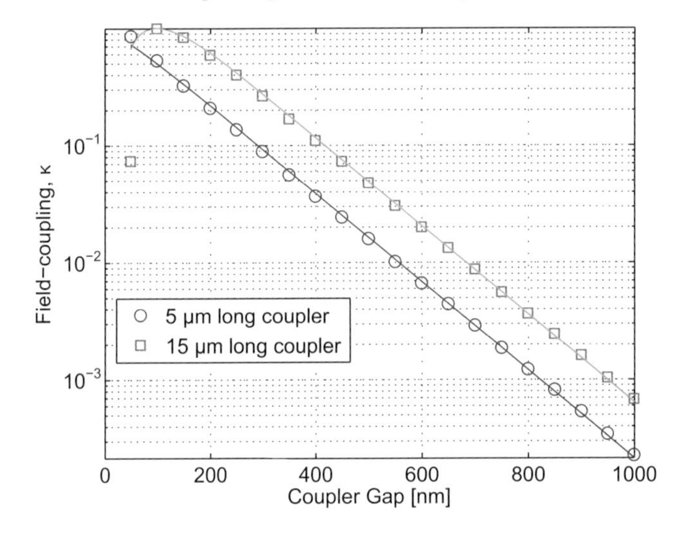

Figure 4.16 Calculations for the field coupling, κ, versus gap, for a strip waveguide directional coupler, 500 × 220 nm.

is that strip waveguides have weaker coupling as compared to rib waveguides. This is visually evident in the field profiles; see Figure 4.13. For the same parameters, the cross-over length and the coupling coefficient are about 2× larger in rib waveguides versus the strip waveguides considered.

Similar to Equation (4.7), we can curve-fit the cross-over length data from Figure 4.14a, with the result for this case being:

$$L_x = 10^{(0.0037645 * g[nm] + 0.799434)} [\mu\text{m}]. \tag{4.14}$$

For example, the coupler with gap $g = 200$ nm has a cross-over length of $L_x = 37.5\,\mu\text{m}$.

4.1.7 Parasitic coupling

For routing waveguides across a chip, it is necessary to determine how much optical cross-talk will exist between parallel waveguides. For example, setting a constraint

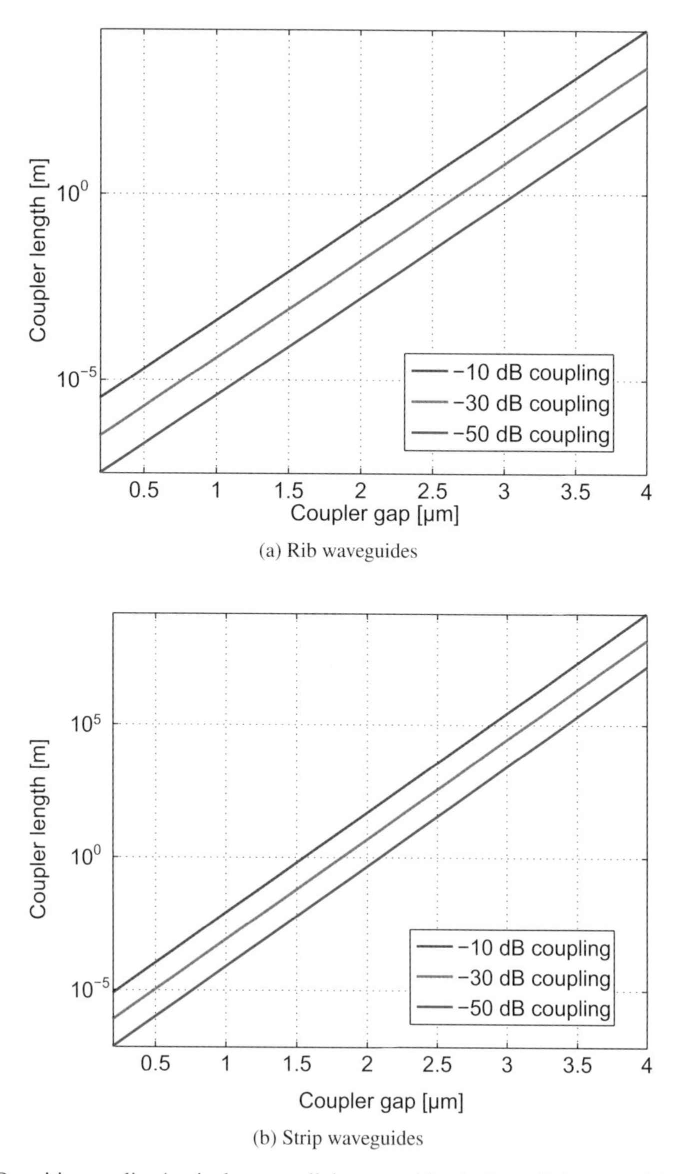

Figure 4.17 Parasitic coupling/optical cross-talk between identical parallel waveguides, calculated using the mode solver. Waveguide width is 500 nm.

for a maximum cross-talk of −50 dB over 1 cm leads to a minimum rib waveguide spacing of 2.3 µm. These results, shown in Figure 4.17a, are obtained using Equations (4.7) and (4.13) and numerically solving for the coupler length. The −10 dB line at 200 nm shows a value of 3 µm, consistent with Figure 4.6. In practice, there will be additional coupling between waveguides due to scattering from the waveguide roughness.

Given that the coupling is weaker between strip waveguides, optical waveguide routing is typically done using strip waveguides since they can be spaced closer together.

Only 1.6 μm waveguide spacing is required to get less than −50 dB cross-talk over 1 cm, as shown in Figure 4.17b.

Delta beta coupling

When routing large arrays of waveguides, and routing density is desired, the parasitic coupling can be reduced by making the waveguides with different propagation constants, thereby reducing the coupling coefficients [6]. This can be done by alternating between waveguides of two widths, e.g. 400 and 500 nm [6]. More elaborate arrangements can be considered by adding to the rotation more than two waveguide geometries, or by alternating the polarization between the waveguides (in circuits where TE and TM polarizations are used).

For the case of coupling between two waveguides, with a propagation constant difference, $\Delta\beta$, the coupling coefficient in Equation (4.1) can be approximated by using coupled mode theory (CMT) to be [2, 7, 8]:

$$\kappa^2 = \frac{\sin^2\left(CL\sqrt{1 + (\Delta\beta/2C)^2}\right)}{1 + (\Delta\beta/2C)^2}. \tag{4.15}$$

Results of this equation are illustrated in Figure 4.18a (CMT). Two strip waveguides, 500 nm and 400 nm in width, separated by a 200 nm gap, are considered. The coupling coefficient, C, is found using the supermodes and Equation (4.3) (as opposed to the perturbative method in CMT). It is seen that maximum coupling is reduced by approximately a factor of two.

However, Equation (4.15) assumes that the two modes are symmetric and asymmetric, which is only the case for very small propagation constant differences and/or waveguides that are very close together. For the high index contrast waveguides considered, the results are not accurate, and a fully vectorial approach needs to be taken

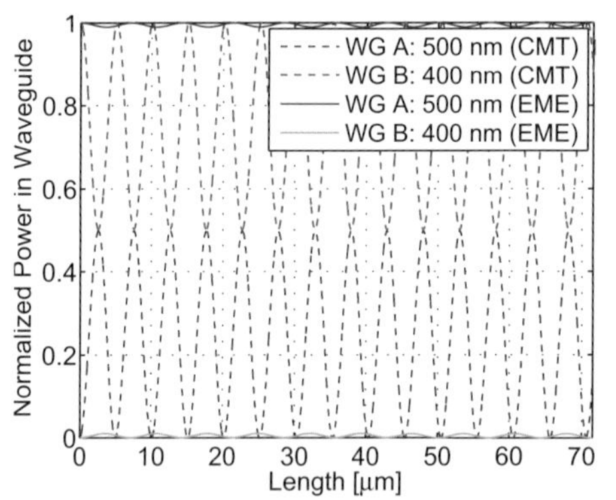

(a) Comparison of coupled mode theory, Equation (4.15), and the eigenmode expansion method.

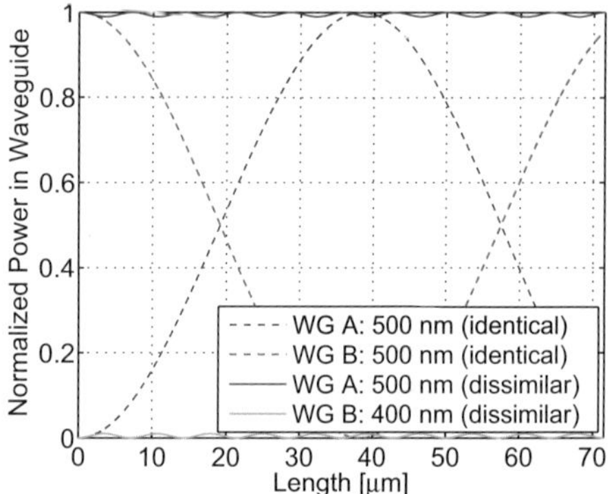

(b) Comparison of coupling between dissimilar (500 nm and 400 nm wide) and identical (500 nm wide) strip waveguides, using eigenmode expansion.

Figure 4.18 Parasitic coupling/optical cross-talk between dissimilar (500 nm and 400 nm wide) strip waveguides, with a 200 nm gap.

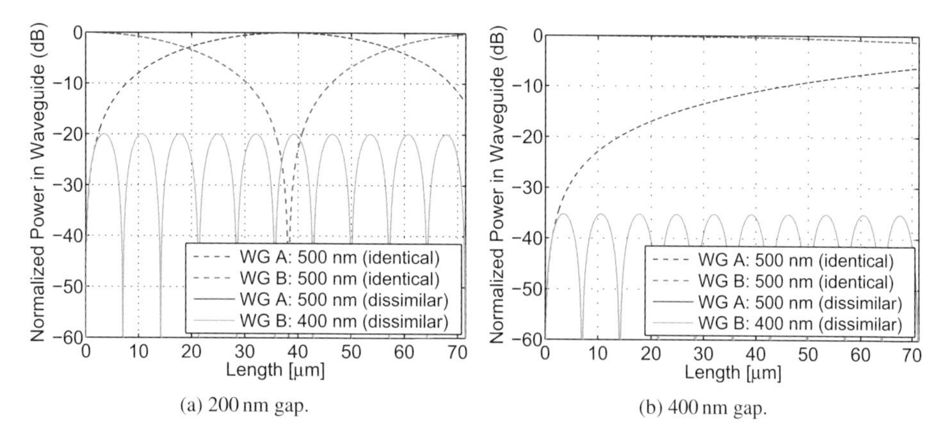

(a) 200 nm gap.

(b) 400 nm gap.

Figure 4.19 Parasitic coupling/optical cross-talk between dissimilar (500 nm and 400 nm wide) and identical (500 nm wide) strip waveguides, calculated using the mode solver and the eigenmode expansion method in Listing 4.11. Results plotted in log scale.

to solve the problem. This is implemented in Listing 4.11. The approach is based on the Eigenmode Expansion Method (EME), described in Section 2.2.3. We calculate the mode profile for the launch waveguide (waveguide A), calculate the two supermodes of the two-waveguide system as in Section 4.1.1, perform overlap integral calculations between the launch waveguide and the two supermodes, propagate the two supermodes, and finally determine how much power is present in each waveguide versus the position. It should be pointed out that this method is the same as in Section 4.1.1, except that overlap integrals are introduced to determine the proportion of light travelling in the two dissimilar waveguides; in identical waveguides, the power is split evenly between the symmetric and asymmetric waveguides, hence overlap integrals were not necessary.

Figure 4.18a compares the results from Coupled Mode Theory versus the Eigenmode Expansion Method. CMT greatly over-estimates the amount of coupling between the dissimilar waveguides, and a fully vectorial technique is necessary to obtain the correct results.

Next, we compare the results to the case of identical waveguides solved using the approach in Section 4.1.1. Shown in Figure 4.18b, the dissimilar waveguides have a much shorter cross-over length since the supermodes have much larger differences; hence, the coupling coefficient, C, is much larger. This is as expected since this approach is based on making waveguides with different propagation constants, hence a more rapid "beating" between the two modes as they propagate down the waveguides. Most importantly, the coupling between the waveguides reaches a maximum that is much less than unity in the case of identical waveguides – in this case, it reaches a maximum coupling of 1%.

Figure 4.19a shows the results plotted on a log scale to better observe the weak coupling found in dissimilar waveguides with a gap of 200 nm. When the gap is increased to 400 nm, the maximum coupling decreases to 0.03% (−35 dB), as shown in Figure 4.19b. For a 600 nm gap, less than −50 dB of cross-talk suppression is predicted

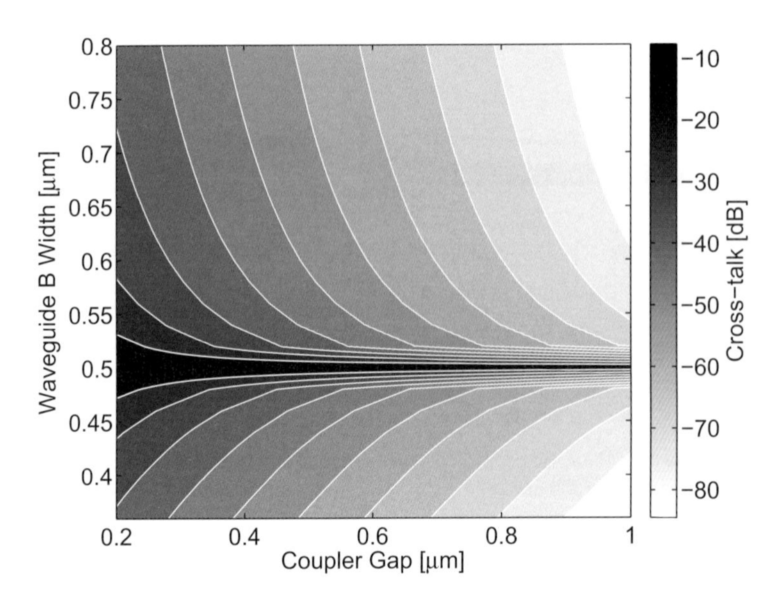

Figure 4.20 Worst-case parasitic coupling between parallel waveguides. Waveguide A is 500 nm and the y-axis is the width of Waveguide B. Results are calculated using the mode solver and the eigenmode expansion method in Listing 4.11.

for all distances; this should be compared to the 1.6 μm spacing required for a length of 1 cm in identical waveguides. Thus, to obtain a similar level of cross-talk with identical waveguides, a spacing three times greater is required; hence dissimilar waveguides can lead to increased space efficiency.

These simulations were repeated by varying the width of the second waveguide (Waveguide B) and the gap between the waveguides. The results are plotted in Figure 4.20. Waveguide A is 500 nm and the y-axis is the width of Waveguide B; the plot corresponds to a waveguide width difference, $B-A$, ranging between -140 nm and 300 nm. The x-axis corresponds to the gap between the waveguides ranging from 200 nm to 1000 nm. The parasitic coupling values are for the worst-case condition, i.e. that which occurs at the cross-over length. As shown in the figure, there is strong coupling between the waveguides when they are the same size. As expected, the strongest coupling occurs when the waveguides are similar in size and close together; conversely, the weakest coupling occurs when the waveguides are mismatched in size and further apart. This plot can serve as a useful guideline for designing waveguide spacing for on-chip routing.

4.2 Y-branch

The role of the Y-branch is to split the light from one waveguide equally into two waveguides (splitter) or to combine light from two waveguides into one (combiner). An example device is shown in Figure 4.21. It is straightforward to understand the splitting functionality – for an input intensity, I_i, the light is split equally, thus each output has

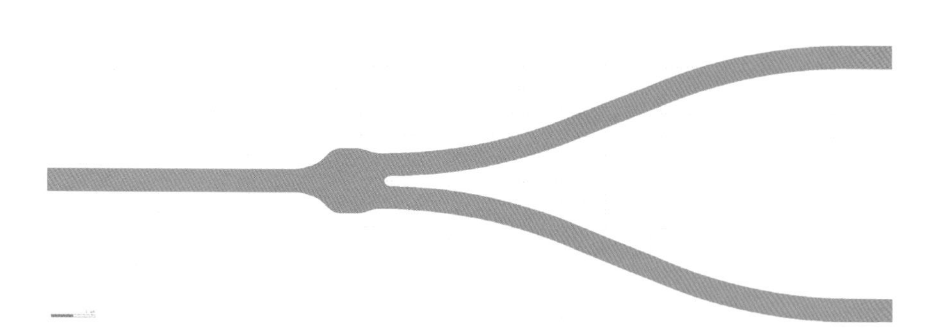

Figure 4.21 Y-branch splitter/combiner, GDS layout file YBranch_Compact.gds; similar to Reference [9].

intensity $I_1 = I_2 = I_i/2$. However, the combiner is less intuitive and will be explained in this section.

We describe a simple model for the Y-branch splitter/combiner, based on the formalism for a 50/50% beam-splitter. The fundamental point is that the Y-branch cannot be thought of as a three-port device. From an eigenmode perspective, it is actually a system with multiple input modes, and multiple output modes (e.g. two and two). In the case of an input waveguide, there are (at least) two modes that must be considered: the fundamental mode of the waveguide, and either the second-order mode or radiation modes.

For the splitter, we begin with an input intensity, I_i, with electric field, E_i. The light is split equally into the two branches. Thus each output has intensity $I_1 = I_2 = I_i/2$, and electric field, $E_1 = E_2 = E_i/\sqrt{2}$ (since $I \propto |E|^2$).

For the combiner, the same equations apply, namely that light input in one waveguide, I_1, is split equally between the fundamental mode of the waveguide and the second-order mode (or radiation mode). Thus the light at the combined port will be $I_i = I_1/2$, and the field will be $E_i = E_1/\sqrt{2}$.

Thus, the Y-branch functions as a 50/50% beam-splitter in both directions. An important point is that it is not possible to combine two incoherent beams of light using a Y-branch in order to increase the power. Also it is important to note that when using a combiner, if light is only present in one port, the output will be reduced to one half of the input.

In reality, Y-branches are not perfect 50/50% splitters and have excess loss. The geometry of the Y-branch needs to be optimized using FDTD. Using a genetic algorithm, and varying the widths of the Y-branch in several locations along the device, the insertion loss was optimized to be less than 0.3 dB [9]. The resulting geometry is shown in Figure 4.21.

Simulations for the Y-branch can be performed with either 2.5D or 3D FDTD. Listing 4.12 provides an example for such simulations, which can be executed in either solver type. The script loads a GDS layout of a Y-branch, and performs several simulations. Movies are also created, which are useful to visualize the results and to understand the loss mechanisms. First, light is launched at the input of the splitter, with results shown in Figure 4.22. It is seen that the light is split evenly between the two ports,

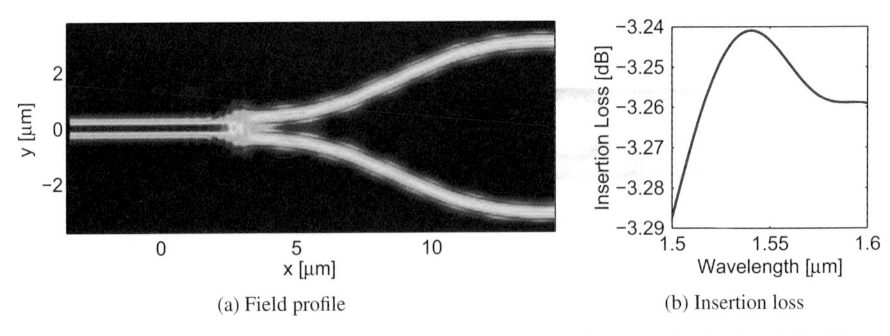

(a) Field profile (b) Insertion loss

Figure 4.22 Simulations of the Y-branch operating as a splitter, simulated by Listing 4.12. The device layout is in Figure 4.21, similar to Reference [9].

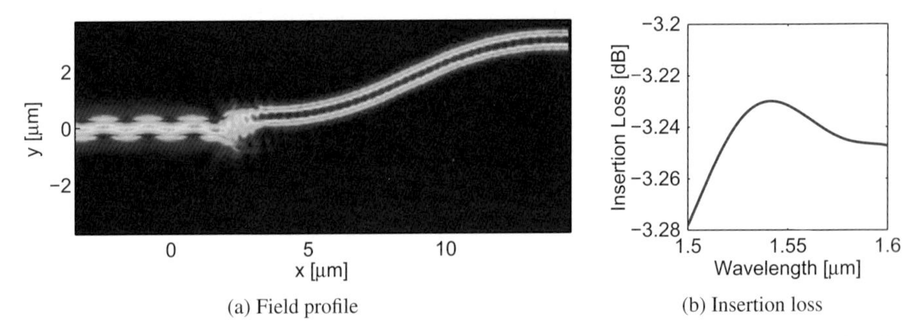

(a) Field profile (b) Insertion loss

Figure 4.23 Simulations of the Y-branch operating as a combiner with a single input, simulated by Listing 4.12.

with an insertion loss of slightly over 3 dB; the insertion loss is as expected, 50%, with excess loss due to the non-ideal device performance. Next, three simulations are performed with the device operating as a combiner. The first is with light on a single port of the combiner, with results shown in Figure 4.23. In this case, two modes are excited at the output, namely the fundamental mode, and the second-order TE mode. The insertion loss is calculated by performing mode overlap integrals (using the mode-expansion monitor) to determine the amount of power in the fundamental mode. Again, the insertion loss is slightly over 3 dB. The second simulation considers two in-phase inputs in the combiner; in this situation, the two sources are coherent hence constructive interference is observed in the fundamental mode of the output port, as shown in Figure 4.24. In this case, the power in the waveguide should approach 100%. The deviation from this is known as the "excess loss" of the device, and plotted in Figure 4.23b as a function of wavelength. Finally, the last simulation considers two out-of-phase inputs in the combiner; in this situation, destructive interference is observed in the fundamental mode of the output port, as shown in Figure 4.25. In this case, the power in the fundamental mode of the waveguide should approach 0%, with all the power going into the second-order TE mode or into radiation modes.

With this understanding of how Y-branches operate, we will treat the interference of coherent beams in Mach–Zehnder interferometers in Section 4.3.

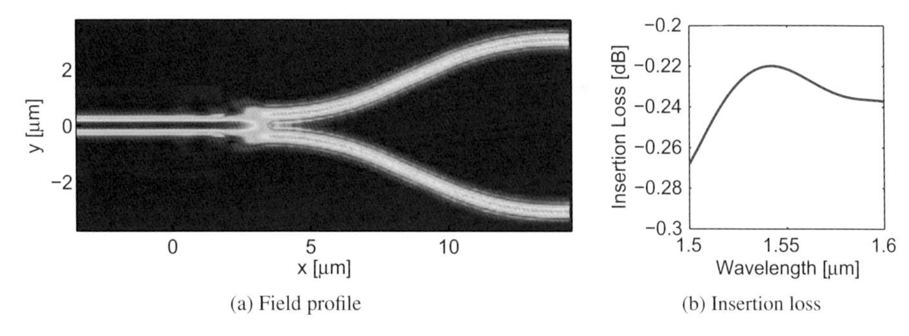

(a) Field profile (b) Insertion loss

Figure 4.24 Simulations of the Y-branch operating as a combiner with two in-phase inputs, simulated by Listing 4.12.

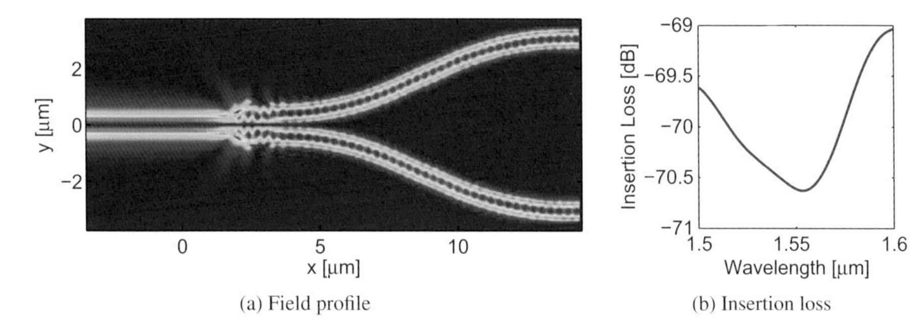

(a) Field profile (b) Insertion loss

Figure 4.25 Simulations of the Y-branch operating as a combiner with two out-of-phase inputs, simulated by Listing 4.12.

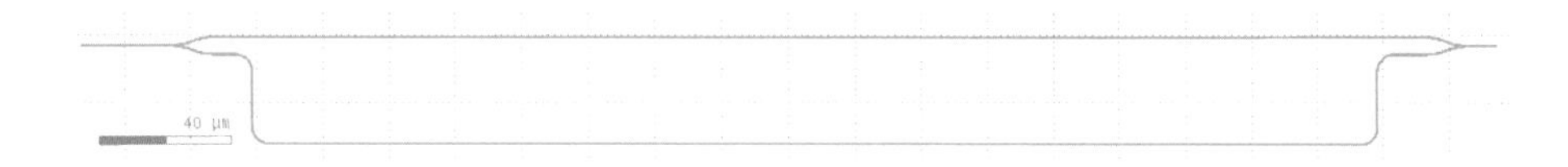

Figure 4.26 Mach–Zehnder interferometer, layout example.

4.3 Mach–Zehnder interferometer

The Mach–Zehnder interferometer is shown in Figure 4.26. It consists of an input split into two branches (here, the upper and lower waveguides), then recombined. The splitting/combining can be achieved using any splitter/combiner, including Y-branches and directional couplers.

In this section, we describe a simple model for the interferometer based on the plane-wave free-space beam-splitter. This model applies to single-mode waveguides, where we consider the total field intensity within each waveguide, and disregard the field distribution inside the waveguides.

Let the input intensity be I_i, with electric field E_i. At the output of the first Y-branch (the splitter), the top branch has field $E_1 = E_i/\sqrt{2}$, and the bottom branch has $E_2 = E_i/\sqrt{2}$. The propagation of light in the waveguides is described by $\beta_1 = \frac{2\pi n_1}{\lambda}$ and $\beta_2 = \frac{2\pi n_2}{\lambda}$, and the waveguides have lengths L_1 and $L_2 = L_1 + \Delta L$, and propagation loss α_1 and α_2 (for intensity, $\alpha/2$ for electric field, see Equation (3.9)), for the top and bottom waveguides respectively. At the end of the two waveguides (at the input to the combiner Y-branch), the fields are:

$$E_{o1} = E_1 e^{-i\beta_1 L_1 - \frac{\alpha_1}{2} L_1} = \frac{E_i}{\sqrt{2}} e^{-i\beta_1 L_1 - \frac{\alpha_1}{2} L_1} \tag{4.16a}$$

$$E_{o2} = E_2 e^{-i\beta_2 L_2 - \frac{\alpha_2}{2} L_2} = \frac{E_i}{\sqrt{2}} e^{-i\beta_2 L_2 - \frac{\alpha_2}{2} L_2}. \tag{4.16b}$$

The output of the second Y-branch (the combiner) is:

$$E_o = \frac{1}{\sqrt{2}} (E_{o1} + E_{o2}) = \frac{E_i}{2} \left(e^{-i\beta_1 L_1 - \frac{\alpha_1}{2} L_1} + e^{-i\beta_2 L_2 - \frac{\alpha_2}{2} L_2} \right). \tag{4.17}$$

The intensity at the output is thus:

$$I_o = \frac{I_i}{4} \left| e^{-i\beta_1 L_1 - \frac{\alpha_1}{2} L_1} + e^{-i\beta_2 L_2 - \frac{\alpha_2}{2} L_2} \right|^2. \tag{4.18}$$

For simplicity, we assume that the total losses are negligible (if losses are to be accounted for, Equation (4.18) can be evaluated numerically). In this case, after some trigonometry, Equation (4.18) simplifies to:

$$I_o = \frac{I_i}{4} \left[2\cos\left(\frac{\beta_1 L_1 - \beta_2 L_2}{2} \right) \right]^2 \tag{4.19a}$$

$$= I_i \cos^2 \left[\frac{\beta_1 L_1 - \beta_2 L_2}{2} \right] \tag{4.19b}$$

$$= \frac{I_i}{2} \left[1 + \cos(\beta_1 L_1 - \beta_2 L_2) \right]. \tag{4.19c}$$

The output of the interferometer is thus a sinusoidally varying function of wavelength (via β_1 and β_2) for an imbalanced interferometer ($L_1 \neq L_2$). The period is termed the free spectral range, and for identical waveguides, is determined by:

$$FSR\,[\text{Hz}] = \frac{c}{n_g \Delta L} \tag{4.20a}$$

$$FSR\,[\text{m}] = \frac{\lambda^2}{n_g \Delta L}, \tag{4.20b}$$

where c is the speed of light in vacuum, and n_g is the waveguide group index (see Equation (3.5)). This oscillation is a convenient method of determining Mach–Zehnder interferometer and modulator parameters such as the group index and the tunability (pm/V).

The intensity in Equation (4.19) also varies sinusoidally with the waveguide effective index (n_1 and n_2), which can be changed by the thermo-optic effect (Section 3.1.1) to implement a thermo-optic switch (Section 6.6), or the plasma dispersion effect (Section 6.1.1) to implement a high-speed Mach–Zehnder modulator, etc.

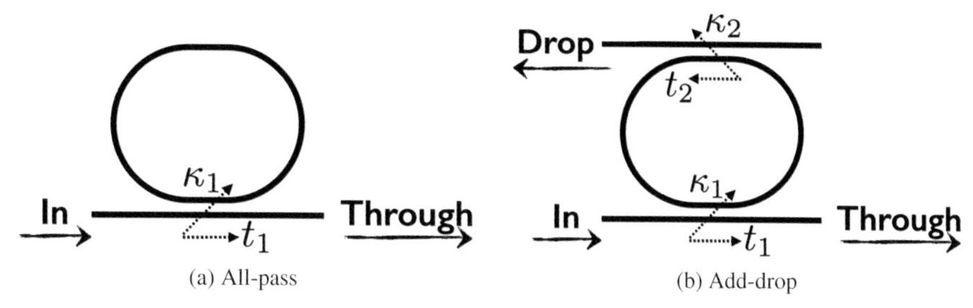

Figure 4.27 Diagram of ring/racetrack resonators.

4.4 Ring resonators

A ring resonator, also known as micro-ring resonator, or racetrack resonator, consists of a loop of optical waveguide, with some coupling to the outside world. The loop is usually a ring (circle), or a racetrack shape. The coupling typically consists of one or two directional couplers. Thus, there are two ring resonator configurations, i.e. all-pass and add-drop, as shown in Figure 4.27. The ring resonator cavity can have a racetrack shape, consisting of two 180° circular waveguides and two straight waveguides (for directional couplers). For a detailed description of ring resonators, please see the review paper by W. Bogaerts *et al.* [10].

4.4.1 Optical transfer function

The roundtrip length is given by:

$$L_{rt} = 2\pi r + 2L_c, \tag{4.21}$$

where r and L_c are the bend radius and the coupler length, respectively. When $L_c = 0$, the micro-ring is point coupled and the cavity becomes a circle.

The through-port response of an all-pass micro-ring resonator, shown in Figure 4.27a, is given by

$$\frac{E_{thru}}{E_{in}} = \frac{-\sqrt{A} + te^{-i\phi_{rt}}}{-\sqrt{A}t^* + e^{-i\phi_{rt}}}, \tag{4.22}$$

where t is the straight-through coupling coefficient of the optical field; * is the complex conjugate; ϕ_{rt} and A are the round-trip optical phase and power attenuation, respectively, and are given by:

$$\phi_{rt} = \beta L_{rt} \tag{4.23a}$$

$$A = e^{-\alpha L_{rt}}. \tag{4.23b}$$

Note that in this model, the phase accumulated by the light propagating in the coupler is included in the total round-trip phase, ϕ_{rt}; t and κ are known as the point-coupling coefficients and there is no phase shift in the through-port coupling, t, as in Equation (4.11a).

The add-drop filter, shown in Figure 4.27b, has two outputs, i.e. the through-port signal, E_{thru}, and the drop-port signal, E_{drop}, which are given by:

$$\frac{E_{thru}}{E_{in}} = \frac{t_1 - t_2^* \sqrt{A} e^{i\phi_{rt}}}{1 - \sqrt{A} t_1^* t_2^* e^{i\phi_{rt}}} \tag{4.24}$$

$$\frac{E_{drop}}{E_{in}} = \frac{-\kappa_1^* \kappa_2 A^{\frac{1}{4}} e^{i\phi_{rt}/2}}{1 - \sqrt{A} t_1^* t_2^* e^{i\phi_{rt}}}, \tag{4.25}$$

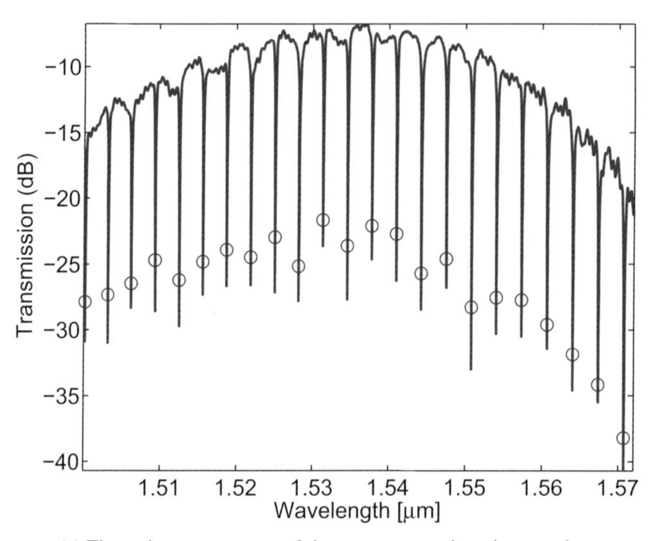

(a) Through-port spectrum of ring resonator and grating coupler

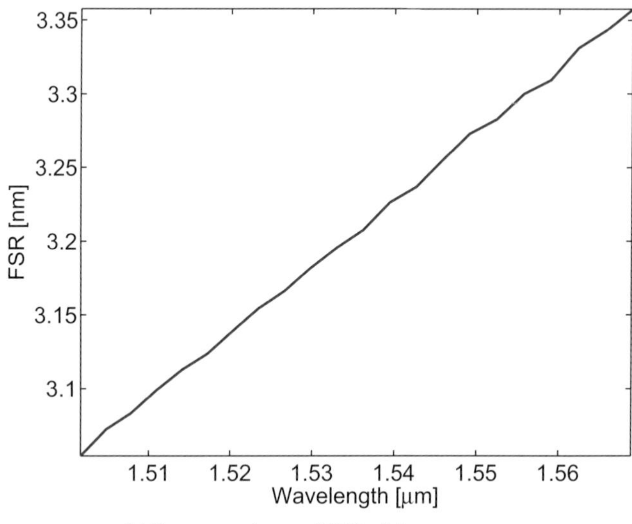

(b) Free spectral range (FSR) of ring resonator

Figure 4.28 Experimental spectra and FSR of a ring resonator modulator. Device fabricated via OpSIS-IME.

where t_1, κ_1, t_2, and κ_2 are the straight-through and cross-over coupling coefficients of the input and drop couplers, respectively, which are typically identical for a symmetric design, i.e. $t_1 = t_2 = t$ and $\kappa_1 = \kappa_2 = \kappa$. Assuming lossless couplers (i.e. the optical losses of the couplers are incorporated into the round trip loss of the entire optical cavity), t and κ have the relationship

$$|\kappa|^2 + |t|^2 = 1. \tag{4.26}$$

The transfer function of the micro-ring resonator is implemented using MATLAB code 6.4.

4.4.2 Ring resonator experimental results

In this section, passive optical characterization results for a fabricated ring modulator are presented. The ring modulator is shown in Figure 6.9. The ring has a radius of 15 μm, uses a rib waveguide with 500 nm width and 90 nm slab, is a double-bus design with straight bus waveguides (a single-bus device is shown in Figure 6.9), and has a small racetrack section (0.1 μm) for the directional couplers. The optical spectrum of the through port is plotted in Figure 4.28a, measured using a pair of fibre grating couplers [11]. The resonant wavelengths are identified using a peak-finding algorithm [12]. The free spectral range (FSR, $\Delta\lambda$), defined as the wavelength difference between resonances, is plotted in Figure 4.28b, and is in general wavelength dependent. The formalism is the same as for the Mach–Zehnder interferometer, see Equation (4.20). Using the FSR, and given the round-trip length of the resonator, L, the group index, n_g, of the waveguide can be found:

$$n_g = \frac{\lambda^2}{L\Delta\lambda}. \tag{4.27}$$

The results are plotted and compared with the waveguide model in Figure 3.21b.

For additional discussion on ring resonators, particularly in the context of modulators, please see Section 6.3.

4.5 Waveguide Bragg grating filters

The Bragg grating, a fundamental component for achieving wavelength-selective functions, has been used in numerous optical devices such as semiconductor lasers and fibres. Over the past few years, the integration of Bragg gratings in silicon waveguides has been attracting increasing research interest. In this section, we describe the theory of gratings, starting from the uniform Bragg grating. We present modelling and design methodologies. We also provide insight into some practical issues and challenges involved with the design and fabrication.

4.5.1 Theory

In the most simple configuration, a Bragg grating is a structure with a periodic modulation of the effective refractive index in the propagation direction of the optical mode, as

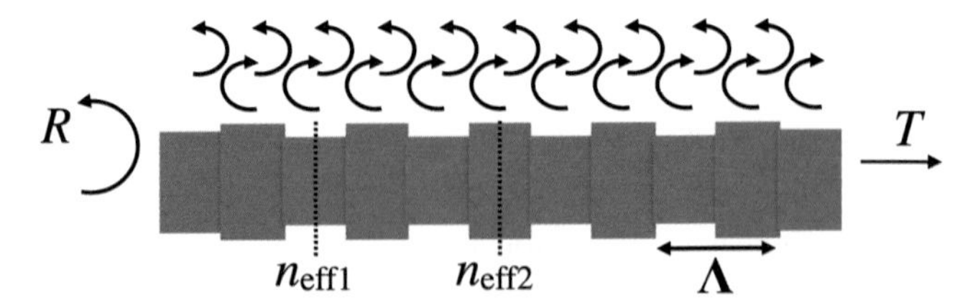

Figure 4.29 Illustration of a uniform Bragg grating. Here n_{eff1} and n_{eff2} are the effective indices of the low and high index sections, respectively, Λ is the grating period, and R and T are the grating's reflection and transmission. The 180° arrows indicate the numerous reflections throughout the grating.

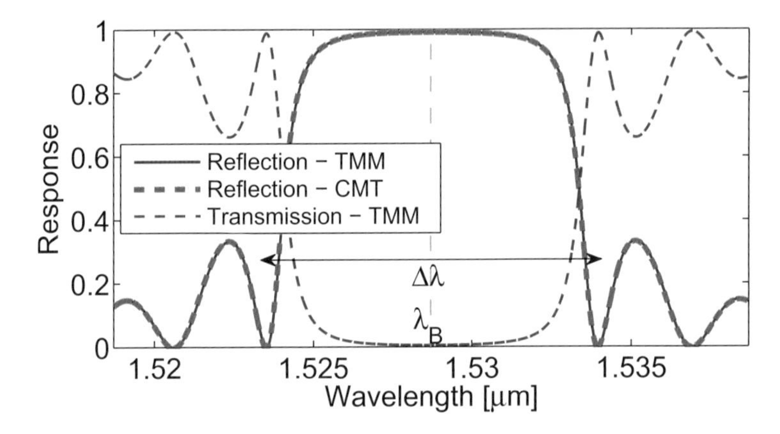

Figure 4.30 Typical spectral response of a uniform Bragg grating. Number of periods = 250; period = 310 nm; 500 × 220 nm strip waveguide geometry, with ΔW = 10 nm corrugations. Grating response calculated using the Transfer Matrix Method (TMM) with Δn = 0.0323, and Coupled Mode Theory (CMT) with a corresponding κ from Equation (4.36).

shown in Figure 4.29. This modulation is commonly achieved by varying the refractive index (e.g. alternating material) or the physical dimensions of the waveguide. At each boundary, a reflection of the travelling light occurs, and the relative phase of the reflected signal is determined by the grating period and the wavelength of the light. The repeated modulation of the effective index results in multiple and distributed reflections. The reflected signals only interfere constructively in a narrow band around one particular wavelength, namely the Bragg wavelength. Within this range, light is strongly reflected. At other wavelengths, the multiple reflections interfere destructively and cancel each other out, and as a result, light is transmitted through the grating.

Figure 4.30 shows the typical spectral transmission and reflection response of a uniform Bragg grating. The spectrum can be calculated using the analytic equations that follow, based on Coupled Mode Theory (CMT), or using numerical methods such as the Transfer Matrix Method (TMM), described in Section 4.5.2.

Comparing the two methods, CMT and TMM, the following observations are made. CMT assumes that the index of refraction perturbation is small. Hence for high index contrasts, TMM is more accurate. However, even for the case of a strip waveguide grating consisting of sections that are 400 nm and 600 nm wide, with a stop-band of approximately 80 nm, the error introduced by CMT is very small: the bandwidth is overestimated by only 0.3%. CMT is also much faster – the CMT code presented here executes nearly 200 faster than the TMM code. The main advantage of TMM, however, is that it allows for an arbitrary index profile, including chirped gratings, apodized gratings, phase-shifted cavities, etc. In both cases, however, the key is to correctly determine the grating coupling coefficient, as described in Section 4.5.1. For comparison purposes, the same coefficient was used in Figure 4.30.

The centre wavelength, which is 1529 nm in Figure 4.30, is known as the Bragg wavelength, and is given as:

$$\lambda_B = 2\Lambda n_{\text{eff}}, \tag{4.28}$$

where Λ is the grating period and n_{eff} is the average effective index. Based on Coupled Mode Theory [13], the reflection coefficient for a uniform grating with length L is described by:

$$r = \frac{-i\kappa \sinh(\gamma L)}{\gamma \cosh(\gamma L) + i\Delta\beta \sinh(\gamma L)} \tag{4.29}$$

with

$$\gamma^2 = \kappa^2 - \Delta\beta^2. \tag{4.30}$$

Here, $\Delta\beta$ is the propagation constant offset from the Bragg wavelength:

$$\Delta\beta = \beta - \beta_0 << \beta_0 \tag{4.31}$$

and κ is often defined as the coupling coefficient of the grating and can be interpreted as the amount of reflection per unit length.

For the case where $\Delta\beta = 0$, Equation (4.29) is written as $r = -i\tanh(\kappa L)$, therefore, the peak power reflectivity at the Bragg wavelength is:

$$R_{peak} = \tanh^2(\kappa L). \tag{4.32}$$

The bandwidth is also an important figure of merit for Bragg gratings. The bandwidth between the first nulls around the resonance can be determined by [13]:

$$\Delta\lambda = \frac{\lambda_B^2}{\pi n_g}\sqrt{\kappa^2 + (\pi/L)^2}, \tag{4.33}$$

where n_g is the group index. It should be noted that this is larger than the 3 dB bandwidth.

For long gratings (relative to the grating strength), i.e. $\kappa >> \pi/L$, the expression can be simplified to be:

$$\Delta\lambda = \frac{\lambda_B^2 \cdot \kappa}{\pi n_g}. \tag{4.34}$$

Please note that the example in Figure 4.30 is a relatively short grating, hence the bandwidth is larger than predicted by the approximation in Equation (4.34); instead Equation (4.33) must be used.

To take the optical propagation losses into account, the above equations are modified by replacing $\Delta\beta$ with $\Delta\beta - i\alpha/2$, where α is the intensity loss coefficient as defined in Equation (3.9).

Grating coupling coefficient

The key task in designing gratings is to determine the grating coupling coefficient parameter, which can be done by several methods: (1) perturbative methods (CMT), (2) based on the reflection coefficients, (3) based on reflection coefficients found using the Fresnel equations and the "plane-wave approximation", (4) based on 3D simulations of the grating, such as using FDTD or EME, (5) 3D simulations with incorporation of fabrication smoothing either via SEM images or by computational lithography, or (6) based on experimental data.

Here, we discuss the approach whereby reflection coefficients are found based on Fresnel equations. For a stepwise effective index variation as shown in Figure 4.29,

$$\Delta n = n_{\text{eff2}} - n_{\text{eff1}}, \tag{4.35}$$

the reflection at each interface can be written as $\Delta n/2n_{\text{eff}}$ according to the Fresnel equations. Each grating period contributes two reflections, therefore the coupling coefficient is:

$$\kappa = 2\frac{\Delta n}{2n_{\text{eff}}}\frac{1}{\Lambda} = \frac{2\Delta n}{\lambda_B}. \tag{4.36}$$

This coupling coefficient is for the case of a rectangular grating profile. Similar equations can be found, e.g. for sinusoidal effective index variations [13]. If it is known that a fabricated grating is sinusoidal,

$$\kappa_{\text{sinusoidal}} = \frac{\pi\,\Delta n}{2\lambda_B}. \tag{4.37}$$

The rest of this chapter will focus on the rectangular grating profile, with Equation (4.36).

4.5.2 Design

In this section, we describe several design methodologies in detail. First we describe the Transfer Matrix Method (TMM), a numerical technique that is very useful for simulating the response of an arbitrary grating profile. MATLAB code is provided. Next we make the connection between the physical structure (waveguide geometry) and the effective index; this allows us to either design a physical structure based on an index profile, or to simulate a physical structure, as shown in the design flow in Figure 4.33. Methods of dealing with the lithography smoothing in gratings is addressed, either via look-up tables derived from experimental results, in Section 4.5.4, or via lithography simulations [14], in Section 4.5.4. In the present section we numerically model the gratings using FDTD. Section 4.5.4 discusses additional fabrication considerations.

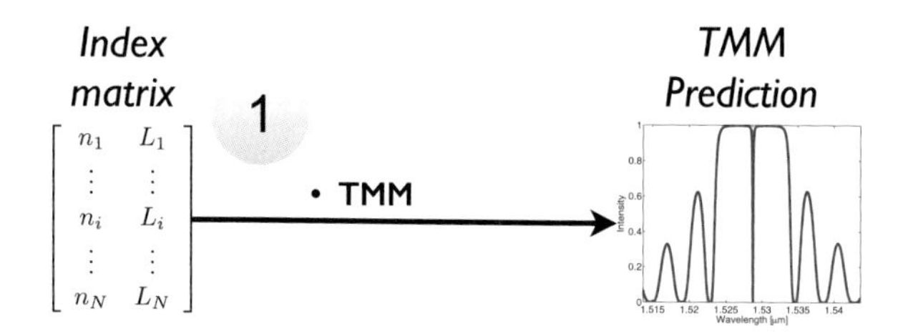

Figure 4.31 Modelling of waveguide gratings.

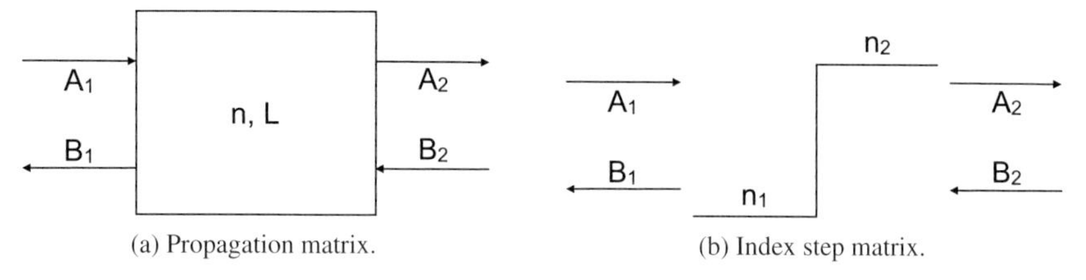

Figure 4.32 Illustration of the transfer matrices.

Transfer Matrix Method

The aim of this section is to calculate the spectrum of a grating defined by an arbitrary index profile, as shown in Figure 4.31. The Transfer Matrix Method, described in detail in Reference [15], is as follows.

(1) Calculates the reflection and transmission coefficients due to index discontinuities, and propagation coefficients for waveguide sections.
(2) Expresses these building blocks as matrices, with a size of 2×2 in the case of a single-mode system.
(3) Multiplies the matrices as a cascaded network, to represent the grating.
(4) Extracts the transmission and reflection values for the overall grating. The calculation is repeated for each wavelength.

The transfer matrix is defined as

$$\begin{bmatrix} A_1 \\ B_1 \end{bmatrix} = \begin{bmatrix} T_{11} & T_{12} \\ T_{21} & T_{22} \end{bmatrix} \begin{bmatrix} A_2 \\ B_2 \end{bmatrix}, \tag{4.38}$$

where the parameters are illustrated in Figure 4.32. The transfer matrix concept is similar to, and related to, the scattering parameter matrix, described in Section 9.3.2.

The transfer matrix for a homogeneous section of a waveguide, as shown in Figure 4.32a, is

$$T_{\text{hw}} = \begin{bmatrix} e^{j\beta L} & 0 \\ 0 & e^{-j\beta L} \end{bmatrix}, \tag{4.39}$$

where β is the complex propagation constant for the field, including the index of refraction and the propagation loss

$$\beta = \frac{2\pi n_{\text{eff}}}{\lambda} - i\frac{\alpha}{2},$$

where α is defined in Equation (3.9a). The MATLAB implementation for the waveguide section is given in Listing 4.1.

Listing 4.1 Transfer Matrix Method – calculate transfer matrix for a section of a homogeneous section of a waveguide.

```
function T_hw=TMM_HomoWG_Matrix(wavelength,l,neff,loss)
% Calculate the transfer matrix of a homogeneous waveguide.

%Complex propagation constant
beta=2*pi*neff./wavelength-1i*loss/2;

T_hw=zeros(2,2,length(neff));
T_hw(1,1,:)=exp(1i*beta*l);
T_hw(2,2,:)=exp(-1i*beta*l);
```

The transfer matrix for an index of refraction step, as defined in Figure 4.32b, is:

$$T_{\text{is-12}} = \begin{bmatrix} 1/t & r/t \\ r/t & 1/t \end{bmatrix} = \begin{bmatrix} \frac{n_1+n_2}{2\sqrt{(n_1 n_2)}} & \frac{n_1-n_2}{2\sqrt{(n_1 n_2)}} \\ \frac{n_1-n_2}{2\sqrt{(n_1 n_2)}} & \frac{n_1+n_2}{2\sqrt{(n_1 n_2)}} \end{bmatrix}, \tag{4.40}$$

where r and t are the reflection coefficients, and here are found based on the Fresnel coefficients. This is the key approximation of this technique, since the Fresnel coefficients are based on plane waves. This can be termed the "plane-wave approximation". More accurate techniques to find the reflection coefficients can be considered, e.g. 3D FDTD simulations. In the section, we choose to live with this approximation for now, and instead make adjustments to find the "effective" or "as-fabricated" Δn values, as described in Section 4.5.4.

The MATLAB implementation for an index step is given in Listing 4.2.

Listing 4.2 Transfer Matrix Method – calculate transfer matrix for a material interface.

```
function T_is=TMM_IndexStep_Matrix(n1,n2)
% Calculate the transfer matrix for a index step from n1 to n2.
T_is=zeros(2,2,length(n1));
a=(n1+n2)./(2*sqrt(n1.*n2));
b=(n1-n2)./(2*sqrt(n1.*n2));
%T_is=[a b; b a];
T_is(1,1,:)=a; T_is(1,2,:)=b;
T_is(2,1,:)=b; T_is(2,2,:)=a;
```

Next, we construct a cascaded network to represent a section of a Bragg grating, by multiplication:

$$T_{\text{p}} = T_{\text{hw-2}} T_{\text{is-21}} T_{\text{hw-1}} T_{\text{is-12}}, \tag{4.41}$$

where the additional subscripts, 1 and 2, represent the two effective index materials, namely regions of low (1) and high (2) index of refraction.

Then, we construct the uniform Bragg grating, with N periods:

$$T = \left(T_{\text{hw-2}} T_{\text{is-21}} T_{\text{hw-1}} T_{\text{hw-12}}\right)^N. \tag{4.42}$$

In the MATLAB example in Listing 4.3, we consider a phase-shifted uniform Bragg grating, i.e. a first-order Fabry–Perot cavity with two Bragg grating reflectors (see Section 4.5.6). The transfer matrix is as follows:

$$T = \left[(T_{\text{p}})^N\right] T_{\text{hw-2}} \left[(T_{\text{p}})^N\right] T_{\text{hw-2}}. \tag{4.43}$$

Listing 4.3 Transfer Matrix Method – calculate transfer matrix for a waveguide Bragg grating cavity consisting of waveguide sections and material interfaces, with a phase-shift in the centre.

```
function T=TMM_Grating_Matrix(wavelength, Period, NG, n1, n2, loss)
% Calculate the total transfer matrix of the gratings

l=Period/2;
T_hw1=TMM_HomoWG_Matrix(wavelength,l,n1,loss);
T_is12=TMM_IndexStep_Matrix(n1,n2);
T_hw2=TMM_HomoWG_Matrix(wavelength,l,n2,loss);
T_is21=TMM_IndexStep_Matrix(n2,n1);

q=length(wavelength);
Tp=zeros(2,2,q); T=Tp;
for i=1:length(wavelength)
  Tp(:,:,i)=T_hw2(:,:,i)*T_is21(:,:,i)*T_hw1(:,:,i)*T_is12(:,:,i);
  T(:,:,i)=Tp(:,:,i)^NG; % 1st order uniform Bragg grating

  % for an FP cavity, 1st order cavity, insert a high index region, n2.
  T(:,:,i)=Tp(:,:,i)^NG * (T_hw2(:,:,i))^1 * Tp(:,:,i)^NG * T_hw2(:,:,i);
end
```

We structure the device to begin and end with an n_2 section. The phase-shift region is implemented using the high index material, n_2.

Finally, the transmission, T, and reflection, R, spectra are generated by Listing 4.4. The calculation is performed across the 1D matrix of wavelength points.

Listing 4.4 Transfer Matrix Method – calculate the reflection and transmission for the grating defined by 4.3.

```
function [R,T]=TMM_Grating_RT(wavelength, Period, NG, n1, n2, loss)
%Calculate the R and T versus wavelength

M=TMM_Grating_Matrix(wavelength, Period, NG, n1, n2, loss);

q=length(wavelength);
T=abs(ones(q,1)./squeeze(M(1,1,:))).^2;
R=abs(squeeze(M(2,1,:))./squeeze(M(1,1,:))).^2;
```

Grating physical structure design

Next, we make the connection between the physical structure (waveguide geometry) and the effective index. In this section, gratings are based on the strip waveguide, 500 × 220 nm with oxide cladding, where the waveguide width is varied as shown in Figure 4.42.

The design methodology is shown in Figure 4.33. It uses eigenmode calculations to determine the effective indices of the grating segments. Table 4.1 calculates the effective index of the waveguide versus the wavelength and waveguide width, which are subsequently parameterized. The simulation results are shown in Figure 4.34.

The data can be summarized by curve fitting to two functions, one for each variation:

$$n_{\text{eff}-\lambda}(\lambda) = a_0 - a_1(\lambda - \lambda_0) - a_2(\lambda - \lambda_0)^2, \tag{4.44}$$

where λ is in units [μm].

$$\Delta n_{\text{eff}-w}(w) = b_1(w - w_0) + b_2(w - w_0)^2 + b_3(w - w_0)^3, \tag{4.45}$$

where w is the width of the waveguide in μm, and $\Delta n_{\text{eff}}(w)$ is the deviation of the effective index relative to its value at λ_0 for a given waveguide width, w.

Table 4.1 Curve fit parameters for the effective index versus waveguide width and wavelength.

	Strip 500 × 220 nm oxide-clad
λ_0	1.554
a_0	2.4379
a_1	1.1193
a_2	0.035
w_0	0.5
b_1	1.6142
b_2	−5.2487
b_3	10.4285

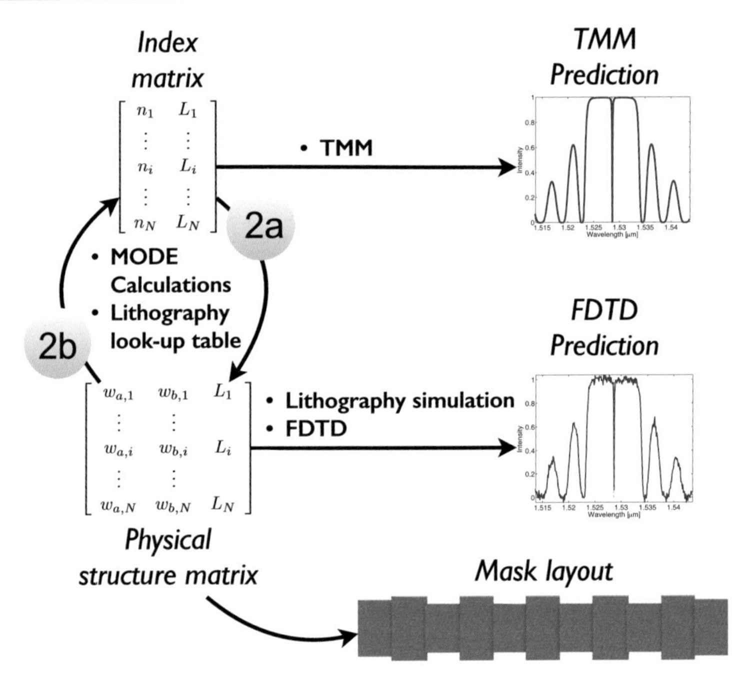

Figure 4.33 Modelling of waveguide gratings.

Considering the effective index versus two dimensions, the simulation results are shown in Figure 4.34c. Assuming the wavelength and width variations are separable, we add the two contributions together:

$$n_{\text{eff}}(\lambda, w) = n_{\text{eff}-\lambda}(\lambda) + \Delta n_{\text{eff}-w}(w). \tag{4.46}$$

The resulting error of this function, relative to the simulation data, is shown in Figure 4.34d, where the data are the difference between the two, and is within ±0.02.

This allows us to either design a physical structure based on an index profile, or to simulate a physical structure, as illustrated in the design flow in Figure 4.33. The MATLAB code in Listing 4.15 defines the grating based on the physical parameters: the

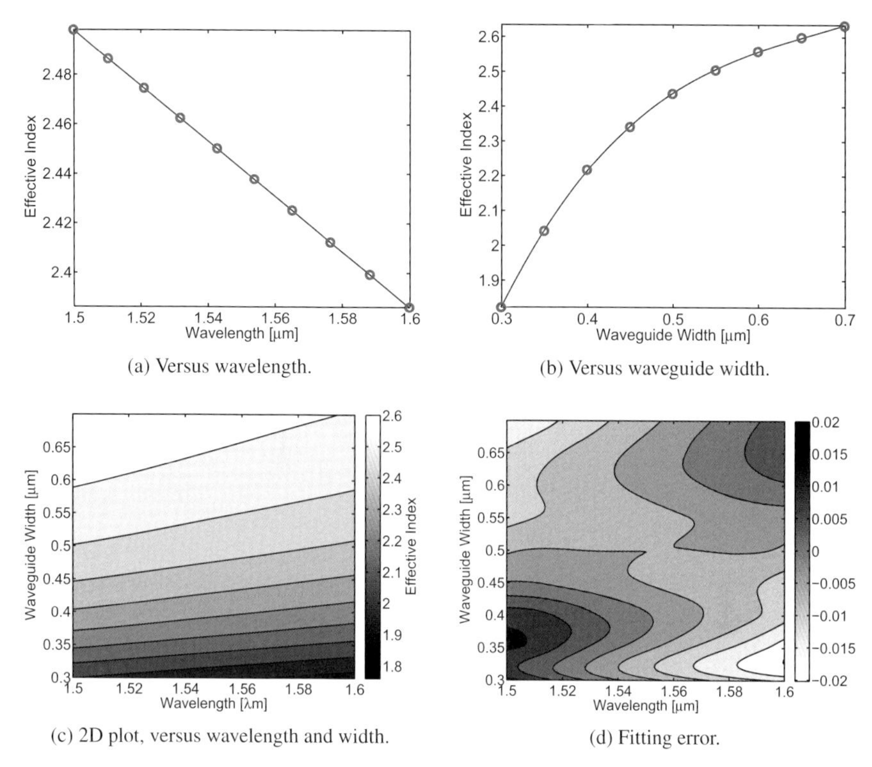

(a) Versus wavelength.

(b) Versus waveguide width.

(c) 2D plot, versus wavelength and width.

(d) Fitting error.

Figure 4.34 Effective index calculations for the strip waveguide with oxide cladding.

grating period, the number of grating periods, the mean waveguide width, the $\pm$ width variation assuming a rectangular profile with a 50–50% duty cycle, and a propagation loss. The simulation parameters are defined: wavelength span and number of points. Next, the relationship between the waveguide width and effective index are defined, as per Equation (4.46). Then the Bragg wavelength is found, with the inclusion of the waveguide dispersion. Then the effect indices are found. Finally, the grating spectra are calculated and plotted, and shown in Figure 4.35.

An example of a simulated result for a Fabry–Perot cavity implemented using Bragg gratings is shown in Figure 4.35. Here, the cavity length is only $\Lambda/2$, which is known as a "phase-shifted" design.

Modelling gratings using FDTD

In this section, we describe the method of simulating a Bragg grating using 3D FDTD. The example provided is a strip waveguide grating with a corrugation width of 50 nm, a 324 nm period, and 280 grating periods, with parameters as defined in Figure 4.42. The principal challenge in simulating periodic structures in FDTD is to ensure that the mesh has the same periodicity as the structure itself. As illustrated in Figure 4.36a, the simulation mesh parameters are chosen to ensure that the mesh points line up with the geometry. This is achieved by using a mesh override. The mesh in the x-direction is forced to be an integer multiple of period (324 nm/8 = 40.5 nm). Similarly, the mesh in

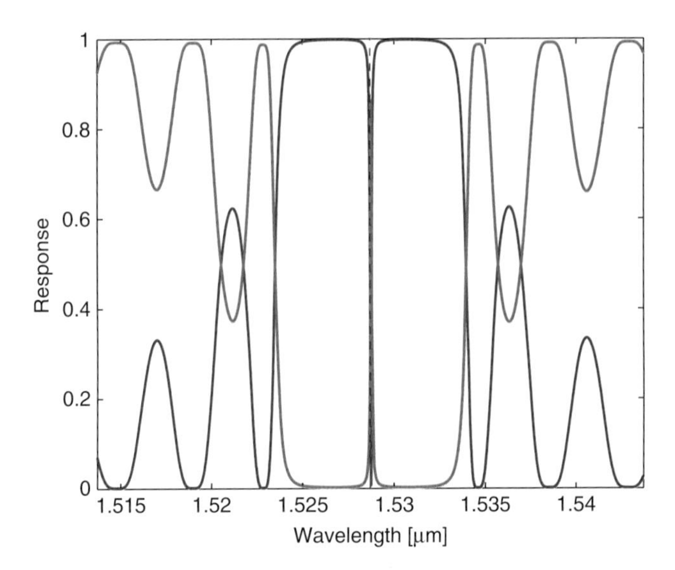

Figure 4.35 Reflection and transmission spectra for a phase-shifted uniform Bragg grating.

the y-direction is chosen to be 50 nm to ensure that simulation points match the grating corrugations.

The simulation script is provided in Listing 4.16, and the simulation results are plotted in Figure 4.36c. For comparison, this device was also fabricated using electron beam lithography (see Section 4.5.4) using a 6 nm shot pitch, with an image of the device shown in Figure 4.36b. An excellent agreement is obtained for the bandwidth of the grating. However, there was a discrepancy in the central wavelengths and the simulated results were shifted by 5 nm to match the experimental results. The discrepancy is attributed to the uncertainty in fabrication: silicon thickness, waveguide width, and sidewall angle (see Section 11.1).

4.5.3 Experimental Bragg gratings

The integration of Bragg gratings in silicon-on-insulator (SOI) waveguides was first demonstrated by Murphy *et al.* [16] in 2001. Typically, the gratings are achieved by physically corrugating the silicon waveguides. This is in contrast to the manufacture of fibre Bragg gratings, in which the fibre is photosensitive and exposed to intense ultraviolet (UV) light to induce refractive index modulation in the core.

Grating corrugations can be the top surface [16–18] or the sidewalls, where the sidewalls can be corrugated either on the rib [19] or on the slab [20]. The top-surface-corrugated configuration usually has a fixed etch depth, therefore, the grating coupling coefficient is constant. The sidewall-corrugated configuration has the advantage that the corrugation width can be easily controlled, which is essential to make complex grating profiles, such as apodized gratings that can suppress reflection side-lobes [20]. Instead of using physical corrugations, there are a few other approaches to form gratings in silicon, such as ion implanted Bragg gratings [21].

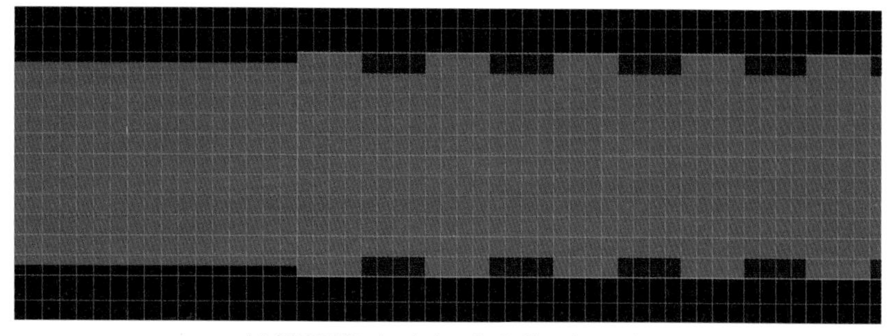

(a) 3D FDTD simulation, including the mesh.

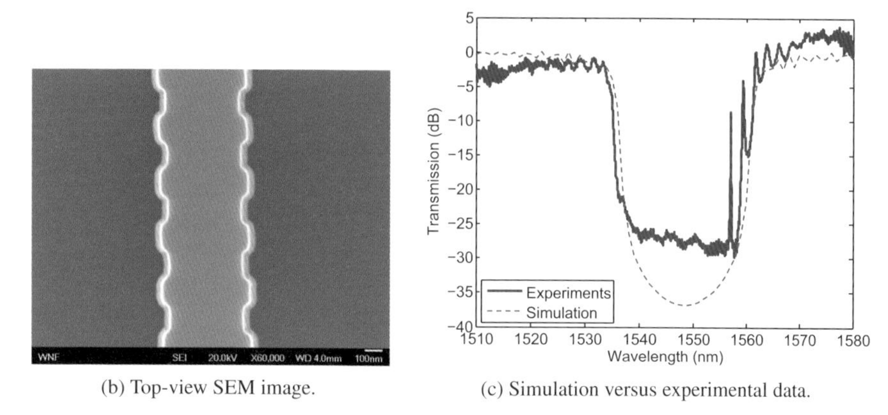

(b) Top-view SEM image.

(c) Simulation versus experimental data.

Figure 4.36 Uniform Bragg grating in a strip waveguide with air cladding. Comparison of 3D FDTD simulations versus a device fabricated by electron beam lithography.

In this section, we will discuss two waveguide structures that are commonly used for integrated Bragg gratings: strip waveguides and rib waveguides, with sidewall corrugations.

Strip waveguide gratings

Figure 4.37a shows the cross-section of a strip waveguide. The grating corrugations are on the waveguide sidewalls, therefore, the grating and the waveguide can be defined in a single lithography step. Owing to the small waveguide geometry and optical mode size, a small perturbation on the sidewalls can cause a considerable grating coupling coefficient, thus resulting in a large bandwidth. The experimentally demonstrated bandwidth is generally on the order of tens of nanometers [23, 24]. The lowest bandwidth reported so far is about 0.8 nm [25]; however, the authors used a very small corrugation width of 10 nm in the design and the fabricated corrugations were even smaller due to the lithography smoothing effect. Therefore, it is quite challenging to fabricate such small corrugations directly on the sidewalls. Tan *et al.* demonstrated another concept by moving the sidewall corrugations outside of the waveguide and placing a periodic array of cylinders near the waveguide to achieve similarly small effective index perturbations [26]. The reported bandwidth using this approach is on the order of a few nanometers; however, this approach remains sensitive to fabrication errors because the cylinders are still small (200 nm diameter) and are isolated structures.

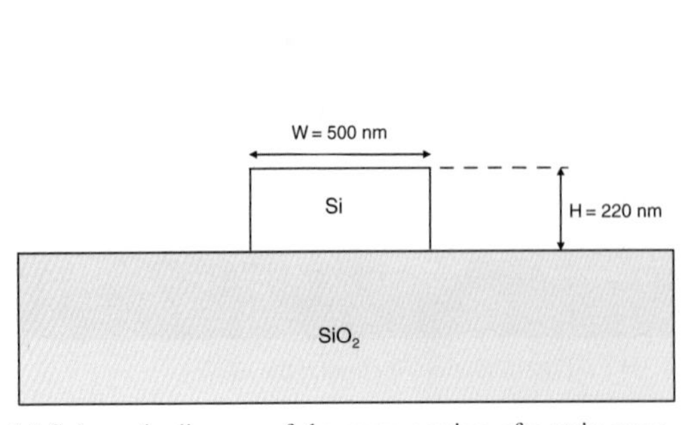

(a) Schematic diagram of the cross-section of a strip waveguide.

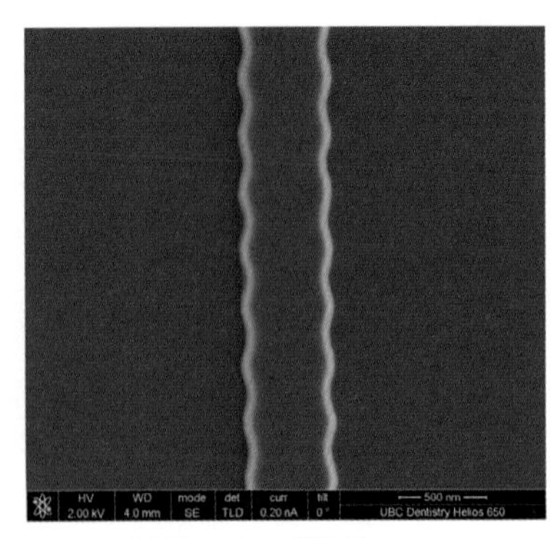

(b) Top-view SEM image.

Figure 4.37 Strip waveguide diagram and SEM image of strip waveguide gratings [22].

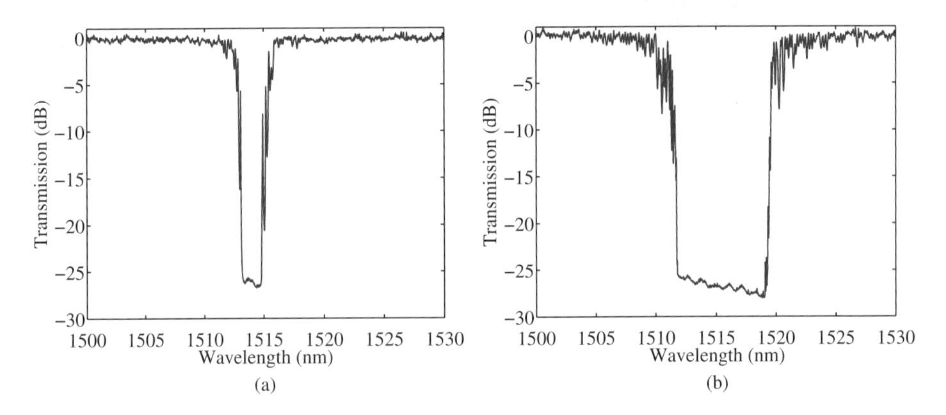

Figure 4.38 Measured transmission spectra of strip waveguide gratings: (a) 10 nm corrugations, (b) 40 nm corrugations.

Figure 4.37b shows the top view SEM image of a fabricated Bragg grating. It is important to keep in mind that, if using optical lithography in the fabrication, the gratings actually fabricated can be severely smoothed. Square corrugations were used in the mask design but the fabricated corrugations are severely rounded and resemble sinusoidal shapes. Therefore, in order to obtain the desired bandwidth, such smoothing effects should be taken into account and compensated. This can be achieved by simply using larger corrugation widths in the mask design than in the simulation. However, this cannot be easily done without look-up tables derived from past experience and/or lithography simulations [14], which is to be discussed in Section 4.5.4.

Figure 4.38 shows the measured transmission spectra for strip waveguide gratings, where the corrugation widths are the design values. As the corrugation width is increased, the coupling coefficient increases, leading to broader bandwidth.

Rib waveguide gratings

As discussed above, strip waveguide gratings have relatively large bandwidths, and are sensitive to fabrication variations. However, numerous applications require narrow

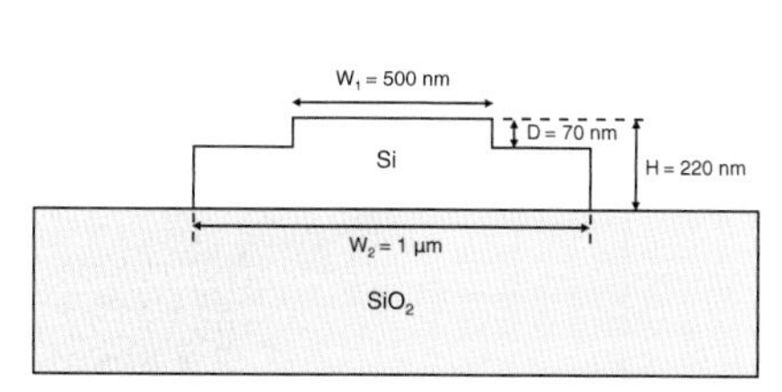

(a) Schematic diagram of the cross-section of a rib waveguide.

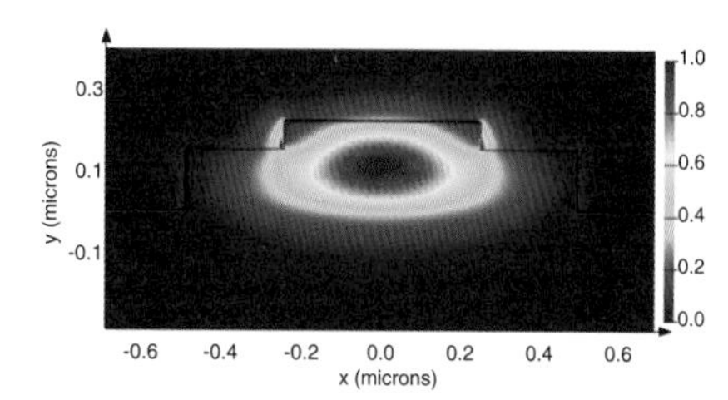

(b) Simulated fundamental transverse-electric (TE) mode profile in a rib waveguide.

Figure 4.39 Rib waveguide.

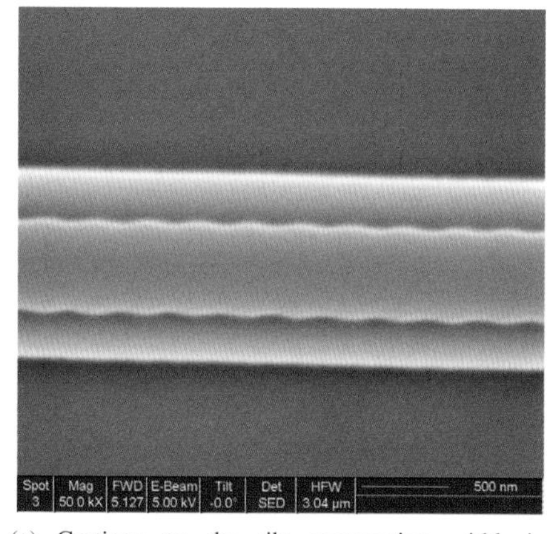

(a) Gratings on the rib: corrugation width is 60 nm (design value)

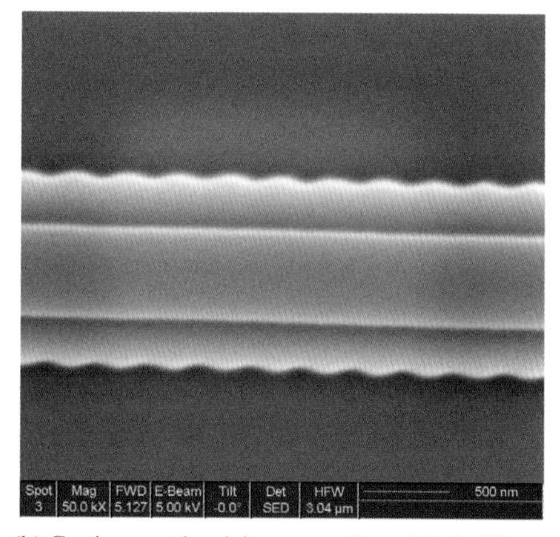

(b) Gratings on the slab: corrugation width is 80 nm (design value)

Figure 4.40 SEM images of rib waveguide gratings. Reprinted with permission from Ref. [22].

bandwidths, such as wavelength-division-multiplexed (WDM) channel filters. An alternative is to use rib waveguides, which typically have larger cross sections and allow higher fabrication tolerance. Figure 4.39a illustrates the waveguide structure in [27]. As shown in Figure 4.39b, most light is confined under the rib and the optical field's overlap with the sidewalls is low around both the rib and slab sidewalls. This overlap reduction makes it possible to introduce weaker effective index perturbations compared to the strip waveguide gratings, thus allowing for smaller coupling coefficients and narrower bandwidths. Two configurations to form gratings on rib waveguides are discussed here: where the sidewalls are corrugated either on the rib [19] or on the slab [20]. Figure 4.40 shows the SEM images of the fabricated devices.

Grating period

Based on the Bragg condition, Equation (4.28), we expect that the Bragg wavelength will increase with increasing grating period. The wavelength shift, taking into account dispersion, is found to be:

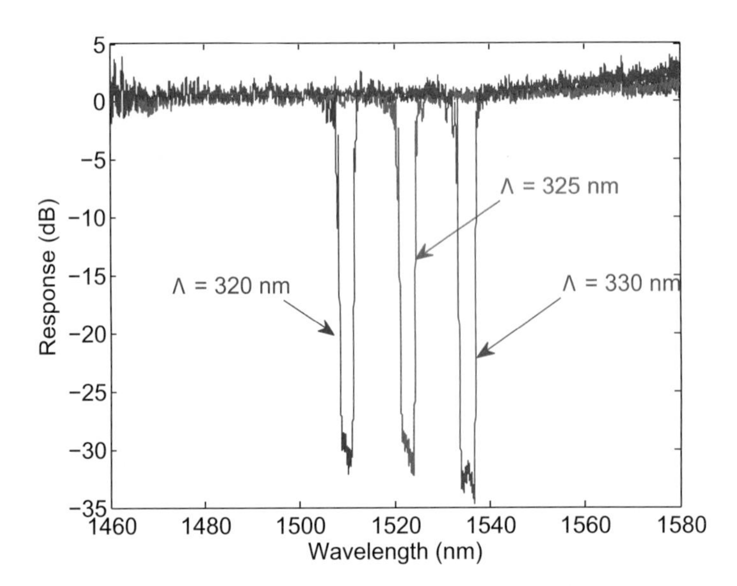

Figure 4.41 Measured transmission spectra for gratings with different grating periods, showing the red shift with increasing grating period. Fixed parameters: air cladding, $W = 500\,\text{nm}, \Delta W = 20\,\text{nm}, N = 1000$.

$$\frac{\partial \lambda_B}{\partial \Lambda} = 2\frac{n_{\text{eff}}^2}{n_g} = \frac{\lambda^2}{2\Lambda^2 n_g}. \tag{4.47}$$

This was observed experimentally. Figure 4.41 shows the measured transmission spectra for three gratings with different periods: 320 nm, 325 nm, and 330 nm. The experimental Bragg wavelength shifts at a rate of $\partial\lambda/\partial\Lambda = 2.53$. This closely matches with the calculated value of 2.515 using Equation (4.47). These data and Equation (4.47) can also be used to extract the effective and group indices of the fabricated waveguides.

4.5.4 Empirical models for fabricated gratings

This section summarizes the fabrication and modelling results for Bragg gratings designed in strip and rib waveguides using sidewall corrugations. The objective of this section is to provide empirical data to designers wishing to synthesize gratings with specific properties, e.g. apodized gratings, using modelling approaches such as the transfer matrix method presented in Section 4.5.2. Uniform gratings are used with the geometric parameters as illustrated in Figure 4.42. The width of the waveguide is thus $W \pm \Delta W$. This configuration, rather than the recessed-only or protruding-only configuration, is chosen because the average effective index is approximately constant as the grating strength changes. This means that the Bragg-wavelength central wavelength will not shift, and this enables these gratings to be used for apodization where the grating corrugation amplitude is varied along the length of the grating.

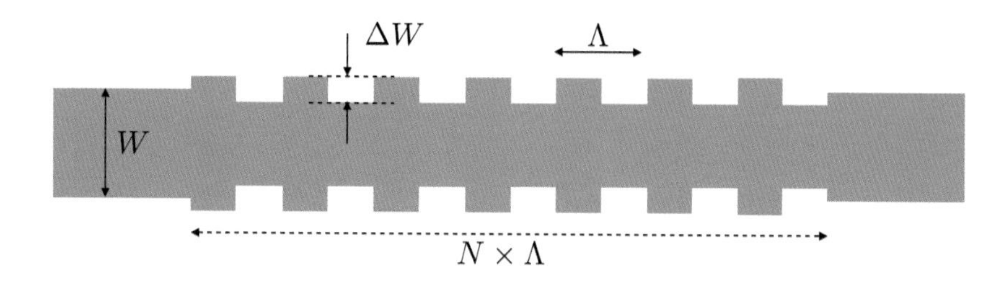

Figure 4.42 Schematic defining the geometric parameters used for the uniform Bragg gratings [22]. The parameters are: W, the average waveguide width; ΔW, the corrugation width on each side wall; Λ, the grating period; N, the number of grating periods; and $L = N \times \Lambda$, the grating length.

Various gratings were fabricated using several fabrication facilities and methods. In all cases, the waveguides had a thickness of 220 nm, width W = 500 nm, were clad with oxide, and the corrugations were applied to the sidewalls. The following is the list of gratings considered in this section:

- 193 nm UV lithography, fabricated at imec: strip and rib waveguides
  - strip waveguides,
  - rib waveguides, with a 1000 nm wide and 150 nm thick slab. Corrugations were applied to: (a) the 150 nm thick slab with an average width of 1000 nm, and (b) the rib of the waveguide (etch depth of 70 nm), with an average width of 500 nm;
- 248 nm UV lithography, fabricated at IME: strip waveguides;
- electron-beam lithography, fabricated at the University of Washington Nanofabrication Facility (WNF) [28]: strip waveguides.

The performance parameters were extracted with the definitions as follows:

- $\Delta\lambda$: the measured Bragg bandwidth, defined from the first-null points in the spectrum, as in Figure 4.30;
- κ: the extracted grating coupling coefficient [m^{-1}]. Equation (4.33) was used to determine the grating coupling coefficient from the experimental results. The simulated values for the group index, n_g, were used;
- Δn: the extracted refractive index modulation, $\Delta n = n_2 - n_1$. This is defined based on the rectangular grating modulation function, Equation (4.36).

Figure 4.43a provides the grating bandwidth (first nulls), $\Delta\lambda$, versus the corrugation width. For experimental results, this was measured from the spectra. For the simulation, it was determined from the Δn values and assuming an infinitely long grating. This graph is useful for the designer in predicting how much optical bandwidth will be attained for the grating geometry and fabrication choices available. To obtain a sub-nanometer bandwidth, if using the strip waveguide, a corrugation width of less than 10 nm is required, whereas, if using the slab region of the rib waveguide, an 80 nm

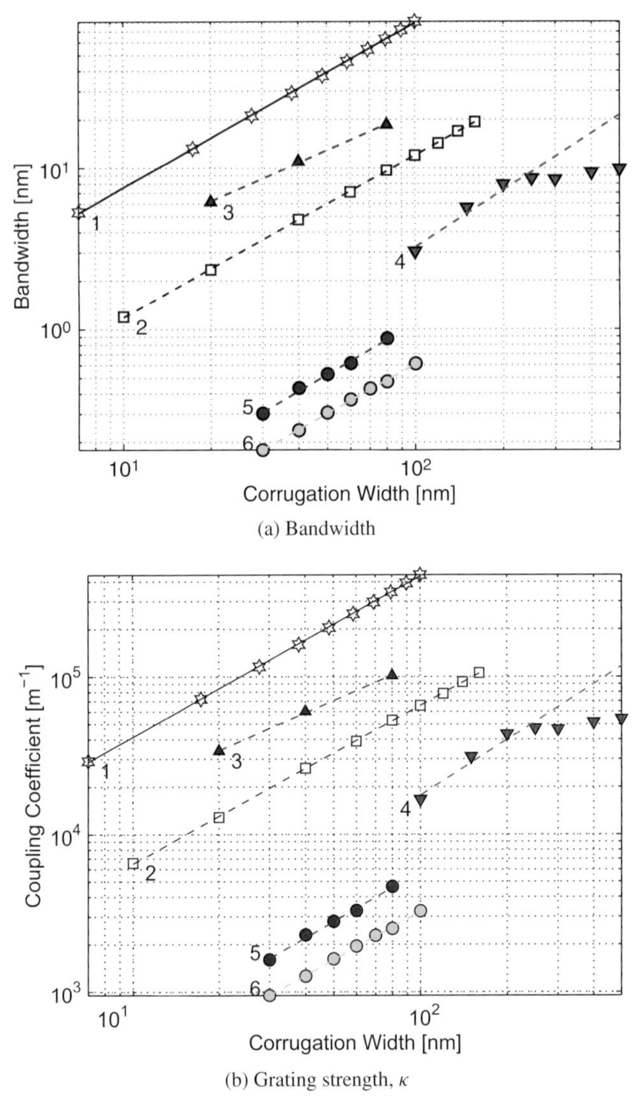

(a) Bandwidth

(b) Grating strength, κ

Figure 4.43 Bragg grating bandwidth (first nulls) and coupling strength, κ, versus as-drawn corrugation size, Δw. (1–4) Oxide-clad strip waveguides, 500 nm $\times$ 220 nm: (1) calculations by the eigensolver, parameterized using Equation (4.45); (2) extracted from devices fabricated using 193 nm lithography; (3) extracted from devices fabricated using electron-beam lithography; (4) extracted from devices fabricated using 248 nm lithography. (5–6) Oxide-clad rib waveguides, 500 nm $\times$ 220 nm, with a slab 1000 nm width and 150 nm thick, fabricated using 193 nm lithography: (5) corrugations applied on the rib; (6) corrugations applied on the slab.

corrugation width should be suitable. This means that the rib waveguide gratings have a relaxed fabrication tolerance. For the rib waveguide gratings, the 3 dB bandwidth ranges from 0.4 nm to 0.8 nm, which is suitable for many narrow-band applications such as WDM channel filters, although apodization may be required to suppress the side-lobes.

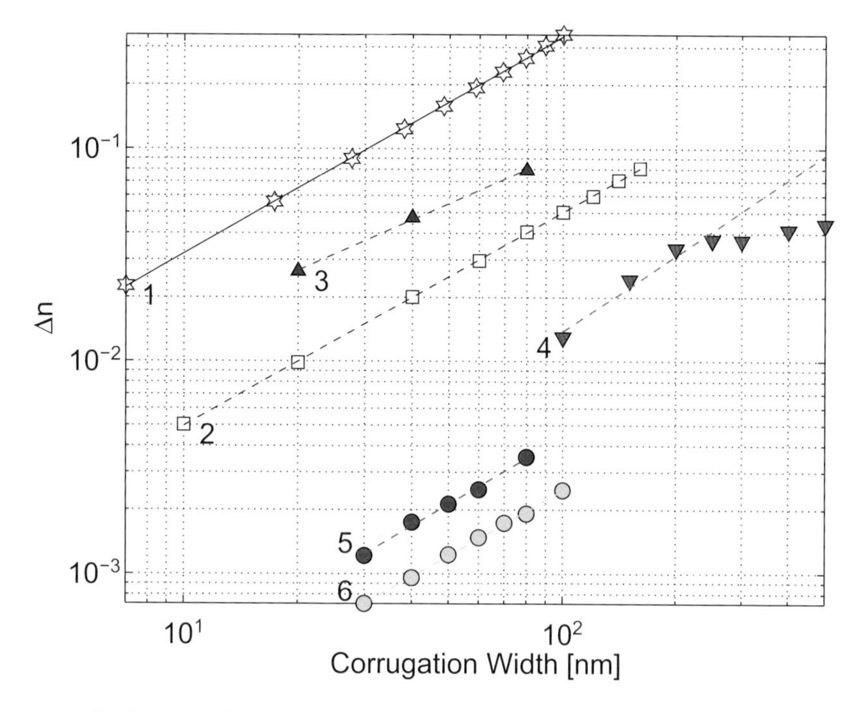

Figure 4.44 Grating strength, Δn, versus as-drawn corrugation size, Δw. (1–4) Oxide-clad strip waveguides, 500 nm × 220 nm: (1) calculations by the eigensolver, parameterized using Equation (4.45); (2) extracted from devices fabricated using 193 nm lithography; (3) extracted from devices fabricated using electron-beam lithography; (4) extracted from devices fabricated using 248 nm lithography; (5–6) Oxide-clad rib waveguides, 500 nm × 220 nm, with a slab 1000 nm wide and 150 nm thick, fabricated using 193 nm lithography: (5) corrugations applied on the rib; (6) corrugations applied on the slab.

The extracted grating coupling coefficient, κ, is plotted in Figure 4.43b. This is useful for designing gratings based on Coupled Mode Theory.

The modelling results presented in Figures 4.43 and 4.44 are based on waveguide eigenmode solutions (Section 4.5.2) and assume a rectangular grating modulation with an index contrast of Δn. The grating coupling coefficient, κ, and bandwidth (first nulls), $\Delta\lambda$, are found using Equations (4.36) and (4.33). The comparison between the experimental results with the waveguide effective index model thereby enables the design of gratings using the transfer matrix method, using the Δn values provided in Figure 4.44.

To facilitate the design, the results shown in Figure 4.44 for Δn versus corrugation width, ΔW, were curve fit using expressions in the log-log scale:

$$\log_{10} \Delta n = c_1 \cdot \log_{10} \Delta W + c_2 \tag{4.48}$$

or equivalently:

$$\Delta n = 10^{c_1 \cdot \log_{10} \Delta W + c_2} \tag{4.49}$$

Table 4.2 Design, fabrication details, and fit parameters for Bragg gratings.

#	1	2	3	4	5	6
Waveguide	Strip				Rib	
Corrugation	Sidewall				Rib	Slab
Method	TMM	193 nm	EBL	248 nm	193 nm	193 nm
Fit slope – c_1	1.0217	1.0044	0.7954	1.1741	1.0557	1.0179
Fit ΔN offset – c_2	−2.5160	−3.3097	−2.5997	−4.2332	−4.4667	−4.6447
Fit BW offset – c_3	−0.1438	−0.9301	−0.2411	−1.8374	−2.0725	−2.2504
Fit κ offset – c_4	3.5947	2.8083	3.4974	1.9011	1.6529	1.4750

The coefficients for the fits are provided in Table 4.2. Similarly, results from Figure 4.43 for the bandwidth (first nulls), BW, and grating strength, κ, were curve-fit using:

$$\log_{10} BW = c_1 \cdot \log_{10} \Delta W + c_3 \tag{4.50}$$

$$\log_{10} \kappa = c_1 \cdot \log_{10} \Delta W + c_4. \tag{4.51}$$

The results in this section indicate that there is a large mismatch between the simulations using the plane-wave approximation implemented in the transfer matrix method, and the fabricated devices. This is due to two sources: (1) lithography smoothing, as discussed in the present section, and (2) the plane-wave approximation used in the Fresnel reflection coefficients. While both of these pose significant challenges in the design of Bragg gratings, the empirical models for Δn and κ presented in this section enable the rapid design of gratings using the Transfer Matrix Method or Coupled Mode Theory. These models account for both sources of mismatch. An alternative approach is to simulate the lithography effects on the geometry, then perform 3D simulations of the virtually fabricated devices. This approach is more involved, and also includes an empirical model for the lithography, and less amenable to the synthesis of gratings.

Computation lithography models

One of the important issues with silicon photonics fabrication is that small features, such as those found in Bragg gratings, are extremely sensitive to fabrication imperfections. For example, the corrugations in Bragg gratings are often designed to be square in the original mask layout. Unsurprisingly, however, the actual fabrication rounds the sharp corners, especially when using optical lithography [29]. As a result, there is a significant performance mismatch between the originally designed (modelled) device and the actually fabricated device, specifically, the experimental bandwidth is usually much narrower than designed [29] (see earlier in Section 4.5.4 and Figure 4.43a).

One solution to this problem is to include the fabrication process in the design flow [30, 31]. Mentor Graphics provides an advanced lithography simulation tool to simulate the fabrication of devices fabricated with deep-ultraviolet (DUV) lithography [31]. After lithography simulation, the virtually fabricated devices can be re-simulated, to obtain a better prediction of the forthcoming experimental results [14].

The lithography model can be constructed using a variety of test structures and is often provided by CMOS foundries. In this example, it was constructed using a 500 nm

Figure 4.45 Lithography simulation for device A: (a) original design, (b) simulation result. Reprinted with permission from Reference [22].

strip waveguide grating designed with 40 nm rectangular corrugations (named as device A). The model parameters were adjusted to match the post-lithography simulation with the experimental data for device A; the parameters were then fixed for all other devices. For the optical system, we use a conventional circular illumination source. The numerical aperture (NA) and the partial coherence factor (σ) are the key parameters that determine the corrugation distortions; we use NA = 0.6 and σ = 0.6 in our simulations. These parameters are within the range defined by the stepper's technical specifications. Figure 4.45 shows the simulation results for device A. We can see that the corrugations are greatly smoothed, and their effective amplitudes are also reduced.

The Bragg grating modelling is based on 3D FDTD. Figure 4.46 shows the simulated transmission spectra for device A, both for the as-drawn structure, and for the virtually fabricated device. These are compared with the measured results for the device, showing a match of the optical bandwidth. It can be seen that the original design has a bandwidth of about 23 nm. In contrast, the post-litho simulation shows a much narrower bandwidth of about 8 nm. Note that the thickness of the waveguide was slightly reduced by a few nanometers in the simulation in order to match the Bragg wavelength, which has little effect on the bandwidth.

Figure 4.47 plots the simulated and measured bandwidths versus the designed corrugation widths. Again, the post-litho simulation agrees very well with the measurement, whereas the mismatch between the original design and the measurement is very large. For the case of 193 nm, as a rule of thumb, the actual bandwidth is approximately three times smaller than the original design value based on FDTD simulations.

The computational lithography technique can be applied to many other silicon photonic devices, especially ones that are sensitive to lithographic distortions, for example, photonic crystals. In photonic crystal cavities, the simulated bulk holes are smaller than the designed ones, therefore, a bias should be applied to the bulk holes in the mask to obtain the desired hole sizes. Owing to the optical proximity effect, the edge holes are smaller than the bulk holes [32], and the displacement of the two cavity side holes

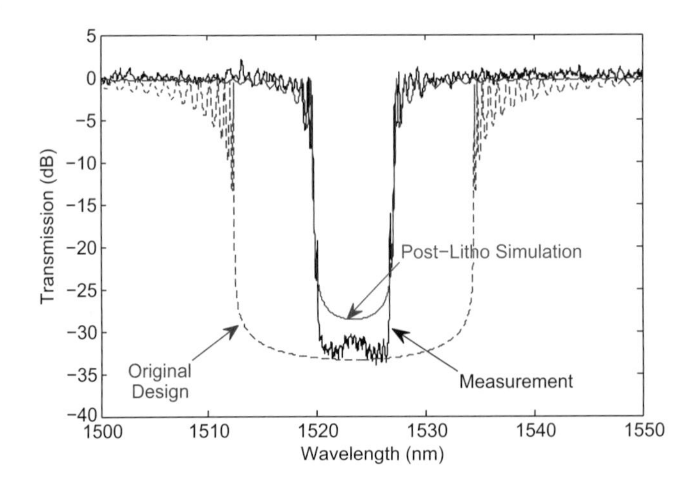

Figure 4.46 Spectral comparison (transmission) for device A. Original design: 3D FDTD simulation using the structure in Figure 4.45a; post-litho simulation: 3D-FDTD simulation using the virtually-fabricated structure in Figure 4.45b. Measurement: device fabricated by imec using 193 nm lithography. Reprinted with permission from Reference [22].

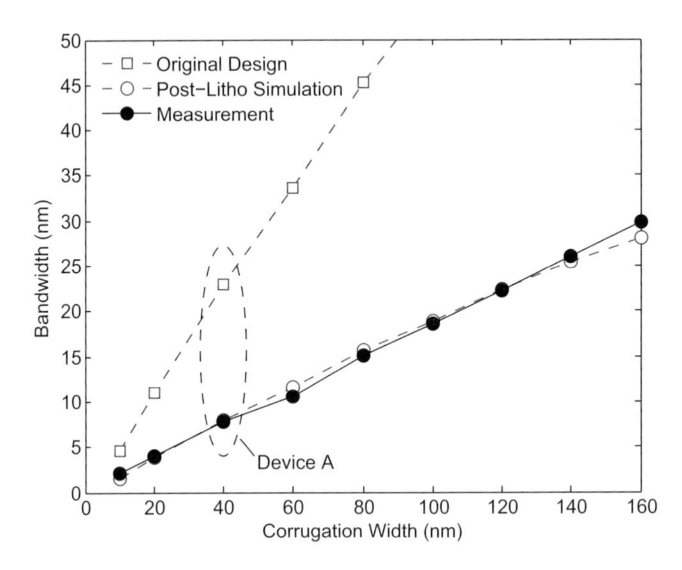

Figure 4.47 Bandwidth versus corrugation width on 500 nm strip waveguides with air cladding. The post-litho simulation agrees well with the measurement result, while the original designs show bandwidths about three times larger.

introduces extra distortions. Therefore, differential bias needs to be applied to the holes next to the cavity, and this cannot be easily done without lithography simulations.

Additional fabrication considerations

Next, we describe additional considerations for Bragg gratings. The first is regarding the challenges of fabricating long devices. Owing to fabrication non-uniformity, as discussed in Section 11.1, long devices (> 100 μm) will experience variations in width and thickness [33]. Thus, non-uniformity is a limiting factor in grating length, as shown in Figure 11.1, whereby a broadening of the spectrum and side-lobes are visible. This problem can be partially alleviated by making compact grating structures, as described in Section 4.5.5. Next, we need to consider the waveguide geometry impact on grating performance. Figure 3.5 shows the tilted cross-section SEM image of a fabricated waveguide [25]. It should be noted that the cross-section profiles are not perfectly rectangular and have slightly sloped sidewalls. Furthermore, the waveguide width and thickness also slightly deviate from the design values. Such geometric imperfections will affect the effective index of the waveguide, and, as a result, usually shift the Bragg wavelength from its design value. The Bragg gratings described in this section typically offer a single Bragg wavelength, thus matching the wavelength to the target application is more critical. Thermal tuning offers limited adjustments (see Section 4.5.5). This is in contrast to ring resonators, which have numerous modes, hence cyclic tuning is possible (see Section 13.1), where the device is heated until the next adjacent peak matches the target wavelength. Finally, the manufacturing grid resolution is a limitation for fabricating chirped or apodized gratings. Specifically, the quantization introduced in the ΔW leads to a quantization of the available grating coupling coefficients available. The quantization in the position of the grating teeth leads can lead to a deterioration of the optical spectrum.

4.5.5 Spiral Bragg gratings

Bragg gratings implemented in rib waveguides can achieve narrow bandwidths. Since the perturbations are very weak, a very long length is required to obtain a high reflectivity. However, this is often not desired from the layout perspective because the high aspect ratio makes it difficult to integrate them efficiently in photonic integrated circuits. More importantly, the performance of long Bragg gratings is more likely to be affected by the thickness and width variations, as discussed in Section 11.1. Therefore, it is important to design long Bragg gratings within a small area [34–36]. This can be implemented in a spiral configuration, as shown in Figures 4.48 and 4.49, where the design achieved a narrow bandwidth of 0.26 nm [36].

As illustrated in Figure 4.48, the gratings are designed on the rib using sidewall corrugations. On each side of the rib, the corrugation width is designed to be 50 nm. The period of the first-order grating is kept constant at 290.9 nm in the whole spiral.

Figure 4.49 shows the optical microscope images of a fabricated device. The spiral consists of a series of half circles with different diameters (D), while the radius of curvature is kept constant within each half circle. As shown in Figure 4.49b, the diameters of the two smallest half circles (D_{min}) in the centre, i.e., the S-shape, are 20 μm. The spacing between two adjacent half circles, or the spiral pitch (P), is 5 μm to ensure low cross-talk. Note that this value is quite conservative and could be reduced (e.g. to 3 μm) to further improve the space efficiency.

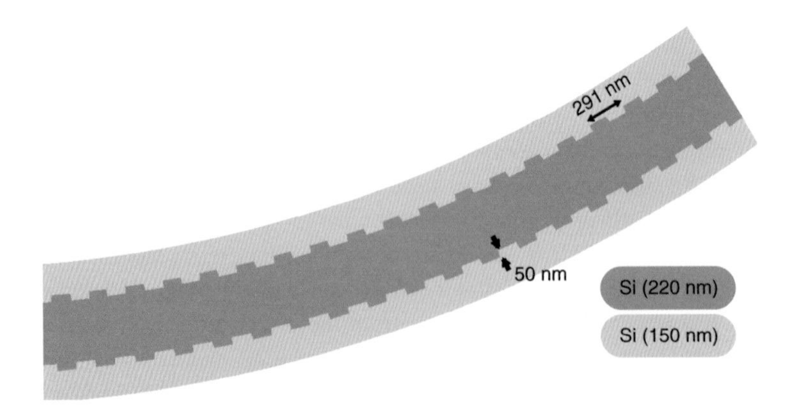

Figure 4.48 Top view of the spiral grating design. Reprinted with permission from Reference [22].

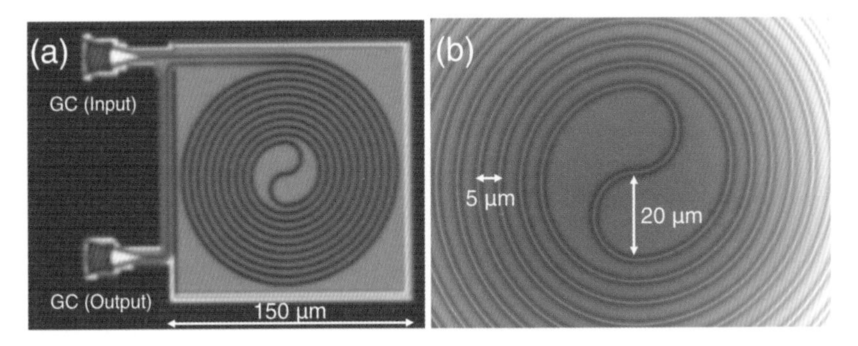

Figure 4.49 Optical microscope images of the fabricated device with N=10. (a) Whole layout. (b) Enlarged image of centre region of the spiral. Reprinted with permission from Reference [22].

Thermal sensitivity

Most silicon photonic devices are highly sensitive to temperature variations on the chip, due to the large thermo-optic coefficient of silicon. In this section, we study the thermal sensitivity of strip waveguide gratings. Figure 4.50 shows the transmission spectra of a strip waveguide grating at different temperatures. The Bragg wavelength shifts to longer wavelengths as the temperature is increased. The linear slope is about 84 pm/°C.

The thermal tuning coefficient for resonant devices (e.g. Bragg gratings, ring resonators, etc.) is:

$$\frac{d\lambda}{dT} = \frac{\lambda}{n_g}\frac{dn_{\text{eff}}}{dT} = \frac{\lambda}{n_g}\frac{dn_{\text{eff}}}{dn}\frac{dn}{dT} \tag{4.52}$$

Using Equation (4.52), the simulated thermo-optic dependence of the Bragg wavelength is about 80 pm/°C, in good agreement with the measurement result.

4.5.6 Phase-shifted Bragg gratings

One application of Bragg gratings is to construct resonators or cavities. These are useful for fabricating lasers (Chapter 8), narrow-band optical filters, or evanescent field

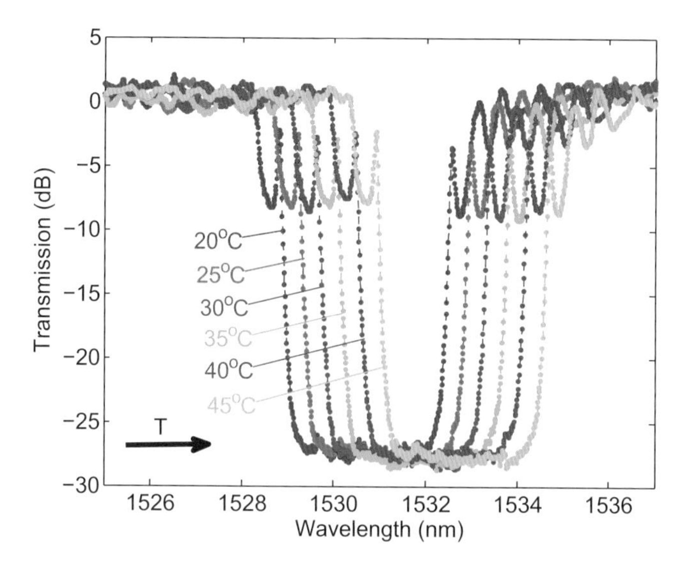

Figure 4.50 Measured transmission spectra of a strip waveguide grating at different temperatures, showing the red shift with increasing temperature. The design parameters are: air cladding, $W = 500$ nm, $\Delta W = 20$ nm, $\Lambda = 330$ nm, and $N = 1000$. Reprinted with permission from Reference [22].

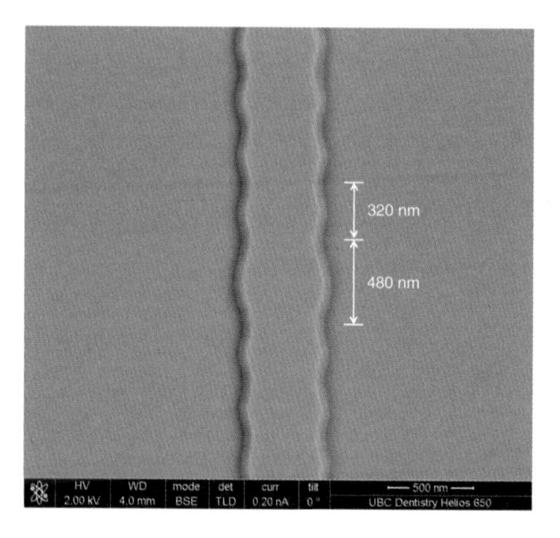

Figure 4.51 Top view SEM image of a fabricated strip waveguide phase-shifted grating. Design parameters: $W = 500$ nm, $\Delta W = 80$ nm, $\Lambda = 320$ nm. Note that the phase shift in the central region can be identified by measuring the spacing between the grating grooves. Here, the spacing with the phase shift is 480 nm, corresponding to 1.5 times the grating period (320 nm). Reprinted with permission from Reference [22].

sensors [37]. Figure 4.51 shows a first-order cavity fabricated using a strip waveguide Bragg grating, whereby a phase-shift is introduced between two identical gratings. The transmission window has a very narrow Lorentzian line shape. An example of a phase-shifted grating response is shown in Figure 4.52, where a FWHM linewidth of about 8 pm (i.e. 1 GHz) is observed. This corresponds to a Q factor of 1.9×10^5. The position

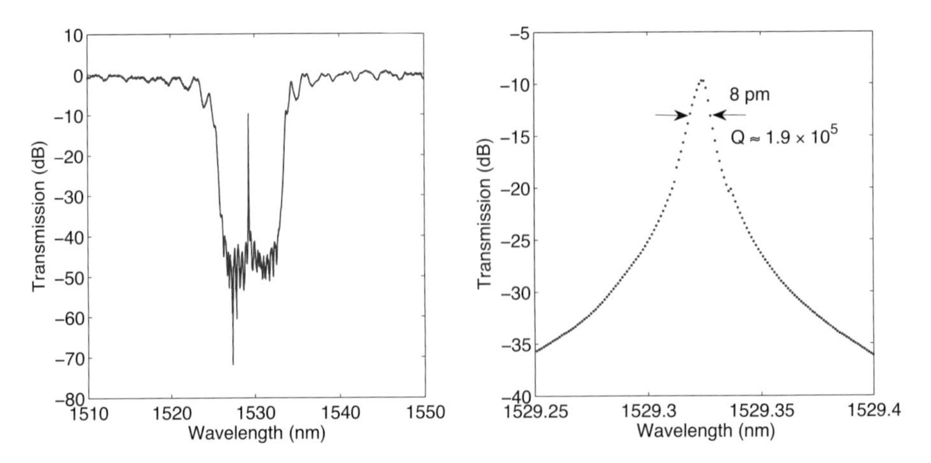

Figure 4.52 Phase-shifted Bragg grating cavity, with a quality factor 1.9×10^5. Design parameters: air cladding, W = 500 nm, ΔW = 40 nm, N =300, and Λ = 330 nm. Reprinted with permission from Reference [22].

and the size of the phase-shift determine the centre wavelength and the sharpness of transmission window. In this example, the phase-shift is placed in the exact centre of the grating so that the resonance peak is the sharpest, and the length of the phase shift is equal to a grating period so that there is always one resonance peak around the centre of the Bragg grating reflection band. If the length of the phase-shift is very long (e.g. 500Λ), it is possible to generate multiple resonance peaks within the stop band [38], which in the case of a laser design can lead to multi-mode operation. Such long gratings are also useful for determining waveguide parameters (effective and group index, propagation loss) [38].

4.5.7 Multi-period Bragg gratings

Since there are two sidewalls in strip waveguides, and four sidewalls in rib waveguides, it is possible to place independent grating profiles on each of these. The resulting spectra will be a superposition of the individual designs. Figure 4.53 shows simulation versus measurement for a dual-period grating. Figure 4.54 illustrates the design of a four-period Bragg grating using the rib waveguide. The grating periods are Λ_1 = 285 nm and Λ_2 = 290 nm, Λ_3 = 295 nm and Λ_4 = 300 nm, respectively. The corrugation widths on the rib and on the slab are ΔW_{rib} = 80 nm (for Λ_2 and Λ_3) and ΔW_{slab} = 100 nm (for Λ_1 and Λ_4), respectively. The grating length is 580 μm, implemented as L =2035Λ_1, 2000Λ_2, 1966Λ_3, and 1933Λ_4.

Figure 4.55 shows the measured spectral responses of a fabricated device. We can clearly observe four Bragg wavelengths, each corresponding to one grating period. Note that the coupling coefficients are reduced by a factor of two since only one side of each waveguide contributes to the grating. In summary, this multi-period grating concept increases the design flexibility and allows for more custom optical functions.

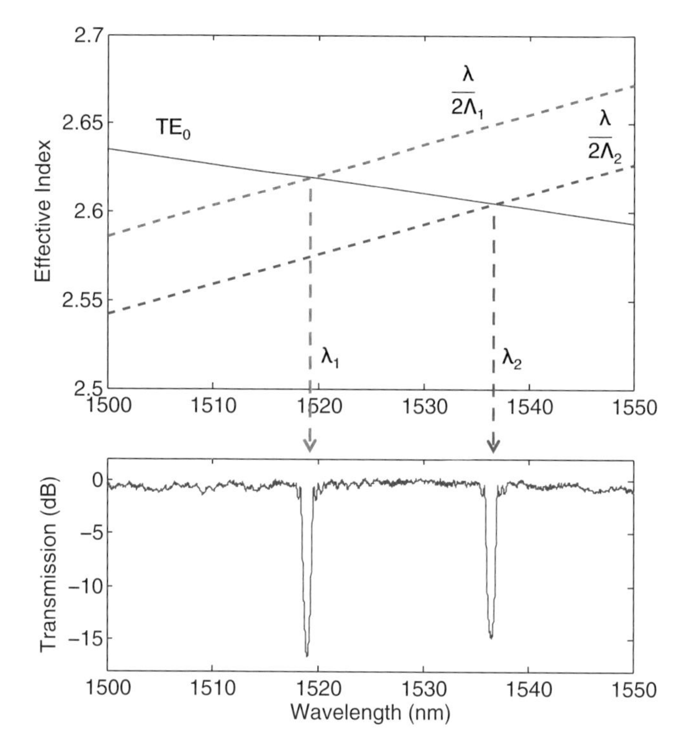

Figure 4.53 Dual-period grating: simulation vs. measurement. In the top graph, the solid blue curve shows the simulated effective index of the fundamental TE mode; the green and red dashed curves correspond to the effective indices that are needed using Λ_1 and Λ_2, respectively; thus, the intersection points correspond to the two Bragg wavelengths. Reprinted with permission from Reference [22].

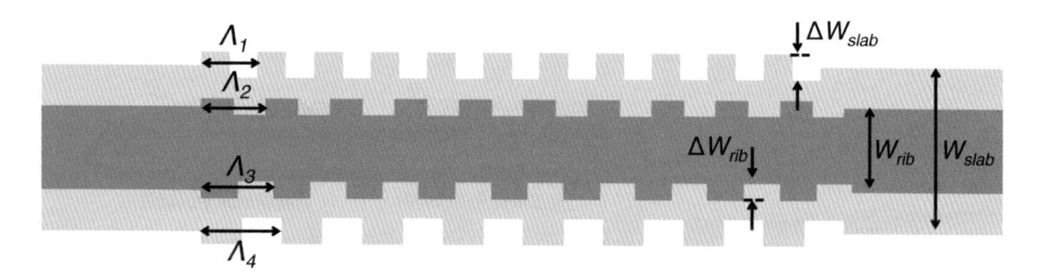

Figure 4.54 Schematic diagram of the four-period rib waveguide grating (not to scale). Design parameters: Λ_1 = 285 nm, Λ_2 = 290 nm, Λ_3 = 295 nm, Λ_4 = 300 nm, ΔW_{rib} = 80 nm, ΔW_{slab} = 100 nm, and the grating length is 580 μm. Reprinted with permission from Reference [22].

4.5.8 Grating-assisted contra-directional couplers

Instead of using a single waveguide, it is possible to construct gratings within directional couplers as illustrated in Figure 4.56. In this approach, the directional coupler is designed to have mis-matched waveguide widths (see Section 4.1.7) resulting in little coupling in the conventional forward propagation direction. The grating is used

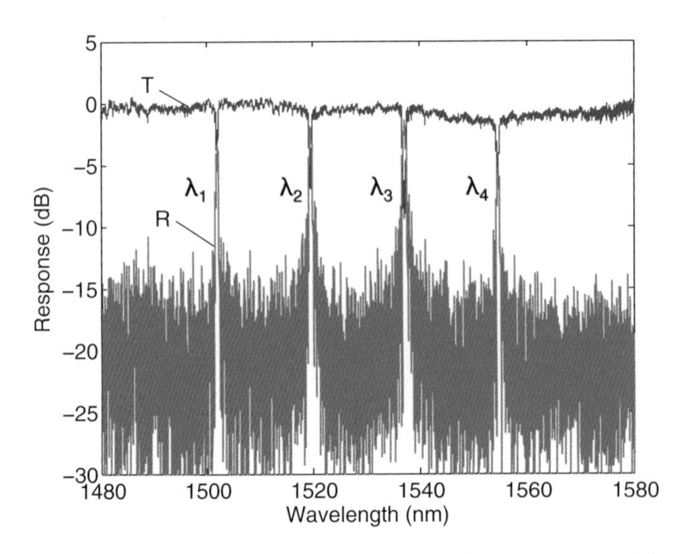

Figure 4.55 Measured spectral responses of a four-period grating. λ_1, λ_2, λ_3, and λ_4 correspond to Λ_1, Λ_2, Λ_3, and Λ_4, respectively. Reprinted with permission from Reference [22].

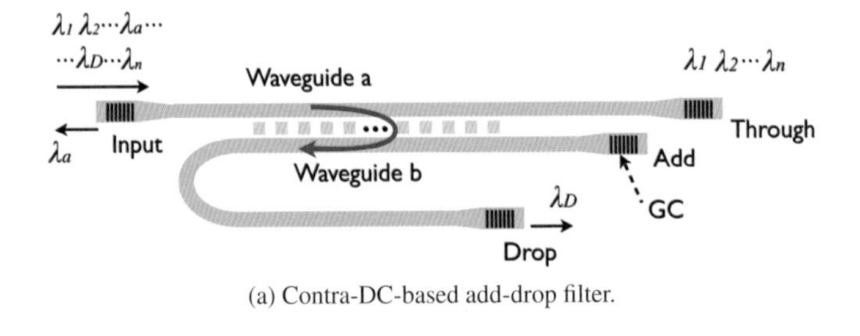

(a) Contra-DC-based add-drop filter.

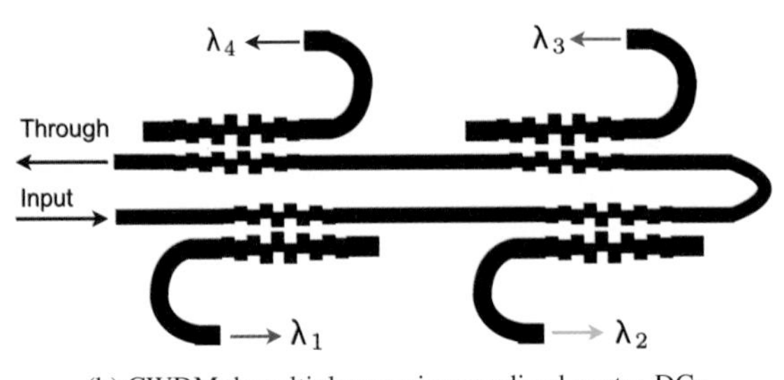

(b) CWDM demultiplexer using apodized contra-DCs

Figure 4.56 Schematic diagram contra-directional couplers for one and four channels. Reference [39].

to provide a coupling mechanism between the waveguides, and in particular, the Bragg condition in Equation (4.28) is modified to be:

$$\lambda_D = \Lambda(n_1 - n_2), \tag{4.53}$$

where λ_D is the Bragg wavelength of the light coupled backwards into the second waveguide.

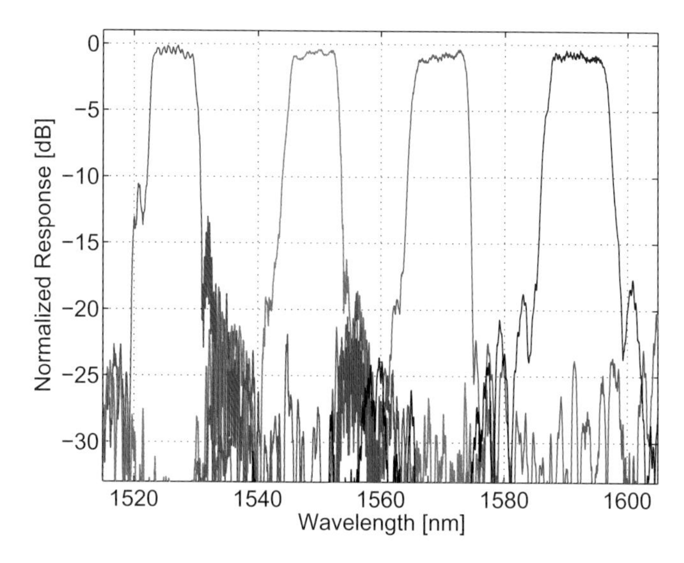

Figure 4.57 Measured transmission spectra of four-channel contra-directional coupler filter designed for coarse-WDM applications. Reference [39].

The advantage of this configuration is that the Bragg grating is no longer a two-port device, but rather a four-port device, as shown in Figure 4.56. The two-port reflection signal is routed to the drop port, which eliminates the need for optical circulators or isolators when used for optical filtering. It is also provides a fourth port, which is used as the "add" port, hence the device can be used as an add-drop multiplexer.

The device can be designed to operate over a large optical bandwidth providing single-channel filtering. In order to obtain a large bandwidth, it is necessary to use an "anti-reflection" design in the waveguides, which eliminate the conventional Bragg reflection as in Equation (4.28). This is accomplished by placing two different gratings on each side of the waveguide, similar to Section 4.5.7. In this case, the two gratings are phase shifted by 180° to completely cancel the Bragg back-reflections [40].

Finally, the multiple contra-directional couplers can be cascaded to implement a multi-channel WDM add-drop multiplexer, as shown in Figure 4.57. More details on these grating devices are described in References [39, 41].

4.6 Problems

4.1 Design a 1% power splitter using a directional coupler operating at 1550 nm.

4.2 Design a Mach–Zehnder interferometer with a free-spectral range of 10 nm operating at a 1310 nm wavelength with 400 nm strip waveguides.

4.3 Design a critically coupled all-pass ring resonator filter with a free-spectral range of 1 nm. In anticipation of designing a ring modulator, assume a waveguide propagation loss of 10 dB/cm (see Section 6.2.2).

4.4 Design an add-drop ring resonator filter with a free-spectral range of 1 nm and a quality factor of 5000. Assume a waveguide propagation loss of 10 dB/cm.

4.5 Design a uniform Bragg grating with a centre wavelength of 1550 nm (first nulls), bandwidth of 5 nm, and peak reflectivity of 99%. Design for both 193 nm and electron-beam lithography using strip waveguides with oxide cladding.

4.7 Code listings

Listing 4.5 Draw the directional coupler geometry, in Lumerical MODE Solutions; DC_wg_draw.lsf

```
# DC_wg_draw.lsf - draw the directional coupler waveguide geometry
new(1);

# define wafer and waveguide structure
thick_Clad = 2.0e-6;
thick_Si  = 0.22e-6;
thick_BOX = 2.0e-6;
#thick_Slab = 0;          # for strip waveguides
thick_Slab = 0.09e-6;         # for strip-loaded ridge waveguides
width_ridge = 0.5e-6;     # width of the waveguide
gap = 100e-9; # Directional coupler gap

# define materials
material_Clad = "SiO2 (Glass) - Palik";
# material_Clad = "H2O (Water) - Palik"; material_Clad = "Air (1)";
material_BOX = "SiO2 (Glass) - Palik";
material_Si  = "Si (Silicon) - Dispersive & Lossless";
materials;    # run script to add materials

# define simulation region
width_margin = 2.5e-6; # space to include on the side of the waveguide
height_margin = 0.5e-6; # space to include above and below the waveguide

# calculate simulation volume
# propagation in the x-axis direction; z-axis is wafer-normal
Xmin = -2e-6; Xmax = 2e-6; # length of the waveguide
Zmin = -height_margin; Zmax = thick_Si + height_margin;
Y_span = 2*width_margin + width_ridge; Ymin = -Y_span/2; Ymax = -Ymin;

# draw cladding
addrect; set("name","Clad"); set("material", material_Clad);
set("y", 0);              set("y span", Y_span+1e-6);
set("z min", 0);          set("z max", thick_Clad);
set("x min", Xmin);       set("x max", Xmax);
set("override mesh order from material database",1);
set("mesh order",3); # similar to "send to back", put the cladding as a background.
set("alpha", 0.05);

# draw buried oxide
addrect; set("name", "BOX"); set("material", material_BOX);
set("x min", Xmin);       set("x max", Xmax);
set("z min", -thick_BOX); set("z max", 0);
set("y", 0);              set("y span", Y_span+1e-6);
set("alpha", 0.05);

# draw silicon wafer
addrect; set("name", "Wafer"); set("material", material_Si);
set("x min", Xmin);       set("x max", Xmax);
set("z max", -thick_BOX); set("z min", -thick_BOX-2e-6);
set("y", 0);              set("y span", Y_span+1e-6);
set("alpha", 0.1);

# draw waveguide 1
addrect; set("name", "waveguide1"); set("material",material_Si);
set("y", -width_ridge/2-gap/2);     set("y span", width_ridge);
set("z min", 0);  set("z max", thick_Si);
set("x min", Xmin); set("x max", Xmax);
```

```
# draw waveguide 2
addrect; set("name", "waveguide2"); set("material",material_Si);
set("y", width_ridge/2+gap/2);    set("y span", width_ridge);
set("z min", 0);  set("z max", thick_Si);
set("x min", Xmin); set("x max", Xmax);

# draw slab for strip-loaded ridge waveguides
addrect; set("name", "slab"); set("material",material_Si);
if (thick_Slab==0) {
 set("y min", 0);   set("y max", 0);
} else {
 set("y", 0);       set("y span", Y_span+1e-6);
}
set("z min", 0);    set("z max", thick_Slab);
set("x min", Xmin); set("x max", Xmax);
set("alpha", 0.2);
```

Listing 4.6 Calculate the directional coupler's even and odd modes, in Lumerical MODE Solutions; DC_modes.lsf

```
# DC_modes.lsf - Calculate directional coupler's even and odd modes, Lumerical MODE
     Solutions

DC_wg_draw;

# define simulation parameters
wavelength = 1.55e-6;
meshsize = 10e-9; # maximum mesh size

# add 2D mode solver (waveguide cross-section)
addfde; set("solver type", "2D X normal");
set("x", 0);
set("y", 0);      set("y span", Y_span);
set("z max", Zmax); set("z min", Zmin);
set("wavelength", wavelength); set("solver type","2D X normal");
set("define y mesh by","maximum mesh step"); set("dy", meshsize);
set("define z mesh by","maximum mesh step"); set("dz", meshsize);
N_modes=2; # modes to output
set("number of trial modes",10);

gap=0.5e-6; switchtolayout;
setnamed("waveguide2","y", -width_ridge/2-gap/2);
setnamed("waveguide1","y", width_ridge/2+gap/2);
n=findmodes;
z=getdata("FDE::data::mode1","z"); y=getdata("FDE::data::mode1","y");
E3=pinch(getelectric("FDE::data::mode1")); image(y,z,E3);
Ey=pinch(getdata("FDE::data::mode1","Ey")); image(y,z,real(Ey));
index=pinch(getdata("FDE::data::material","index_x"));
matlabsave("DC_mode1",y,z,E3,Ey,index);
E3=pinch(getelectric("FDE::data::mode2")); image(y,z,E3);
Ey=pinch(getdata("FDE::data::mode2","Ey")); image(y,z,real(Ey));
matlabsave("DC_mode2",y,z,Ey,E3);
```

Listing 4.7 Calculate the directional coupler's cross-over length versus gap, in Lumerical MODE Solutions; DC_gap.lsf

```
# DC_gap.lsf - Calculate directional coupler's gap dependence, Lumerical MODE Solutions

gap_list=[.1:.1:1]*1e-6; # sweep waveguide width
neff = matrix (length(gap_list), N_modes );
L_cross= matrix (length(gap_list));

for(jj=1:length(gap_list)) {
  switchtolayout;
  setnamed("waveguide2","y", -width_ridge/2-gap_list(jj)/2);
  setnamed("waveguide1","y", width_ridge/2+gap_list(jj)/2);
  n=findmodes;
  for (m=1:N_modes) { # extract mode data
```

```
    neff (jj,m) =abs( getdata ("MODE::data::mode"+num2str(m),"neff") );
  }
  L_cross(jj) = wavelength / 2 / abs( neff (jj,1)-neff (jj,2));
}

plot (gap_list*1e9, L_cross*1e6, "Gap [nm]", "Cross-over length [micron]", "Cross-over
     length versus gap");
plot (gap_list*1e9, L_cross*1e6, "Gap [nm]", "Cross-over length [micron]", "Cross-over
     length versus gap","logy");
matlabsave ("DC_gap", L_cross,neff,gap_list);
```

Listing 4.8 Calculate the directional coupler's wavelength dependence, in Lumerical MODE Solutions; DC_wavelength.lsf

```
# DC_wavelength.lsf - Calculate directional coupler's wavelength dependence, Lumerical
     MODE Solutions

wavelength_start=1.5e-6;
wavelength_stop=1.6e-6;
wavelength_num=5;
gap=0.2e-6;

switchtolayout;
setnamed("waveguide2","y", -width_ridge/2-gap/2);
setnamed("waveguide1","y", width_ridge/2+gap/2);

setanalysis("wavelength",wavelength_start);
findmodes;
selectmode(1);
setanalysis("track selected mode",1);
setanalysis("number of test modes",3);
setanalysis("number of points",wavelength_num);
setanalysis("stop wavelength",wavelength_stop);
frequencysweep;
f=getdata("frequencysweep","f"); wavelengths=c/f;
neff1 = getdata("frequencysweep","neff");

selectmode(2);
setanalysis("track selected mode",2);
frequencysweep;
neff2= getdata("frequencysweep","neff");

plot(wavelengths*1e6,real(neff1),real(neff2), "Wavelength [micron]", "Effective Index","");
legend("Symmetric mode","Antisymmetric mode");
matlabsave ("DC_wavelength", neff1, neff2,wavelengths);
```

Listing 4.9 Load the directional coupler from a mask layout file (GDS), in Lumerical FDTD Solutions; DC_GDS_import.lsf

```
# DC_GDS_import.lsf - Script to import GDS for 3D FDTD simulations in Lumerical Solutions

newproject;
filename = "DC.gds";  cellname = "DC_0";

Material_Clad = "SiO2 (Glass) - Const";
Material_Ox = "SiO2 (Glass) - Const";
Material_Si  = "Si (Silicon) - Dispersive & Lossless";
materials;    # run script to add materials

Thickness_Si=0.22e-6; Etch2=130e-9;

FDTD_above=300e-9; # Extra simulation volume added
FDTD_below=300e-9;

minvxWAFER=1e9; minvyWAFER=1e9;
maxvxWAFER=-1e9; maxvyWAFER=-1e9; # design extent
maxvzWAFER=Thickness_Si;
```

```
# Waveguide   Si 220nm
n = gdsimport(filename, cellname, 1, Material_Si, 0, Thickness_Si);
if (n==0) { delete; } else {
  groupscope("::model::GDS_LAYER_1");
  set("script","");
  selectall;
  set('material', Material_Si);
  set('z span',Thickness_Si);  set('z',0);
  selectpartial("poly");
  minvx=1e9; minvy=1e9; maxvx=-1e9; maxvy=-1e9;
  for (i=1:getnumber) { # find the extent of this GDS layer.
    v=get("vertices",i);  a=size(v);
    minvx = min ( [minvx, min( v(1:a(1), 1 ))]);
        minvy = min ( [minvy, min( v(1:a(1), 2 ))]);
    maxvx = max ( [maxvx, max( v(1:a(1), 1 ))]);
    maxvy = max ( [maxvy, max( v(1:a(1), 2 ))]);
  }
  minvxWAFER = min ( [minvx, minvxWAFER]); # save design extent
  minvyWAFER = min ( [minvy-2.25e-6, minvyWAFER]);
  maxvxWAFER = max ( [maxvx, maxvxWAFER]);
  maxvyWAFER = max ( [maxvy+2.25e-6, maxvyWAFER]);
  groupscope("::model");
}

# Waveguide âĂŞ Rib Si partial etch 2 (130~\SI{}{\nano\meter} deep)
addrect; set("name", "Slab");
set("x min", minvxWAFER); set("y min", minvyWAFER);
set("x max", maxvxWAFER); set("y max", maxvyWAFER);
set("z min", 0); set("z max", Thickness_Si-Etch2);
set("material", Material_Si);
set("alpha",0.2);

addrect; set("name", "Oxide"); # Buried Oxide
set("x min", minvxWAFER); set("y min", minvyWAFER);
set("x max", maxvxWAFER); set("y max", maxvyWAFER);
set("z min", -2e-6); set("z max", 0);
set("material", Material_Ox); set("alpha",0.2);

addrect; set("name", "Cladding"); # Cladding
set("x min", minvxWAFER); set("y min", minvyWAFER);
set("x max", maxvxWAFER); set("y max", maxvyWAFER);
set("z min", 0); set("z max", 2.3e-6);
set("material", Material_Clad); set("alpha",0.1);
set("override mesh order from material database", 1);
set("mesh order", 4); # make the cladding the background

addfdtd; # FDTD simulation volume
set("x min", minvxWAFER+2e-6); set("y min", minvyWAFER+1.5e-6);
set("x max", maxvxWAFER-2e-6); set("y max", maxvyWAFER-1.5e-6);
set("z min", -FDTD_below); set("z max", maxvzWAFER+FDTD_above);
set("mesh accuracy", 3);
set("x min bc", "PML"); set("x max bc", "PML");
set("y min bc", "metal"); set("y max bc", "PML");
set("z min bc", "metal"); set("z max bc", "metal");

addmesh; # mesh override in the coupler gap.
set("x min", minvxWAFER+2e-6); set("y min", -100e-9);
set("x max", maxvxWAFER-2e-6); set("y max", 100e-9);
set("z min", 0);    set("z max", Thickness_Si);
set("override y mesh",1); set("override z mesh",0); set("override x mesh",0);
set("set equivalent index",1); set("equivalent y index",5);
```

Listing 4.10 Setup and perform 3D FDTD simulations for the directional coupler, in Lumerical FDTD Solutions; DC_FDTD_sweeps.lsf

```
# Perform 3D FDTD simulations for the directional coupler

DC_GDS_import;
```

```
DC_length_list=[0:1:25]*1e-6;
#DC_length_list=[5]*1e-6;
DC_length=0;

setglobalsource("wavelength start",1500e-9);
setglobalsource("wavelength stop",1600e-9);
setglobalmonitor("use source limits",0);
setglobalmonitor("frequency points",1);
setglobalmonitor("minimum wavelength",1550e-9);
setglobalmonitor("maximum wavelength",1550e-9);

# add mode source:
addmode; set("name", "source");
set("injection axis", "x-axis");
set("direction", "forward");
set("y", 1e-6); set("y span", 1.5e-6);
set("x", -5e-6 - DC_length/2);
set("z min", -FDTD_below); set("z max", maxvzWAFER+FDTD_above);
updatesourcemode;

addpower;
set("name", "through");
set("monitor type", "2D X-normal");
set("y", 1e-6); set("y span", 1.4e-6);
set("x", 5e-6 + DC_length/2);
set("z min", -FDTD_below); set("z max", maxvzWAFER+FDTD_above);

addpower;
set("name", "cross");
set("monitor type", "2D X-normal");
set("y", -1e-6); set("y span", 1.4e-6);
set("x", 5e-6 + DC_length/2);
set("z min", -FDTD_below); set("z max", maxvzWAFER+FDTD_above);


for (i=1:length(DC_length_list))
{
  switchtolayout;
  DC_length=DC_length_list(i);
  # stretch the coupler both to the left and right (keep symmetric at x=0)
  select("source"); set("x", -5e-6 - DC_length/2);
  select("through"); set("x", 5e-6 + DC_length/2);
  select("cross");  set("x", 5e-6 + DC_length/2);
  select("Oxide");
  set("x min", minvxWAFER-DC_length/2); set("x max", maxvxWAFER+ DC_length/2);
  select("Slab");
  set("x min", minvxWAFER-DC_length/2); set("x max", maxvxWAFER+ DC_length/2);
  select("Cladding");
  set("x min", minvxWAFER-DC_length/2); set("x max", maxvxWAFER+ DC_length/2);
  select("FDTD");
  set("x min", minvxWAFER+2e-6-DC_length/2); set("x max", maxvxWAFER-2e-6+DC_length/2);
  groupscope("GDS_LAYER_1"); selectall;
  set("x",-DC_length/2,1); set("x",-DC_length/2,2);
  set("x", DC_length/2,3); set("x", DC_length/2,4);
  groupscope("::model");
  select("wg1");
  if (getnumber==0) { addrect; set("name", "wg1");}
  set("x min", -DC_length/2); set("y min", 0.1e-6);
  set("x max", DC_length/2); set("y max", 0.6e-6);
  set("z min", 0); set("z max", Thickness_Si);
  set("material", Material_Si);
  select("wg2");
  if (getnumber==0) { addrect; set("name", "wg2");}
  set("x min", -DC_length/2); set("y max", -0.1e-6);
  set("x max", DC_length/2); set("y min", -0.6e-6);
  set("z min", 0); set("z max", Thickness_Si);
  set("material", Material_Si);

  save("DC_"+num2str(DC_length)+"_FDTD.fsp");
  run;
```

```
}

Tthrough=matrix(length(DC_length_list));
Tcross=matrix(length(DC_length_list));
for (i=1:length(DC_length_list))
{
  DC_length=DC_length_list(i);
  load("DC_"+num2str(DC_length)+"_FDTD.fsp");
  Tthrough(i)=transmission("through");
  Tcross(i)=transmission("cross");
}
plot(DC_length_list,[Tthrough,Tcross]);
matlabsave("DC_FDTD_mesh" +num2str(MESH_ACCURACY) +".mat", DC_length_list, Tthrough,
     Tcross);
```

Listing 4.11 Calculations for the coupling between dissimilar waveguides, in Lumerical MODE Solutions; DC_DeltaBeta.lsf

```
# Calculations for coupling between dissimilar waveguides, in Lumerical MODE

new(1); clear; cleardcard;
materials;

# Simulation Parameters
meshsizex = 0.02e-6;
meshsizey = 0.02e-6;
xrange = 5e-6;
yrange = 2.75e-6;
wavelength = 1.55e-6;

# Process Parameters
material_Clad = "SiO2 (Glass) - Palik";
material_BOX = "SiO2 (Glass) - Palik";
material_Si = "Si (Silicon) - Palik";

ridge_thick = 0.22e-6;
slab_thick = 0;
wg_width1 = 0.5e-6;
wg_width2 = 0.4e-6;
gap = 0.4e-6;

# Draw Cladding
addrect; set("name","Clad");
set("material",material_Clad);
set("y",0); set("y span",yrange+1);
set("x",0); set("x span",xrange+1);

#Draw Waveguides
addrect; set("name","WG1");
set("x min",-gap/2-wg_width1);set("x max",-gap/2);
set("y min",-ridge_thick/2);set("y max",ridge_thick/2);
set("material",material_Si);

addrect; set("name","WG2");
set("x min",gap/2);set("x max",gap/2+wg_width2);
set("y min",-ridge_thick/2);set("y max",ridge_thick/2);
set("material",material_Si);

#Mode Solver
addfde; set("solver type","2D Z Normal");
set("x",0); set("y",0); set("z",0);
set("x span",xrange); set("y span",yrange);
set("wavelength",wavelength);
set("define x mesh by","maximum mesh step");
set("define y mesh by","maximum mesh step");
set("dx",meshsizex); set("dy",meshsizey);
modes = 2;
set("number of trial modes",modes);
```

```
#Find Mode of Input Waveguide (isolated)
select("WG1"); set("enabled",1);
select("WG2"); set("enabled",0);
findmodes;
copydcard( "mode1", "modeA");
BetaA = 1e-6*(2*pi/wavelength)*real(getdata("mode1","neff"));

#Find Mode of 2nd Waveguide (isolated)
switchtolayout;
select("WG1"); set("enabled",0);
select("WG2"); set("enabled",1);
findmodes;
copydcard( "mode1", "modeB");
BetaB = 1e-6*(2*pi/wavelength)*real(getdata("mode1","neff"));

#Find Supermodes and Propagation Constants of dissimilar waveguide system
switchtolayout;
select("WG1"); set("enabled",1);
select("WG2"); set("enabled",1);
findmodes;
Beta1 = 1e-6*(2*pi/wavelength)*real(getdata("mode1","neff"));
Beta2 = 1e-6*(2*pi/wavelength)*real(getdata("mode2","neff"));

#Assume two waveguides are adiabaticaly brought together (or abrupt transition?)
#Perform Overlap Integrals
AB1A = overlap("mode1","modeA");
AB2A = overlap("mode2","modeA");

coeff1 = sqrt(AB1A(2))/sqrt((AB1A(2)+AB2A(2)));
coeff2 = sqrt(AB2A(2))/sqrt((AB1A(2)+AB2A(2)));

# Power In Each Waveguide vs. Distance, Eigenmode Expansion Method
L = ((2*pi)/abs((Beta2-Beta1)))*[0:0.001:10];
ones = matrix(length(L))+1;
P1 = ones*(abs(coeff1)^4 +
     abs(coeff2)^4)+2*abs(coeff1)^2*abs(coeff2)^2*cos((Beta2-Beta1)*L);
P2 = ones-P1;
plot(L,P1,P2,"Distance (microns)","Transmission");
legend("Waveguide A (EME)","Waveguide B (EME)");

##############################################
#Coupled-Mode Equation, dissimilar waveguides
C = abs(Beta2-Beta1)/2;
temp1=1 + ((BetaA-BetaB)/2/C)^2;
kappa2 = sin(C*L*sqrt(temp1))^2/temp1;
t2=1-kappa2;

plotxy(L,P1,L,P2,L,kappa2,L,t2, "Distance (microns)","Transmission");
legend("Waveguide A " +num2str(wg_width1*1e9)+" nm (EME)","Waveguide B "
     +num2str(wg_width2*1e9)+" nm (EME)","Waveguide A " +num2str(wg_width1*1e9)+" nm
     (CMT)","Waveguide B " +num2str(wg_width2*1e9)+" nm (CMT)");

##################################################
#Compare to coupling between identical waveguides:
switchtolayout;
select("WG2"); set("x max",gap/2+wg_width1);

#Find Supermodes and Propagation Constants
findmodes;
Beta1i = 1e-6*(2*pi/wavelength)*real(getdata("mode1","neff"));
Beta2i = 1e-6*(2*pi/wavelength)*real(getdata("mode2","neff"));
switchtolayout;

#Coupled-Mode Equation, dissimilar waveguides
C = abs(Beta2i-Beta1i)/2;
kappa2i = sin(C*L)^2;
t2i=1-kappa2i;
```

```
plot(L,P1,P2,kappa2,t2,kappa2i,t2i, "Distance (microns)","Transmission","Gap = " +
     num2str(gap*1e9)+" nm");
legend("Waveguide A " +num2str(wg_width1*1e9)+" nm (EME)","Waveguide B "
     +num2str(wg_width2*1e9)+" nm (EME)","Waveguide A " +num2str(wg_width1*1e9)+" nm
     (CMT)","Waveguide B " +num2str(wg_width2*1e9)+" nm (CMT)", "Waveguide A "
     +num2str(wg_width1*1e9)+" nm (CMT)","Waveguide B " +num2str(wg_width1*1e9)+" nm
     (CMT)");

matlabsave('DeltaBeta_gap'+num2str(gap*1e9)+'_wgB'+num2str(wg_width2*1e9));

# Fig 4.19:
plot(L,P1+1e-6,P2+1e-6,kappa2i+1e-6,t2i+1e-6, "Distance (microns)","Transmission","Gap = "
     + num2str(gap*1e9)+" nm","log10y");
legend("Waveguide A " +num2str(wg_width1*1e9)+" nm (EME)","Waveguide B "
     +num2str(wg_width2*1e9)+" nm (EME)", "Waveguide A " +num2str(wg_width1*1e9)+" nm
     (CMT)","Waveguide B " +num2str(wg_width1*1e9)+" nm (CMT)");
```

Listing 4.12 Simulations for the Y-branch, in Lumerical MODE (2.5D FDTD) or FDTD Solutions; YBranch_FDTD.lsf

```
# Script to import YBranch GDS into Lumerical MODE or FDTD and simulate

clear;
# perform four simulations:
# 1: splitter
# 2: combiner with one input
# 3: combiner with two inputs in phase
# 4: combiner with two inputs out of phase
for (r=1:4) {
  if (r==1) {
   SIM_DIRECTION = 1; # 1 = splitter, 2 = combiner
  } else {
   SIM_DIRECTION = 2; # 1 = splitter, 2 = combiner
  }
  if (r<3) {
   SOURCE2 = 0; # for combiner only. 0 = one source
  }
  if (r==3) {
   SOURCE2 = 1; # for combiner only. 1 = 2nd source in phase
  }
  if (r==4) {
   SOURCE2 = 2; # for combiner only. 2 = 2nd source pi phase
  }

  newproject;

  filename = "YBranch_Compact.gds";
  cellname = "y";

  save('YBranch');
  fileout=filebasename(currentfilename) + '_Dir' + num2str(SIM_DIRECTION) + '_Source2_' +
       num2str(SOURCE2);

  setglobalsource("wavelength start",1500e-9);
  setglobalsource("wavelength stop",1600e-9);
  setglobalmonitor("frequency points",100);

  # define materials
  Material_Clad = "SiO2 (Glass) - Palik";
  Material_Ox = "SiO2 (Glass) - Palik";
  Material_Si  = "Si (Silicon) - Dispersive & Lossless";
  materials;   # run script to add materials

  Thickness_Si=0.22e-6;

  FDTD_above=200e-9; # Extra simulation volume added
  FDTD_below=200e-9;

  minvxWAFER=1e9; minvyWAFER=1e9; maxvxWAFER=-1e9; maxvyWAFER=-1e9;
  maxvzWAFER=Thickness_Si;
```

```
n = gdsimport(filename, cellname, 1, Material_Si, 0, Thickness_Si);
if (n==0) { delete; } else {
 groupscope("::model::GDS_LAYER_1");
 set("script","");
 selectall;
 set('material', Material_Si);
 set('z span',Thickness_Si);
 set('z',0);
 selectpartial("poly");
 minvx=1e9; minvy=1e9; maxvx=-1e9; maxvy=-1e9;
 for (i=1:getnumber) { # find the extent of this GDS layer.
  v=get("vertices",i);
  a=size(v);
  minvx = min ( [minvx, min( v(1:a(1), 1 ))]);
  minvy = min ( [minvy, min( v(1:a(1), 2 ))]);
  maxvx = max ( [maxvx, max( v(1:a(1), 1 ))]);
  maxvy = max ( [maxvy, max( v(1:a(1), 2 ))]);
 }
 minvxWAFER = min ( [minvx, minvxWAFER]); # save the extent of overall design.
 minvyWAFER = min ( [minvy-2.25e-6, minvyWAFER]);
 maxvxWAFER = max ( [maxvx, maxvxWAFER]);
 maxvyWAFER = max ( [maxvy+2.25e-6, maxvyWAFER]);
 groupscope("::model");
}

# Oxide
addrect; set("name", "Oxide");
set("x min", minvxWAFER); set("y min", minvyWAFER);
set("x max", maxvxWAFER); set("y max", maxvyWAFER);
set("z min", -2e-6);
set("z max", 0);
set("material", Material_Ox);
set("alpha",0.2);

# Cladding
addrect; set("name", "Cladding");
set("x min", minvxWAFER); set("y min", minvyWAFER);
set("x max", maxvxWAFER); set("y max", maxvyWAFER);
set("z min", 0);
set("z max", 2.3e-6);
set("material", Material_Clad);
set("alpha",0.1);
set("override mesh order from material database", 1);
set("mesh order", 4); # make the cladding the background, i.e., "send to back".

if (fileextension(currentfilename) == "lms") {
 addpropagator;
 set("x min", minvxWAFER+0.5e-6); set("y min", minvyWAFER+1.5e-6);
 set("x max", maxvxWAFER-0.5e-6); set("y max", maxvyWAFER-1.5e-6);
 set("z min", -FDTD_below);
 set("z max", maxvzWAFER+FDTD_above);
 set("mesh accuracy", 3);
 set('x0',-get('x span')/2+0.1e-6);
}
 else {
 addfdtd;
 set("x min", minvxWAFER+0.5e-6); set("y min", minvyWAFER+1.5e-6);
 set("x max", maxvxWAFER-0.5e-6); set("y max", maxvyWAFER-1.5e-6);
 set("z min", -FDTD_below);
 set("z max", maxvzWAFER+FDTD_above);
 set("mesh accuracy", 2);
}

PointsX=get('mesh cells x');
PointsY=get('mesh cells y');
addmovie;
set('lock aspect ratio',1);
set('horizontal resolution',PointsX*2);
set('min sampling per cycle', 2);
```

```
set('name',fileout);

addmodeexpansion; set('name', 'expansion_v');
if (fileextension(currentfilename) == "fsp") {
 set('monitor type', '2D X-normal');
 set('y', 0); set("y span",1.5e-6);
 set("z min", -FDTD_below); set("z max", maxvzWAFER+FDTD_above);
 set('mode selection','fundamental TE mode');
} else {
 set('monitor type', 'Linear Y');
 set('y', 0); set("y span",1.5e-6);
 set('mode selection','fundamental mode');
}
set("x", minvxWAFER+0.6e-6);

if (SIM_DIRECTION==1) { # simulate splitter
 # add mode source:
 if (fileextension(currentfilename) == "fsp") {
  addmode;
  set("z min", -FDTD_below); set("z max", maxvzWAFER+FDTD_above);
 } else { addmodesource; }
 set("name", "source");
 set("injection axis", "x-axis");
 set("direction", "forward");
 set("y", 0e-6); set("y span", 1.5e-6);
 set("x", minvxWAFER+0.6e-6);
 updatesourcemode;

 addpower;
 set("name", "port1");
 set("monitor type", "2D X-normal");
 set("y", 2.75e-6); set("y span", 1.4e-6);
 set("x", maxvxWAFER-0.6e-6);
 if (fileextension(currentfilename) == "fsp") {
  set("z min", -FDTD_below); set("z max", maxvzWAFER+FDTD_above);
 }
 select('expansion_v');
 setexpansion('expansion_monitor','port1');

 addpower;
 set("name", "port2");
 set("monitor type", "2D X-normal");
 set("y", -2.75e-6); set("y span", 1.4e-6);
 set("x", maxvxWAFER-0.6e-6);
 if (fileextension(currentfilename) == "fsp") {
  set("z min", -FDTD_below); set("z max", maxvzWAFER+FDTD_above);
 }

}
else { # simulate the combiner
 # add mode source:
 if (fileextension(currentfilename) == "fsp") {
  addmode;
  set("z min", -FDTD_below); set("z max", maxvzWAFER+FDTD_above);
 } else { addmodesource; }
 set("name", "source1");
 set("injection axis", "x-axis");
 set("direction", "backward");
 set("y", 2.75e-6); set("y span", 1.4e-6);
 set("x", maxvxWAFER-0.6e-6);
 updatesourcemode;

 if (SOURCE2>0) {
  if (fileextension(currentfilename) == "fsp") {
   addmode;
   set("z min", -FDTD_below); set("z max", maxvzWAFER+FDTD_above);
  } else { addmodesource; }
  set("name", "source2");
  set("injection axis", "x-axis");
  set("direction", "backward");
```

```
    set("y", -2.75e-6); set("y span", 1.4e-6);
    set("x", maxvxWAFER-0.6e-6);
    updatesourcemode;
    if (SOURCE2==2) { # pi out of phase 2nd source for destructive interference
     set('phase',180);
    }
   }

   addpower;
   set("name", "port0");
   set("monitor type", "2D X-normal");
   set("y", 0e-6); set("y span", 1.5e-6);
   set("x", minvxWAFER+0.6e-6);
   if (fileextension(currentfilename) == "fsp") {
    set("z min", -FDTD_below); set("z max", maxvzWAFER+FDTD_above);
   }
   select('expansion_v');
   setexpansion('expansion_monitor','port0');
  }

  addpower; # surface power monitor
  set("name", "surface");
  set("monitor type", "2D Z-normal");

  run;

  # Insertion loss vs. wavelength
  Port=getresult("expansion_v","expansion for expansion_monitor");
  wavelengths=c/Port.f;
  T=Port.T_net;
  plot(wavelengths*1e6,10*log10(abs(T)),'Wavelength [micron]','Transmission [dB]');

  # Plot field profile in the device
  x=pinch(getdata('surface','x'));
  y=pinch(getdata('surface','y'));
  z=pinch(abs(getdata('surface','Ey')));
  z=pinch(z,3,50);
  image(x,y,z);

  matlabsave(fileout, x,y,z,wavelengths,T);
}
```

Listing 4.13 Ring resonator spectrum, MATLAB model, RingResonator.m

```
% RingResonator.m: Ring Resonator spectrum
% Usage, e.g.,
%   lambda = (1540:0.001:1550)*1e-6
%   [Ethru Edrop Qi Qc]=RingMod(lambda, 'add-drop', 10e-6, 0 );
%   plot (lambda, [abs(Ethru); abs(Edrop)])
% Wei Shi, UBC, 2012, weis@ece.ubc.ca

function [Ethru, Edrop, Qi, Qc]=RingResonator(lambda, Filter_type, r, Lc)
% lambda: wavelength (can be a 1D array) in meters
% type: "all-pass" or "add-drop"
% r: radius
% Lc: coupler length
%
k=0.2; t=sqrt(1-k^2);      %coupling coefficients

neff = neff_lambda(lambda);
if lambda(1)==lambda(end)
  ng=neff - lambda(1) * (neff-neff_lambda(lambda(1)+0.1e-9)/0.1e-9);
else
  ng = neff - mean(lambda) * diff(neff)./diff(lambda); % for Q calculations
  ng = [ng(1) ng];
end

alpha_wg_dB=10;  % optical loss of optical waveguide, in dB/cm
alpha_wg=-log(10^(-alpha_wg_dB/10));% converted to /cm
L_rt=Lc*2+2*pi*r;
```

```
phi_rt=(2*pi./lambda).*neff*L_rt;
A=exp(-alpha_wg*100*L_rt); % round-trip optical power attenuation
alpha_av=-log(A)/L_rt;     % average loss of the cavity
Qi=2*pi*ng./lambda/alpha_av; % intrinsic quality factor

if (Filter_type=='all-pass')
  Ethru=(-sqrt(A)+t*exp(-1i*phi_rt)) ./ (-sqrt(A)*conj(t)+exp(-1i*phi_rt));
  Edrop=zeros(1,length(lambda));
  Qc=-(pi*L_rt*ng)./(lambda*log(abs(t)));
elseif (Filter_type=='add-drop') % symmetrically coupled
  Ethru=(t-conj(t)*sqrt(A)*exp(1i*phi_rt)) ./ (1-sqrt(A)*conj(t)^2*exp(1i*phi_rt));
  Edrop=-conj(k)*k*sqrt(sqrt(A)*exp(1i*phi_rt)) ./ (1-sqrt(A)*conj(t)^2*exp(1i*phi_rt));
  Qc=-(pi*L_rt*ng)./(lambda*log(abs(t)))/2;
else
  error(1, 'The''Filter_type'' has to be ''all-pass'' or ''add-drop''.\n');
end

function [neff]=neff_lambda(lambda)
neff = 2.57 - 0.85*(lambda*1e6-1.55);
```

Listing 4.14 Calculate the wavelength and width dependence of waveguide's effective index, Lumerical MODE Solutions

```
# wg_2D_neff_sweep_wavelength_width.lsf - Calculate the wavelength and width dependence of
     waveguide's neff

wg_2D; # draw waveguide

# define parameters to sweep
width_ridge_list=[.4:.05:.61]*1e-6; # sweep waveguide width
Nf=10; # number of wavelength points

neff = matrix (length(width_ridge_list), Nf );
ng = matrix (length(width_ridge_list), Nf );
for(ii=1:length(width_ridge_list)) {
  switchtolayout;
  setnamed("waveguide","y span", width_ridge_list(ii));

  run; mesh;
  setanalysis('wavelength',1.6e-6);
  findmodes; selectmode(1); # find the fundamental mode

  setanalysis("track selected mode",1);
  setanalysis("number of test modes",5);
  setanalysis("number of points",Nf);
  setanalysis("detailed dispersion calculation",1);
  setanalysis('stop wavelength',1.5e-6);
  frequencysweep;  # perform sweep of wavelength and plot
  f=getdata("frequencysweep","f");
  neff1=getdata("frequencysweep","neff");
  ng1=c/getdata("frequencysweep","vg");
  wavelengths=c/f;
  for (m=1:Nf) { # extract mode data
    neff (ii,m) = abs( neff1(m) );
    ng (ii,m) = abs( ng1(m) );
  }
}
matlabsave ('wg_2D_neff_sweep_wavelength_width.mat', f, neff, ng, wavelengths,
     width_ridge_list);
```

Listing 4.15 Transfer Matrix Method – define the grating based on physical parameters, and plot the spectra.

```
function Grating
%This file is used to plot the reflection/transmission spectrum.

% Grating Parameters
Period=310e-9; % Bragg period
NG=200;    % Number of grating periods
```

```
L=NG*Period;  % Grating length
width0=0.5;   % mean waveguide width
dwidth=0.01;  % +/- waveguide width
width1=width0 - dwidth;
width2=width0 + dwidth;
loss_dBcm=3;  % waveguide loss, dB/cm
loss=log(10)*loss_dBcm/10*100;
% Simulation Parameters:
span=30e-9; % Set the wavelength span for the simultion
Npoints = 10000;

% from MODE calculations
switch 1
    case 1 % Strip waveguide; 500x220 nm
        neff_wavelength = @(w) 2.4379 - 1.1193 * (w*1e6-1.554) - 0.0350 * (w*1e6-1.554).^2;
            % 500x220 oxide strip waveguide
        dneff_width = @(w) 10.4285*(w-0.5).^3 - 5.2487*(w-0.5).^2 + 1.6142*(w-0.5);
end

% Find Bragg wavelength using lambda_Bragg = Period * 2neff(lambda_bragg);
% Assume neff is for the average waveguide width.
f = @(lambda) lambda - Period*2*
      (neff_wavelength(lambda)+(dneff_width(width2)+dneff_width(width1))/2);
wavelength0 = fzero(f,1550e-9);

wavelengths=wavelength0 + linspace(-span/2, span/2, Npoints);
n1=neff_wavelength(wavelengths)+dneff_width(width1); % low index
n2=neff_wavelength(wavelengths)+dneff_width(width2); % high index

[R,T]=TMM_Grating_RT(wavelengths, Period, NG, n1, n2, loss)
figure;
plot (wavelengths*1e6,[R, T],'LineWidth',3); hold all
plot ([wavelength0, wavelength0]*1e6, [0,1],'--'); % calculated bragg wavelength
xlabel('Wavelength [\mum]')
ylabel('Response');
axis tight;
%printfig ('PS-WBG')

function printfig (pdf)
FONTSIZE=20;
set ( get(gca, 'XLabel'),'FontSize',FONTSIZE)
set ( get(gca, 'YLabel'),'FontSize',FONTSIZE)
set (gca, 'FontSize',FONTSIZE); box on;
print ('-dpdf', pdf); system([ 'pdfcrop ' pdf ' ' pdf '.pdf' ]);
```

Listing 4.16 Simulations for the Bragg grating, in Lumerical FDTD Solutions (3D simulation); Bragg_FDTD.lsf

```
##############################################
# script file: Bragg_FDTD.lsf
#
# Create and simulate a basic Bragg grating
# Copyright 2014 Lumerical Solutions
##############################################

# DESIGN PARAMETERS
##############################################
thick_Si = 0.22e-6;
thick_BOX = 2e-6;
width_ridge = 0.5e-6; # Waveguide width
Delta_W = 50e-9; # Corrugation width
L_pd = 324e-9; # Grating period
N_gt = 280;     # Number of grating periods
L_gt = N_gt*L_pd;# Grating length
W_ox = 3e-6; L_ex = 5e-6; # simulation size margins
L_total = L_gt+2*L_ex;
material_Si = 'Si (Silicon) - Dispersive & Lossless';
material_BOX = 'SiO2 (Glass) - Const';
# Constant index materials lead to more stable simulations
```

```
# DRAW
################################################
newproject; switchtolayout;
materials;

# Oxide Substrate
addrect;
set('x min',-L_ex); set('x max',L_gt+L_ex);
set('y',0e-6); set('y span',W_ox);
set('z min',-thick_BOX); set('z max',-thick_Si/2);
set('material',material_BOX);
set('name','oxide');

# Input Waveguide
addrect;
set('x min',-L_ex); set('x max',0);
set('y',0);  set('y span',width_ridge);
set('z',0);  set('z span',thick_Si);
set('material',material_Si);
set('name','input_wg');

# Bragg Gratings
addrect;
set('x min',0); set('x max',L_pd/2);
set('y',0);  set('y span',width_ridge+Delta_W);
set('z',0);  set('z span',thick_Si);
set('material',material_Si);
set('name','grt_big');
addrect;
set('x min',L_pd/2); set('x max',L_pd);
set('y',0);  set('y span',width_ridge-Delta_W);
set('z',0);  set('z span',thick_Si);
set('material',material_Si);
set('name','grt_small');
selectpartial('grt');
addtogroup('grt_cell');
select('grt_cell');
redrawoff;
for (i=1:N_gt-1) {
 copy(L_pd);
}
selectpartial('grt_cell');
addtogroup('bragg');
redrawon;

# Output WG
addrect;
set('x min',L_gt); set('x max',L_gt+L_ex);
set('y',0);  set('y span',width_ridge);
set('z',0);  set('z span',thick_Si);
set('material',material_Si);
set('name','output_wg');


# SIMULATION SETUP
################################################
lambda_min = 1.5e-6;
lambda_max = 1.6e-6;
freq_points = 101;
sim_time = 6000e-15;
Mesh_level = 2;
mesh_override_dx = 40.5e-9; # needs to be an integer multiple of the period
mesh_override_dy = 50e-9;
mesh_override_dz = 20e-9;

# FDTD
addfdtd;
set('dimension','3D');
set('simulation time',sim_time);
set('x min',-L_ex+1e-6); set('x max',L_gt+L_ex-1e-6);
```

```
set('y', 0e-6);  set('y span',2e-6);
set('z',0);   set('z span',1.8e-6);
set('mesh accuracy',Mesh_level);
set('x min bc','PML');  set('x max bc','PML');
set('y min bc','PML');  set('y max bc','PML');
set('z min bc','PML');  set('z max bc','PML');

#add symmetry planes to reduce the simulation time
#set('y min bc','Anti-Symmetric'); set('force symmetric y mesh', 1);

# Mesh Override
if (1){
addmesh;
set('x min',0e-6);  set('x max',L_gt);
set('y',0); set('y span',width_ridge+Delta_W);
set('z',0); set('z span',thick_Si+2*mesh_override_dz);
set('dx',mesh_override_dx);
set('dy',mesh_override_dy);
set('dz',mesh_override_dz);
}

# MODE Source
addmode;
set('injection axis','x-axis');
set('mode selection','fundamental mode');
set('x',-2e-6);
set('y',0); set('y span',2.5e-6);
set('z',0); set('z span',2e-6);
set('wavelength start',lambda_min);
set('wavelength stop',lambda_max);

# Time Monitors
addtime;
set('name','tmonitor_r');
set('monitor type','point');
set('x',-3e-6); set('y',0); set('z',0);
addtime;
set('name','tmonitor_m');
set('monitor type','point');
set('x',L_gt/2); set('y',0); set('z',0);
addtime;
set('name','tmonitor_t');
set('monitor type','point');
set('x',L_gt+3e-6); set('y',0); set('z',0);

# Frequency Monitors
addpower;
set('name','t');
set('monitor type','2D X-normal');
set('x',L_gt+2.5e-6);
set('y',0); set('y span',2.5e-6);
set('z',0); set('z span',2e-6);
set('override global monitor settings',1);
set('use source limits',1);
set('use linear wavelength spacing',1);
set('frequency points',freq_points);

addpower;
set('name','r');
set('monitor type','2D X-normal');
set('x',-2.5e-6);
set('y',0); set('y span',2.5e-6);
set('z',0); set('z span',2e-6);
set('override global monitor settings',1);
set('use source limits',1);
set('use linear wavelength spacing',1);
set('frequency points',freq_points);

#Top-view electric field profile
if (0) {addprofile;
```

```
set('name','field');
set('monitor type','2D Z-normal');
set('x min',-2e-6); set('x max',L_gt+2e-6);
set('y', 0); set('y span',1.2e-6);
set('z', 0);
set('override global monitor settings',1);
set('use source limits',1);
set('use linear wavelength spacing',1);
set('frequency points',21);
}

# SAVE AND RUN
################################################
save('Bragg_FDTD');
run;

# Analysis
################################################
transmission_sim=transmission('t');
reflection_sim=transmission('r');
wavelength_sim=3e8/getdata('t','f');
plot(wavelength_sim*1e9, 10*log10(transmission_sim), 10*log10(abs(reflection_sim)),
     'wavelength (nm)', 'response');
legend('T','R');
matlabsave('Bragg_FDTD');
```

References

[1] Tom Baehr-Jones, Ran Ding, Ali Ayazi, *et al.* "A 25 Gb/s silicon photonics platform". *arXiv:1203.0767v1* (2012) (cit. on pp. 93, 101).

[2] Amnon Yariv. "Coupled-mode theory for guided-wave optics". *IEEE Journal of Quantum Electronics* **9**.9 (1973), pp. 919–933 (cit. on pp. 92, 108).

[3] Amnon Yariv and Pochi Yeh. *Photonics: Optical Electronics in Modern Communications (The Oxford Series in Electrical and Computer Engineering)*. Oxford University Press, Inc., 2006 (cit. on p. 92).

[4] N. Rouger, L. Chrostowski, and R. Vafaei. "Temperature effects on silicon-on-insulator (SOI) racetrack resonators: a coupled analytic and 2-D finite difference approach". *Journal of Lightwave Technology* **28**.9 (2010), pp. 1380–1391. DOI: 10.1109/JLT.2010.2041528 (cit. on p. 95).

[5] Fengnian Xia, Lidija Sekaric, and Yurii A. Vlasov. "Mode conversion losses in silicon-on-insulator photonic wire based racetrack resonators". *Optics Express* **14**.9 (2006), pp. 3872–3886 (cit. on p. 102).

[6] Valentina Donzella, Sahba Talebi Fard, and Lukas Chrostowski. "Study of waveguide crosstalk in silicon photonics integrated circuits". *Proc. SPIE 8915, Photonics North 2013* (2013), 89150Z (cit. on p. 108).

[7] Herwig Kogelnik and R V Schmidt. "Switched directional couplers with alternating $\Delta\beta$". *IEEE Journal of Quantum Electronics* **12**.7 (1976), pp. 396–401 (cit. on p. 108).

[8] R. V. Schmidt and R. Alferness. "Directional coupler switches, modulators, and filters using alternating $\Delta\beta$ techniques". *IEEE Transactions on Circuits and Systems* **26**.12 (1979), pp. 1099–1108 (cit. on p. 108).

[9] Yi Zhang, Shuyu Yang, Andy Eu-Jin Lim, Guo-Qiang Lo, Christophe Gal-land, Tom Baehr-Jones, and Michael Hochberg. "A compact and low loss Y-junction for submicron silicon waveguide". *Optics Express* 21.1 (2013), pp. 1310–1316 (cit. on pp. 111, 112).

[10] W. Bogaerts, P. De Heyn, T. Van Vaerenbergh, *et al.* "Silicon microring resonators". *Laser & Photonics Reviews* (2012) (cit. on p. 115).

[11] A. Mekis, S. Gloeckner, G. Masini, *et al.* "A grating-coupler-enabled CMOS photonics platform". *IEEE Journal of Selected Topics in Quantum Electronics* **17**.3 (2011), pp. 597–608. DOI: 10.1109/JSTQE.2010.2086049 (cit. on p. 117).

[12] *Peak Finding and Measurement.* [Accessed 2014/04/14]. URL: http://terpconnect.umd.edu/~toh/spectrum/PeakFindingandMeasurement.htm (cit. on p. 117).

[13] Jens Buus, Markus-Christian Amann, and Daniel J. Blumenthal. *Tunable Laser Diodes and Related Optical Sources, 2nd Edn.* John Wiley & Sons, Inc., 2005 (cit. on pp. 119, 120).

[14] Xu Wang, Wei Shi, Michael Hochberg, *et al.* "Lithography simulation for the fabrication of silicon photonic devices with deep-ultraviolet lithography". *IEEE International Conference on Group IV Photon.* 2012, ThP 17 (cit. on pp. 120, 128, 134).

[15] L. A. Coldren, S. W. Corzine, and M. L. Mashanovitch. *Diode Lasers and Photonic Integrated Circuits.* Wiley Series in Microwave and Optical Engineering. John Wiley & Sons, 2012. ISBN: 9781118148181 (cit. on p. 121).

[16] Thomas Edward Murphy, Jeffrey Todd Hastings, and Henry I. Smith. "Fabrication and characterization of narrow-band Bragg-reflection filters in silicon-on-insulator ridge waveguides". *Journal of Lightwave Technology* **19**.12 (2001), pp. 1938–1942 (cit. on p. 126).

[17] Ivano Giuntoni, David Stolarek, Harald Richter, *et al.* "Deep-UV technology for the fabrication of Bragg gratings on SOI rib waveguides". *IEEE Photonics Technology Letters* **21**.24 (2009), pp. 1894–1896 (cit. on p. 126).

[18] Ivano Giuntoni, Andrzej Gajda, Michael Krause, *et al.* "Tunable Bragg reflectors on silicon-on-insulator rib waveguides". *Optics Express* **17**.21 (2009), pp. 18 518–18 524 (cit. on p. 126).

[19] J. T. Hastings, Michael H. Lim, J. G. Goodberlet, and Henry I. Smith. "Optical waveguides with apodized sidewall gratings via spatial-phase-locked electron-beam lithography". *Journal of Vacuum Science & Technology B* **20**.6 (2002), pp. 2753–2757 (cit. on pp. 126, 129).

[20] Guomin Jiang, Ruiyi Chen, Qiang Zhou, *et al.* "Slab-modulated sidewall Bragg gratings in silicon-on-insulator ridge waveguides". *IEEE Photonics Technology Letters* **23**.1 (2011), pp. 6–9 (cit. on pp. 126, 129).

[21] Renzo Loiacono, Graham T. Reed, Goran Z. Mashanovich, *et al.* "Laser erasable implanted gratings for integrated silicon photonics". *Optics Express* **19**.11 (2011), pp. 10728–10734. DOI: 10.1364/OE.19.010728 (cit. on p. 126).

[22] Xu Wang. "Silicon photonic waveguide Bragg gratings". PhD thesis. University of British Columbia, 2013 (cit. on pp. 128, 129, 131, 135, 136, 138, 139, 140, 141, 142).

[23] D. T. H. Tan, K. Ikeda, R. E. Saperstein, B. Slutsky, and Y. Fainman. "Chip-scale dispersion engineering using chirped vertical gratings". *Optics Letters* **33**.24 (2008), pp. 3013–3015 (cit. on p. 127).

[24] A. S. Jugessur, J. Dou, J. S. Aitchison, R. M. De La Rue, and M. Gnan. "A photonic nano-Bragg grating device integrated with microfluidic channels for bio-sensing applications". *Microelectronic Engineering* **86**.4-6 (2009), pp. 1488–1490 (cit. on p. 127).

[25] Xu Wang, Wei Shi, Raha Vafaei, Nicolas A. F. Jaeger, and Lukas Chrostowski. "Uniform and sampled Bragg gratings in SOI strip waveguides with sidewall corrugations". *IEEE Photonics Technology Letters* **23**.5 (2011), pp. 290–292 (cit. on pp. 127, 137).

[26] D. T. H. Tan, K. Ikeda, and Y. Fainman. "Cladding-modulated Bragg gratings in silicon waveguides". *Optics Letters* **34**.9 (2009), pp. 1357–1359 (cit. on p. 127).

[27] Xu Wang, Wei Shi, Han Yun, *et al.* "Narrow-band waveguide Bragg gratings on SOI wafers with CMOS-compatible fabrication process". *Optics Express* **20**.14 (2012), pp. 15 547–15 558. DOI: 10.1364/OE.20.015547 (cit. on p. 129).

[28] R. J. Bojko, J. Li, L. He, *et al.* "Electron beam lithography writing strategies for low loss, high confinement silicon optical waveguides". *Journal of Vacuum Science & Technology B: Microelectronics and Nanometer Structures* **29**.6 (2011), 06F309–06F309 (cit. on p. 131).

[29] X. Wang, W. Shi, R. Vafaei, N. A. F. Jaeger, and L. Chrostowski. "Uniform and sampled Bragg gratings in SOI strip waveguides with sidewall corrugations". *IEEE Photonics Technology Letters* **23**.5 (2010), pp. 290–292 (cit. on p. 134).

[30] W. Bogaerts, P. Bradt, L. Vanholme, P. Bienstman, and R. Baets. "Closed-loop modeling of silicon nanophotonics from design to fabrication and back again". *Optical and Quantum Electronics* **40**.11 (2008), pp. 801–811 (cit. on p. 134).

[31] *Calibre Computational Lithography – Mentor Graphics.* [Accessed 2014/04/14]. URL: http://www.mentor.com/products/ic-manufacturing/computational-lithography (cit. on p. 134).

[32] S. K. Selvaraja, P. Jaenen, W. Bogaerts, *et al.* "Fabrication of photonic wire and crystal circuits in silicon-on-insulator using 193-nm optical lithography". *Journal of Lightwave Technology* **27**.18 (2009), pp. 4076–4083 (cit. on p. 135).

[33] Alexandre D. Simard, Guillaume Beaudin, Vincent Aimez, Yves Painchaud, and Sophie LaRochelle. "Characterization and reduction of spectral distortions in silicon-on-insulator integrated Bragg gratings". *Optics Express* **21**.20 (2013), pp. 23 145–23 159 (cit. on p. 137).

[34] Steve Zamek, Dawn T. H. Tan, Mercedeh Khajavikhan, *et al.* "Compact chip-scale filter based on curved waveguide Bragg gratings". *Optics Letters* **35**.20 (2010), pp. 3477–3479. DOI: 10.1364/OL.35.003477 (cit. on p. 137).

[35] Alexandre D. Simard, Yves Painchaud, and Sophie LaRochelle. "Integrated Bragg gratings in spiral waveguides". *Optics Express* **21**.7 (2013), pp. 8953–8963. DOI: 10.1364/OE.21.008953 (cit. on p. 137).

[36] Xu Wang, Han Yun, and Lukas Chrostowski. "Integrated Bragg gratings in spiral waveguides". In *Conference on Lasers and Electro-Optics, San Jose, CA, paper CTh4F.8* (2013) (cit. on p. 137).

[37] Xu Wang, Samantha Grist, Jonas Flueckiger, Nicolas A. F. Jaeger, and Lukas Chrostowski. "Silicon photonic slot waveguide Bragg gratings and resonators". *Optics Express* **21** (2013), pp. 19 029–19 039 (cit. on p. 139).

[38] Y. Painchaud, M. Poulin, C. Latrasse, and M. Picard. "Bragg grating based Fabry–Perot filters for characterizing silicon-on-insulator waveguides". *Group IV Photonics (GFP).* IEEE. 2012, pp. 180–182 (cit. on p. 140).

[39] Wei Shi, Venkat Veerasubramanian, David V. Plant, Nicolas A. F. Jaeger, and Lukas Chrostowski. "Silicon photonic Bragg-grating couplers for optical communications". *Proc. SPIE.* 2014 (cit. on pp. 142, 143).

[40] Wei Shi, Han Yun, Charlie Lin, *et al.* "Ultra-compact, flat-top demultiplexer using anti-reflection contradirectional couplers for CWDM networks on silicon". *Optics Express* **21**.6 (2013), pp. 6733–6738 (cit. on p. 143).

[41] Lukas Chrostowski and Krzysztof Iniewski. *High-speed Photonics Interconnects, Chapter 3 Silicon Photonic Bragg Gratings.* Vol. 13. CRC Press, 2013 (cit. on p. 143).

5 Optical I/O

In this chapter, we describe the design of these two types of optical input/output coupling techniques: fibre grating couplers in Section 5.2, and edge couplers in Section 5.3. A method of creating a mask layout for a focusing grating coupler is presented. Methods for polarization management are also discussed in Section 5.4.

5.1 The challenge of optical coupling to silicon photonic chips

Owing to the large refractive index contrast between the silicon core (n = 3.47 at 1550 nm) and the silicon dioxide cladding (n = 1.444 at 1550 nm), propagation modes are highly confined within the waveguide with dimensions on the order of a few hundred nanometers (see Sections 3.1 and 3.2). Although a benefit for large-scale integration, the small feature size of the waveguide raises the problem of a huge mismatch between the optical mode within an optical fibre and the mode within the waveguide. The cross-sectional area of an optical fibre core (with a diameter of 9 μm) is almost 600 times larger than that of a silicon waveguide (with dimensions of 500 nm × 220 nm), hence requires components that adjust the mode-field diameter accordingly.

Several approaches have been demonstrated to tackle the problem of the aforementioned mode mismatch. Edge coupling using spot-size converters and lensed fibres is one solution used to address this, and high-efficiency coupling with an insertion loss below 0.5 dB has been demonstrated [1]. In addition, both TE and TM polarizations can be efficiently coupled. However, this approach can only be used at the edge of the chips, and the implementation of such designs requires complicated post-processes and high-resolution optical alignment, which increase the packaging cost.

Grating couplers are an alternative solution to tackle the issue of mode mismatch. Compared to the edge coupling, grating couplers have several advantages: alignment to grating couplers during measurement is much easier than alignment to edge couplers; the fabrication of grating couplers does not require post-processing, which reduces the fabrication cost; grating couplers can be put anywhere on a chip, which provides flexibility in the design as well as enabling wafer-scale automated testing. Both academic and industrial research groups have demonstrated high efficiency grating couplers; below 1 dB is possible [2, 3]. Polarization splitting grating couplers have also been demonstrated [4].

5.2 Grating coupler

What is grating coupler?

A grating coupler is a periodic structure that can diffract light from propagation in the waveguide (in plane) to free-space (out of plane). It is normally used as an I/O device to couple light between fibre (or free-space) and sub-micrometer SOI waveguides. Figure 5.1 is a cross-section diagram of a shallow-etched grating coupler design in a silicon-on-insulator wafer. The thickness of the functional Si layer and the thickness of the buried oxide (BOX) layer are determined by the wafer type (Section 3.1). A cladding layer is generally employed to protect the functional silicon layer and to enable the fabrication of multi-layer electrical interconnects. In some applications (e.g. evanescent field sensors), the cladding is air or liquid.

Figure 5.1 describes the design parameters.

- The coupler consists of a silicon waveguide core, a top cladding (oxide or air), a bottom cladding (buried oxide, BOX), and a substrate (silicon). The effective index of the slab waveguide is n_{eff};
- Λ is the period of the grating;
- W is the width of the grating teeth (assuming uniform grating);
- ff is the fill factor (or duty cycle), and is defined as ff $= W/\Lambda$;
- ed is the etch depth of the grating; and
- θ_c is the angle between surface normal and the propagation direction of the diffracted light, in the cladding.
- θ_{air} is the angle between surface normal and the propagation direction of the diffracted light, in the air.
- θ_{fibre} is the angle between surface normal and the propagation direction of the diffracted light, in fibre. This corresponds to the fibre polish angle.

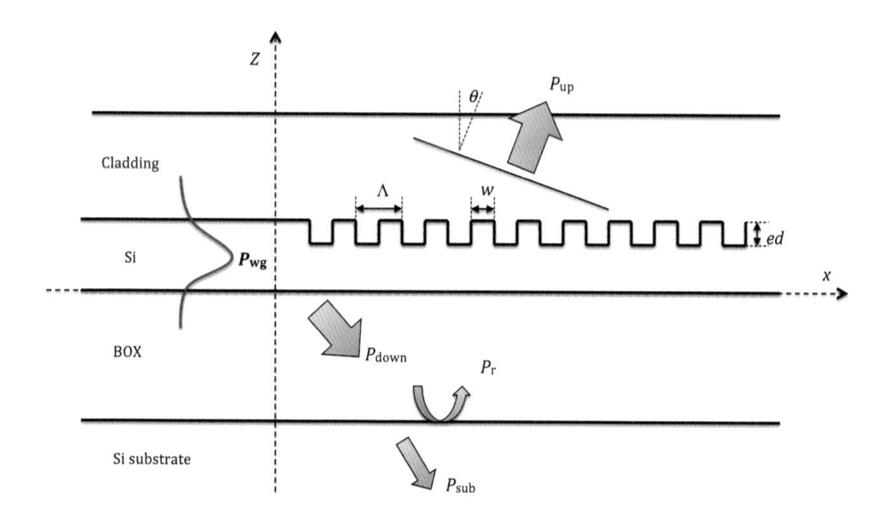

Figure 5.1 Schematic cross-section diagram of a grating coupler.

In consideration of the output grating coupler, P_{wg} is the optical power of the input light; P_{up} and P_{down} indicate the power that goes up and the power that penetrates down into the wafer. Not shown is the optical fibre, and the power that is coupled into its fundamental mode, P_{fibre}.

The performance of the grating coupler can be described by the following parameters.

(1) Directionality: the ratio between the power diffracted upwards (P_{up}) and the input power from the waveguide (P_{wg}), which is usually expressed in decibels (dB) as $10\log_{10}\left(P_{up}/P_{wg}\right)$.

(2) Insertion loss (coupling efficiency): the ratio between the power coupled into the fundamental mode of the fibre (P_{fibre}) and the input power from the waveguide (P_{wg}). It is usually expressed in decibels (dB), and insertion loss can be expressed as $IL = 10\log_{10}\left(P_{fibre}/P_{wg}\right)$.

(3) Penetration loss: the ratio between the power lost in the substrate (P_{sub}) and the input power from the waveguide, which is $10\log_{10}\left(P_{down}/P_{wg}\right)$.

(4) Reflection to the waveguide: owing to the refractive index contrast between the silicon wire waveguide and the grating, part of the input light from the waveguide will be reflected back into the waveguide. The ratio between the reflected power and the input power from the waveguide is called back reflection to the waveguide, or optical return loss. It is usually expressed in dB, as $10\log_{10}\left(P_{back\text{-}wg}/P_{wg}\right)$. This back reflection is unwanted because it will cause Fabry–Perot oscillations by reflecting back and forth between the input and output grating couplers [5, 6]; typically suppression in the range of 20–30 dB is desired.

(5) Reflection to the fibre: for the input grating couplers, part of the input light will be reflected back to the fibre. It is also referred to as optical return loss (dB), and calculated by $10\log_{10}\left(P_{back\text{-}fibre}/P_{in\text{-}fibre}\right)$ to measure it. This reflection is unwanted because it may affect the stability of light source.

(6) 1 dB or 3 dB bandwidth: the wavelength range at which insertion loss is 1 dB or 3 dB lower than the peak coupling efficiency.

5.2.1 Performance

There has been considerable effort by the international community to improve the efficiency of grating couplers [1–3, 7–12]. There are three main factors that contribute to a reduced efficiency of a grating coupler: penetration loss, mode mismatch, and back reflection.

(1) For a shallow-etch process, there is about 30% of energy lost in the substrate; and the penetration loss can be more than 50% in a full-etch process. This can be improved by using reflectors imbedded in the substrate [3, 12]. These can be either a metal layer or a multilayer distributed Bragg reflector film.

(2) In addition, another 10% energy loss results from the mode mismatch between grating coupler and fibre; this can be improved by apodizing or chirping the grating [8, 9, 11].

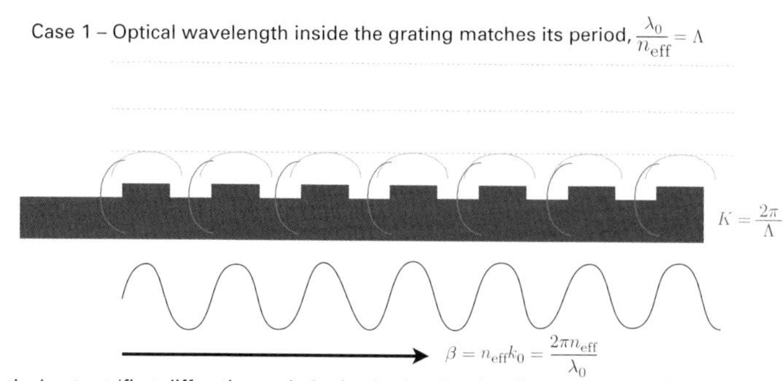

(a) Vertical output (first diffraction order) and back-reflection (second diffraction order)

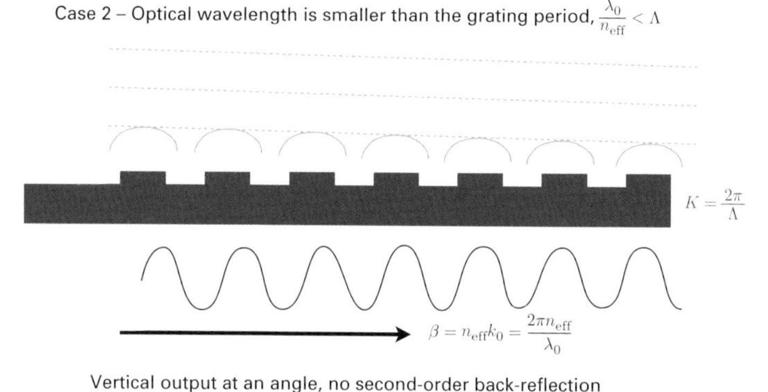

(b) Vertical output at a detuned angle, no second-order back-reflection

Figure 5.2 Diagram illustrating the concept of output grating coupler. Light is incident in the waveguide from the left, with output upwards (in this case, in air).

(3) For well-designed shallow-etched grating couplers, the back-reflections are small (e.g. −30 dB), so this does not represent a significant source of loss. However, it can be as high as 30% in a full-etch process. It is also necessary to couple the light under a small angle to eliminate the first-order Bragg reflection (termed "detuning grating", Figure 5.2b); hence grating couplers seldom are designed for perfectly vertical operation (Figure 5.2a).

5.2.2 Theory

Figure 5.2 illustrates the grating operation. It can be understood in terms of the Huygens–Fresnel principle, namely the constructive and destructive interference arising from the wavefronts created by the diffraction of light from the grating teeth.

Figure 5.2a shows the case when the optical wavelength inside the grating matches its period. In this case, the first-order diffraction will propagate vertically (indicated by the green line) and the second-order diffraction will propagate back to the waveguide

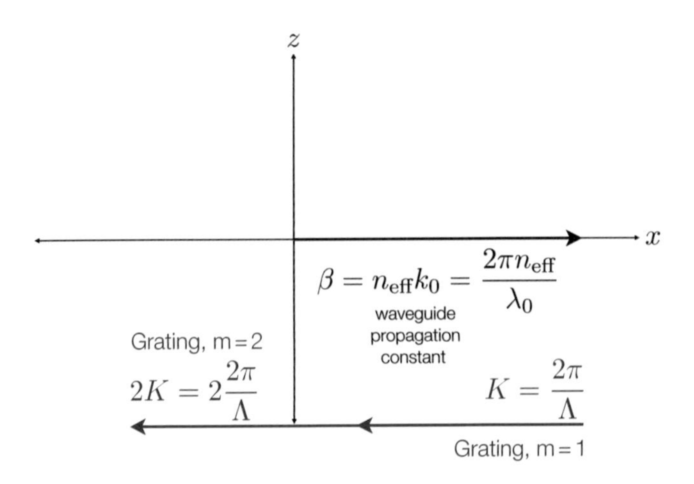

Figure 5.3 Graphical illustration of the Bragg condition for the grating coupler (Part 1).

(indicated by the red line). In practice, the back-reflection is undesired because this can lead to Fabry–Perot oscillation between the input coupler and output coupler. In order to avoid the second-order diffraction back into the waveguide, the grating is detuned and the fibre has a small angle to the normal of the grating surface. Figure 5.2b is the case when the optical wavelength inside the grating is smaller than the grating period. Then the output wave will propagate at an angle (indicates by the green line) and no second-order reflection exists in this case.

The grating couplers discussed in this section are one-dimensional periodic structures, which are well described by the Bragg Law. This is illustrated in Figures 5.3–5.5. Even the focusing grating couplers described in Section 5.2.3 can be considered as one-dimensional, but with an additional built-in focusing element. The wave incident on the grating is a guided wave propagating in a slab waveguide (see Section 3.2.2), with a direction of propagation in the same plane as the grating, and is normal to the grating teeth, as shown in Figure 5.3. The waveguide propagation constant is:

$$\beta = \frac{2\pi n_{\text{eff}}}{\lambda_0}, \tag{5.1}$$

where λ_0 is the optical wavelength, and n_{eff} is the effective index of the slab waveguide.

As shown in Figure 5.3, the periodicity of the grating is described by $K = 2\pi/\Lambda$, where Λ is the grating period. Higher-order diffraction gratings can be considered by using $m \cdot K$.

The general form of the Bragg condition can be expressed as:

$$\beta - k_x = m \cdot K, \tag{5.2}$$

where k_x is the component of the wave vector of the diffracted wave in the direction of the incident wave (Figure 5.4a). The diffracted wave is travelling in the cladding with an index of refraction, n_c (assumed to be air in the figure). The diffracted light has a

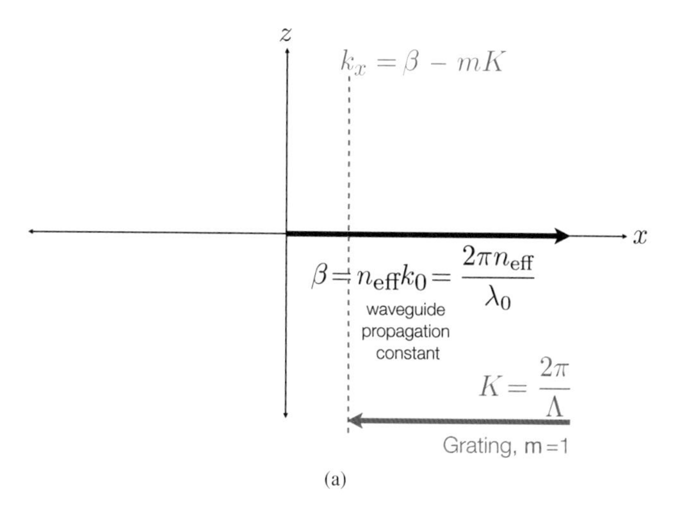

(a)

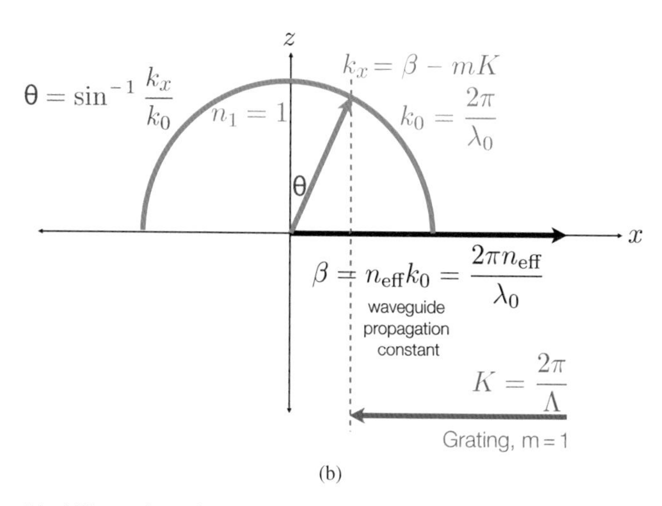

(b)

Figure 5.4 Graphical illustration of the Bragg condition for the grating coupler (Part 2).

wave vector:

$$k = \frac{2\pi n_c}{\lambda}. \tag{5.3}$$

Given the difference between the wave vector of the diffracted light, k, and the horizontal component, k_x, this leads to a diffracted angle of:

$$\sin\theta_c = \frac{k_x}{k} = n_{\text{eff}} \frac{\lambda}{\Lambda}. \tag{5.4}$$

This is illustrated in Figure 5.4b (assumed to be air in the figure, i.e. $k = k_0$). Thus the Bragg condition can be simplified to be:

$$n_{\text{eff}} - n_c \cdot \sin\theta_c = \frac{\lambda}{\Lambda} \tag{5.5}$$

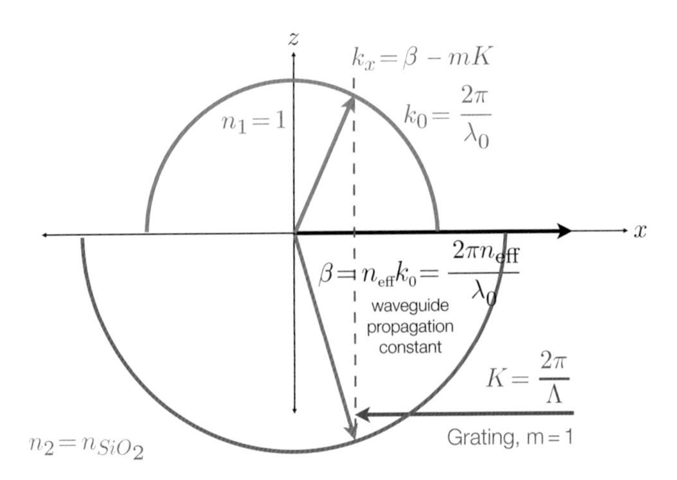

Figure 5.5 Graphical illustration of the Bragg condition for the grating coupler (Part 3).

and for the angle in the air, θ_{air}, using Snell's Law, this becomes:

$$n_{eff} - \sin\theta_{air} = \frac{\lambda}{\Lambda} \tag{5.6}$$

Finally, diffraction into the substrate can also be considered, as shown in Figure 5.5. From the diagram, the angle of the light in the oxide is smaller (closer to perpendicular to the surface normal), as compared to that in the air.

5.2.3 Design methodology

The design methodology presented in this section follows Reference [13]. The first step is identifying the fabrication process limitations and defining the design objectives. Next, a design based on analytic calculations from the Bragg condition presented in Section 5.2.2 is found. The simulation of the performance of the design, and its optimization, is then performed using 2D and 3D FDTD simulations. We then generate the mask layout of the desired grating couplers using the physical parameters. According to the idea demonstrated by Van Laere *et al.* [14], and described here, a linear grating coupler can be transformed into a compact shape using focusing gratings without an efficiency penalty. This whole design flow, including the Bragg condition calculations, has been implemented in Mentor Graphics Pyxis and Lumerical FDTD. The mask layout of the grating coupler can be directly generated in Pyxis given the process-determined and design-intent parameters. The script is described later in this section.

Prior to performing simulations on a specific grating coupler, we need to understand the fabrication process limitations and define the design objectives. For a specific grating coupler design, there are two types of input parameters: process-determined parameters and design-intent parameters. The process-determined parameters include the etch depth, the cladding material, the thicknesses of various layers, and the minimum feature sizes, which are determined by the wafers used by a foundry in its fabrication process. Fabrication process details are typically described in the design rules (see Sections 10.1.1 and 10.1.6). The design-intent parameters include central wavelength, λ, incident angle, and the optical polarization, which are specified by the designer.

The design objectives and optimization criteria are identified as follows.

- The polarization of the incident wave (s or p) and the corresponding target polarization in the waveguide (quasi-TE or quasi-TM).
- The desired central wavelength.
- The 3 dB bandwidth (a narrow-bandwidth or a wide-bandwidth grating coupler?).
- The desired incident angle (e.g. typically in the range of 10°–30°). This should take the experimental setup into consideration because the mechanical stages may have limitations, e.g. the angle must be smaller than 40° (see Section 12.2). However, it is the optimal performance that typically dictates the angle that will be chosen for the physical setup.
- Specification on the optical return loss.

In this section, we present an example of how to design a grating coupler with the following characteristics:

- quasi-TE polarization,
- 1550 nm central wavelength,
- 20° angle of incidence in air. Using Snell's Law, this corresponds to a fibre polish angle, and the angle in the oxide cladding, of $\arcsin(\sin(20)/1.44) = 13.7°$.

For the fabrication process, we assume the process described in Table 10.1, except with a shallow etch of $ed = 70$ nm for the grating. The other parameters are a full etch to define the strip waveguide and the focusing element; a silicon thickness of 220 nm; oxide cladding; and a minimum feature size of 200 nm.

Analytic grating coupler design

Based on the Bragg condition described in the previous part, we can design the desired grating coupler. We calculate the effective index of the two silicon thicknesses using the approach in Section 3.2.2. The assumption is that the grating has an infinite width. This assumption holds because the width of the grating coupler is typically 10 µm, which is much larger than the central wavelength; an example layout of a 1D coupler is shown in Figure 5.6. If we denote the effective index of the grating teeth as n_{eff1}, and the effective index of the grating slots as n_{eff2}, then the effective index of the grating region can be expressed as:

$$n_{\text{eff}} = \text{ff} \cdot n_{\text{eff1}} + (1 - \text{ff}) \cdot n_{\text{eff2}}. \tag{5.7}$$

For the example, the effective index of the 220 nm slab is about $n_{\text{eff1}} = 2.848$, and the effective index of the shallow etched region with thickness 150 nm is about $n_{\text{eff2}} = 2.534$, at $\lambda_0 = 1550$ nm. As an initial design, we choose the fill factor as ff = 50%. The weighted-average effective index of the grating region, n_{eff}, is then 2.691.

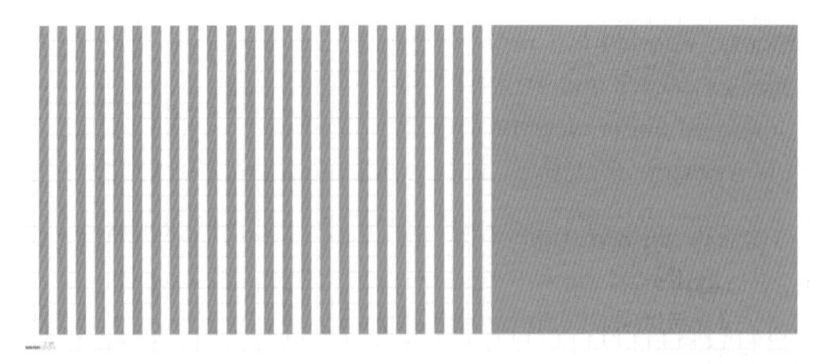

(a) Layout of the grating coupler

(b) Layout of a grating coupler with a linear taper

Figure 5.6 Mask layout for a 1D grating coupler and a linear taper several hundred microns in length

The Bragg condition, Equation (5.6), is employed to calculate the grating period:

$$\Lambda = \frac{\lambda}{n_{\text{eff}} - \sin\theta_{\text{air}}}. \tag{5.8}$$

For an incident angle of $\theta_{\text{air}} = 20°$, the period of the grating is calculated to be approximately $\Lambda = 660\,\text{nm}$.

As described in Reference [13], it was found that the analytic calculations yield a design that is very close to the optimal for a uniform grating coupler optimized by FDTD. Namely, the insertion loss of the design does not improve significantly after FDTD optimization. The central wavelength was found to deviate slightly from the design target, typically by 0–10 nm from the target, but this could be compensated by slightly changing input wavelength. Thus, this approach is very suitable for designers requiring a grating coupler for a specific fabrication process, wavelength, and polarization. These physical design parameters can be used directly for creating the mask layout later in the present section, or simulated and optimized as in the next section.

Design using 2D FDTD simulations

The analytic calculations presented above have yielded a design, but its performance is not known. Thus, the next tasks are to evaluate the performance, and optimize the design, using FDTD. There are several objectives that can be considered: (1) determining efficiency of the analytic design including determining the optimal input beam location, (2) understanding the impact of the various physical parameters (buried oxide thickness, fill factor, fibre angle, etch depths), (3) evaluating other performance parameters such as back-reflections, sensitivity to process variations (e.g. etch depth), and (4) optimizing the design.

Two-dimensional (2D) FDTD simulations are generally used to simulate grating couplers because it takes much less computational memory and simulation time. After designing a grating using 2D simulations, 3D simulations are used to verify the

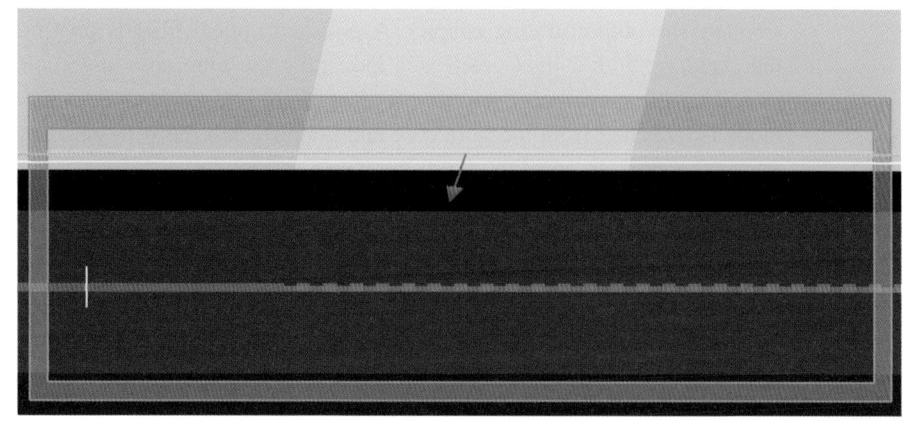

(a) Simulation configuration for an input grating coupler

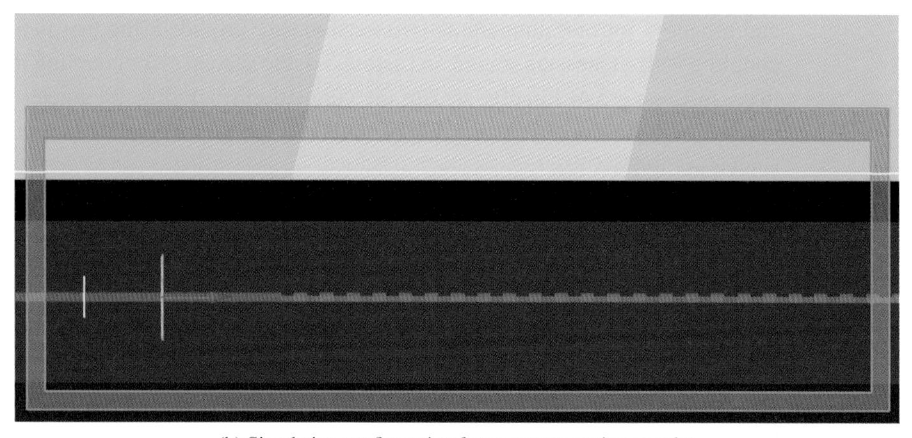

(b) Simulation configuration for an output grating coupler

Figure 5.7 2D FDTD simulation configuration for the grating coupler, including the optical fibre. Mode sources are used to inject light into the simulation either from the optical fibre (a) or from the waveguide (b). Power monitors and mode expansion monitors are used to measure the power in the waveguide (a) and optical fibre (b).

behaviour of the final design. The illustration in Figure 5.7 depicts a grating coupler structure: the Si wafer on the bottom, a functional Si layer on top of a 2 µm buried oxide, and a top oxide cladding layer for protection. The orange rectangle defines the simulation region and Perfectly Matched Layer (PML) boundary is used so that radiation appears to propagate out of the computational area and, therefore, does not interfere with the fields inside. The yellow lines shown represent frequency-domain power monitors which collect power flow information in the frequency domain from simulation results across spatial regions within the simulation. The green area denotes the optical fibre.

Two types of simulation structures are often employed to simulate a grating coupler. Figure 5.7a is used to simulate the input grating coupler and Figure 5.7b is used to simulate the output grating coupler. A polished optical fibre is presented on top of the cladding, with a light green area indicating the fibre core and a dark green area indicating the cladding of the fibre. For an input grating coupler, a fundamental TE mode is launched from the fibre core and coupled into the waveguide by the grating. Power monitors are used to record the insertion loss and reflection to the fibre of the grating coupler. For an output grating coupler, a fundamental TE mode is launched from the waveguide and out-coupled into the fibre. A mode expansion monitor is used to calculate the power that goes into the fundamental mode of the fibre. This is essential for a coupler because not all of the power is coupled into the fundamental mode due to the mode mismatch.

This simulation is implemented via FDTD simulation scripts consisting of four parts: initial settings, structure drawing, simulation setting, and simulation running. First, we need to set the initial parameters, in Listing 5.1. Secondly, we need to draw the simulation structures, in Listing 5.2. Then, we need to set up the simulation area, source, and monitors for our simulation. Two versions are provided: the first is for free-space coupling with a Gaussian source, in Listing 5.3; the second is for coupling into an optical fibre, in Listing 5.4. Finally, we run the simulation (including parameter sweeps) and plot the transmission, in Listing 5.5. Simulations are performed for an input coupler, as shown in Figure 5.7a.

Finally, as presented in Section 9.6, we later perform simulations in both directions (input and output coupler) and determine the back reflections. These results are used to determine the compact S-parameters (see Section 9.3.2) for circuit modelling.

Position of the beam

The position of the incident beam needs to be swept to find the optimal fibre position. The maximum transmission of this grating coupler, for various positions, is shown in Figure 5.8. This graph indicates that the highest efficiency is obtained with the optical fibre positioned approximately 5 μm away from the start of the grating. This plot is also indicative of the alignment sensitivity. In this case, the 1 dB alignment sensitivity is $\pm$ 2.3 μm.

Results

The simulation result is shown in Figure 5.9. From the simulation results we can see that the Bragg condition theory gives a good prediction of the grating coupler's central wavelength. For this design, we obtain an insertion loss of -2.7 dB and a central wavelength of 1548 nm. Figure 5.9 also shows the simulated back reflections, or optical return loss, of the grating coupler. In this case, the return loss is below -10 dB over the grating bandwidth. The large reflection is due in part to the optical fibre–air interface; index matching (e.g. glue) would reduce the return loss.

Polarization

Grating couplers are intrinsically polarization sensitive, since the waveguide effective index has very different values for TE and TM polarized modes. Thus, the TE-designed grating will suppress light in the opposite polarization, as shown in Figure 5.10.

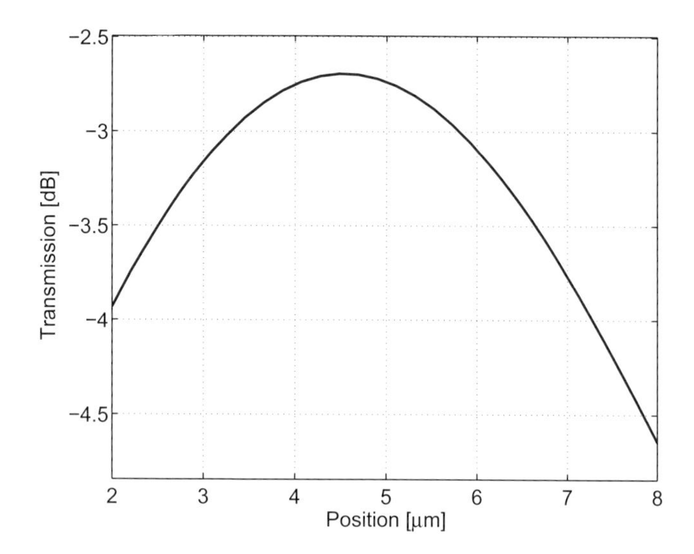

Figure 5.8 Grating coupler efficiency (insertion loss) versus optical fibre position, for the analytic design in the present section. The distance is defined relative to the start of the grating ($x = 0$ is the centre of the fibre at the start of the grating).

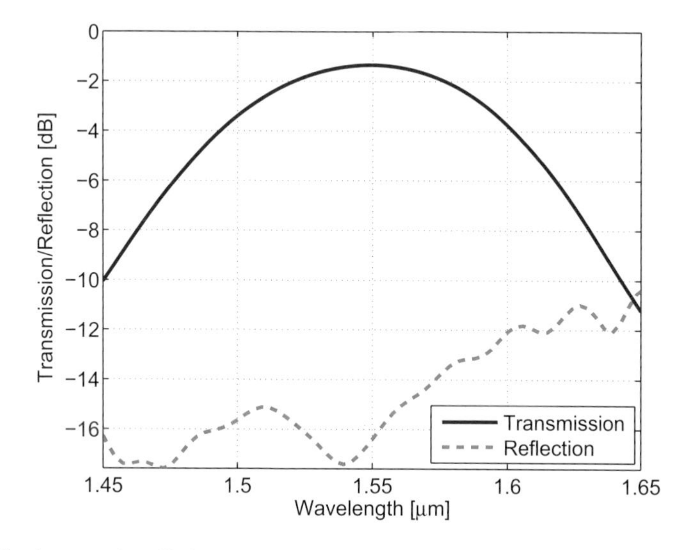

Figure 5.9 Grating coupler efficiency (insertion loss) and reflection back into the waveguide (optical return loss) versus wavelength, for the design in the present section.

Design parameters

There are several parameters that decide the performance of a grating coupler: period, fill factor, incident angle, incident position, etch depth, thickness of the SiO_2 substrate, thickness of the SiO_2 cladding, number of grating periods. Among them, etch depth and SiO_2 thickness are decided by the fabrication process.

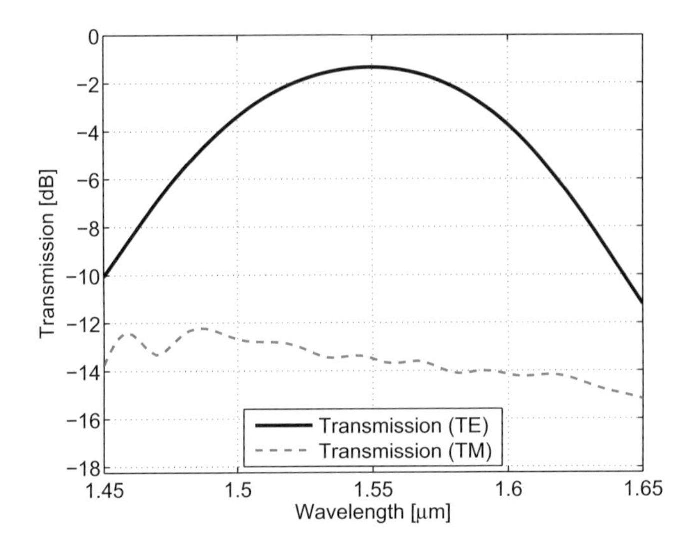

Figure 5.10 Grating coupler efficiency (insertion loss) versus wavelength for TE and TM polarized light. The coupler is designed for TE polarization, and acts as a polarizer for the TM polarization.

The primary impact of these parameters is that they change the central wavelength. This can be found from the Bragg condition, Equation (5.6), to find the central wavelength:

$$\lambda = \Lambda \left(n_{\text{eff}}(\lambda) - n_{\text{c}} \cdot \sin\theta_{\text{c}}\right). \tag{5.9}$$

One should keep in mind that when the grating coupler central wavelength changes, so too does the effective index, which is wavelength dependent (dispersion), $n_{\text{eff}}(\lambda)$.

Grating period

The grating period is the parameter with the most impact on the grating wavelength, as seen by Equation (5.9). A parameter sweep simulation is implemented in Listing 5.5. All parameters are constant except for the period. As the grating period is varied from 620 nm to 700 nm, the central wavelength of the grating coupler shifted from 1483 nm to 1608 nm, as shown in Figure 5.11. From the simulated spectra, the tuning coefficient for the grating period, which is defined as $\delta\lambda/\delta\Lambda$, was determined to be 1.56 nm/nm.

Fill factor

The fill factor (or duty cycle) affects the performance of a grating coupler. Its impact is by changing the effective index, via Equation (5.7). Hence, as the fill factor increases, the effective index increases, and the grating response shifts to longer wavelength as per Equation (5.9). Figure 5.12 shows the simulation results for varied fill factor, with all other parameters kept constant. The fill factors were varied from ff = 0.3 to ff = 0.6. The central wavelength shifted from 1522 nm to 1560 nm. The tuning coefficient for the fill factor can be defined as $\delta\lambda/\delta W$, and is determined to be 0.215 nm/nm. We note that

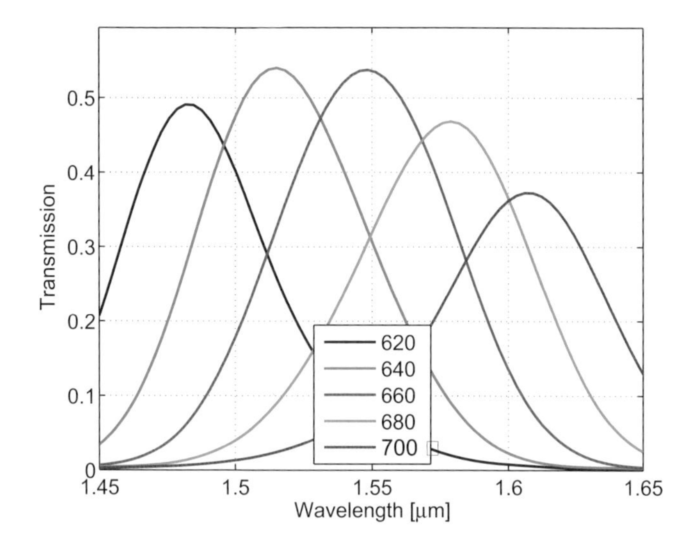

Figure 5.11 Grating coupler – sweep of the grating period (from 620 nm to 700 nm).

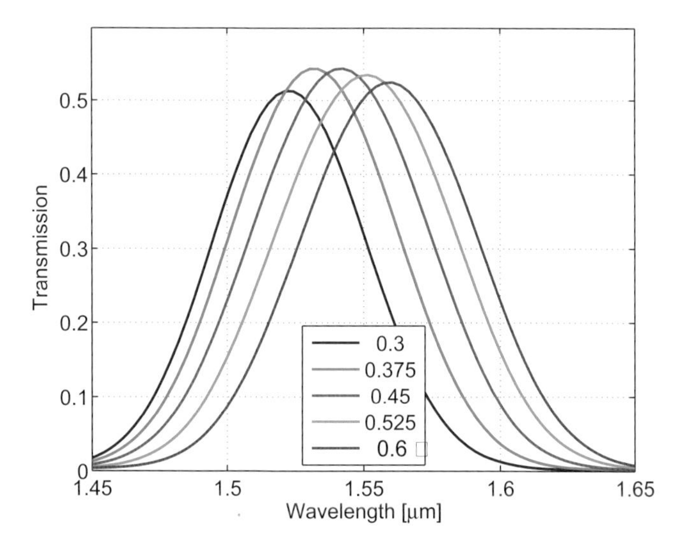

Figure 5.12 Grating coupler – sweep of the grating fill factor (from 30% to 60%).

grating period has a stronger impact on the central wavelength of the grating than the fill factor.

Etch depth

Etch depth of a grating coupler also influences the performance of the grating coupler through its impact on the effective index of refraction of the grating. As the etch depth increases, the effective index of refraction of the shallow etched area decreases, thus n_{eff} decreases. As per Equation (5.9), the central wavelength of the grating is proportional to the effective index; hence the central wavelength of the grating is inversely proportional to the etch depth. The grating is similarly sensitive to the SOI silicon thickness.

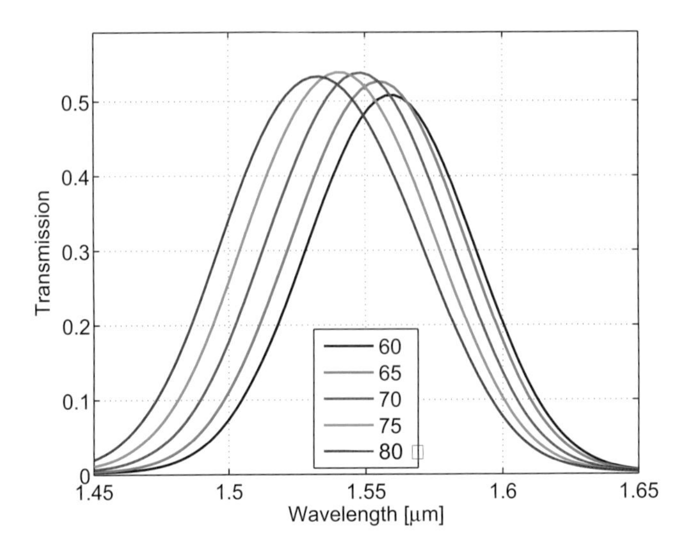

Figure 5.13 Grating coupler – sweep of the etch depth (*ed* varied from 60 nm to 80 nm).

Figure 5.13 shows the simulation results for varying the etch depth from 60 nm to 80 nm, with all other parameters fixed. The central wavelength shows a blueshift as the etch depth increases, which is consistent with our analytical calculation. The tuning coefficient of the etch depth, which is defined as $\delta\lambda/\delta ed$, was determined to be 1.9 nm/nm.

Incident angle

The incident angle of a grating coupler is defined as the angle between the incident wave (or out-coupled wave) and the normal to the grating surface. A positive angle indicates the case in which the incident wave and the coupled wave in the waveguide propagate in the same direction and a negative angle indicates the case in which the incident wave and the coupled wave in the waveguide propagate in opposites directions. The angle can be defined as the free-space angle (useful for free-space measurements, see Section 12.1.1), or as the angle of light in the cladding (e.g. oxide). When coupling light using a lensed fibre, the angle of light in air is the same as the angle of the fibre. When using polished fibres or fibre arrays, Snell's Law is used, and the fibre polish angle is the same as the angle of light in the cladding (assuming the same index in both the fibre and the cladding).

The incident angle influences the central wavelength of the grating coupler via Equation (5.9). Simulation results for varied incident angle (in air), with other parameters constant, are shown in Figure 5.14. As the incident angle varies from 15° to 25°, the central wavelength shifts from from 1586 nm to 1512 nm. Thus, the tuning coefficient, defined as $\delta\lambda/\delta\theta$, is determined to be 7 nm per degree.

Parameter sensitivity

From the simulation results shown above, the grating period, fill factor (or width of the grating tooth), etch depth, and incident angle all have impacts on the central wavelength

Table 5.1 Grating coupler sensitivity to geometry parameters.

Parameter	Sensitivity coefficient
Period, Λ	1.56 nm/nm
Width, W	0.215 nm/nm
Etch depth, ed	1.9 nm/nm
SOI thickness	1.82 nm/nm
Incident angle, θ	7 nm/°

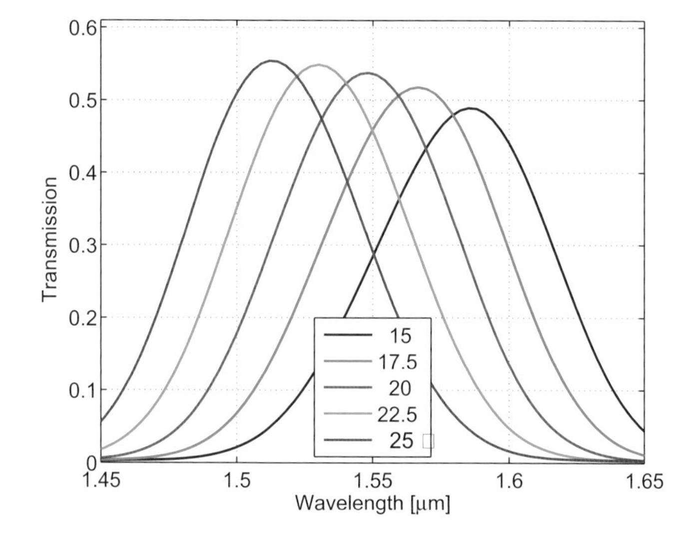

Figure 5.14 Grating coupler – sweep of angle (θ_{air} varied from 15° to 25°).

of the grating coupler. Table 5.1 shows the comparison of the tuning or sensitivity coefficients for these parameters. These parameters are also useful for understanding the impact of fabrication variations on the grating coupler performance; see Section 11.1.3.

Cladding and buried oxide

The thickness of the buried oxide and the thickness of the cladding are two important factors that have impacts on the insertion loss of the grating coupler. An illustration of different reflections at various interfaces of the grating coupler is shown in Figure 5.15. The interference between these reflections contributes the insertion loss of the grating coupler. Minimum insertion loss can be achieved when $P_{\text{reflection1}}$ and $P_{\text{reflection2}}$ result in destructive interference, and when $P_{\text{reflection3}}$ and $P_{\text{reflection4}}$ result in constructive interference.

Simulation results for varied buried oxide are shown in Figure 5.16, from 1 μm to 3 μm. The insertion loss of the grating coupler oscillates in a sinusoidal manner, which is determined by the interference between $P_{\text{reflection3}}$ and $P_{\text{reflection4}}$. The thickness of the buried oxide for a particular wafer type is chosen to achieve constructive interference between $P_{\text{reflection3}}$ and $P_{\text{reflection4}}$, therefore, low insertion loss can be obtained. There

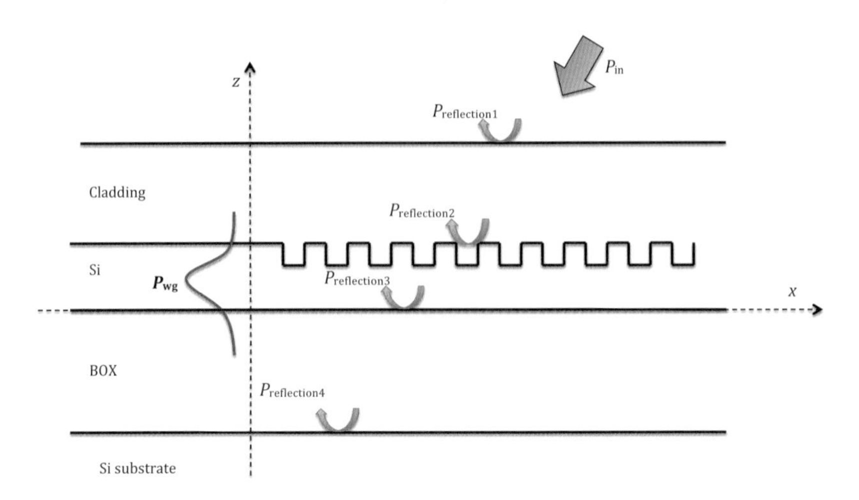

Figure 5.15 Schematic diagram of reflections at different interfaces in the grating coupler.

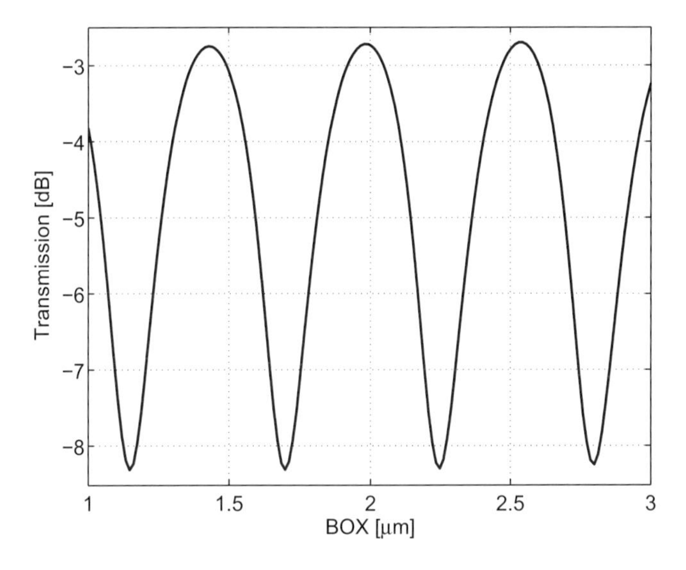

Figure 5.16 Grating coupler efficiency versus the buried oxide (BOX) thickness.

is a peak at an oxide thickness of 2 μm. It also turns out that a 2 μm oxide provides high coupling efficiency for both 1310 nm and 1550 nm. It is a common standard among silicon photonics foundries.

Similarly, the phase condition of $P_{reflection1}$ and $P_{reflection2}$ varies as the thickness of the cladding changes. Here, we define the thickness of the cladding to be the height from the interface of the silicon and buried oxide to the top surface of the cladding. Figure 5.17 shows the simulation results for varying the cladding thickness. Minimum insertion loss is achieved where destructive interference occurs and maximum insertion loss achieved where constructive interference occurs. Depending on the incident angle

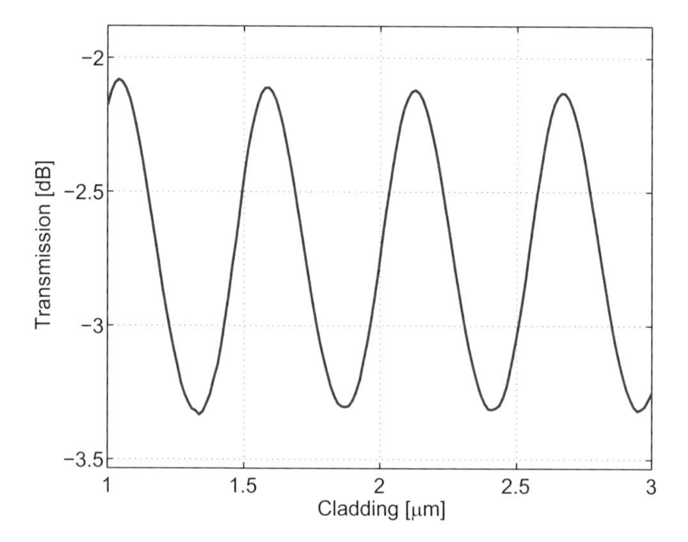

Figure 5.17 Grating coupler efficiency versus the cladding thickness.

and the central wavelength, the optimal thickness of cladding changes. By comparing Figure 5.16 and Figure 5.17 we note that the thickness of the buried oxide has a larger impact on the insertion loss of the grating. This is the case because the reflection coefficients of $P_{\text{reflection3}}$ and $P_{\text{reflection4}}$ are larger than those of $P_{\text{reflection1}}$ and $P_{\text{reflection2}}$.

Compact design – focusing

So far we have addressed how to design a grating coupler with straight gratings. With this approach, a taper is required to convert the approximately 9 μm diameter fibre mode to a 0.5 μm waveguide mode. In order to have high conversion efficiency, the length of taper needs to be more than 100 μm, which is undesirable for compact circuits [7, 8]. Instead of using straight gratings, we can use confocal gratings to make the grating coupler more compact [14,15]. The layout of a confocal grating is shown in Figure 5.18. The grating lines form an ellipse with a common focal point, which coincides with the optical focal point of the coupler.

According to [14], compact grating structure can be obtained by curving the grating lines following the equation:

$$q \cdot \lambda_0 = n_{\text{eff}}\sqrt{y^2 + z^2} - z \cdot n_t \cdot cos\,(\theta_c)\,, \tag{5.10}$$

where q is an integer number for each grating line, θ_c is the angle between the fibre and the chip surface, n_t is the refractive index of the environment, λ_0 is the vacuum wavelength, and n_{eff} is the effective index of the grating. The above equation can be expressed as (where t is the angle variable in a polar coordinate system):

$$x = \sqrt{\frac{q \cdot \lambda_0 \cdot n_t \cdot \cos(\theta_c) + q^2 \cdot \lambda_0^2}{n_{\text{eff}}^2 - n_t^2 \cdot \cos^2(\theta_c)}} \cdot \cos(t) + \frac{q \cdot \lambda_0 \cdot n_t \cdot \cos(\theta_c)}{n_{\text{eff}}^2 - n_t^2 \cdot \cos^2(\theta_c)} \tag{5.11}$$

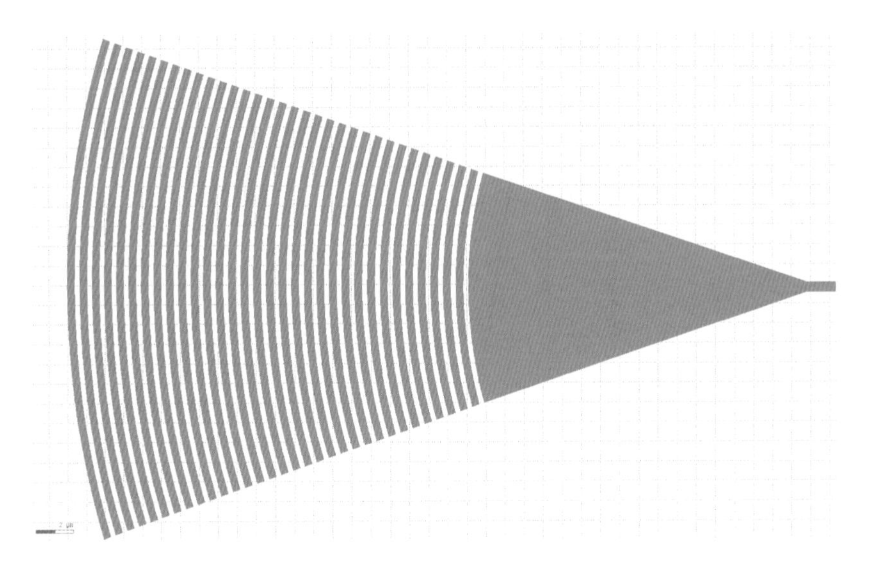

Figure 5.18 Mask layout for a focusing grating coupler.

$$y = \sqrt{\frac{q^2 \cdot \lambda_0^2}{n_{\text{eff}} - n_t^2 \cdot \cos^2(\theta_c)}} \cdot \sin(t) \tag{5.12}$$

The expression of Bragg condition for the curved gratings can also be converted to a polar system [11]:

$$n_{\text{eff}} \cdot k_0 \cdot r = n_c \cdot k \cdot r \sin(\theta) \cos(\phi) + 2\pi N \tag{5.13}$$

where k_0 is the wave vector in free space, and ϕ denotes the angle subtended by an arbitrary point on the grating and to the z-axis (in this case, the direction of propagation in the waveguide).

If a straight-lines grating coupler has the same period and fill factor as an elliptical grating coupler, then the coupling efficiency of the two couplers should be nearly the same. In other words, the curvature of the grating line does not affect the coupling efficiency of a grating coupler. Thus, the design of a focal grating coupler begins with 2D simulations of the straight-lines grating coupler. With a given 2D design, the focusing grating coupler can be drawn using the same period and fill factor, and 3D FDTD simulations are used to double check the result as described in Section 5.2.4.

Mask layout

The grating coupler layout is drawn using a script implementing the equations in the present section. The design is implemented in Mentor Graphics mask layout package "Pyxis," in the form of a parameterized cell (PCell). The layout is shown in Figure 5.18, with the script in Listing 5.8. This script can be used to generate grating couplers with different period and fill factor, and can be used to design grating couplers for any wavelength, polarization, and angle of incidence. The function requires the following

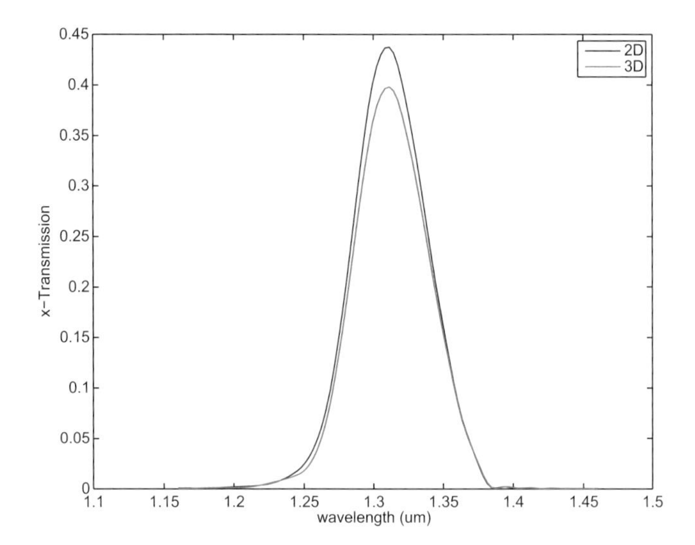

Figure 5.19 Grating coupler 2D and 3D efficiency versus wavelength.

parameters: wavelength, period, fill factor, cladding index of refraction, incident angle (defined in air), waveguide width, slab waveguide effective index, and number of segments in the curves.

Using the analytic design methodology here, we can implement the basic optimal calculations (effective index) using Listing 5.7, and design the resulting grating coupler via Listing 5.6. This script implements the grating coupler based on physical parameters (wafer thickness, etc.), as well as design-intent parameters (wavelength, polarization, angle). The required parameters are: wavelength, etch depth, silicon thickness, incident angle (defined in air), waveguide width, cladding index of refraction, polarization, and fill factor. The default fill factor of 0.5 is not necessarily optimal, and should be optimized using FDTD simulations as described in Section 5.2.3.

3D simulation

2D simulations are used in the initial optimization and provide a good approximation. However, 3D simulations are required to verify the design. In general, a 3D simulation will require much more memory and time than running a 2D simulation. The results comparing the 2D versus 3D simulations are shown in Figure 5.19. The 3D simulations were performed by exporting the mask layout and importing into the FDTD solver. The 3D simulation result matches very well with the 2D simulation, and the difference is due to the dimension limitation of the 3D simulation. In 2D simulation, the third dimension is considered to be infinite.

5.2.4 Experimental results

Optimization of a grating coupler involves simulation sweeps of various parameters such as grating period, duty cycle, and incident angle, etc. The optimization can be

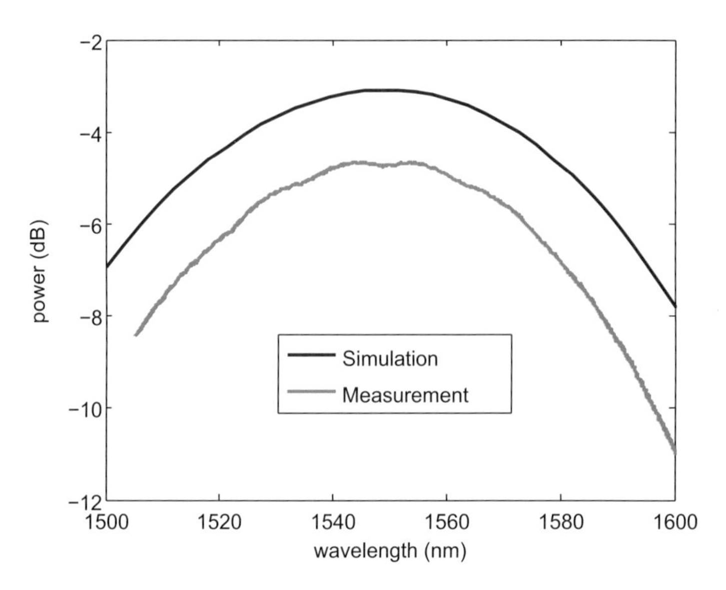

Figure 5.20 Experimental and simulation results for a grating coupler design fabricated via OpSIS-IME. Reprinted with permission from Reference [16].

performed using methods such as the genetic algorithm. Figure 5.20 shows the simulation and measurement results of the grating coupler designed for the fabrication process provided by OpSIS-IME [17]. The designed grating coupler has a grating period of 650 nm with a duty cycle of 350 nm. The simulated results show an insertion loss of −2.74 dB with a 3 dB bandwidth of 79.8 nm, and the measurement shows an insertion loss of −4.64 dB with a 3 dB bandwidth of 74.9 nm [16]. The simulation results shown in Figure 5.20 are obtained from the model with the assumption that the distance between the fibre tip and the chip is negligible. The measured insertion loss is higher than the simulated insertion loss. In addition, the bandwidth of the measured spectrum is narrower than that of the simulated one. The mismatches are due to the gap between the fibre ribbon and the photonics chip.

5.3 Edge coupler

Edge coupling, also known as end fire coupling, is the standard technique for coupling to and from a single-mode fibre for most photonic devices such as DFB lasers, modulators, and high-speed detectors. Edge coupling generally offers a very broadband response and features low insertion loss (lower than 0.5 dB [18, 19]). It also couples both TE and TM polarizations. The challenges with edge coupling are related to the need for precision alignment, polishing/etching the facet, beam astigmatism, and the need for anti-reflection coatings.

There are several methods of implementing an edge coupler in silicon photonics. The main requirement is to convert the highly confined mode in the waveguide to a much larger fibre mode, using a spot-size converter. This requires addressing both the

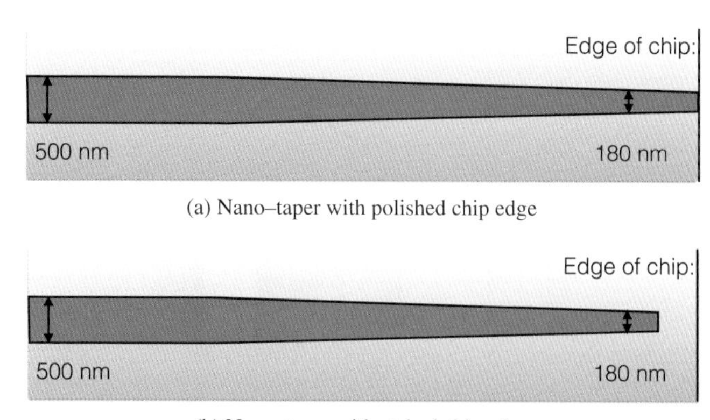

(a) Nano–taper with polished chip edge

(b) Nano–taper with etched chip edge

Figure 5.21 Illustration of the nano-taper edge coupler. Light is incident in the waveguide from the left, with output out of the right side of the chip.

mode-size mismatch as well as the effective index mismatch of the fibre and waveguide modes. The typical approach, known as an inverse taper, or nano-taper, is one where the waveguide is tapered down to a very small size such that the light is barely guided by the waveguide and the evanescent field of the waveguide becomes large [20, 21], as shown in Figure 5.21. This approach can also be implemented by sub-wavelength structures [22] or a knife-edge structure [23]. Both approaches lead to a mode field diameter expanding in the taper. More discussion on edge coupler test and packaging is included in Section 12.1.1. An additional technique is also to expand the waveguide in both in-plane and out-of-plane directions, for example, by depositing a polymer overlay [21, 24].

In this section, we first discuss the nano-taper (surrounded by a cladding oxide), then describe the polymer overlay approach. Three modelling approaches are presented: (1) simple mode calculations, (2) 3D FDTD simulations, and (3) eigenmode expansion method.

5.3.1 Nano-taper edge coupler

In this section, we discuss the design and modelling of the nano-taper edge coupler geometry in Figure 5.21b.

Mode overlap calculation approach

The first method to determine the coupling efficiency is to estimate it using mode calculations at the tip of the nano-taper. This is justified by the assumption that the taper is adiabatic, and that the mode at the tip of the taper will represent the mode profile after propagating down the taper. Specifically, we find the mode profile for a 180 nm wide waveguide, 220 nm thick, surrounded by oxide. The simulated mode profile of the nano-taper is shown in Figure 5.22. We wish to consider the coupling from the edge coupler to a Gaussian beam (e.g. a lensed fibre, a lens assembly with a fibre attached, a high numerical aperture lens). Hence, in the simulation, Gaussian beams are created

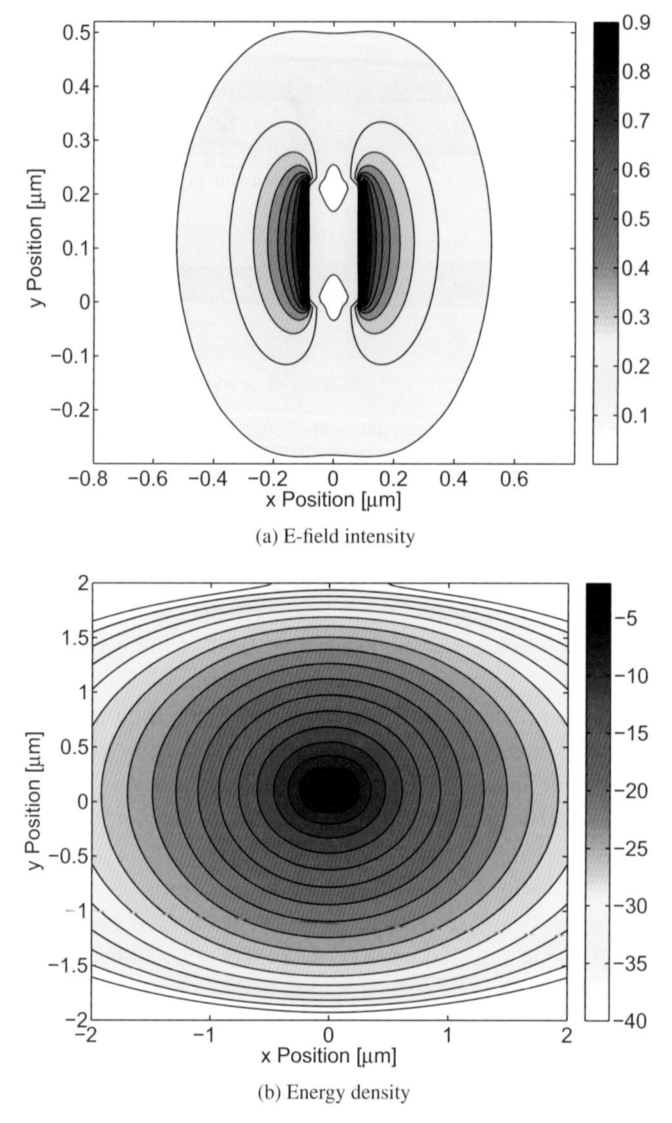

(a) E-field intensity

(b) Energy density

Figure 5.22 Mode profile of a 180 × 220 nm nano-taper, $\lambda = 1.55\,\mu\text{m}$, quasi-TE polarization. Effective index is 1.46 (very close to the SiO_2 material index).

by ideal lenses with various numerical apertures, NA. The profile for a Gaussian beam with NA = 0.4 is shown in Figure 5.23. From these images, it is clear that the mode profiles do not match, hence it is anticipated that there will be mode-mismatch loss.

We then perform overlap calculations with the Gaussian beams with varied numerical apertures. This is accomplished using Listing 5.9. Overlap calculations allow us to determine the best coupling efficiency between the nano-taper and the Gaussian beam, and to find the best numerical aperture for this beam. As shown in Figure 5.24, the best coupling is obtained for relatively large numerical apertures (0.4 for TE, and 0.55

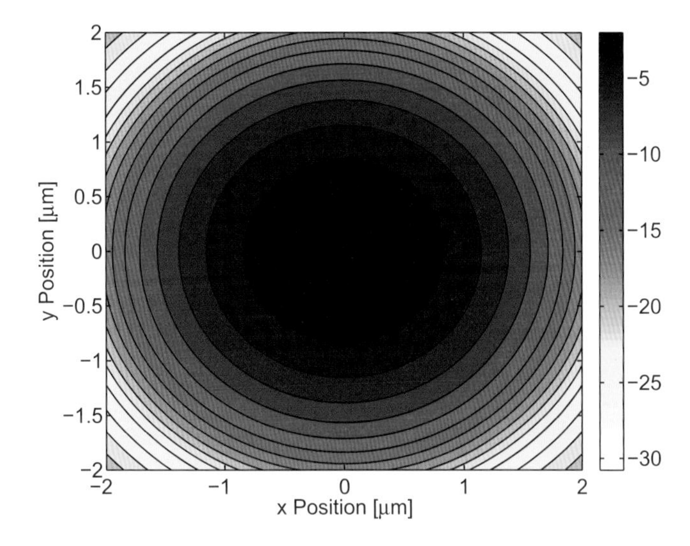

Figure 5.23 Mode profile Gaussian beam, created using a lens with an optimized numerical aperture of 0.4 for TE coupling.

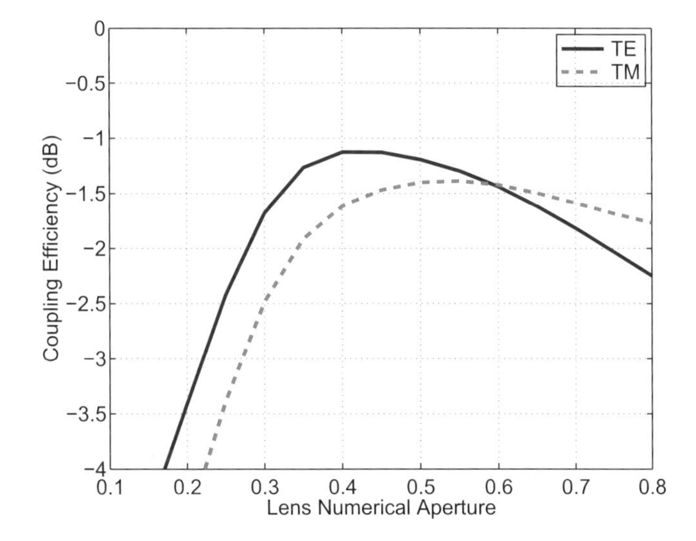

Figure 5.24 Mode overlap calculations showing power coupling efficiency versus the lens numerical aperture. 180 × 220 nm nano-taper, $\lambda = 1.55\,\mu$m, quasi-TE polarization.

for TM), especially as compared to conventional single-mode fibres (NA = 0.14). Fibres with these values are available, and are termed high-NA fibres.

Using mode overlap calculations, the alignment sensitivity can be simulated. This is done by translating one mode relative to the other, and performing overlap calculations. The results for misalignment in the two dimensions, (x, y), are shown in Figure 5.25. The results indicate that a 1 dB excess loss is introduced for a misalignment of 0.6 μm.

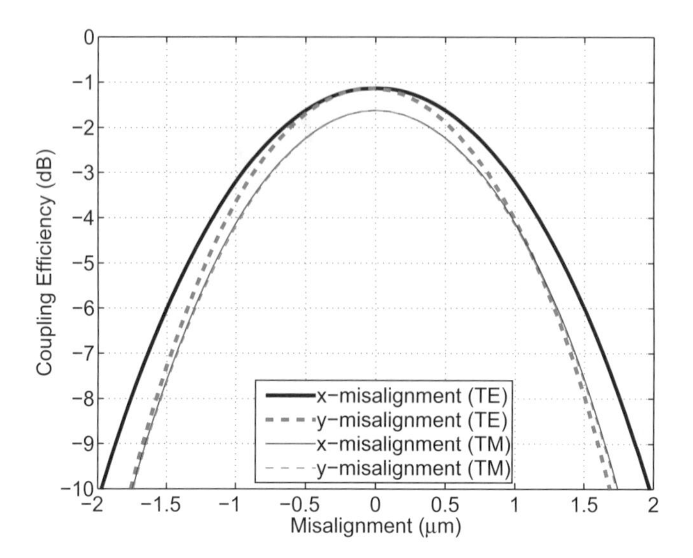

Figure 5.25 Mode overlap calculations showing power coupling efficiency versus misalignment (x, y): 180 × 220 nm nano-taper, $\lambda = 1.55\,\mu$m, quasi-TE polarization.

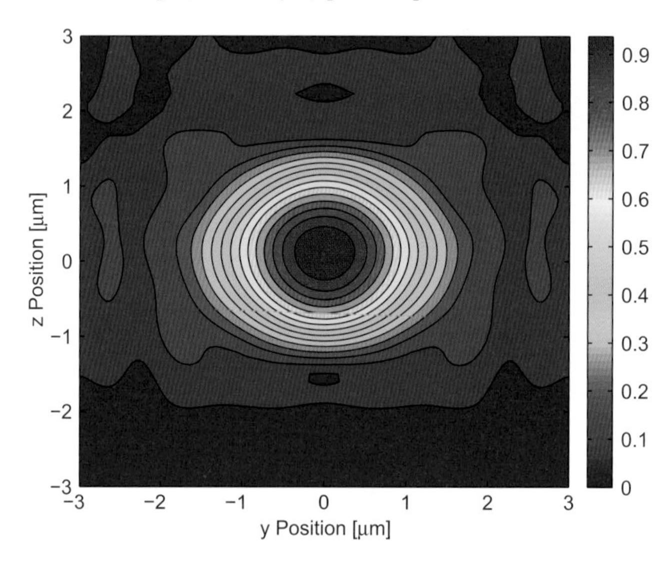

Figure 5.26 FDTD simulation of the nano-taper edge coupler – field profile at the output, in air: x-axis tangential waveguide; y-axis, surface normal to wafer. Nano-taper length 200 μm, with 180 × 220 nm tip, $\lambda = 1.55\,\mu$m, quasi-TE polarization.

There are limitations in the mode overlap calculation approach. Namely, it neglects:

- reflections at the waveguide–oxide and the oxide–air interfaces;
- propagation of the field in the short oxide region between the tip and the etched oxide facet;
- propagation of the field in the gap between the facet and the fibre;
- the taper itself, which needs to be long enough to ensure that it is adiabatic and loss-less.

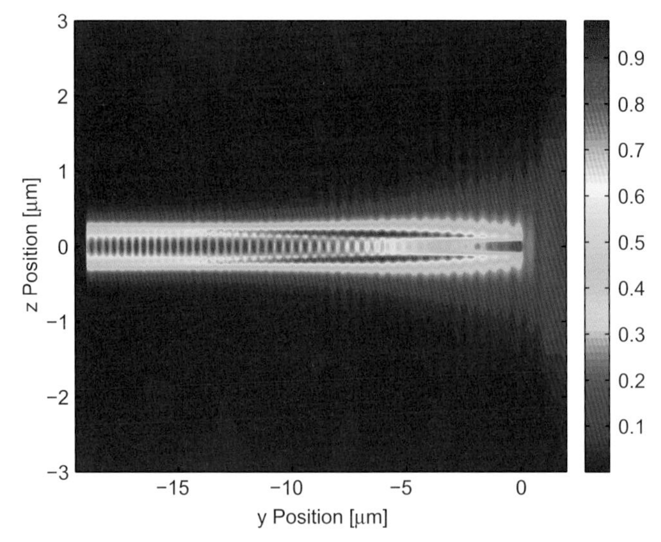

(a) Field profile (top view): x-axis, direction of waveguide; y-axis, tangential to waveguide.

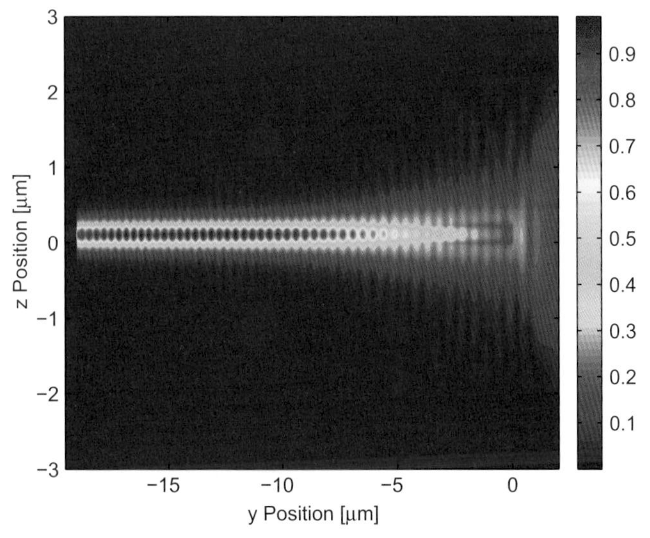

(b) Field profile (side view): x-axis, direction of waveguide; y-axis, surface normal to wafer.

Figure 5.27 FDTD simulation of the nano-taper edge coupler – field profiles. Nano-taper length 20 μm, with 180 × 220 nm tip, $\lambda = 1.55$ μm, quasi-TE polarization.

FDTD approach

The second method to determine the coupling efficiency is to perform 3D FDTD calculations (see Figures 5.26 and 5.27). Specifically, we model in FDTD the adiabatic taper, the tip, oxide, and air, in Listing 5.10. We then perform overlap calculations with Gaussian beams created by ideal lenses with various numerical apertures, in Listing 5.11. As in the previous method, this allows us to determine the best coupling efficiency and the best numerical aperture. This simulation could be extended to include the optical fibre itself.

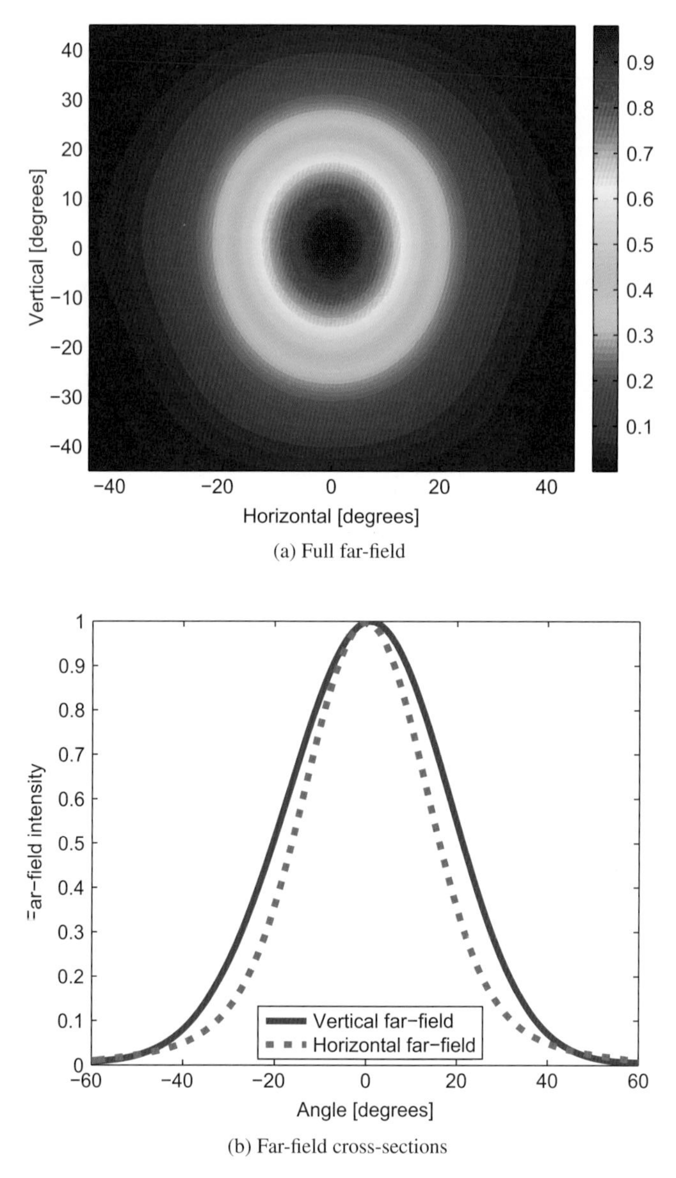

(a) Full far-field

(b) Far-field cross-sections

Figure 5.28 FDTD simulation of the nano-taper edge coupler – far-field projection at the output. Nano-taper length 200 μm, with 180 × 220 nm tip, $\lambda = 1.55\,\mu m$, quasi-TE polarization.

Figure 5.27 shows the simulated results for the field propagation down the taper, for the top and cross-section views. In the simulation, the oxide–air interface is at $y = 0$. The simulation shows some spatial oscillations, which are due to the reflection at the oxide–air interface. The field is seen to expand as it propagates from left to right.

The field profile at the output of the taper (in the air) is shown in Figure 5.26. This is the near-field profile, as would be seen by a fibre in direct contact with (or very close to) the chip.

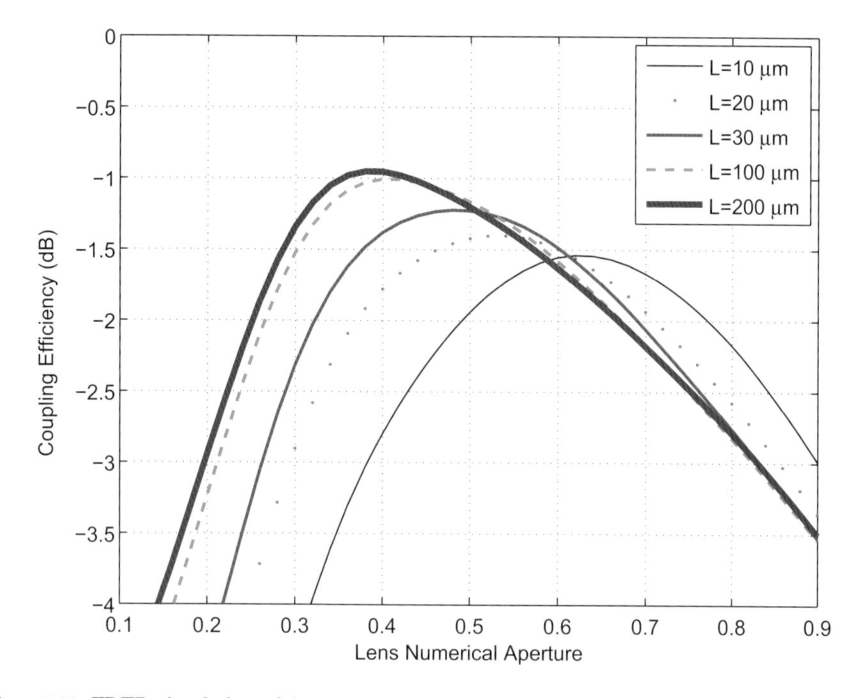

Figure 5.29 FDTD simulation of the nano-taper edge coupler – mode overlap calculations showing power coupling efficiency versus the lens numerical aperture. Several taper lengths are considered, $L = [10, 20, 30, 100, 200]$ μm; 180 × 220 nm nano-taper, $\lambda = 1.55$ μm, quasi-TE polarization.

For coupling to lenses, the far-field profile is also important, and is shown in Figure 5.28a. From the far-field pattern, the divergence angles of the taper can be determined. The far-field are shown along one dimension in Figure 5.28b. These show a full-width half-maximum far-field intensity of approximately 40°.

Finally, overlap calculations are performed between the FDTD-calculated near-field profile and the Gaussian beam from an ideal lens. The results are shown in Figure 5.29. Calculations were performed for several lengths of nano-tapers. It is seen that the taper needs to be at least 100 μm long to avoid introducing loss due to the taper itself. The simulated insertion loss reaches a best-case value of 1 dB.

5.3.2 Edge coupler with overlay waveguide

In this section, we consider the edge coupler that consists of a nano-taper connected to a second and much larger waveguide, which is then connected to an optical fibre. This waveguide can be made using a polymer or an inorganic material such as SiON [24] or an oxide [21].

Eigenmode expansion method

The simulation for the nano-taper with overlay waveguide is performed using the eigenmode expansion method. The structure consists of a 400 nm wide waveguide tapered

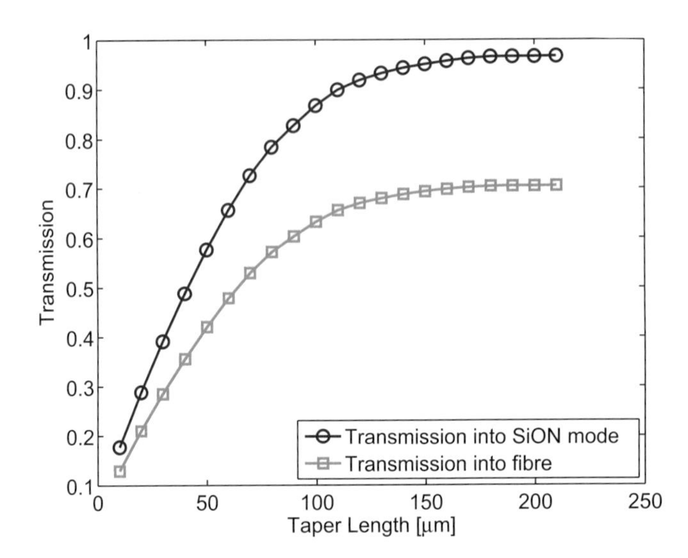

Figure 5.30 Eigenmode expansion method simulation of the nano-taper edge coupler with overlay waveguide, and considering fibre coupling efficiency; $\lambda = 1.5\,\mu m$, quasi-TE polarization.

down to 80 nm. Above the waveguide, a 3 μm square waveguide with a 3% index contrast, as described in Reference [24], is deposited and continues to the edge of the chip. A parameter sweep is performed by varying the length of the nano-taper, to determine the length required for efficient coupling. Finally, coupling to an Gaussian beam with a numerical aperture of 0.4 is performed, to simulate the insertion loss to a high-NA optical fibre. The simulation script is provided in Listing 5.12. The results are shown in Figure 5.30. The results indicate that nearly 100% power transfer into the overlay waveguide occurs for a length of 200 μm. Most of the loss occurs from the mode-mismatch between the fibre and the overlay waveguide. Similar conclusions were obtained for the nano-taper considered previously, via the mode overlap and the 3D FDTD simulations.

5.4 Polarization

Silicon photonic waveguides, particularly those based on the 220 nm thickness platform, are highly polarization dependent (birefringent). In theory it is possible to make the waveguides polarization insensitive; however, this is neither technically nor economically feasible due to the nanometer-level sensitivity of these high-contrast waveguides [25]. An alternative is to use much thicker silicon layers, e.g. 1.5 μm, in which polarization insensitive components can be build, as developed by Kotura and others.

Given this challenge with the 220 nm thickness platform, this means that photonic circuits typically need to be designed to operate on a single polarization. For application such as transmitters, where the laser is closely integrated with the circuit (see Chapter 8), the polarization of the light source can be maintained to a single polarization, and the output can be effectively coupled to a single-mode fibre. For the receiver, however, the

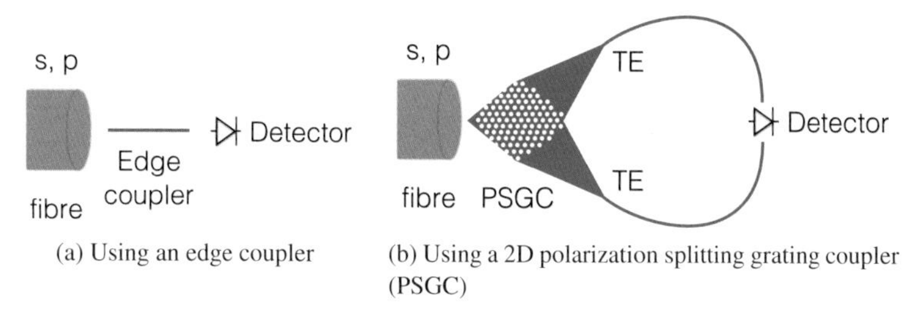

(a) Using an edge coupler (b) Using a 2D polarization splitting grating coupler (PSGC)

Figure 5.31 Detection of arbitrary polarized light.

light will have typically travelled through a single-mode fibre where the polarization is not preserved and is typically slowly time varying due to temperature and strain variations in the fibre. This is because the fibre has random birefringence (where the axes are also randomly spatially varying) due to bends and strain that leads to coherent cross-talk and coupling between the polarization modes. Thus, the receiver circuit needs to be polarization insensitive. There are several possible approaches. The first is to solve the polarization issue in the fibre by using polarization-maintaining fibre [26]. Another approach is to have a data format that allows for a polarization insensitive receiver. This is particularly straightforward for single wavelength communication systems using on–off modulation, where the light needs only to be directly coupled to a photodetector. In one approach, a polarization splitting grating coupler (PSGC) is used to separate the *s* and *p* polarized light from the fibre into two waveguides, each in the TE polarization [27]. A detector with two input ports is used to collect the light from the two waveguides, as shown in Figure 5.31. This can be implemented either with edge or grating couplers.

Polarization diversity

For applications which require more components than one detector, such as wavelength division multiplexing (WDM) or coherent detection, the common approach has been two split the *s* and *p* polarized light into two separate waveguides, both in the TE polarization. Then, two identical circuits are implemented. The light can be recombined into an optical fibre, if necessary. This can be implemented using either polarization splitting grating couplers (PSGC) [28] or using edge couplers with on-chip polarization splitter-rotators (PSR) [25]. The polarization diversity approach is illustrated in Figure 5.32 for both optical interface types. The need for two identical circuits has disadvantages, namely an increased chip area, and the need to have the two circuits matched (e.g. ring resonators or filters), which can be achieved by thermal tuning but this increases the power consumption. For certain circuits (e.g. passive filters), it is possible to use bi-directional propagation to re-use the same circuit instead of having two identical copies [28].

Active polarization management

An alternative to dealing with the polarization challenge is based on active polarization management. In this approach, linear optical transformations [29] are used to convert an arbitrary polarization of light in the fibre into a single TE polarization in a waveguide.

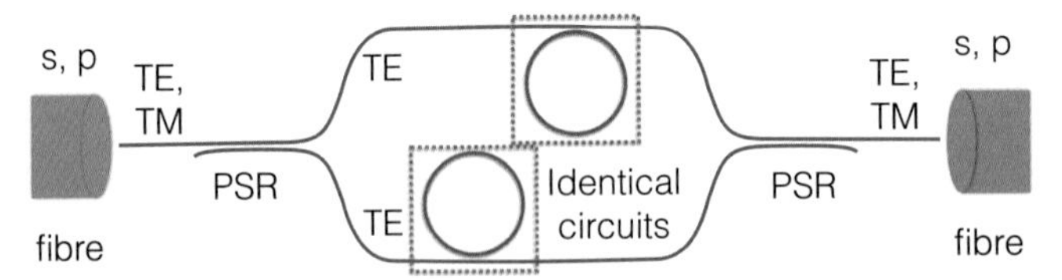

(a) Polarization diversity using edge couplers, polarization splitters, and polarization rotators

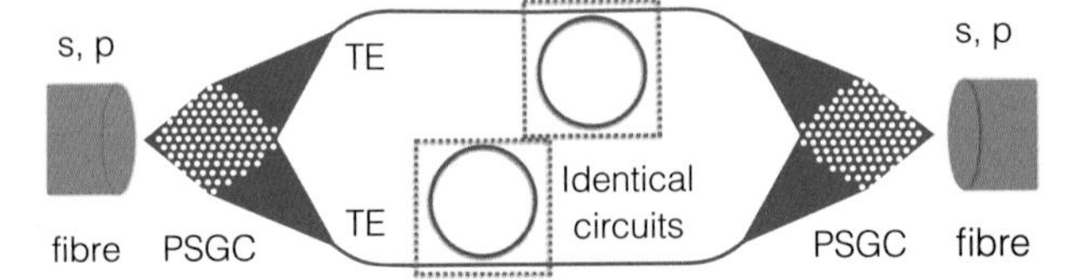

(b) Polarization diversity using 2D polarization splitting grating couplers (PSGC)

Figure 5.32 Polarization diversity

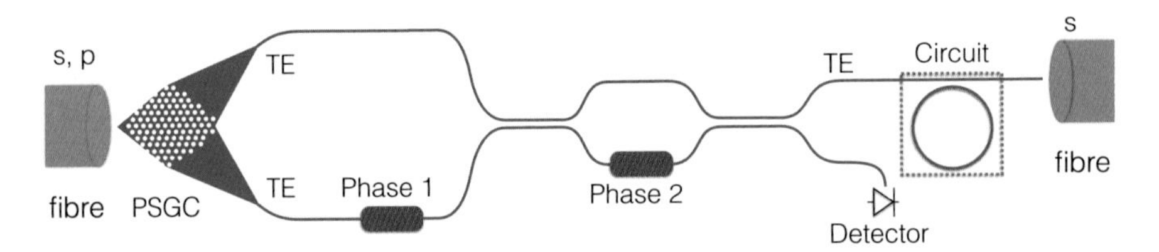

Figure 5.33 Active polarization management using two phase shifters to convert to/from an arbitrary polarization in the fibre, and a single TE polarization in a waveguide [30]. The input fibre is coupled to a polarization splitting grating coupler (PSGC) to capture both fibre polarizations; the output has a single polarization, and can be coupled to a fibre using an edge coupler (shown) or a grating coupler.

The advantage is that only a single copy of the photonic circuit is required, whereas two copies are required in the polarization diversity scheme. This approach requires an active polarization controller with two phase shifters (see Section 6.5) to adjust the phase between the s and p fibre polarizations and to match their amplitudes. This has recently been implemented using two approaches: the first is based on polarization splitting grating couplers (PSGC) [30], and the second uses edge couplers with polarization splitter-rotators (PSRs) [31]. The realization of this polarization controller concept, using grating couplers, is illustrated in Figure 5.33. The 2D grating coupler (PSGC) converts the two fibre polarization states into the TE mode of two separate silicon waveguides; at this point, the two waveguides have light with different phases and amplitudes. Two phase shifters are used to constructively interfere all the light into the top waveguide at the output. This light can be used in a photonic circuit (e.g. WDM filter, coherent receiver, optical switch), and optionally coupled back into a fibre. The bottom output port is used to monitor and implement a feedback control loop for the polarization controller; the phase shifters are adjusted to minimize the detected signal, thereby ensuring all the power is coupled to the top waveguide.

This active polarization controller can also be used to measure the state of polarization, or can be used in the opposite direction to generate an arbitrary output state of polarization.

5.5 Problems

5.1 Design a grating coupler for TM 1310 nm (straight fingers), for an incident angle of 10°.

5.2 Design a grating coupler for TM 1310 nm (focusing).

5.3 Design and optimize an edge coupler for TE 1310 nm operation, coupled to an ideal lens assembly.

5.6 Code listings

Listing 5.1 Grating coupler 2D FDTD simulation – parameters; GC_init.lsf

```
# define grating coupler parameters
period=0.66e-6;         # grating period
ff=0.5;                 # fill factor
gc_number=50;           # number of gratings

# define wafer and waveguide structure
thick_Si=0.22e-6;       # thickness of the top silicon layer
etch_depth=0.07e-6;     # etch depth;
thick_BOX=2e-6;         # thickness of the oxide cladding
thick_Clad=2e-6;        # thickness of the cladding material
Si_substrate=4e-6;      # thickness of the silicon substrate
materials;              # creates a dispersive Si material model
material='Si (Silicon) - Dispersive & Lossless';
width_wg=0.5e-6;        # width of the waveguide

# define input optical source parameters
theta0=20;              # incident angle
polarization='TE';      # TE or TM
lambda=1.55e-6;         # desired central wavelength
Position=4.5e-6;        # position of the optical source on GC

# define simulation parameters
wl_span=0.3e-6;         # wavelength span
mesh_accuracy=3;        # FDTD simulation mesh accuracy
frequency_points=100;   # global frequency points
simulation_time=1000e-15; # maximum simulation time [s]

# define optical fibre parameters
core_index=1.4682;
cladding_index=1.4629;
core_diameter=8.2e-6;
cladding_diameter=100e-6;
```

Listing 5.2 Grating coupler 2D FDTD simulation – structure drawing; http://siepic.ubc.ca/files/GC_draw.lsf

```
# 2D Grating Coupler Model

# Draw GC
redrawoff;
gap=period*(1-ff); # etched region of the grating
```

```
# add GC base
addrect;
set('name','GC_base');
set('material',material);
set('x max',(gc_number+1)*period);
set('x min',0);
set('y',0.5*(thick_Si-etch_depth));
set('y span',thick_Si-etch_depth);

# add GC teeth;
for(i=0:gc_number)
{
  addrect;
  set('name','GC_tooth');
  set('material',material);
  set('y',0.5*thick_Si);
  set('y span',thick_Si);
  set('x min',gap+i*period);
  set('x max',period+i*period);
}
selectpartial('GC');
addtogroup('GC');


# draw silicon substrate;
addrect;
set('name','Si_sub');
set('material','Si (Silicon) - Dispersive & Lossless');
set('x max',30e-6);
set('x min', -20e-6);
set('y',-1*(thick_BOX+0.5*Si_substrate));
set('y span',Si_substrate);
set('alpha',0.2);

#draw burried oxide;
addrect;
set('name','BOX');
set('material','SiO2 (Glass) - Const');
set('x max',30e-6);
set('x min',-20e-6);
set('y min',-thick_BOX);
set('y max',thick_Clad);
set('override mesh order from material database',true);
set('mesh order',3);
set('alpha',0.3);

#draw waveguide;
addrect;
set('name','WG');
set('material','Si (Silicon) - Dispersive & Lossless');
set('x min',-20e-6);
set('x max', 0);
set('y',0.11e-6);
set('y span',0.22e-6);
```

Listing 5.3 Grating coupler 2D FDTD simulation – simulation setup for Gaussian beam; GC_setup_Gaussian.lsf

```
# 2D Grating Coupler Model with Gaussian input

GC_draw; # Draw GC

# add simulation region;
addfdtd;
set('dimension','2D');
set('x max',15e-6);
set('x min',-6e-6);
set('y min',-(thick_BOX+0.2e-6));
```

```
set('y max',thick_Clad+2e-6);
set('mesh accuracy',mesh_accuracy);
set('simulation time',simulation_time);

# add monitor;
addpower;
set('name','T');
set('monitor type','2D X-normal');
set('x',-5e-6);
set('y',0.5*thick_Si);
set('y span',1e-6);

# add waveguide mode expansion monitor
addmodeexpansion;
set('name','waveguide');
set('monitor type','2D X-normal');
setexpansion('T','T');
set('x',-5e-6);
set('y',0.5*thick_Si);
set('y span',1e-6);

# add Gaussian mode
addgaussian;
set('name','fibre');
set('injection axis','y');
set('x',Position);
set('x span', 16e-6);
set('direction','Backward');
set('y',thick_Clad+1e-6);

if(polarization=='TE'){
  set('polarization angle',90);
}
else{
  set('polarization angle',0);
}

set('angle theta',-theta0);
set('center wavelength',lambda);
set('wavelength span',wl_span);
set('waist radius w0',4.5e-6);
set('distance from waist',10e-6);

# global properties
setglobalmonitor('frequency points',frequency_points);
setglobalmonitor('use linear wavelength spacing',1);
setglobalmonitor('use source limits',1);
setglobalsource('center wavelength',lambda);
setglobalsource('wavelength span',wl_span);

save('GC_Gaussian');
```

Listing 5.4 Grating coupler 2D FDTD simulation – simulation setup for optical fibre; GC_setup_fibre.lsf

```
# 2D Grating Coupler Model with Fibre

GC_draw; # Draw GC

# add simulation region;
addfdtd;
set('dimension','2D');
set('x max',15e-6);
set('x min',-3.5e-6);
set('y min',-(thick_BOX+0.2e-6));
set('y max',thick_Clad+2e-6);
set('mesh accuracy',mesh_accuracy);
set('simulation time',simulation_time);
```

```
#add waveguide mode source;
addmode;
set('name','waveguide_source');
set('x',-3e-6);
set('y',0.5*thick_Si);
set('y span',2e-6);
set('direction','Forward');
set('use global source settings',true);
set('enabled',false);


#add fibre;
theta=asin(sin(theta0*pi/180)/core_index)*180/pi;
r1 = core_diameter/2;
r2 = cladding_diameter/2;
if(theta > 89) { theta = 89; }
if(theta < -89) { theta = -89; }

thetarad = theta*pi/180;
L = 20e-6/cos(thetarad);

V1 = [ -r1/cos(thetarad), 0;
       r1/cos(thetarad), 0;
       r1/cos(thetarad)+L*sin(thetarad), L*cos(thetarad);
       -r1/cos(thetarad)+L*sin(thetarad), L*cos(thetarad)
     ];

V2 = [ -r2/cos(thetarad), 0;
       r2/cos(thetarad), 0;
       r2/cos(thetarad)+L*sin(thetarad), L*cos(thetarad);
       -r2/cos(thetarad)+L*sin(thetarad), L*cos(thetarad)
     ];

addpoly;
set('name','fibre_core');
set('x',0); set('y',0);
set('vertices',V1);
set('index',core_index);

addpoly;
set('name','fibre_cladding');
set('override mesh order from material database',1);
set('mesh order',3);
set('x',0); set('y',0);
set('vertices',V2);
set('index',cladding_index);

addmode;
set('name','fibre_mode');
set('injection axis','y-axis');
set('direction','Backward');
set('use global source settings',1);
set('theta',-theta);
span = 15*r1;
set('x span',span);
d = 0.4e-6;
set('x',d*sin(thetarad));
set('y',d*cos(thetarad));
set('rotation offset',abs(span/2*tan(thetarad)));

addpower;
set('name','fibre_top');
set('x span',span);
d = 0.2e-6;
set('x',d*sin(thetarad));
set('y',d*cos(thetarad));

addmodeexpansion;
set('name','fibre_modeExpansion');
set('monitor type','2D Y-normal');
```

```
setexpansion('fibre_top','fibre_top');
set('x span',span);
set('x',d*sin(thetarad));
set('y',d*cos(thetarad));
set('theta',-theta);
set('rotation offset',abs(span/2*tan(thetarad)));
set('override global monitor settings',false);

selectpartial('fibre');
addtogroup('fibre');
selectpartial('::fibre_modeExpansion');
setexpansion('fibre_top','::model::fibre::fibre_top');

unselectall;
select('fibre');
set('x',Position);
set('y',thick_Clad+1e-6);

# add monitor;
addpower;
set('name','T');
set('monitor type','2D X-normal');
set('x',-2.8e-6);
set('y',0.5*thick_Si);
set('y span',1e-6);

# add waveguide mode expansion monitor
addmodeexpansion;
set('name','waveguide');
set('monitor type','2D X-normal');
setexpansion('T','T');
set('x',-2.9e-6);
set('y',0.5*thick_Si);
set('y span',1e-6);


if (polarization=='TE'){
   select('fibre::fibre_mode'); set('mode selection','fundamental TM');
   select('fibre::fibre_modeExpansion'); set('mode selection','fundamental TM');
   select('waveguide_source'); set('mode selection','fundamental TM');
   select('waveguide'); set('mode selection','fundamental TM');
} else {
   select('fibre::fibre_mode'); set('mode selection','fundamental TE');
   select('fibre::fibre_modeExpansion'); set('mode selection','fundamental TE');
   select('waveguide_source'); set('mode selection','fundamental TE');
   select('waveguide'); set('mode selection','fundamental TE');
}
# global properties
setglobalmonitor('frequency points',frequency_points);
setglobalmonitor('use linear wavelength spacing',1);
setglobalmonitor('use source limits',1);
setglobalsource('center wavelength',lambda);
setglobalsource('wavelength span',wl_span);

save('GC_fibre');
```

Listing 5.5 Grating coupler 2D FDTD simulation – parameter sweeps; GC_sweeps.lsf

```
# Sweep various parameters of the grating coupler

newproject;

# Choose one of the following:
Sweep_type = 'Period';     # Period of the grating
#Sweep_type = 'FillFactor'; # Fill factor of the grating
#Sweep_type = 'Position'; # Position of the optical source on the grating
#Sweep_type = 'Angle';     # Angle of the gaussian beam
#Sweep_type = 'BOX';       # Thickness of the buried oxide
#Sweep_type = 'Cladding'; # Thickness of the cladding
#Sweep_type = 'EtchDepth'; # Etch depth on the silicon grating
```

```
GC_init;

if (Sweep_type == 'Period') {
    sweep_start = 0.62e-6; sweep_end = 0.7e-6; loop = 5;
}
if (Sweep_type == 'FillFactor') {
    sweep_start = 0.3; sweep_end = 0.6; loop = 5;
}
if (Sweep_type == 'Position') {
    sweep_start = 2e-6; sweep_end = 8e-6; loop = 10;
}
if (Sweep_type == 'Angle') {
    sweep_start = 15; sweep_end = 25; loop = 5;
}
if (Sweep_type == 'BOX') {
    sweep_start = 1e-6; sweep_end = 3e-6; loop = 50;
}
if (Sweep_type == 'Cladding') {
    sweep_start = 1e-6; sweep_end = 3e-6; loop = 50;
}
if (Sweep_type == 'EtchDepth') {
    sweep_start = 0.06e-6; sweep_end = 0.08e-6; loop = 5;
}


M_sweep = linspace(sweep_start, sweep_end, loop);
M_Tlambda = matrix(loop,1); # matrix to store transmission at central wavelength
M_T = matrix(frequency_points,loop); # matrix to store transmission for all wavelengths

for(ii=1:loop)
{
    ? ii;
    if (Sweep_type == 'Period') {
        period = M_sweep(ii,1);
    }
    if (Sweep_type == 'FillFactor') {
        ff = M_sweep(ii,1);
    }
    if (Sweep_type == 'Position') {
        Position = M_sweep(ii,1);
    }
    if (Sweep_type == 'Angle') {
        theta0 = M_sweep(ii,1);
    }
    if (Sweep_type == 'BOX') {
        thick_BOX = M_sweep(ii,1);
    }
    if (Sweep_type == 'Cladding') {
        thick_Clad = M_sweep(ii,1);
    }
    if (Sweep_type == 'EtchDepth') {
        etch_depth = M_sweep(ii,1);
    }

    switchtolayout; selectall; delete; redrawoff;

    #GC_setup_Gaussian;
    # or:
    GC_setup_fibre;

    run;
    T = transmission('T');
    M_T(1:frequency_points,ii) = T;
    M_Tlambda(ii,1) = T(floor(frequency_points/2));
    switchtolayout;
}

WL=linspace(lambda-0.5*wl_span,lambda+0.5*wl_span,frequency_points);
```

```
for(jj=1:loop)
{
    plot(WL, abs(M_T(1:frequency_points,jj)));
    holdon;
}
?10*log10(max(abs(M_T))); # lowest insertion loss
holdoff;
plot(M_sweep, abs(M_Tlambda));

matlabsave('GC_sweep_'+Sweep_type);
```

Listing 5.6 Design and layout of a focusing grating coupler, in Mentor Graphics Pyxis; UGC.ample

```
// Design a grating coupler using analytic equations, effective index calculations
// Create the layout for the grating coupler
function UGC(query: optional boolean {default = false}),INVISIBLE
{
  if(query) { return[@point, @block, []] }
  local device = $get_device_iobj();
  local wl = $get_property_value(device,"wl"); // wavelength
  local etch_depth = $get_property_value(device,"etch_depth");
  local Si_thickness = $get_property_value(device,"Si_thickness");
  local incident_angle = $get_property_value(device,"incident_angle");
  local wg_width = $get_property_value(device,"wg_width");
  local n_cladd = $get_property_value(device,"n_cladd");
  local pl = $get_property_value(device,"pl"); // polarization "TE" or "TM"
  local ff = $get_property_value(device,"ff"); // fill factor
  build_UGC(wl,etch_depth,Si_thickness,incident_angle,wg_width,n_cladd,pl, ff);
}

function build_UGC(wl, etch_depth, Si_thickness, incident_angle, wg_width, n_cladd, pl, ff)
{
  local neff=effective_index(wl, etch_depth, Si_thickness, n_cladd, pl, ff);
  local ne_fiber=1; // effective index of the mode in the air
  local period= wl/ (neff- sin(rad(incident_angle)) *ne_fiber);
  $writes_file($stdout,"Grating_Period is:", period, "\n");
  $add_point_device("Draw_GC", @block, [], [@to, [0, 0], @rotation, 0.0, @flip, "none"],
       [["ff", ff], ["incident_angle", incident_angle], ["n_clad", n_cladd], ["neff",
       neff], ["period", period], ["segnum", "100"], ["wg_width", wg_width], ["wl", wl]],
       @placed);
  //output port
  $unselect_all(@nofilter);
  $add_shape([[-1, -wg_width/2], [0, wg_width/2]], "Si", @both);
  $make_port(@signal, @bidirectional,"opt_in");
}

function UGC_parameters(
  layer:optional number {default=1},
  wl:optional number {default=1.55},
  etch_depth:optional number {default=0.07},
  Si_thickness:optional number {default=0.22},
  incident_angle:optional number {default=20},
  wg_width:optional number {default=0.5},
  n_cladd:optional number {default=1.44},
  pl:optional string {default="TE"},
  ff:optional number {default=0.5} )
{ return [ ["wl",$g(wl)], ["etch_depth",$g(etch_depth)],
      ["Si_thickness",$g(Si_thickness)], ["incident_angle",$g(incident_angle)],
      ["wg_width",$g(wg_width)], ["n_cladd",$g(n_cladd)], ["pl",pl], ["ff",$g(ff)] ]; }
```

Listing 5.7 Effective index calculation, Mentor Graphics Pyxis; Effective_ Index.ample

```
 // calculate the effective index (neff) of a grating coupler
function Effective_index(
  wl:number {default=1.55},
  etch_depth:number {default=0.07},
  Si_thickness:number {default=0.22},
  n_cladd:number {default=1.444},
```

```
    pl:string {default="TE"},
    ff:number {default=0.5} )
  {
    local point=1001, ii=0, jj=0, kk=0, mm=0, nn=0, n0=n_cladd, n1=0, n3=1.444;
    // Silicon wavelength-dependant index of refraction:
    local n2=sqrt(7.9874+(3.68*pow(3.9328,2)*pow(10,30)) /
          ((pow(3.9328,2)*pow(10,30)-pow(2*3.14*3*pow(10,8) /(wl*pow(10,-6)),2))));
    local delta=n0-n3, t=Si_thickness, t_slot=t-etch_depth;
    local k0 = 2*3.14159/wl;
    local b0 = $create_vector(point-1), te0 = $create_vector(point-1), te1 =
         $create_vector(point-1), tm0 = $create_vector(point-1), tm1 =
         $create_vector(point-1), h0 = $create_vector(point-1), q0 =
         $create_vector(point-1), p0 = $create_vector(point-1), qbar0 =
         $create_vector(point-1), pbar0 = $create_vector(point-1);
    local mini_TE=0, index_TE=0, mini_TE1=0, index_TE1=0, mini_TM=0, index_TM=0, mini_TM1=0,
         index_TM1=0, nTE=0, nTE1=0, nTM=0, nTM1=0, ne=0;

    //------ calculating neff for the silicon layer ----
    if( delta<0)  {
      n1=n3;
    } else  {
      n1=n0;
    }

    for(ii=0;ii<point-1;ii=ii+1)  {
      b0[ii]= n1*k0+(n2-n1)*k0/(point-10)*ii;
    }

    for(jj=0;jj<point-1;jj=jj+1) {
      h0[jj] = sqrt( abs(pow(n2*k0,2) - pow(b0[jj],2)));
      q0[jj] = sqrt( abs(pow(b0[jj],2) - pow(n0*k0,2)));
      p0[jj] = sqrt( abs(pow(b0[jj],2) - pow(n3*k0,2)));
    }

    for(kk=0; kk<point-1; kk=kk+1) {
      pbar0[kk] = pow(n2/n3,2)*p0[kk];
      qbar0[kk] = pow(n2/n0,2)*q0[kk];
    }

  //----calculating neff for TE mode--------

     if (pl=="TE") {
      for (nn=0;nn<point-1;nn=nn+1) {
        te0[nn] = tan( h0[nn]*t )-(p0[nn]+q0[nn])/h0[nn]/(1-p0[nn]*q0[nn]/pow(h0[nn],2));
        te1[nn] = tan( h0[nn]*t_slot
             )-(p0[nn]+q0[nn])/h0[nn]/(1-p0[nn]*q0[nn]/pow(h0[nn],2));
      }

      local abs_te0=abs(te0);
      local abs_te1=abs(te1);
      mini_TE=$vector_min(abs_te0);
      mini_TE1=$vector_min(abs_te1);
      index_TE=$vector_search(mini_TE,abs(te0),0);
      index_TE1=$vector_search(mini_TE1,abs(te1),0);
      nTE=b0[index_TE]/k0;
      nTE1=b0[index_TE1]/k0;

      do  {
        abs_te0[index_TE]=100;
        mini_TE=$vector_min(abs_te0);
        index_TE=$vector_search(mini_TE,abs(te0),0);
        nTE=b0[index_TE]/k0;
      }
      while ( nTE<2 || nTE>3); // && is logic and || is logic or

      do  {
        abs_te1[index_TE1]=100;
        mini_TE1=$vector_min(abs_te1);
        index_TE1=$vector_search(mini_TE1,abs(te1),0);
        nTE1=b0[index_TE1]/k0;
```

```
    }
    while ( nTE1<2 || nTE1>3);

    ne=ff*nTE+(1-ff)*nTE1;
  }

//-----calculating neff for TM mode----

  else if (pl=="TM") {
    for (mm=0;mm<point-1;mm=mm+1) {
      tm0[mm] = tan(h0[mm]*t)- h0[mm]*(pbar0[mm]+qbar0[mm]) /
            (pow(h0[mm],2)-pbar0[mm]*qbar0[mm]);
      tm1[mm] = tan(h0[mm]*t_slot)- h0[mm]*(pbar0[mm]+qbar0[mm]) /
            (pow(h0[mm],2)-pbar0[mm]*qbar0[mm]);
    }

    local abs_tm0=abs(tm0);
    local abs_tm1=abs(tm1);
    mini_TM=$vector_min(abs(tm0));
    mini_TM1=$vector_min(abs(tm1));
    index_TM=$vector_search(mini_TM,abs(tm0),0);
    index_TM1=$vector_search(mini_TM1,abs(tm1),0);
    nTM=b0[index_TM]/k0;
    nTM1=b0[index_TM1]/k0;

    do  {
      abs_tm0[index_TM]=100;
      mini_TM=$vector_min(abs_tm0);
      index_TM=$vector_search(mini_TM,abs(tm0),0);
      nTM=b0[index_TM]/k0;
    }
    while ( nTM<1.5 || nTM>3);

    do  {
      abs_tm1[index_TM1]=100;
      mini_TM1=$vector_min(abs_tm1);
      index_TM1=$vector_search(mini_TM1,abs(tm1),0);
      nTM1=b0[index_TM1]/k0;
    }
    while ( nTM1<1.5 || nTM1>3);

    ne=ff*nTM+(1-ff)*nTM1;
  } else {
    $writes_file("Please type TE or TM for pl(polarization)");
  }

  $writes_file($stdout,"ne=",ne,"\n");
  return ne;
}
```

Listing 5.8 Grating coupler mask layout generation script, Mentor Graphics Pyxis; Draw_GC.ample

```
// Create the layout for the grating coupler
function Draw_GC(query: optional boolean {default = false}),INVISIBLE
{
  if(query) { return[@point, @block, []] }
  local device = $get_device_iobj();
  local wl = $get_property_value(device,"wl"); // wavelength
  local period = $get_property_value(device,"period");
  local ff = $get_property_value(device,"ff"); // fill factor
  local n_clad = $get_property_value(device,"n_clad");
  local incident_angle = $get_property_value(device,"incident_angle");
  local wg_width = $get_property_value(device,"wg_width");
  local neff = $get_property_value(device,"neff");
  local segnum = $get_property_value(device,"segnum"); // number of points in the curves
  build_Draw_GC(wl, period, ff, n_clad, incident_angle, wg_width, neff, segnum);
}

function build_Draw_GC( wl,period,ff, n_clad, incident_angle, wg_width, neff,segnum)
{
```

```
  //Save Original user settings
  local selectable_types_orig = $get_selectable_types();
  local selectable_layers_orig = $get_selectable_layers();
  local autoselect_orig = $get_autoselect();

  //Set up selection settings
  $set_selectable_types(@replace, [@shape, @path, @pin, @overflow,@row, @property_text,
       @instance, @array, @device, @via_object, @text, @region, @bisector, @channel,
       @slice], @both);
  $set_selectable_layers(@replace, ["0-4096"]);
  $set_autoselect(@true);

  local seg_points = segnum+1;
  local arc_vec = $create_vector(2*seg_points),taper_vec = $create_vector(seg_points+2);
  local nf=1.44, e =nf*sin(rad(incident_angle))/neff,angle_e=38,
       gc_number=$round(18/period);
  local i=0,j=0,k=0,x_r=0,y_r=0,x_l=0,y_l=0,t_lx=0,t_ly=0,phi0=0,phi1=0,phi2=0,r=0,
       r_taper=0;
  local grating_width=period*ff;
  local N=$round((18+period-grating_width)*(1+e)*neff/wl);

  for(k=0;k<seg_points;k=k+1)
  {
      phi0=rad(180-angle_e/2+angle_e/segnum*k);
      r_taper=(N*wl/neff)/(1-e*cos(phi0));
      t_lx=r_taper*cos(phi0);
      t_ly=r_taper*sin(phi0);

      taper_vec[k] = [t_lx,t_ly];
      taper_vec[seg_points] = [0,0-1/2*wg_width];
      taper_vec[seg_points+1] = [0,0+1/2*wg_width];
  }
  $add_shape(taper_vec,"Si");

  for(j=0;j<gc_number;j=j+1)
  {
    for(i=0;i<seg_points;i=i+1)
    {
      phi1=rad(180-angle_e/2+angle_e/segnum*i);
      r=(N*wl/neff)/(1-e*cos(phi1));
      x_r=(r-grating_width+(j+1)*period)*cos(phi1);
      y_r=(r-grating_width+(j+1)*period)*sin(phi1);
      arc_vec[i] = [x_r,y_r];

      phi2=rad(180+angle_e/2-angle_e/segnum*i);
      x_l=(r+(j+1)*period)*cos(phi2);
      y_l=(r+(j+1)*period)*sin(phi2);
      arc_vec[seg_points+i] = [x_l,y_l];
    }
    $add_shape(arc_vec,"Si");
  }

  // generate the shallow etch area
  $add_shape([[-42,15],[-12,15],[-12,-15],[-42,-15]],'SiEtch1');

  //output port
  $unselect_all(@nofilter);
  $add_shape([[-1, -wg_width/2], [0, wg_width/2]], "Si", @both);
  $make_port(@signal, @bidirectional,"opt_in");

  //Restore original user settings
  $set_selectable_types(@replace,
       (selectable_types_orig[0]==void)?[]:selectable_types_orig[0],
        selectable_types_orig[1]);
  $set_selectable_layers(@replace, selectable_layers_orig);
  $set_autoselect(autoselect_orig);
}
function Draw_GC_parameters(
  layer:optional number {default=1},
  wl:optional number {default=1.55},
```

```
    period:optional number {default=0.66},
    ff:optional number {default=0.5},
    n_clad:optional number {default=1.44},
    incident_angle:optional number {default=20},
    wg_width:optional number {default=0.5},
    neff:optional string {default=2.8},
    segnum:optional string {default=100})
  { return [ ["wl",$g(wl)], ["period",$g(period)], ["ff",$g(ff)], ["n_clad",$g(n_clad)],
        ["incident_angle",$g(incident_angle)], ["wg_width",$g(wg_width)], ["neff",$g(neff)],
        ["segnum",$g(segnum)] ]; }
```

Listing 5.9 Mode overlap calculations for the edge coupler with a Gaussian beam, using Lumerical MODE; edgecoupler_mode.lsf

```
# Lumerical MODE script to estimate edge coupling efficiency for the nanotaper with
    different gaussian beams
cleardcard;
closeall; redrawoff;

wg_list=[180e-9];
t_list=[220e-9];
lambda_list=[1.31e-6, 1.55e-6];
NA_list=0.1:0.02:0.8;
PLOT_misalignment = 1;
PLOT_modeprofiles = 1;

write ('edgecoupler_mode.txt','WG_w, WG_t, Wavelength, Best NA TE, Best coupling TE, Best
    NA TM, Best coupling TM');

for (l=1:length(t_list)) {
  for (k=1:length(wg_list)) {
    switchtolayout; new;

    # Draw the silicon nano-taper
    addrect; set('name','Si waveguide');
    set('x span',wg_list(k));
    set('y min',0); set('y max',t_list(l));
    set('material','Si (Silicon) - Palik');

    addrect; set('name','Oxide');
    set('x span',10e-6);
    set('y min',-2e-6); set('y max',2e-6);
    set('material','SiO2 (Glass) - Palik');
    set('override mesh order from material database',1);
    set('mesh order',3);

    addmode; # create simulate mesh
    set('x span',6e-6); set('y span',4e-6);
    set('mesh cells x',100); set('mesh cells y',200);

    addmesh; # mesh override, higher resolution in the waveguide.
    set('x span',0.5e-6); set('y min',-0.1e-6); set('y max',t_list(l)+0.1e-6);
    set('dx',10e-9); set('dy',10e-9);

    run;

    for (j=1:length(lambda_list)) {

      # for energy density calculation
      # find the material dispersion (using 2 frequency points)
      switchtolayout; select('MODE');
      set("wavelength", lambda_list(j)*(1 + .001) );
      run; mesh;
      f1 = getdata("MODE::data::material","f");
      eps1 = pinch(getdata("MODE::data::material","index_x"))^2;
      switchtolayout; set("wavelength", lambda_list(j)*(1 - .001) );
      run; mesh;
      f3 = getdata("MODE::data::material","f");
      eps3 = pinch(getdata("MODE::data::material","index_x"))^2;
```

```
re_dwepsdw = real((f3*eps3-f1*eps1)/(f3-f1));

FILE='EdgeCoupling_'+num2str(lambda_list(j)*1e9) +'nm_W='+ num2str(wg_list(k)*1e9) +
     'nm_t=' +num2str(t_list(1)*1e9)+'nm';

setanalysis('wavelength',lambda_list(j) );
setanalysis('search',1);
setanalysis('use max index',0);
setanalysis('n',1.45);
setanalysis('number of trial modes',20);

n=findmodes;

# find out which mode is TE and which is TM
pol1=getdata('mode1','TE polarization fraction');
pol2=getdata('mode2','TE polarization fraction');
if (pol1 > 0.8) { TEmode='mode1'; } if (pol2 > 0.8) { TEmode='mode2'; }
if (pol1 < 0.2) { TMmode='mode1'; } if (pol2 < 0.2) { TMmode='mode2'; }

# save the mode profiles
if (PLOT_modeprofiles) {
  x = getdata(TEmode,"x"); y=getdata(TEmode,"y");
  E1_TE = pinch(getelectric(TEmode)); H1 = pinch(getmagnetic(TEmode));
  W_TE = 0.5*(re_dwepsdw*eps0*E1_TE+mu0*H1);
  E1_TM = pinch(getelectric(TMmode)); H1 = pinch(getmagnetic(TMmode));
  W_TM = 0.5*(re_dwepsdw*eps0*E1_TM+mu0*H1);
}
setanalysis('sample span',6e-6);
edge_coupling_TE=matrix(length(NA_list),1);
edge_coupling_TM=matrix(length(NA_list),1);
gaussianbeams=matrix(length(NA_list)*2,1);

setanalysis('polarization angle',0); # TE
for (i=1:length(NA_list)) {
  setanalysis('NA',NA_list(i));
  beam_name=createbeam;
  cou=overlap(TEmode,'gaussian'+num2str(i),0,t_list(1)/2,0);
  ?edge_coupling_TE(i)=cou(2); # power coupling
}

setanalysis('polarization angle',90); # TM
for (i=1:length(NA_list)) {
  setanalysis('NA',NA_list(i));
  beam_name=createbeam;
  cou=overlap(TMmode,'gaussian'+num2str(i+length(NA_list)), 0,t_list(1)/2,0);
  ?edge_coupling_TM(i)=cou(2); # power coupling
}

plot(NA_list, edge_coupling_TE, edge_coupling_TM, 'Lens NA', 'Coupling efficiency',
     num2str(lambda_list(j)*1e6)+ 'um,W='+ num2str(wg_list(k)*1e9)+'nm,t='+
     num2str(t_list(1)*1e9)+'nm');
legend('TE','TM');
setplot ('y min', max ([ min( [edge_coupling_TE, edge_coupling_TM]), 0.2] ) );
setplot ('y max',1);
exportfigure(FILE+'(linear).jpg');
plot(NA_list, 10*log10(edge_coupling_TE), 10*log10(edge_coupling_TM), 'Lens NA',
     'Coupling efficiency (dB)', num2str(lambda_list(j)*1e6)+'um,W='+
     num2str(wg_list(k)*1e9)+'nm,t='+ num2str(t_list(1)*1e9)+'nm');
legend('TE','TM');
setplot ('y min', max ([ min( 10*log10([edge_coupling_TE, edge_coupling_TM])), -4] )
     );
setplot ('y max',0);
exportfigure(FILE+'(dB).jpg');

best_coupling_TE = max(10*log10(edge_coupling_TE));
     posTE=find(10*log10(edge_coupling_TE), best_coupling_TE); best_NA_TE =
     NA_list(posTE);
best_coupling_TM = max(10*log10(edge_coupling_TM));
     posTM=find(10*log10(edge_coupling_TM), best_coupling_TM); best_NA_TM =
     NA_list(posTM);
```

```
        # save the gaussian mode profiles
        if (PLOT_modeprofiles) {
          x_g = getdata('gaussian'+num2str(posTE),"x");
          y_g=getdata('gaussian'+num2str(posTE),"y");
          E1 = pinch(getelectric('gaussian' +num2str(posTE)));
          H1 = pinch(getmagnetic('gaussian' +num2str(posTE)));
          W_g_TE = 0.5*(1*eps0*E1+mu0*H1);
          E1 = pinch(getelectric('gaussian' +num2str(posTM+length(NA_list))));
          H1 = pinch(getmagnetic('gaussian' +num2str(posTM+length(NA_list))));
          W_g_TM = 0.5*(1*eps0*E1+mu0*H1);
        }

        # calculate the fibre misalignment sensitivity
        if (PLOT_misalignment) {
          xlist=[-2:.1:2]*1e-6;
          xTE_misalign=matrix(length(xlist)); xTM_misalign=matrix(length(xlist));
          yTE_misalign=matrix(length(xlist)); yTM_misalign=matrix(length(xlist));
          for (m=1:length(xlist)) {
            cou=overlap(TEmode,'gaussian'+num2str(i), xlist(m), t_list(1)/2,0);
                 xTE_misalign(m)=cou(2);
            cou=overlap(TMmode,'gaussian'+num2str(i+length(NA_list)),
                 xlist(m),t_list(1)/2,0); xTM_misalign(m)=cou(2);
            cou=overlap(TEmode,'gaussian'+num2str(i), 0, t_list(1)/2+xlist(m),0);
                 yTE_misalign(m)=cou(2);
            cou=overlap(TMmode,'gaussian'+num2str(i+length(NA_list)), 0,
                 t_list(1)/2+xlist(m),0); yTM_misalign(m)=cou(2);
          }
          plot (xlist, xTE_misalign, yTE_misalign, xTM_misalign, yTM_misalign);
          legend('xTE_misalign', 'yTE_misalign', 'xTM_misalign', 'yTM_misalign');
        }
        matlabsave ( FILE + '.mat' );
        write ('edgecoupler_mode.txt', num2str(wg_list(k)*1e9)+', '+
             num2str(t_list(1)*1e9)+', '+ num2str(lambda_list(j)*1e6) +', ' +
             num2str(best_NA_TE)+', '+ num2str(best_coupling_TE) +', '+
             num2str(best_NA_TM)+', '+ num2str(best_coupling_TM) );
      }
    }
  }
```

Listing 5.10 FDTD simulation of the nano-taper edge coupler, using Lumerical FDTD; nanotaper_fdtd.lsf

```
# FDTD Solutions script
# Draw the silicon nano-taper, setup 3D FDTD simulations
# inputs example:
wg_w1=500e-9; wg_w2=180e-9; wg_t=220e-9; wg_l=20e-6;
BC='Metal'; # Metal is faster than PML, error introduced is < 0.02 dB in coupling.
Wavelength=1.55e-6;
PLOT_FIELD=1;
MESH=2;
# Mesh coupling dB  (convergence test at 1550 nm, L=10um)
# 1 -1.62
# 2 -1.55
# 3 -1.55

newproject; redrawoff;

addpyramid; set('name','Si taper');
set('x span bottom',wg_t); set('x span top',wg_t);
set('y span bottom',wg_w1); set('y span top',wg_w2);
set('z span',wg_l);
set('first axis','y'); set('rotation 1',90);
set('x',-wg_l/2); set('z',wg_t/2); set('y',0);
set('material','Si (Silicon) - Palik');

addrect; set('name','Oxide');
set('x min',-wg_l); set('x max',1e-6);
set('z min',-2e-6); set('z max',2.1e-6);
```

```
set('y span', 10e-6);
set('material','SiO2 (Glass) - Palik');
set('override mesh order from material database',1);
set('mesh order',3); set('alpha',0.2);

addrect; set('name','SiSubstrate');
set('x min',-wg_l); set('x max',1e-6);
set('z min',-3e-6); set('z max',-2e-6);
set('y span', 10e-6);
set('material','Si (Silicon) - Palik');
set('override mesh order from material database',1);
set('mesh order',3); set('alpha',0.2);

addfdtd; # create simulate mesh
set('x min',-wg_l+0.5e-6); set('x max',2e-6);
set('z span',6e-6); set('y span', 6e-6);
set('simulation time', 200e-15+wg_l/c*4.5);
set('mesh type','auto non-uniform'); set('mesh accuracy',MESH);
set('x min bc', 'PML'); set('x max bc', 'PML');
#set('y min bc', 'Anti-Symmetric'); set('y max bc', BC); # problem with field import into
     MODE
set('y min bc', BC); set('y max bc', BC);
set('z min bc', BC); set('z max bc', BC);


if(1) {
addmesh; # mesh override, higher resolution in the waveguide.
set('y span',wg_w1); set('x span',0); set('z span', wg_t*2); set('z',wg_t/2);
set('set equivalent index', 1);
set('override y mesh',1); set('equivalent y index',5);
set('override z mesh',1); set('equivalent z index',5);
}

addmode;
set('injection axis','x-axis'); set('set wavelength',1);
set('center wavelength', Wavelength); set('wavelength span', 200e-9);
set('x', -wg_l+1e-6); set('y',0); set('y span', 2e-6);
set('z', wg_t/2); set('z span', 2e-6);
updatesourcemode;

if (PLOT_FIELD) {
  addpower; set('name','XY');
  set('override global monitor settings',1);
  set('use source limits',0); set('frequency points',1);
  set('wavelength center', Wavelength);
  set('monitor type', '2D Z-normal');
  set('x min',-wg_l+0.5e-6); set('x max',2e-6);
  set('y span', 6e-6); set('z',wg_t/2);

  addpower; set('name','XZ');
  set('override global monitor settings',1);
  set('use source limits',0); set('frequency points',1);
  set('wavelength center', Wavelength);
  set('monitor type', '2D Y-normal');
  set('x min',-wg_l+0.5e-6); set('x max',2e-6);
  set('z span', 6e-6);

  addpower; set('name','reflected'); # about 5% reflection observed from the nano-tip &
      oxide.
  set('monitor type', '2D X-normal');
  set('x', -wg_l+0.8e-6); set('z span',6e-6); set('y span', 6e-6);
  set('override global monitor settings',1);
  set('use source limits',0); set('frequency points',1);
  set('wavelength center', Wavelength);
}

addpower; set('name','output');
set('monitor type', '2D X-normal');
set('x', 1.05e-6); set('z span',6e-6); set('y span', 6e-6);
```

```
# Necessary to export only one wavelength point; difficult in multi-wavelength overlap
     integrals
set('override global monitor settings',1);
set('use source limits',0); set('frequency points',1);
set('wavelength center', Wavelength);


save('nanotaper');
run;


if (PLOT_FIELD) {
  Exy=sqrt(abs(getdata('XY','Ex'))^2 + abs(getdata('XY','Ey'))^2 +
       abs(getdata('XY','Ez'))^2);
  x=getdata('XY','x'); y=getdata('XY','y');
  image(x,y,Exy/max(Exy)); # plot |E|

  Exz=sqrt(abs(getdata('XZ','Ex'))^2 + abs(getdata('XZ','Ey'))^2 +
       abs(getdata('XZ','Ez'))^2);
  z=getdata('XZ','z');
  image(x,z,Exz/max(Exz)); # plot |E|

  Eo=sqrt(abs(getdata('output','Ex'))^2 + abs(getdata('output','Ey'))^2 +
       abs(getdata('output','Ez'))^2);
  y=getdata('output','y'); z=getdata('output','z');
  image(y,z,Eo/max(Eo)); # plot |E|

  Eff = farfield3d("output",1);
  ux = farfieldux("output",1); uy = farfielduy("output",1);
  image(ux,uy,Eff,"","","Far field of output","polar");

  plot(ux*90, Eff(75,1:150)/max(Eff(75,1:150)),
       Eff(1:150,75)/max(Eff(1:150,75)),"Angle","Far field intensity");
  legend ('Vertical far-field','Horizontal far-field');
}

savedcard("nanotaper","::model::output");
matlabsave ("nanotaper_output");
```

Listing 5.11 Mode overlap calculations for the edge coupler with FDTD simulations of the nano-taper edge coupler, using Lumerical MODE; nanotaper_overlap.lsf

```
# MODE Solutions script
#   do mode overlap calculations of gaussian beams with FDTD field

new; cleardcard;
FILE = "nanotaper";
loaddata(FILE);
addfde;

NA_list=0.1:0.02:0.9;
PLOT_misalignment = 0;
PLOT_modeprofiles = 0;
Polarization = 0; # 0=TE, 90=TM;

wg_list=[180e-9];
t_list=[220e-9];
lambda_list=c/getdata('output','f');

write ('edgecoupler_FDTD.txt','FILE, WG_w, WG_t, Wavelength, Best NA, Best coupling');
for (l=1:length(t_list)) {
  for (k=1:length(wg_list)) {
    for (j=1:length(lambda_list)) {
      setanalysis('wavelength',lambda_list(j) );
      setanalysis('sample span',6e-6);
      setanalysis('number of plane waves', 200);
      edge_coupling=matrix(length(NA_list),1);
      gaussianbeams=matrix(length(NA_list)*2,1);
      setanalysis('polarization angle',Polarization);
```

```
      setanalysis('beam direction','2D X normal');
      for (i=1:length(NA_list)) {
        setanalysis('NA',NA_list(i));
        beam_name=createbeam;
        cou=overlap('output','gaussian'+num2str(i),0,t_list(1)/2,0);
        ?edge_coupling(i)=cou(2); # power coupling
      }
      plot(NA_list, 10*log10(edge_coupling),'Lens NA', 'Coupling efficiency (dB)', FILE);
      setplot ('y min', max ([ min( 10*log10(edge_coupling)), -4] ) );
      setplot ('y max',0);
      exportfigure(FILE+'(dB).jpg');

      best_coupling = max(10*log10(edge_coupling)); pos=find(10*log10(edge_coupling),
           best_coupling); best_NA = NA_list(pos);

      # calculate the fibre misalignment sensitivity
      if (PLOT_misalignment) {
        xlist=[-2:.1:2]*1e-6;
        x_misalign=matrix(length(xlist));
        y_misalign=matrix(length(xlist));
        for (m=1:length(xlist)) {
          cou=overlap('output','gaussian'+num2str(i), xlist(m), t_list(1)/2,0);
               x_misalign(m)=cou(2);
          cou=overlap('output','gaussian'+num2str(i), 0, t_list(1)/2+xlist(m),0);
               y_misalign(m)=cou(2);
        }
        plot (xlist, x_misalign,y_misalign);
        legend('x_misalign','y_misalign');
      }
      matlabsave ( FILE + '.mat' );
      write (FILE+'.txt',FILE+ ', '+num2str(wg_list(k)*1e9)+', '+num2str(t_list(1)*1e9)+',
           '+ num2str(lambda_list(j)*1e6) +', ' + num2str(best_NA)+',
           '+num2str(best_coupling) );
    }
  }
}
```

Listing 5.12 Eigenmode expansion method simulation for the edge coupler consisting of a nano-taper, overlay waveguide, and optical fibre, using Lumerical MODE; spot_size_converter.lsf

```
#####################################################################
#
# spot size converter
#
# Copyright 2014 Lumerical Solutions
#####################################################################
switchtolayout; newproject;
clear; deleteall; cleardcard;

filename = "spot_size_converter";

#Define materials ################################################
mat_sub = "SiO2 (Glass) - Palik" ;
mat_Si = "Si (Silicon) - Palik";
mat_Ox = "SiO2 (Glass) - Palik";
SiON_index = 1.5;

# ******************************************************************************
## Geometry
# x-axis: propagation
# y-z: cross section of wg
# ******************************************************************************
#add substrate
addrect;
select("rectangle");
set("name","substrate");
set("x span",20e-6);
set("x",0); #period on either side.
set("y", 0);
```

```
set("y span", 10e-6);
set("z",-2.5e-6);
set("z span", 5e-6);
set("material", mat_sub);
unselectall;

#add input waveguide
addrect;
select("rectangle");
set("name","input");
set("x span",5e-6);
set("x",-7.5e-6);
set("y", 0);
set("y span",0.4e-6);
set("z",0.1e-6);
set("z span", 0.2e-6);
set("material",mat_Si);
unselectall;


#add taper
lx_top = 0.4e-6;
lx_base = 0.08e-6;
y_span = 10e-6;
z_span = 0.2e-6;
z = 0.1e-6;
x = 0;
y = 0;

V=matrix(4,2);
V(1,1:2)=[-lx_base/2,-y_span/2];
V(2,1:2)=[-lx_top/2,y_span/2];
V(3,1:2)=[lx_top/2,y_span/2];
V(4,1:2)=[lx_base/2,-y_span/2];
addpoly;
  set("x",0);
  set("y",0);
  set("z",0.1e-6);
  set("z span",z_span);
  set("vertices",V);
  set("material",mat_Si);
set("name","taper");
set("first axis", "z");
set("rotation 1", 90);


#add low index polymer
addrect;
select("rectangle");
set("name","SiON");
set("x span",15e-6);
set("x",2.5e-6);
set("y", 0);
set("y span",3e-6);
set("z",1.5e-6);
set("z span", 3e-6);
set("index",SiON_index);
set("override mesh order from material database",1);
set("mesh order",3);
unselectall;

# ********************************************************************************
# Add EME solver
# ********************************************************************************
addeme;
set("solver type", "3D: X Prop");
set("background index", 1.465);
set("wavelength", 1.5e-6);
set("z", 0.5e-6);
set("z span", 7e-6);
```

```
set("y",0);
set("y span",5.5e-6);
set("x min", -8e-6);
set("number of cell groups", 3);
set("display cells", 1);
set("number of modes for all cell groups", 20);
set("number of periodic groups", 1);
set("energy conservation", "make passive"); # or "none", "conserve energy"
set("subcell method", [0;1;0]);
set("cells", [1;19; 1]);
set("group spans",[3e-6; 10e-6; 3e-6]);

#update port configuration
setnamed("EME::Ports::port_1", "y", 0);
setnamed("EME::Ports::port_1", "y span", 5.5e-6);
setnamed("EME::Ports::port_1", "z", 0);
setnamed("EME::Ports::port_1", "z span", 7e-6);
setnamed("EME::Ports::port_1", "mode selection", "fundamental mode");

setnamed("EME::Ports::port_2", "y", 0);
setnamed("EME::Ports::port_2", "y span", 5.5e-6);
setnamed("EME::Ports::port_2", "z", 0);
setnamed("EME::Ports::port_2", "z span", 7e-6);
setnamed("EME::Ports::port_2", "mode selection", "fundamental mode");


addmesh; #mesh override.
set("x",0); set("x span", 10e-6);
set("y", 0); set("y span", 0.45e-6);
set("z", 0.1e-6); set("z span", 0.2e-6);
set("set mesh multiplier",1);
set("y mesh multiplier",5);
set("z mesh multiplier",5);


addemeindex;
set("name", "index");
set("x",0); set("x span", 20e-6);
set("y", 0); set("y span", 6e-6);
set("z", 0.1e-6);

addemeprofile;
set("name", "profile");
set("monitor type", "2D Y-normal");
set("x",0); set("x span", 20e-6);
set("y", 0);
set("z", 0.5e-6);set("z span",8e-6);


# ****************************************************************************
# Run: calculate modes
# ****************************************************************************
save(filename);
run;

# ****************************************************************************
# Propagate fields
# ****************************************************************************

setemeanalysis("source port", "port 1");

setemeanalysis("Propagation sweep", 1);
setemeanalysis("parameter", "group span 2");
setemeanalysis("start", 10e-6);
setemeanalysis("stop", 200e-6);

step=10;
setemeanalysis("interval", step);

emesweep;
```

```
S = getemesweep('S');

# ********************************************************************************
# Account for fiber overlap
# ********************************************************************************
switchtolayout;
cleardcard;
addfde;
set("solver type","2D X normal");
set("background index", 1.465);
set("z", 0.5e-6);
set("z span", 7e-6);
set("y",0);
set("y span",5.5e-6);
set("x", 8e-6);

setanalysis("wavelength", 1.5e-6);
findmodes;

# Create fiber mode
setanalysis("NA",0.4);
setanalysis("beam direction","2D X normal");
createbeam;

setanalysis("shift d-card center",1);
out = overlap("mode1","gaussian1",0,0,getnamed("SiON","z"));
power = out(2);

# ********************************************************************************
# Plot result
# ********************************************************************************
plot(S.group_span_2,abs(S.s21)^2,(abs(S.s21)^2)*out(2),"taper length (um)","Transmission");
legend('Transmission into SiON mode', 'Transmission into fibre');
```

References

[1] Sharee McNab, Nikolaj Moll, and Yurii Vlasov. "Ultra-low loss photonic integrated circuit with membrane-type photonic crystal waveguides". *Optics Express* **11**.22 (2003), pp. 2927–2939 (cit. on pp. 162, 164).

[2] A. Mekis, S. Abdalla, D. Foltz, *et al.* "A CMOS photonics platform for high-speed optical interconnects". *Photonics Conference (IPC)*. IEEE. 2012, pp. 356–357 (cit. on pp. 162, 164).

[3] Wissem Sfar Zaoui, Andreas Kunze, Wolfgang Vogel, *et al.* "Bridging the gap between optical fibers and silicon photonic integrated circuits". *Optics Express* **22**.2 (2014), pp. 1277–1286. DOI: 10.1364/OE.22.001277 (cit. on pp. 162, 164).

[4] Dirk Taillaert, Harold Chong, Peter I. Borel, *et al.* "A compact two-dimensional grating coupler used as a polarization splitter". *IEEE Photonics Technology Letters* **15**.9 (2003), pp. 1249–1251 (cit. on p. 162).

[5] N. Na, H. Frish, I. W. Hsieh, *et al.* "Efficient broadband silicon-on-insulator grating coupler with low back-reflection". *Optics Letters* **36**.11 (2011), pp. 2101–2103 (cit. on p. 164).

[6] D. Vermeulen, Y. De Koninck, Y. Li, *et al.* "Reflectionless grating coupling for silicon-on-insulator integrated circuits". *Group IV Photonics (GFP)*. IEEE. 2011, pp. 74–76 (cit. on p. 164).

[7] D. Taillaert, F. Van Laere, M. Ayre, *et al.* "Grating couplers for coupling between optical fibers and nanophotonic waveguides". *Japanese Journal of Applied Physics* **45**.8A (2006), pp. 6071–6077 (cit. on pp. 164, 179).

[8] D. Vermeulen, S. Selvaraja, P. Verheyen, *et al.* "High-efficiency fiber-to-chip grating couplers realized using an advanced CMOS-compatible silicon-on-insulator platform". *Optics Express* **18**.17 (2010), pp. 18278–18283 (cit. on pp. 164, 179).

[9] X. Chen, C. Li, C. K. Y. Fung, S. M. G. Lo, and H. K. Tsang. "Apodized waveguide grating couplers for efficient coupling to optical fibers". *Photonics Technology Letters, IEEE* **22**.15 (2010), pp. 1156–1158 (cit. on p. 164).

[10] G. Roelkens, D. Vermeulen, S. Selvaraja, *et al.* "Grating-based optical fiber interfaces for silicon-on-insulator photonic integrated circuits". *IEEE Journal of Selected Topics in Quantum Electronics* **99** (2011), pp. 1–10 (cit. on p. 164).

[11] A. Mekis, S. Gloeckner, G. Masini, *et al.* "A grating-coupler-enabled CMOS photonics platform". *IEEE Journal of Selected Topics in Quantum Electronics* **17**.3 (2011), pp. 597–608. DOI: 10.1109/JSTQE.2010.2086049 (cit. on pp. 164, 180).

[12] Attila Mekis, Sherif Abdalla, Peter M. De Dobbelaere, *et al.* "Scaling CMOS photonics transceivers beyond 100 Gb/s". *SPIE OPTO.* International Society for Optics and Photonics. 2012, 82650A–82650A (cit. on p. 164).

[13] Yun Wang, Jonas Flueckiger, Charlie Lin, and Lukas Chrostowski. "Universal grating coupler design". *Proc. SPIE* 8915 (2013), 89150Y. DOI: 10. 1117/12.2042185 (cit. on pp. 168, 170).

[14] F. Van Laere, T. Claes, J. Schrauwen, *et al.* "Compact focusing grating couplers for silicon-on-insulator integrated circuits". *IEEE Photonics Technology Letters* **19**.23 (2007), pp. 1919–1921 (cit. on pp. 168, 179).

[15] R. Waldhusl, B. Schnabel, P. Dannberg, *et al.* "Efficient coupling into polymer waveguides by gratings". *Applied Optics* **36**.36 (1997), pp. 9383–9390 (cit. on p. 179).

[16] Yun Wang. "Grating coupler design based on silicon-on-insulator". MA thesis. University of British Columbia, 2013 (cit. on p. 182).

[17] Tom Baehr-Jones, Ran Ding, Ali Ayazi, *et al.* "A 25 Gb/s silicon photonics platform". *arXiv:1203.0767v1* (2012) (cit. on p. 182).

[18] Na Fang, Zhifeng Yang, Aimin Wu, *et al.* "Three-dimensional tapered spot-size converter based on (111) silicon-on-insulator". *IEEE Photonics Technology Letters* **21**.12 (2009), pp. 820–822 (cit. on p. 182).

[19] Minhao Pu, Liu Liu, Haiyan Ou, Kresten Yvind, and Jorn M Hvam. "Ultra-low-loss inverted taper coupler for silicon-on-insulator ridge waveguide". *Optics Communications* **283**.19 (2010), pp. 3678–3682 (cit. on p. 182).

[20] V. R. Almeida, R. R. Panepucci, and M. Lipson. "Nanotaper for compact mode conversion". *Optics Letters* **28**.15 (2003), pp. 1302–1304 (cit. on p. 183).

[21] B. Ben Bakir, A. V. de Gyves, R. Orobtchouk, *et al.* "Low-loss (< 1 dB) and polarization-insensitive edge fiber couplers fabricated on 200-mm silicon-on-insulator wafers". *IEEE Photonics Technology Letters* **22**.11 (2010), pp. 739–741. DOI: 10.1109/LPT.2010.2044992 (cit. on pp. 183, 189).

[22] Jens H. Schmid, Przemek J. Bock, Pavel Cheben, *et al.* "Applications of subwavelength grating structures in silicon-on-insulator waveguides". *OPTO.* International Society for Optics and Photonics. 2010, 76060F–76060F (cit. on p. 183).

[23] R. Takei, M. Suzuki, E. Omoda, *et al.* "Silicon knife-edge taper waveguide for ultralow-loss spot-size converter fabricated by photolithography". *Applied Physics Letters* **102**.10 (2013), p. 101108 (cit. on p. 183).

[24] Tai Tsuchizawa, Koji Yamada, Hiroshi Fukuda, *et al.* "Microphotonics devices based on silicon microfabrication technology". *IEEE Journal of Selected Topics in Quantum Electronics*, **11**.1 (2005), pp. 232–240 (cit. on pp. 183, 189, 190).

[25] Tymon Barwicz, Michael R. Watts, Milos A. Popovi, *et al.* "Polarization-transparent microphotonic devices in the strong confinement limit". *Nature Photonics* **1**.1 (2007), pp. 57–60 (cit. on pp. 190, 191).

[26] Thierry Pinguet, Steffen Gloeckner, Gianlorenzo Masini, and Attila Mekis. "CMOS photonics: a platform for optoelectronics integration". In *Silicon Photonics II.* Ed. David J. Lockwood and Lorenzo Pavesi. Vol. 119. Topics in Applied Physics. Springer Berlin Heidelberg, 2011, pp. 187–216. ISBN: 978-3-642-10505-0. DOI: 10.1007/978-3-642-10506-7_8 (cit. on p. 191).

[27] Daniel Kucharski, Drew Guckenberger, Gianlorenzo Masini, *et al.* "10Gb/s 15mW optical receiver with integrated Germanium photodetector and hybrid inductor peaking in 0.13μm SOI CMOS technology". *Solid-State Circuits Conference Digest of Technical Papers (ISSCC), 2010 IEEE International.* IEEE. 2010, pp. 360–361 (cit. on p. 191).

[28] Wim Bogaerts, Dirk Taillaert, Pieter Dumon, *et al.* "A polarization-diversity wavelength duplexer circuit in silicon-on-insulator photonic wires". *Optics Express* **15**.4 (2007), pp. 1567–1578 (cit. on p. 191).

[29] David A. B. Miller. "Self-configuring universal linear optical component". *Photonics Research* **1**.1 (2013), pp. 1–15 (cit. on p. 191).

[30] Jan Niklas Caspers, Yun Wang, Lukas Chrostowski, and Mohammad Mo-jahedi. "Active polarization independent coupling to silicon photonics circuit". *Proc. SPIE.* 2014, pp. 9133–9217 (cit. on p. 192).

[31] Wesley D. Sacher, Tymon Barwicz, Benjamin J. F. Taylor, and Joyce K. S. Poon. "Polarization rotator-splitters in standard active silicon photonics platforms". *Optics Express* **22**.4 (2014), pp. 3777–3786 (cit. on p. 192).

Part III

Active components

6 Modulators

This chapter describes how optical modulation and tuning are achieved using carrier depletion in a pn-junction, carrier injection in a PIN junction, and the thermo-optic effect. We discuss the modelling and design considerations for a ring modulator, variable optical attenuator, active tuning techniques, and a thermo-optic switch.

6.1 Plasma dispersion effect

6.1.1 Silicon, carrier density dependence

In 1987, Soref and Bennett predicted the change in silicon's refraction index due to carriers [1]. This is termed the plasma dispersion effect, and is used for silicon modulators where the concentration of carriers is varied either by injecting or removing carriers from the device. This effect will be used for high-speed pn-junction modulators, Section 6.2, and PIN phase shifters and variable attenuators, Section 6.4. The commonly used (e.g. [2]) phenomenological expressions are as follows:

The change in index of refraction is phenomenologically described by:

$$\begin{aligned} \Delta n \text{ (at 1550 nm)} &= -8.8 \times 10^{-22} \Delta N - 8.5 \times 10^{-18} \Delta P^{0.8} \\ \Delta n \text{ (at 1310 nm)} &= -6.2 \times 10^{-22} \Delta N - 6 \times 10^{-18} \Delta P^{0.8}. \end{aligned} \tag{6.1}$$

The change in absorption is described by:

$$\begin{aligned} \Delta \alpha \text{ (at 1550 nm)} &= 8.5 \times 10^{-18} \Delta N + 6 \times 10^{-18} \Delta P \ [\text{cm}^{-1}] \\ \Delta \alpha \text{ (at 1310 nm)} &= 6 \times 10^{-18} \Delta N + 4 \times 10^{-18} \Delta P \ [\text{cm}^{-1}], \end{aligned} \tag{6.2}$$

where ΔN, ΔP are the carrier densities of electrons and holes [cm^{-3}].

It should be noted that holes have a smaller absorption as compared with electrons, whereas holes have a larger index shift. Thus, holes are most effective for providing an index shift with minimal absorption, hence modulators typically use holes for offset-junction designs (e.g. in Mach–Zehnder or ring modulators).

Equations (6.1) and (6.2) were updated in 2011 using more recent experimental data as follows [3]:

$$\begin{aligned} \Delta n \text{ (at 1550 nm)} &= -5.4 \times 10^{-22} \Delta N^{1.011} - 1.53 \times 10^{-18} \Delta P^{0.838} \\ \Delta n \text{ (at 1310 nm)} &= -2.98 \times 10^{-22} \Delta N^{1.016} - 1.25 \times 10^{-18} \Delta P^{0.835}. \end{aligned} \tag{6.3}$$

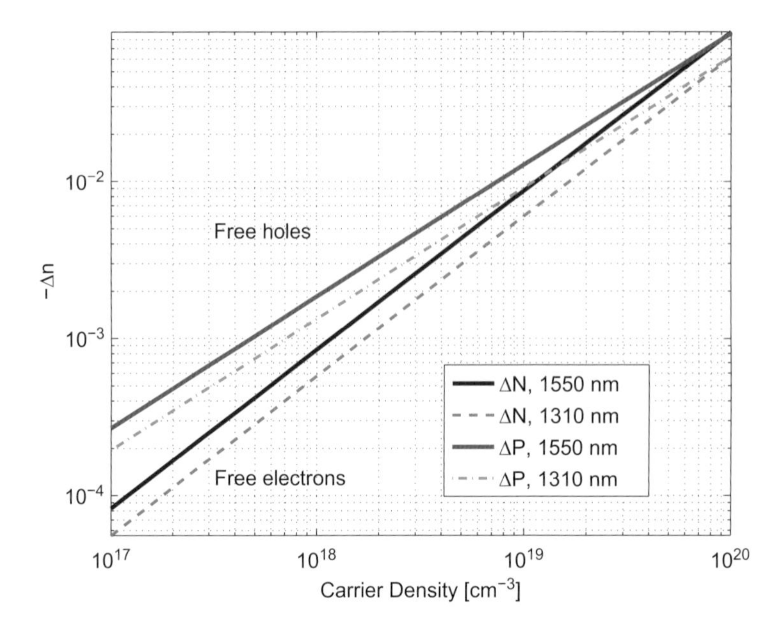

Figure 6.1 Change in index of refraction versus carrier density, Equations (6.3).

The change in absorption is described by:

$$\begin{aligned}\Delta\alpha\ (1550\,\text{nm}) &= 8.88 \times 10^{-21}\Delta N^{1.167} + 5.84 \times 10^{-20}\Delta P^{1.109} \\ \Delta\alpha\ (1310\,\text{nm}) &= 3.48 \times 10^{-22}\Delta N^{1.229} + 1.02 \times 10^{-19}\Delta P^{1.089}.\end{aligned} \tag{6.4}$$

The wavelength dependence can be introduced by considering the theoretical free-carrier (Drude) model, in which both the index and absorption vary as λ^2. See also Figures 6.1 and 6.2. Taking into account the wavelength dependence, Equations (6.1) and (6.2) are extended with wavelength-dependent fitting parameters:

$$\begin{aligned}\Delta n(\lambda) &= -3.64 \times 10^{-10}\lambda^2\Delta N - 3.51 \times 10^{-6}\lambda^2\Delta P^{0.8} \\ \Delta\alpha(\lambda) &= 3.52 \times 10^{-6}\lambda^2\Delta N + 2.4 \times 10^{-6}\lambda^2\Delta P\ \ [\text{cm}^{-1}],\end{aligned} \tag{6.5}$$

where λ is the wavelength [m]. The wavelength dependence is plotted in Figure 6.3.

6.2 pn-Junction phase shifter

6.2.1 pn-Junction carrier distribution

The impurity and carrier distributions in a carrier-depletion phase modulator are illustrated in Figure 6.4. We have the following assumptions or approximations for the pn-junction.

- The diffused pn-junction is approximated by a step or abrupt junction where the impurity profile changes abruptly across the mask-defined doping boundary.

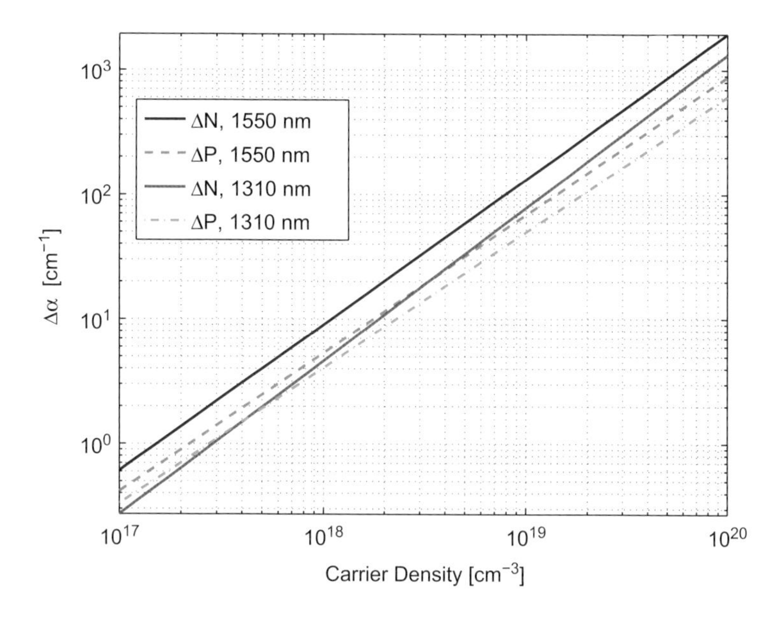

Figure 6.2 Change in absorption versus carrier density, Equations (6.4).

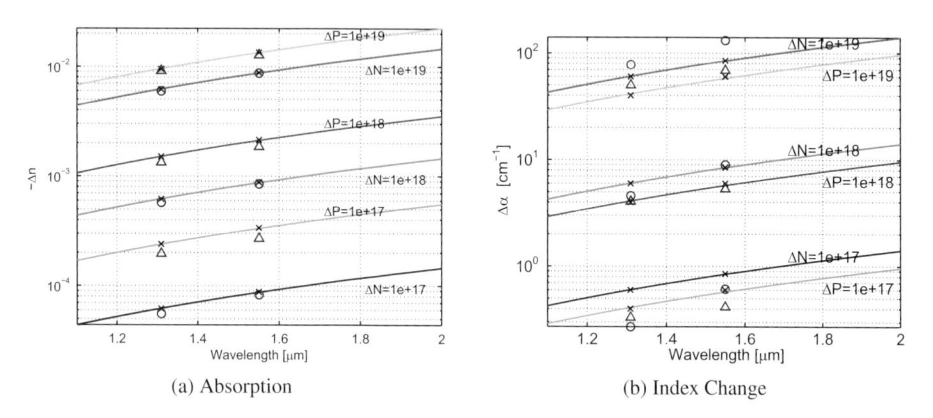

Figure 6.3 (a) Change in index of refraction versus wavelength. (b) Change in absorption versus wavelength. Equation (6.5) with data points (X) taken from Equations (6.1) and (6.2). Update data from Reference [3] is included (circles, triangles).

- The width of the pn-junction is much shorter than the diffusion length, therefore, a linear distribution in minority carrier densities is assumed between the depletion region and the heavily doped region.

The width of the depletion region, W_d, is determined by the impurity densities (N_A and N_D), as well as the applied voltage (V), and is given by:

$$W_d = \sqrt{\frac{2\epsilon_0\epsilon_s(N_A + N_D)(V_{bi} - V)}{qN_AN_D}}, \tag{6.6}$$

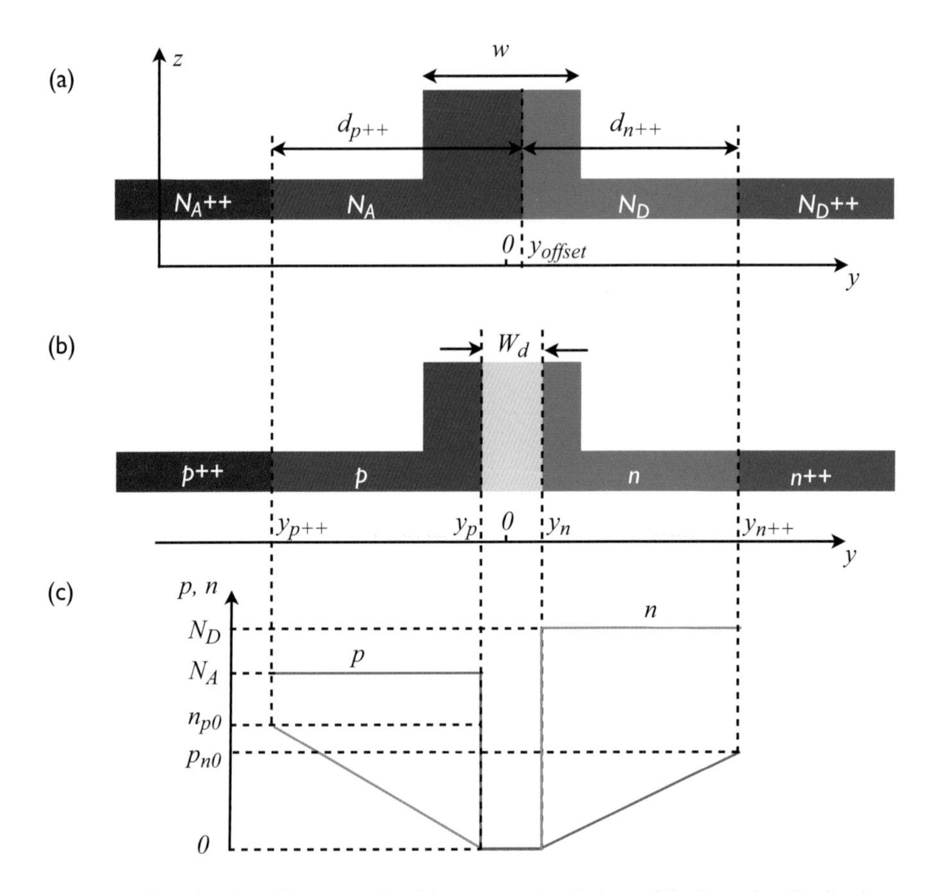

Figure 6.4 pn-Junction in a rib waveguide: (a) cross-sectional view of the impurity distribution assumed in the abrupt junction model; (b) cross-sectional view of the carrier distribution; (c) 1D free-carrier distributions.

where ϵ_s is the relative permittivity and V_{bi} is the built-in or diffusion potential of the junction given by:

$$V_{bi} = \frac{k_B T}{q} \ln \frac{N_A N_D}{n_i^2}. \tag{6.7}$$

The boundaries of the depletion region are given by:

$$y_p = y_{\text{offset}} - \frac{W_d}{1 + N_A/N_D} \tag{6.8a}$$

$$y_n = y_{\text{offset}} + \frac{W_d}{1 + N_D/N_A} \tag{6.8b}$$

$$n(y, V) = \begin{cases} n_{p0}\left[1 + (1 - \frac{y_p - y}{y_p - y_{p++}})e^{(\frac{qV}{k_B T} - 1)}\right] & \text{for} \quad y_{p++} < y < y_p \\ 0 & \text{for} \quad y_p < y < y_n \\ N_D & \text{for} \quad y_{n++} > y > y_n \end{cases} \tag{6.9}$$

$$p(y,V) = \begin{cases} N_A & \text{for} \quad y_{p++} < y < y_p \\ 0 & \text{for} \quad y_p < y < y_n \\ p_{n0}\left[1 + \left(1 - \frac{y - y_n}{y_{p++} - y_p}\right) e^{\left(\frac{qV}{k_B T} - 1\right)}\right] & \text{for} \quad y_{n++} > y > y_n \end{cases}. \tag{6.10}$$

The carrier densities, $\Delta N = n(y,V)$, and $\Delta P = p(y,V)$, are given in Equations (6.9) and (6.10), where n_{p0} and p_{n0} are given by

$$n_{p0} = \frac{n_i^2}{N_A} \tag{6.11a}$$

$$p_{n0} = \frac{n_i^2}{N_D}. \tag{6.11b}$$

Using the above equations, the carrier distributions in the waveguide can be solved using the MATLAB code 6.1.

6.2.2 Optical phase response

The changes in the silicon refractive index, n_{co}, and optical loss, at $\lambda = 1.55\,\mu\text{m}$, due to the free carriers are given by Equations (6.1) and (6.2).

Then the effective index, n_{eff}, and the optical loss due to the free-carrier absorption, α_{pn}, as functions of applied voltage are given by:

$$\begin{aligned} n_{\text{eff}}(V) &= n_{\text{eff},i} + \frac{\int E^*(y) \cdot \Delta n(y,V)\, E(y)\, dy}{\int E^*(y) \cdot E(y)\, dy} \cdot \frac{dn_{\text{eff}}}{dn_{co}} \\ \alpha_{pn}(V) &= \frac{\int E^*(y) \cdot \Delta\alpha(y,V)\, E(y)\, dy}{\int E^*(y) \cdot E(y)\, dy}, \end{aligned} \tag{6.12}$$

where $n_{\text{eff},i}$ is the effective index of the waveguide without any doping and dn_{eff}/dn_{co} (change of mode effective index versus change in the waveguide core effective index) is typically very close to 1 (including silicon strip and rib waveguides). $E(y)$ is the 1D field profile found using the effective index method, see MATLAB function 3.11. Then the voltage-dependent changes in effective index and phase are given by:

$$\begin{aligned} \Delta n_{\text{eff}}(V) &= n_{\text{eff}}(V) - n_{\text{eff}}(0) \\ \Delta\phi(V)\ [\pi \cdot \text{cm}^{-1}] &= \frac{0.02\, \Delta\, n_{\text{eff}}(V)}{\lambda}. \end{aligned} \tag{6.13}$$

Using the above equations (MATLAB code 6.2), we calculate the changes in effective index and optical loss due to the free-carrier absorption for a design with the following waveguide parameters: rib width w = 500 nm, rib thickness, t = 220 nm, and slab thickness, t_{slab} = 90 nm. Holes have stronger effect on the effective index than electrons do, as revealed by Equations (6.1), therefore, an offset of the junction to the waveguide centre can be used to optimize the modulation efficiency. A 50 nm doping offset is used in the calculation.

The calculated results are shown in Figure 6.5 (MATLAB code 6.3). The effective index increases, while the optical loss reduces, as the voltage increases, because the

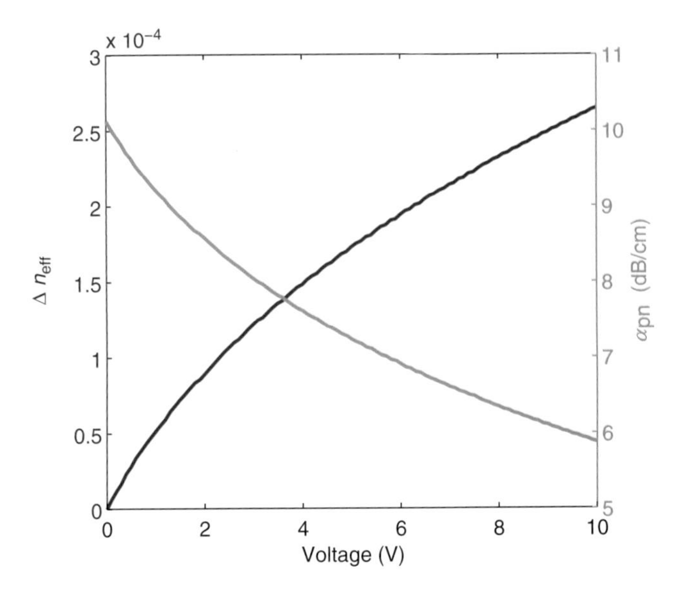

Figure 6.5 Changes in effective index and free-carrier-caused optical loss as functions of applied voltage (reverse biased).

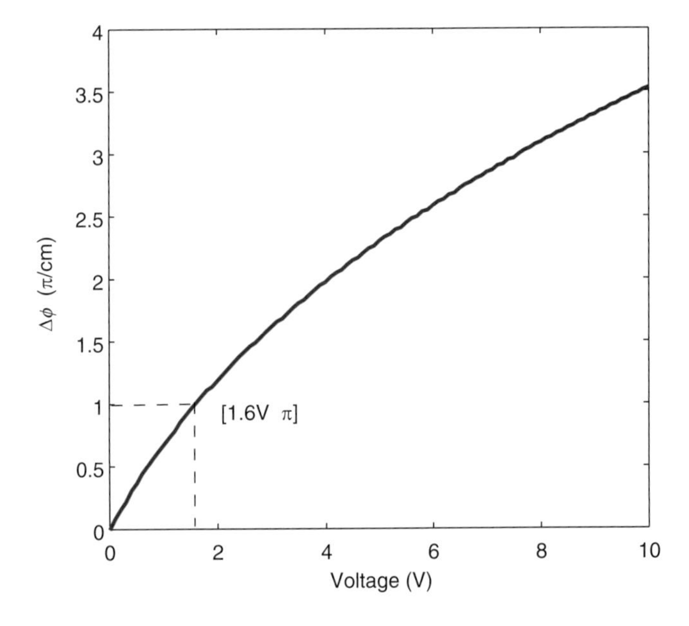

Figure 6.6 Phase change as a function of applied voltage (reverse biased).

carriers are removed from the waveguide by the applied voltage. The change in the phase is plotted in Figure 6.6. A voltage of 1.6 V is needed to make a phase shift of π for a waveguide 1 cm long, indicating a $V_\pi \cdot L$ product of 1.6 V·cm for a phase shifter 1 cm long.

6.2.3 Small-signal response

The resistance and capacitance of the pn-junction are given by:

$$\begin{aligned} R_j\,[\Omega \cdot \mathrm{m}] &= \left(\frac{w}{2} + y_p\right) R_{srp} + \left(\frac{w}{2} - y_n\right) R_{srn} \\ &\quad - \left(\frac{w}{2} + y_{p++}\right) R_{ssp} + \left(y_{n++} - \frac{w}{2}\right) R_{ssn} \\ C_j\,[\mathrm{F/m}] &= t_{rib}\sqrt{\frac{q\epsilon_0\epsilon_s}{2(1/N_D + 1/N_A)(V_{bi} - V)}} \end{aligned} \tag{6.14}$$

where R_{srn}, R_{srp}, R_{ssn}, and R_{ssp} are the sheet resistances of the n-doped rib, p-doped rib, n-doped slab, and p-doped slab, respectively.

Then, the 3 dB cutoff frequency determined by the RC time constant can be found by:

$$f_c = \frac{1}{2\pi R_j C_j}. \tag{6.15}$$

Using the parameters given above, f_c is calculated to be 35 GHz and 51 GHz at 0 and 1 V, respectively. As shown in Figure 6.7, f_c increases as the applied DC voltage increases due to the simultaneously reduced R_j and C_j, as a result of the expanded depletion region. We can see that the frequency response of the pn-junction can easily go beyond tens of GHz, therefore, the intrinsic RC of the junction is typically not a limiting factor of silicon optical modulators. The RC becomes a limitation in cases

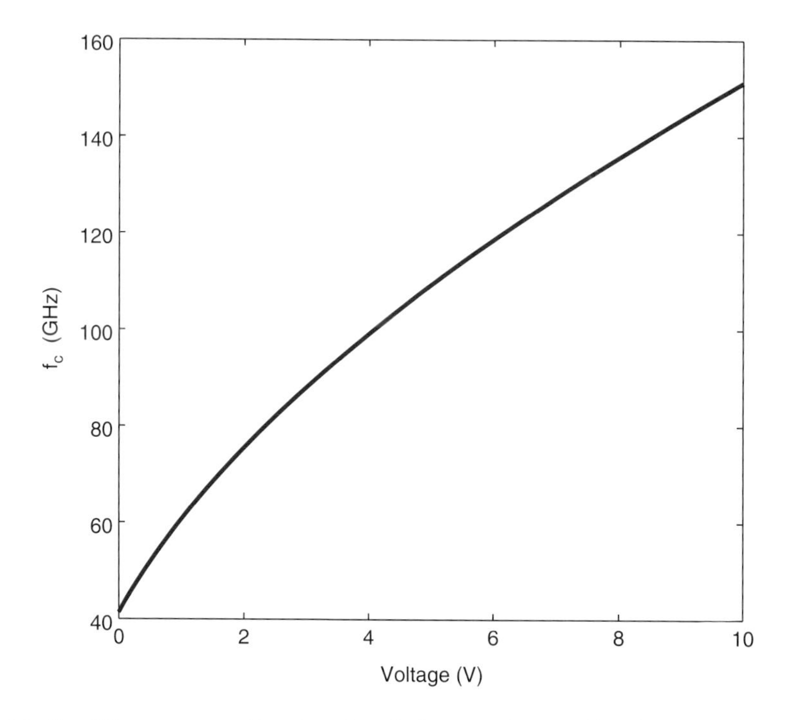

Figure 6.7 Cutoff frequency as a function of applied voltage (reverse biased).

where a long junction is used (hence a large capacitance) together with a large source impedance, e.g. 50 Ω. This is the case for a basic Mach–Zehnder modulator or phase shifter; to avoid this RC limit, structures such as travelling wave electrodes need to be employed.

The resistance of the junction is an important parameter that can be reduced by optimizing the doping concentrations and the distances of the dopants from the junction. Specifically, minimizing the distance from the contact to the junction will lead to a reduced RC time constant. However, the optical losses of the doping contacts need to be considered. This is described in the context of the PIN junction in Section 6.4, with results in Figure 6.16. Similar results can be obtained for the pn-junction.

6.2.4 Numerical TCAD modelling of pn-junctions

In the preceding sections, specifically in Section 6.2.1, we assumed a one-dimensional model for the pn-junction, and used the effective index method in Section 6.2.2 to analytically calculate the optical mode. The 1D model is computationally efficient and provides significant insight into the functioning of the phase modulator. However, it makes several approximations (e.g. neglects the carrier distribution in the vertical dimension), and relies on numerous parameters (e.g. sheet resistance, contact resistance). In this section, we present 2D simulations using TCAD tools for both the pn-junction and the optical mode. These models are expected to be more accurate.

In the following example, the dimensions of the waveguide reflect the structures in the publication by T. Baehr-Jones *et al.* [4]. The structure is a rib waveguide with a pn-junction in the centre. The peak doping concentrations and dimensions of the doping profile in the referenced paper are used to define the pn-junction profile in the waveguide. Analytic models are used to construct the doping profile; however, doping profiles determined from a process simulation could also be used. To establish an accurate correspondence between the simulations and measured results, the measured capacitance versus voltage characteristic was fit by setting the pn-junction profile in the waveguide. The size and position of the analytic models were adjusted until a match between simulation and experiment was obtained.

The simulation is similar to that presented in Sections 6.2.1 and 6.2.2, as follows.

(1) Parameters are defined, including the waveguide geometry, doping parameters, contacts, and simulation regions; Listing 6.8.
(2) Electrical simulations are performed. The electrical simulation is used to calculate the distribution of electrons and holes in response to the applied voltage, here from −0.4 to 4 V. The spatial charge density (2D profile) is exported for each voltage; Listing 6.9.
(3) Additional electrical simulations are performed to determine the junction capacitance versus voltage. The pn-junction capacitance is simulated at DC by calculating the numerical derivative, $C = dQ/dV$. The total charge can be calculated using a charge monitor. The monitor will integrate the carrier density (electrons

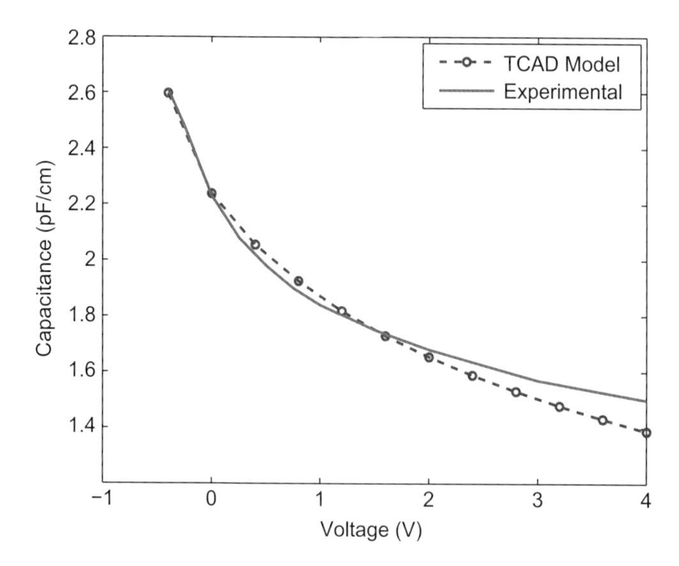

Figure 6.8 Capacitance of a pn-junction phase modulator. Experimental data are from Reference [4]. Simulation is from Listing 6.8.

and holes) over the volume of simulation. By running two simulations at bias voltage V and $V + \Delta V$, the capacitance can be estimated as

$$C_{n,p} = \frac{Q_{n,p}(V + \Delta V) - Q_{n,p}(V)}{\Delta V},$$

where $C_{n,p}$ and $Q_{n,p}$ are the capacitance and charge due to the electrons (n) and holes (p), at an applied voltage V. These two values should be numerically the same, if the simulation has correctly converged. In this example, the capacitance is ~2 pF. It is voltage dependent, as plotted in Figure 6.8, due to the depletion region being voltage dependent, as per Equation (6.6).

The total charge integration is sensitive to the mesh density, particularly in the region containing the pn-junctions. Therefore, a mesh override is used in the junction to improve the accuracy.

(4) Additional electrical simulations are performed to determine the resistance of the n and the p sides. This is done by splitting the simulation into two parts by putting a metal contact in the pn-junction region. Two simulations are performed to determine the resistance of each side, by measuring the current in response to the voltage. In this example, the total series resistance is 2 Ω.

(5) Optical simulations are performed. The simulation loads the carrier density from the electrical simulations, calculates the corresponding plasma dispersion effect using Equation (6.5) (or similar), performs optical mode calculations for each voltage, and determines the n_{eff} versus voltage, similar to results in Figure 6.5; Listing 6.10. This includes both the real and imaginary part of the refractive index change.

(6) Finally, the results are exported to a file. These results are used to create a compact model for the phase modulator, which can then be used for photonic circuit modelling, such as the ring modulator shown in Figure 9.13 or a travelling wave modulator. The compact model is created in Listing 9.3.

6.3 Micro-ring modulators

A high-Q optical resonator, such as a ring resonator described in Section 4.4 has strong wavelength selectivity, performing as a narrow-bandwidth filter. The resonant wavelength is determined by the round-trip phase of the resonator. Therefore, when operating at a wavelength close to the resonance of an optical resonator, the transmission is very sensitive to the phase change of the cavity. Based on this effect, one can obtain a very efficient modulator by integrating a pn-junction into the resonator cavity and modulating the phase through the plasma effect as described in the previous section. The reader is also referred to numerous papers on ring modulators, e.g. [5, 6], and two review papers [7, 8]. See also Figure 6.9.

Two ring resonator configurations, i.e., all-pass and add-drop filters, are commonly used to obtain a micro-ring modulator, as shown in Figure 6.10. The micro-ring cavity can have a racetrack shape, consisting of two 180° circular waveguides and two straight waveguides (for directional couplers).

Because the micro-ring modulator can only operate within a narrow spectral window around its resonant wavelength, wavelength stabilization is usually needed for practical applications. For example, as shown in Figure 6.10a, a quarter of the optical cavity (in the directional coupler region) is integrated with a resistor heater for thermal tuning and wavelength stability. As a result of this compromise, the modulation efficiency is lower as compared to the fully modulated cavity.

The optical transfer function of the micro-ring modulator is implemented using MATLAB code 6.4.

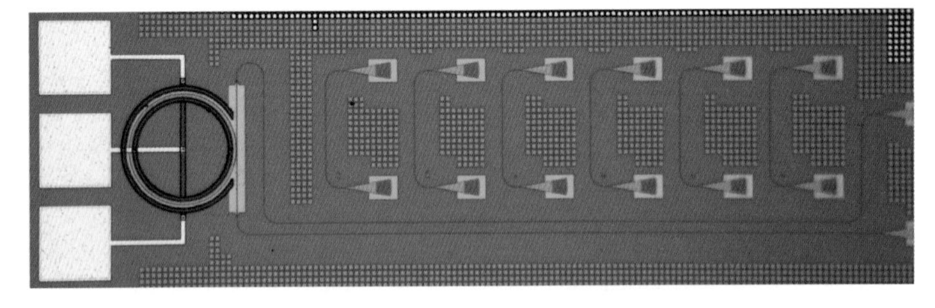

Figure 6.9 Microscope image of a ring resonator modulator [6]. Three electrical pads are shown on the left, for ground–signal–ground (GSG); the ring is adjacent; it is optically characterized using two grating couplers on the far right side. The couplers are separated by 0.5 mm from the microwave probes for ease of testing. Additional pairs of grating couplers are included in the layout as characterization test structures.

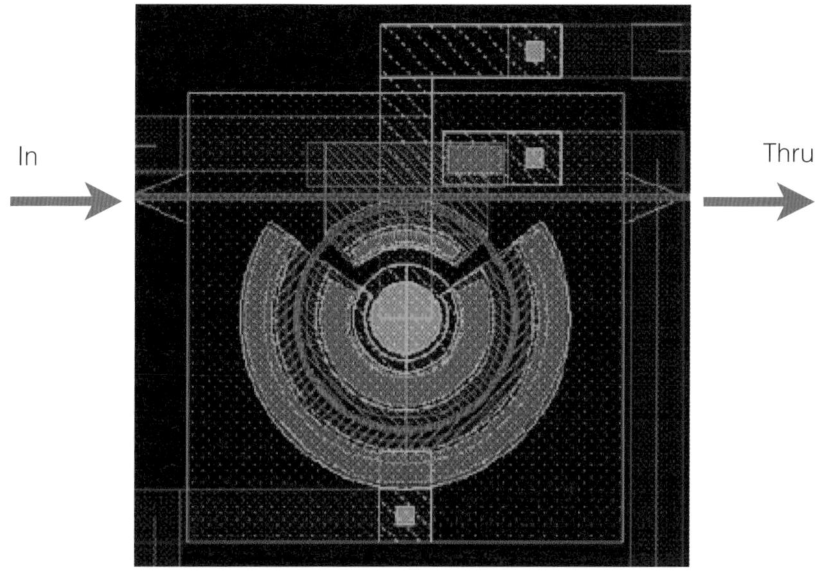

(a) All-pass (integrated with a heater for wavelength tuning)

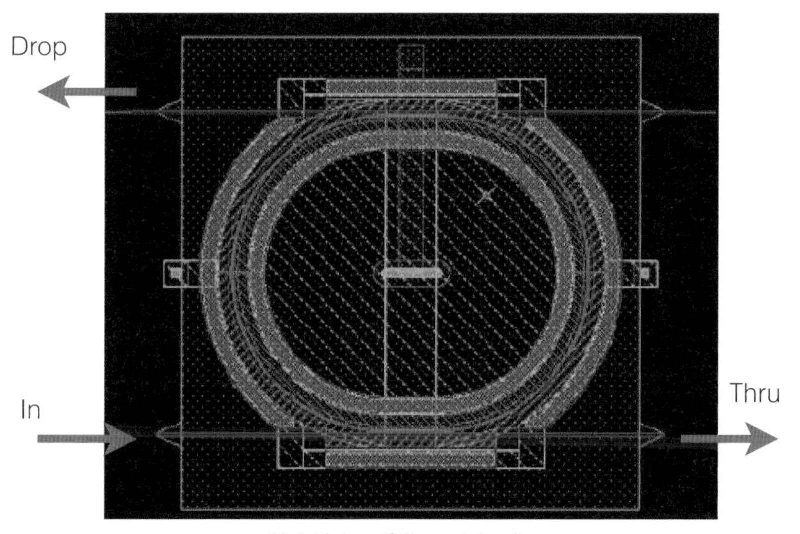

(b) Add-drop (fully modulated)

Figure 6.10 Mask layouts of micro-ring modulators.

6.3.1 Ring tuneability

Here we discuss the micro-ring modulator with a reverse-biased pn-junction. Incorporating the pn-junction model (MATLAB codes 6.2 and 6.1) with the micro-ring resonator transfer function (Equations (4.24) and (4.25)), we can simulate its spectrum as a function of applied voltage (MATLAB codes 6.6 and 6.7). This example micro-ring modulator is implemented based on the structure shown in Figure 6.11.

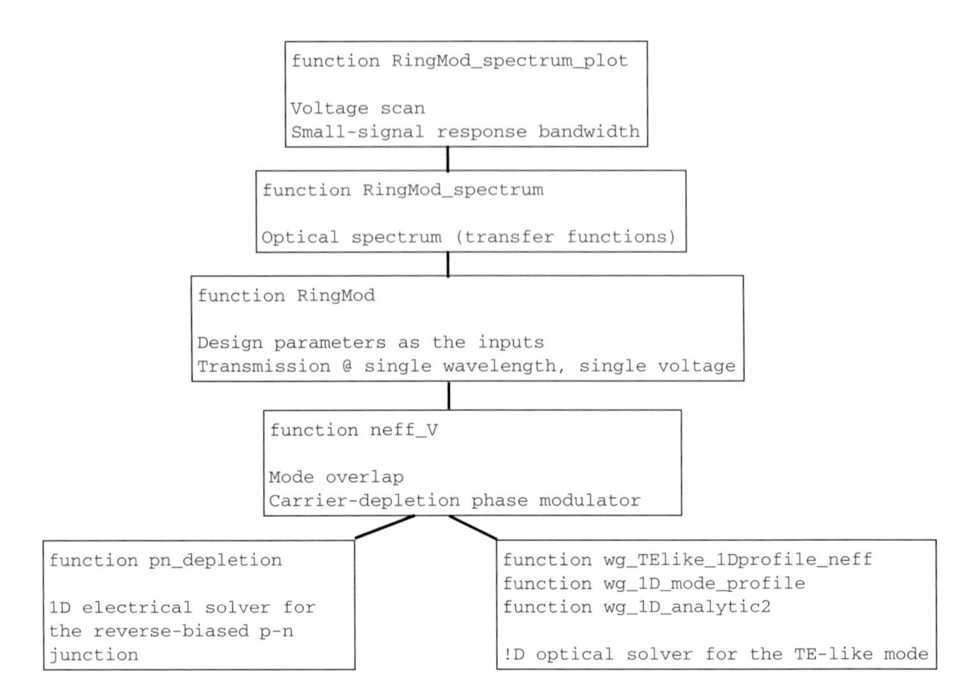

Figure 6.11 Structure diagram of a 1D analytic model for micro-ring modulators.

We consider a fully modulated (i.e. no thermal tuning is used), add-drop micro-ring modulator using point couplers (i.e. $L_c = 0$) with $r = 10\,\mu\text{m}$ and default values for other parameters given in MATLAB code 6.7. The calculated through-port and drop-port responses are shown in Figure 6.12. We can see that the through-port power transmission is more sensitive to the change of the roundtrip phase and, therefore, should be used as the modulator output. As shown in Figure 6.12a, the central wavelength shifts by 0.016 nm/V. Owing to the high quality factor (about 10 000), this relatively small spectral shift results in a considerable change in the power transmission. For example, for the 3 dB insertion loss wavelength at zero bias (~1540.9 nm), when the applied voltage (reverse biased) changes from 0 to 4V, the transmission drops by about 8 dB. In order to improve the modulation efficiency, we can increase the quality factor, for example, by reducing the coupling (i.e. reducing κ) to make the transmission notch narrower. However, a higher Q means a longer photon lifetime, which will limit the frequency response of the modulator, as we will see next.

6.3.2 Small-signal modulation response

The cutoff frequency (3 dB), f_c, of the small-signal response of a micro-ring modulator is determined by both the RC constant of the reverse-biased pn-junction and the photon lifetime, τ_p, of the optical cavity:

$$\frac{1}{f_c^2} = \frac{1}{f_{\tau_p}^2} + \frac{1}{f_{RC}^2}. \tag{6.16}$$

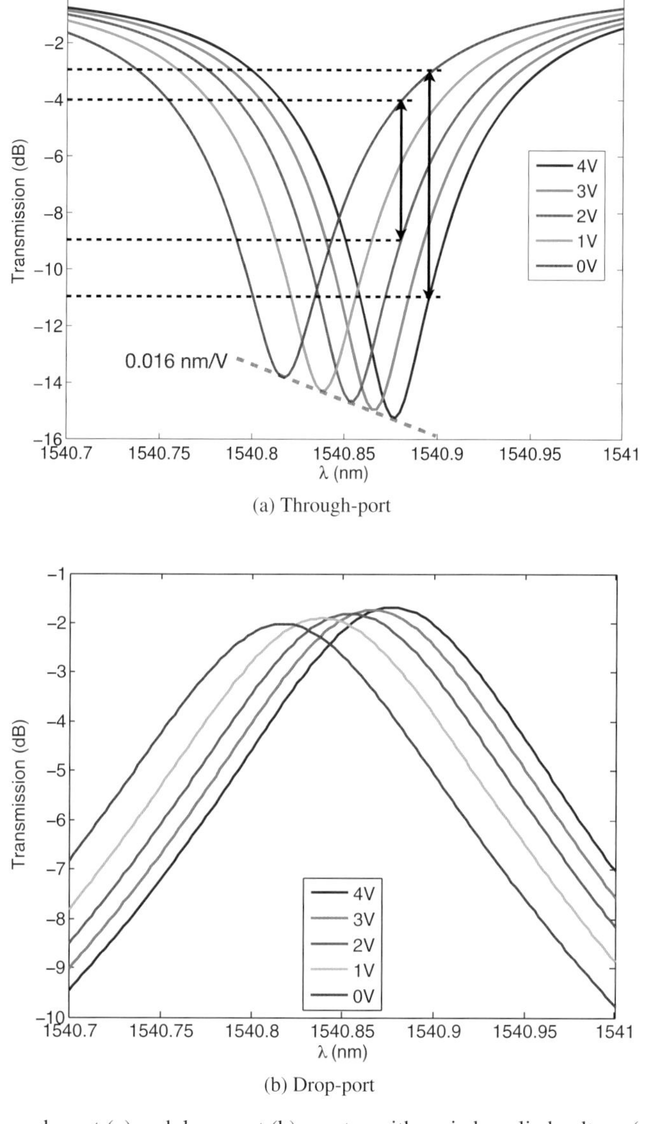

(a) Through-port

(b) Drop-port

Figure 6.12 Through-port (a) and drop-port (b) spectra with varied applied voltage (reverse biased).

The τ_p determined cutoff frequency is given by:

$$f_{\tau_p} = \frac{1}{2\pi\,\tau_p}, \tag{6.17}$$

where τ_p is related to the total quality factor, Q_t, of the optical cavity and is given by:

$$\tau_p = \frac{Q_t}{\omega_o}, \tag{6.18}$$

where ω_o is the optical frequency and Q_t is determined by both the coupling and propagation losses:

$$\frac{1}{Q_t} = \frac{1}{Q_c} + \frac{1}{Q_i}, \tag{6.19}$$

where the intrinsic quality factor is given by [9]:

$$Q_i = \frac{2\pi n_g}{\lambda \alpha}. \tag{6.20}$$

For the all-pass filter, the coupling-determined quality factor, Q_c, is given by:

$$Q_c = -\frac{\pi L_{rt} n_g}{\lambda \log_e |t|} \tag{6.21}$$

If the add-drop configuration is used, the coupling-determined quality factor should be divided by 2 since two couplers are used in this case.

For the same design as described for the DC performance, we can predict the cutoff frequency of the micro-ring modulator using the above equations. The calculated results are shown in Figure 6.13. In this case, the RC constant of the pn-junction has a cutoff frequency (i.e. f_{RC}) of over 40 GHz, while the τ_p determined cutoff frequency (i.e. f_{τ_p})

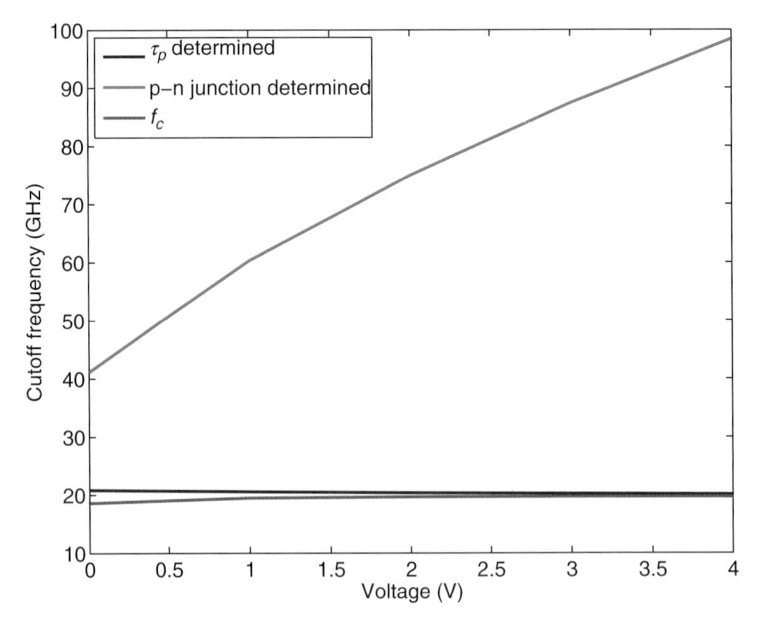

Figure 6.13 Ring modulator small-signal modulation bandwidth versus applied voltage. Two factors are considered: the photon lifetime, τ_p, and the pn junction RC time constant.

is about 20 GHz. As a result, the total cutoff frequency, f_C, is 15 GHz at the bias of 1 V, mainly limited by the photon lifetime.

6.3.3 Ring modulator design

In this section, we provide an overview for a design methodology for the ring modulator. The first step is to identify the desired modulator characteristics: modulation bandwidth, FSR, extinction ratio, drive voltage, architecture choice between double-bus and single-bus, and operation in the drop port or through port.

A common design target is to design a ring for critical coupling. The highest extinction ratio is obtained in this case. This condition leads to 0 transmission on resonance (i.e., all power is absorbed within the ring, or goes to the drop port); this is the complete destructive interference condition. Critical coupling is obtained when the input coupler matches the other losses of the resonator (internal losses in the ring, and output coupler losses).

The double-bus ring modulator is useful because the second waveguide is used to "load" the resonator (add extra loss). This leads to balanced losses, enabling critical coupling and high extinction ratio. This approach is particularly useful when the losses of the dopants are not known *a priori*. It allows for the control of the Q and hence the modulation bandwidth. A disadvantage of this approach is that this design is sub-optimal. Namely, if the modulator requires the Q to be lowered, it is preferable to increase the pn-junction propagation losses, thereby increasing the junction efficiency (pm/V).

One important consideration is the pn-junction phase shifter fill-factor. This fill-factor determines the fraction of the ring circumference that contains the pn-junction. This is illustrated in the ring modulator in Figure 6.10a. A fill-factor of less than 100% may be used to include thermal tuning in the ring. The fill-factor is also determined by mask layout and manufacturing constraints.

Next, parameters from the fabrication process are identified: waveguide propagation loss (due to scattering, doping absorption, metal absorption, and bend radiation and mismatch losses, which are larger for smaller devices with large FSR); slab thickness for the rib waveguide (e.g. 150, 90, and 50 nm); characteristics of the pn-junction, particularly the RC time constant; and considerations for fabrication process variations and any required fabrication bias. The pn-junction itself can also be optimized (e.g. choice of doping concentrations, junction offset in the waveguide, etc.)

Next, we calculate the quality factor of the ring for the target modulation bandwidth. Then we calculate the design parameters to match the required Q and FSR: radius, and coupling coefficients. At this point, the optical transfer function can be verified and optimized, and a time-domain model can be constructed; see Section 9.5. Such a model can be used to predict eye diagrams, extinction ratio, energy efficiency, and to investigate the impact of the bias point.

Finally, the physical structure needs to be calculated, specifically the directional coupler for a desired coupling coefficient (i.e. directional coupler gap), typically done

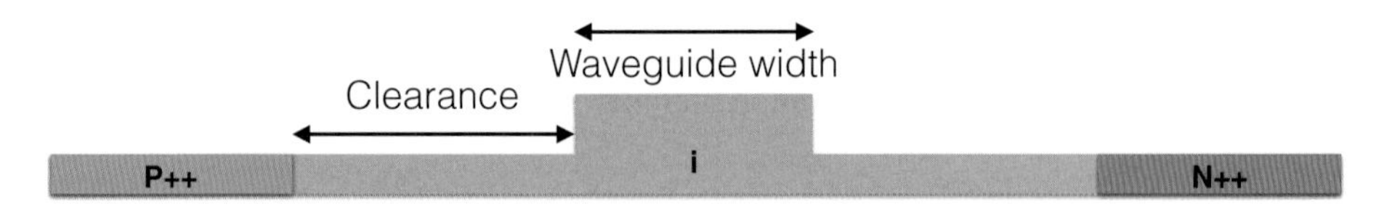

Figure 6.14 Cross-section of a PIN junction in a rib waveguide. The contacts connected to the "P++" and "N++" regions, and the intrinsic region is "I".

using 3D FDTD. Only the directional coupler by itself needs to be simulated in 3D FDTD; see Sections 4.1.4 and 9.4.2. With all physical parameters in place, the mask layout can be implemented.

6.4 Forward-biased PIN junction

Similar to the reverse-biased pn-junctions considered above, active silicon photonics devices can be constructed using forward bias. In this case, it is useful to increase the size of the junction by introducing an un-doped (intrinsic) region between the P and N regions, as shown in Figure 6.14. This eliminates the excess optical losses (Figure 6.2) in the waveguide when the device is unbiased; thus, the device functions as a conventional rib waveguide. By applying a forward bias, carriers are injected. These carriers change the index of refraction (Figure 6.1) and introduce optical absorption (Figure 6.2). Hence, this is useful to construct variable optical attenuators (VOA) using the absorption property, and phase tuners using the index of refraction change. It should be noted that both pn and PIN junctions operate on the same fundamental principle, namely the plasma-dispersion effect, and that both the real and imaginary parts of the index of refraction change simultaneously; namely, amplitude modulation occurs simultaneously with phase modulation. One important difference is that PIN junctions have much longer time constants, dominated by the carrier recombination lifetime in the intrinsic region. For example, a 90 MHz lifetime was measured in the device in Reference [10]. This is in contrast with the tens of GHz that are obtained in pn-junctions (Section 6.2.3).

6.4.1 Variable optical attenuator

Experimental results for a variable optical attenuator consisting of a 1 cm long PIN junction waveguide are shown in Figure 6.15. The waveguide geometry is shown in Figure 6.14, where the width of the waveguide is 500 nm, and the clearance is 800 nm. The experimental VOA efficiency is approximately 0.35 dB/mA. The device is easiest to model and understand when operated in constant current mode. Assuming that every injected electron and hole recombine in the intrinsic region, and the dominant mechanism is non-radiative recombination with a time constant τ_n, a simple model for the carrier density in the device is:

$$N = \frac{I\tau_n}{qV}. \tag{6.22}$$

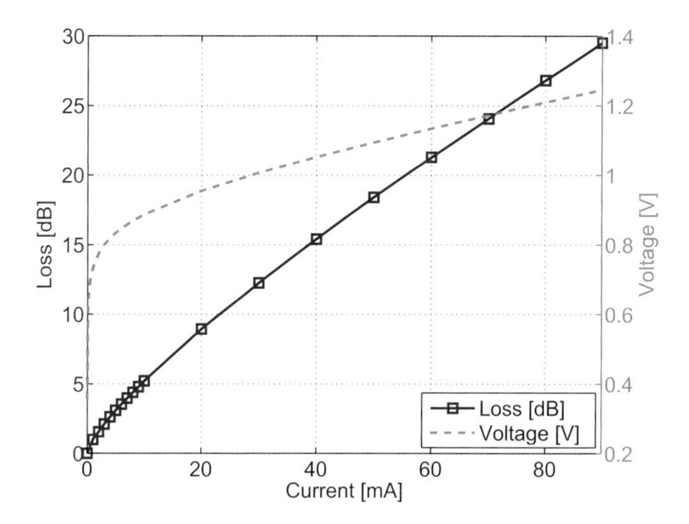

Figure 6.15 Performance of a variable optical attenuator (VOA) consisting of a 1 cm long PIN junction in a rib waveguide: (left) optical attenuation; (right) voltage.

Considering the volume of the intrinsic region, a carrier lifetime of $\tau_n = 1/2\pi f_{3dB} =$ 1.8 ns, and a current of 90 mA, using Equations (6.2) and (3.9), this corresponds to a propagation loss of 25 dB/cm, close to the results in Figure 6.15.

One of the main design considerations for PIN junction waveguide devices is the distance between the doping regions; referred to here as "doping offset," or "clearance," it is 2.1 μm in Figure 6.14. To optimize the device efficiency, the volume of the device needs to be reduced according to Equation (6.22); this can be achieved by bringing the dopants closer to the waveguide, leading to a larger carrier density for a given current. However, the dopants introduce excess optical loss. This can be simulated using a mode solver with the inclusion of doped regions with a complex index of refraction with absorption as per Equations (6.5). In the simulation Listing 6.11, the two doped regions are defined, and mode calculations are repeated for several doping offset distances. The results are shown in Figure 6.16. The high doping density is typically used for these devices, in which case the doping offset needs to be greater than 1.6 μm to ensure that the excess loss due to the dopants is less than 1 dB/cm, and 2.1 μm to ensure an excess loss less than 0.1 dB/cm.

The diode current–voltage (IV) relation should also be considered. The Shockley diode equation, with an ideality factor, n, is:

$$I = I_s \left(e^{V/nV_T} - 1 \right), \tag{6.23}$$

where $V_T = kT/q$, I_s is the reverse bias saturation current, and I is the bias current. Re-arranging this equation, and adding a series resistance due to the probes, contacts, etc., the voltage observed when a current source is used is:

$$V = \ln \frac{I}{I_s} nV_T + nV_T + IR. \tag{6.24}$$

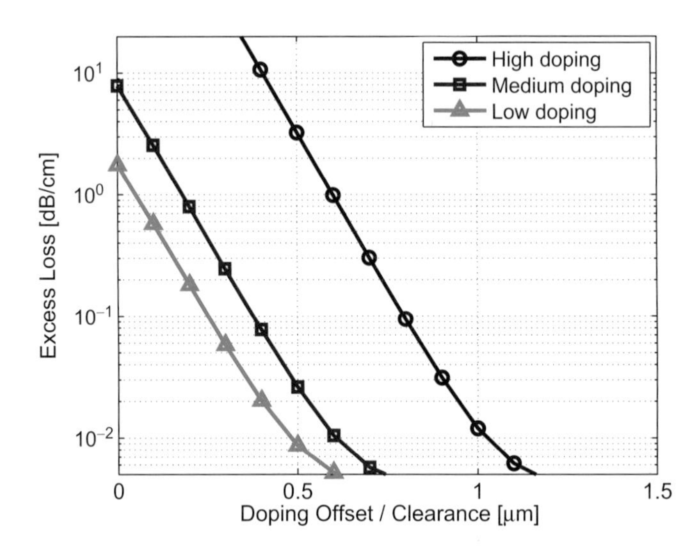

Figure 6.16 Optical excess loss due to the P and N dopants for the PIN junction rib waveguide. Three levels of doping density are considered, as per References [6, 11].

For the device in Figure 6.15, the fit to the IV data has parameters $n = 2.1$, $I_s = 3.8 \times 10^{-9}$, and $R = 3\,\Omega$, as shown in Figure 6.15. In such measurements, it is common to see variations in the resistance (e.g. from 3 to 15 Ω) primarily due to the probe contact.

Given the IV curve, the power consumption of the device can be calculated. This allows one to determine the optical absorption in relation to the power consumed. For applications where power consumption needs to be minimized, an additional important geometry parameter can be adjusted – the length. There are additional trade-offs to consider given the nonlinear behaviour of the diode and of the measured absorption data. For very long devices, the propagation loss (scattering, e.g. 3 dB/cm) will dominate, negatively contributing to the insertion loss. For very short devices, a large voltage will be required leading to a large current; in such a regime, a large portion of the current is "wasted", namely the carrier density does not increase linearly with current as per Equation (6.22). Thus the efficiency is reduced. With these considerations, typical PIN junction devices have lengths in the range of 0.1–10 mm.

More detailed simulations and device optimization taking into account the diode behaviour, contacts, device length, waveguide geometry, etc., can be performed using TCAD simulations, using similar methods as described in Section 6.2.4.

6.5 Active tuning

In this section, we describe two common approaches of tuning photonic circuits by introducing an electrically controlled phase shift. The first is the PIN junction waveguide, which offers moderately high-speed operation and efficiency, with a trade-off of variable insertion loss. The second is a thermal phase shifter, which offers pure phase shifting with no amplitude change, but operates at low speed.

6.5.1 PIN phase shifter

The forward-biased PIN junction waveguide from Section 6.4 can also be used as a phase shifter. This is useful for tuning the phase in a Mach–Zehnder modulator or other circuits that require phase tuning. When this waveguide is placed inside a Mach–Zehnder interferometer, the phase shift can be measured. Experimental results are shown for waveguide width of 500 nm and a clearance of 800 nm. Shown in Figure 6.17 are the spectra for a PIN junction inside an imbalanced interferometer with a free-spectral range of 55 nm. The measurement was performed using grating couplers, hence the spectrum has a signature Gaussian-like insertion-loss. At 0 bias, the extinction ratio of the interferometer is large (40 dB) signifying that there is no excess loss in the PIN waveguide. At a 5 mA current, a π phase shift is obtained, noting that the spectrum has shifted by half of the FSR. The figure of merit for PIN junction phase tuners can be expressed in mA per FSR, and in this case is approximately 10 mA/FSR. It is also observed that there is excess optical loss present in the PIN waveguide, thus the extinction ratio of the unbalanced interferometer is degraded (approximately 15 dB). For a large current of 90 mA, the losses are so large that the interference is no longer observed. Given that the attenuation in one arm of the interferometer is nearly 30 dB, as per Figure 6.15, this corresponds to light being transmitted only in one arm of the interferometer.

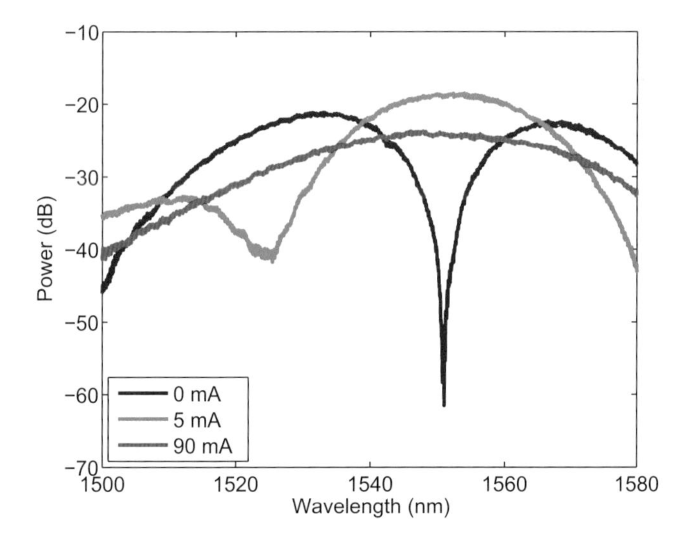

Figure 6.17 Optical transmission spectra of a Mach–Zehnder with each arm consisting of a 3 mm long PIN junction in a rib waveguide.

6.5.2 Thermal phase shifter

In this section, we consider the implementation of a phase shifter using a thermal heater. There are numerous methods of implementing a thermal phase shifter.

- A (metal) resistor above an optical waveguide, such that the heat generated in the heater descends towards the substrate past the waveguide. The metal is typically located far enough above the waveguide so as not to introduce optical loss: typically 1–2 μ above the waveguide.
- Placing the resistor inside the waveguide, so that the heat is generated directly within the waveguide. This would be more efficient; however, it incurs optical loss from the doping. This is typically implemented as an N++/N/N++ structure, where the two N++ regions are used to make contact with the silicon and are placed on opposite sides of the waveguide, as in Figure 6.18. The current flows across the waveguide, perpendicular to the light propagation. One of the limitations is that this requires access to both sides of the waveguide, i.e. requires a rib waveguide. This approach is used in the system presented in Section 13.1.3.
- Placing the resistor on the side of the waveguide with the current running parallel to the waveguide. This can be implemented either in the silicon slab of a rib waveguide, or a nearby metal.

Heater efficiency is an important issue for silicon photonic systems, and the common figure of merit is the tuning efficiency, expressed as mW/FSR, namely the power required to obtain a 2π optical phase-shift in a waveguide. This figure of merit suggests that the efficiency is nearly independent of the length of the phase shifter. A short thermal phase shifter will require a higher temperature of operation to achieve the same phase shift as a long phase shifter; however, it will consume the same power. This length independence is true for straight phase-shifter waveguides, as considered in this section, where the structure can be considered as a 2D cross-section. Compact heater structures (e.g., folding the waveguide), which would require 3D thermal modelling, can achieve improved heater efficiency since the heat is more concentrated on the waveguides. Other techniques of improving thermal phase shifters include reducing the thermal-conduction pathways by removing material, e.g. selectively undercutting the back-side of the substrate under the heater [12] (3.9 mW/FSR demonstrated), etching vertical trenches next to the thermal phase shifter [13] (0.8 mW/FSR demonstrated), or under-etching to improve the thermal isolation of the waveguide [14] (0.49 mW/FSR demonstrated).

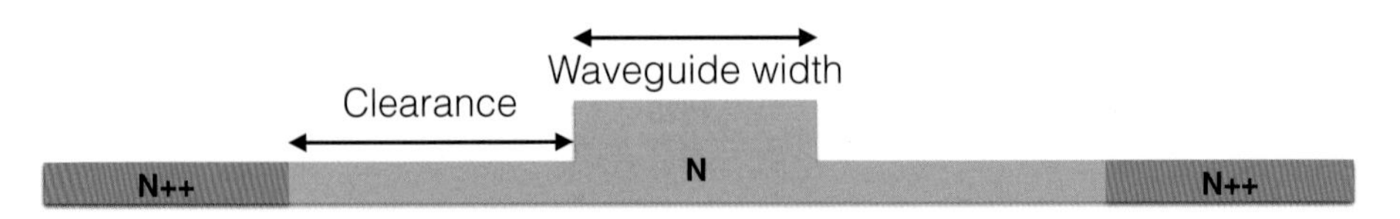

Figure 6.18 Cross-section of a N++/N/N++ resistor in a rib waveguide. The contacts are connected to the "N++" regions, and the resistance is dominated by the lightly doped region "N".

Thermal modelling of the thermal phase shifter uses the steady-state heat equation (Poisson's equation):

$$-\nabla \cdot (k\nabla T) = Q, \tag{6.25}$$

where k is the coefficient of heat conduction, Q is the heat source [W/m^3], and T is the temperature.

The example in Listing 6.12 was generated by the MATLAB Partial Differential Equation (PDE) Toolbox [15] graphic user interface. It was subsequently modified to extract the relevant information. Using the PDE Toolbox, we consider the geometry as follows, drawn in the order listed to ensure that the waveguide and metal are "on-top" of the oxide.

- Define $y = 0$ as the bottom of the waveguide.
- Oxide extends from $y = -2$ to 2, and $x = -50$ to 50.
- Metal is 1 µm above the waveguide, i.e. extends from $y = 1.22$ to 1.72, and $x = -0.5$ to 0.5.
- Waveguide is a 500 nm × 220 nm strip, i.e. extends from $y = 0$ to 0.22, and $x = -0.25$ to 0.25.
- Silicon substrate is 100 µm thick, i.e. extends from $y = 0$ to -100, and $x = -50$ to 50.

Two boundary conditions are used.

(1) Dirichlet type, where the temperature on the boundary is specified: this is used for the bottom of the substrate, assuming that it is on a heat sink. The temperature is set to 0 degrees, and the results are plotted relative to this heat sink temperature.
(2) Neumann type, where the heat flux, $-\mathbf{n} \cdot (k\nabla T)$, is specified: this is used on all the other boundaries to assume they are insulating, i.e. no heat passes through. We are neglecting convection and radiation.

Heat is generated in a metal conductor above a waveguide. Assume the metal is 500 nm thick, 1000 nm wide, and 100 µm long, and that this resistive heater is dissipating Q_{tot}=10 mW of power. Thus,

$$Q = \frac{Q_{tot}}{V} = \frac{0.01\text{W}}{0.5 \cdot 1 \cdot 100 \times 10^{-18}\text{m}^3}.$$

The material properties are

- $kSi = 149$; thermal conductivity of silicon [W/m·K],
- $kSiO_2 = 1.4$; thermal conductivity of SiO2 [W/m·K],
- $kAl = 250$; thermal conductivity of aluminum [W/m·K].

To be consistent with the geometry definition in microns, the thermal conductivities are also provided in micron units, i.e. $kSi = 149 \times 10^{-6}$ W/m·K.

The thermal map, zoomed in to the cross-section near the waveguide and heater, is plotted in Figure 6.19a. Observations from the simulations are as follows.

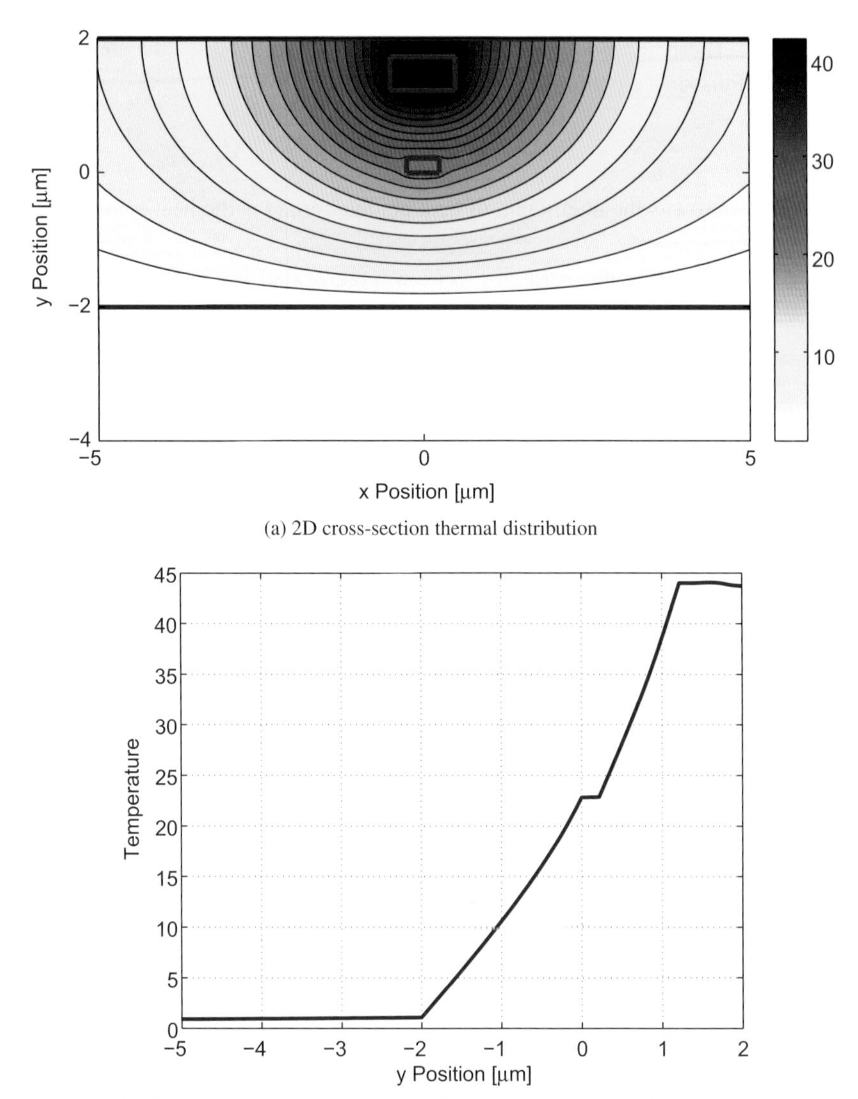

(a) 2D cross-section thermal distribution

(b) 1D cross-section temperature distribution through the metal and waveguide, as position $x = 0$ in Figure 6.19a. The waveguide is located at y position 0, and the metal is at position 1.22.

Figure 6.19 Temperature distribution in the cross-section of a wafer. The simulation includes a strip waveguide at position (0,0), a metal heater above it at position (0,1.5); both are surrounded by oxide. The silicon substrate is below the interface at $y = -2$. The bottom of the substrate is fixed at a temperature 0 °C, and the other boundaries are assumed insulating.

- Because the silicon and metal thermal conductivities are much higher than the surrounding oxide (100×), the temperature is uniform within the metal and waveguide.
- Similarly, because silicon thermal conductivity is significantly higher (100×) than the oxide, the substrate is nearly uniform in temperature (1.1 °C higher

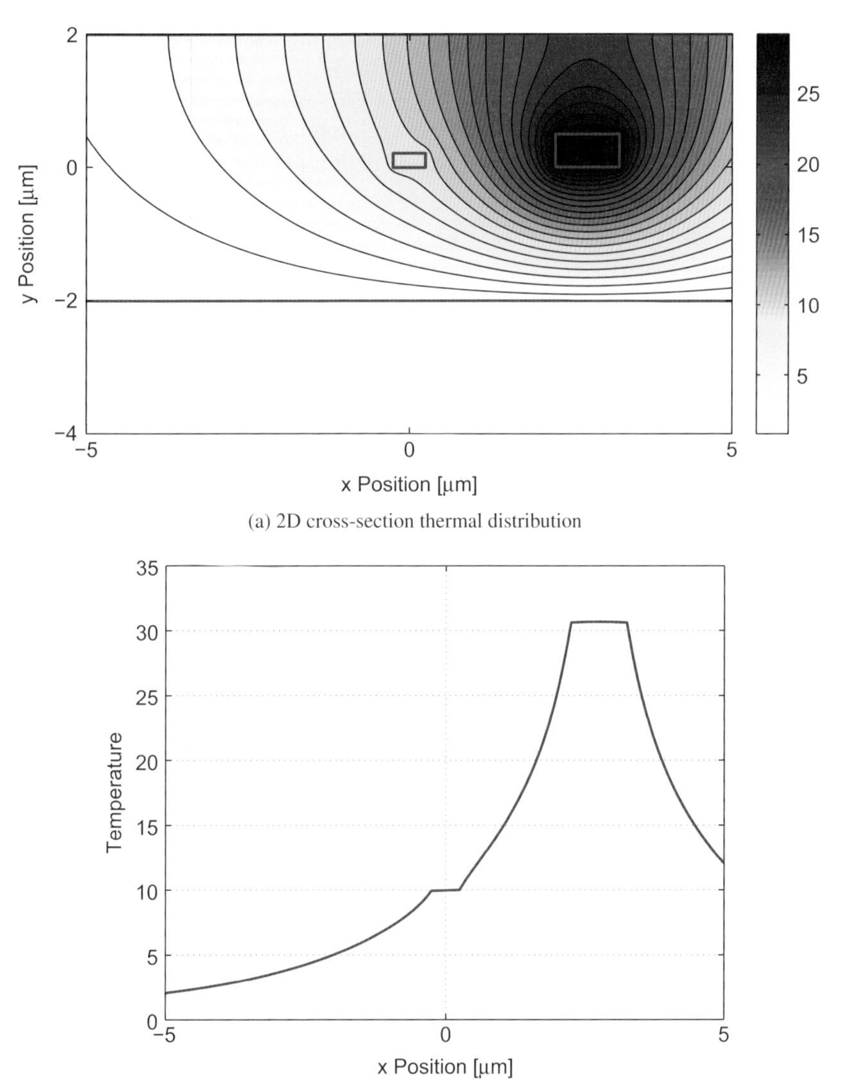

(a) 2D cross-section thermal distribution

(b) 1D cross-section temperature distribution through the metal and waveguide, as position $y = 0.11$ in Figure 6.20a. The waveguide is located at x position 0, and the metal is at position 2.25–3.25.

Figure 6.20 Temperature distribution in the cross-section of a wafer. The simulation includes a strip waveguide at position (0,0), a metal heater next to it at position (2.75,0.25); both are surrounded by oxide. The silicon substrate is below the interface at $y = -2$. The bottom of the substrate is fixed at a temperature 0 °C, and the other boundaries are assumed insulating.

at the silicon–oxide interface than the heat sink), and the thermal gradient is almost entirely in the oxide. Although the silicon substrate is very thick relative to the oxide and waveguide (700 µm versus 2 µm and 0.22 µm, respectively), the thickness of the substrate does not significantly impact the thermal results. Namely, a substrate that is 10 µm or 700 µm thick gives very similar results.

- The simulation predicts a temperature rise of 44 °C in the metal, and 23 °C in the waveguide.

We can calculate the thermal tuning efficiency for this simulation. From Equation (6.25), and based on the definition that one FSR is equivalent to a 2π phase shift, we can write that:

$$\text{efficiency [mW/FSR]} = \frac{Q_{tot}\lambda}{\frac{dn}{dT}\Delta T}. \tag{6.26}$$

For this simulation, the efficiency is found to be 36 mW/FSR.

Similar calculations can be performed for the heater on the side of the waveguide. The results for a metal heater 2 μm away from the strip waveguide are shown in Figure 6.20. In this case, the efficiency is 83 mW/FSR.

6.6 Thermo-optic switch

The Mach–Zehnder inteferometer (MZI), described in Section 4.3, can function as a thermo-optic switch if a temperature difference is applied between the arms [13, 16]. This can be implemented by using a resistive heater on one of the arms [17]. Let us consider a temperature increase of ΔT applied to the lower arm, and a thermo-optic coefficient of $\frac{dn}{dT} = 1.87 \times 10^{-4} K^{-1}$ (Section 3.1.1). The propagation constant in the

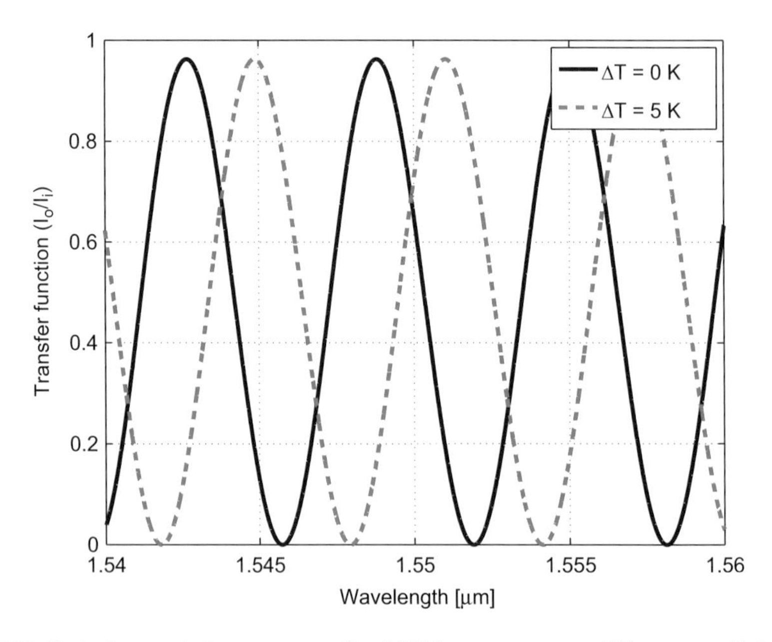

Figure 6.21 Optical transmission spectrum of an MZI for a temperature difference applied on the second arm (L_2). Parameters: $L_1 = 500\,\mu\text{m}$, $\Delta L = 100\,\mu\text{m}$, $\alpha = 3$ dB/cm, lossless *y*-branch. Wavelength-dependent effective index for both waveguides as per Equation (3.7).

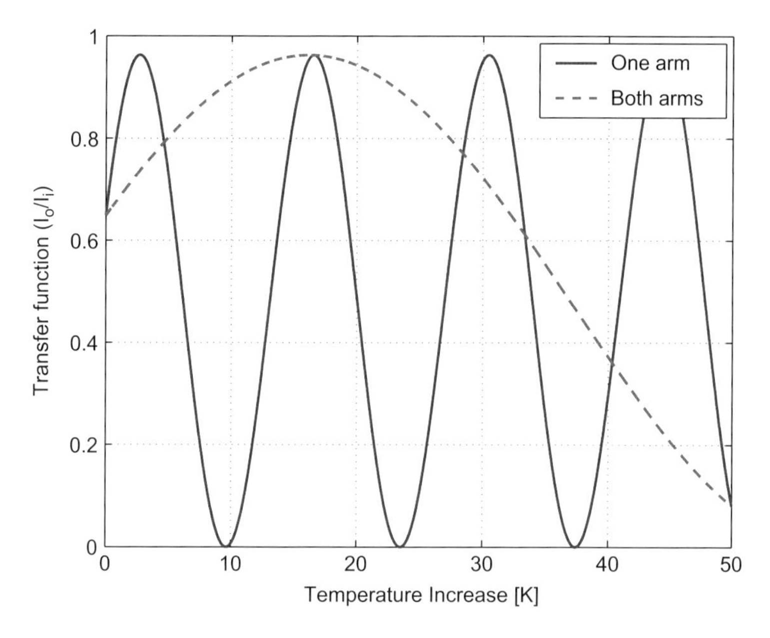

Figure 6.22 Optical transmission versus temperature of the MZI in Figure 6.21. Two cases are considered: thermally tuning one arm, and thermally tuning the substrate (both arms).

lower arm is:

$$\beta_2 = \frac{2\pi\left(n_2 + \frac{dn}{dT}\Delta T\right)}{\lambda}. \tag{6.27}$$

Inserting this in Equation (4.19), the optical output is thus temperature dependent:

$$I_o(\Delta T) = \frac{I_i}{2}\left[1 + \cos\left(\beta_1 L_1 - \frac{2\pi\left(n_2 + \frac{dn}{dT}\Delta T\right)}{\lambda} L_2\right)\right]. \tag{6.28a}$$

For identical waveguide cross-sections ($n_1 = n_2$), this becomes

$$I_o(\Delta T) = \frac{I_i}{2}\left[1 + \cos\left(\frac{2\pi n}{\lambda}\Delta L - \frac{2\pi\left(\frac{dn}{dT}\Delta T\right)}{\lambda} L_2\right)\right] \tag{6.29a}$$

Here, we explicitly see that the output of the interferometer is a sinusoidally varying function of wavelength and temperature. This is illustrated in Figures 6.21 and 6.22 (with loss included).

6.7 Problems

6.1 Determine the thermal tuning efficiency for a thermal phase shifter where the heater is embedded in the waveguide via a N++/N/N++ region. Assume the waveguide is a rib waveguide with a 10 μm slab width, 90 nm slab thickness, with a ridge that is 0.5 μm width and 220 nm thick.

6.2 Consider a ring resonator with 220 nm × 500 nm strip waveguides that is uniformly heated. Determine an expression for the wavelength shift versus temperature, i.e. $d\lambda/dT$, at 1550 nm. How does this vary with the radius of the ring resonator?

6.8 Code listings

Listing 6.1 pn-Junction depletion, MATLAB model, pn_depletion.m

```
% pn_depletion.m: 1D pn junction model for carrier-depletion phase modulation
% Wei Shi, UBC, Nov. 2012
%
% usage, e.g.:
%   [n, p, x, xn, xp, Rj, Cj]=pn_depletion(500e-9, 50e-9, 1e-6, 1e-6, 25, -1, 100)
%
function [n, p, x, xn, xp, Rj, Cj]=pn_depletion(wg_width, pn_offset, ds_npp, ds_ppp, T, V,
     pts)
%
% N_D, N_A: doping densities
% V: applied voltage; positive for forward bias; negative for reverse bias
% ds_npp: distance of the n++ boundary to the pn junction centre
% ds_ppp: distance of the p++ boundary to the pn junction centre
% Rj: junction resistance in ohms
% Cj: junction capacitance in F/m

epsilon0 = 8.854187817620e-12; % [F/m]
epsilon_s = 11.8; % relative dielectric constant for Si
q = 1.60217646e-19; % electronic charge [Coulumbs]
kB = 1.3806503e-23; % Boltzmann constant in J/K
T=T+273.15; % Temperature [K]
VT=kB*T/q;

%material constants
NA_plus=4.4e20*1e6;% cm^-3*1e6
ND_plus=4.4e20*1e6;
NA=5e17*1e6;% cm^-3*1e6
ND=3e17*1e6;

Rs_rib_n=2.5e3;
Rs_rib_p=4.0e3;
Rs_slab_n=0.6e4;
Rs_slab_p=1e4;

% waveguide height
h_rib=220e-9; h_slab=90e-9;

h=4.135e-15; % Plank's constant [eV-s]
m_0=9.11e-31; % electron mass [kg]
m_n=1.08*m_0; % Density-of-states effective mass for electrons
m_p=1.15*m_0; % Density-of-states effective mass for holes
Nc=2*(2*pi*m_n*(kB/q)*T/h^2)^(3/2)/(q)^(3/2); % Effective Density of states for Conduction
     Band
Nv=2*(2*pi*m_p*(kB/q)*T/h^2)^(3/2)/(q)^(3/2); % Effective Density of states for Valence
     Band
Eg=1.1242; % band gap for Si [eV]
% ni=1e10*1e6;
ni=sqrt(Nc*Nv).*exp(-Eg/(2*(kB/q)*T)); % intrinsic charge carriers in m^-3

Vbi=VT*log(NA*ND/ni^2); % built-in or diffusion potential
Wd=sqrt(2*epsilon0*epsilon_s*(NA+ND) / (q*NA*ND) *(Vbi-V)); % depletion width
xp=-Wd/(1+NA/ND)+pn_offset;
xn=Wd/(1+ND/NA)+pn_offset;
```

```
del_x=wg_width/(pts-1);
x_ppp=-ds_ppp+pn_offset; x_npp=ds_npp+pn_offset;
x_min=x_ppp-500e-9; x_max=x_npp+500e-9;
%
x_NA_plus=x_min:del_x:x_ppp-del_x;
x_NA=x_ppp:del_x:xp-del_x;
x_dep=xp:del_x:xn;% for the depletion region
x_ND=xn+del_x:del_x:x_npp;
x_ND_plus=x_npp+del_x:del_x:x_max;
x=[x_NA_plus, x_NA, x_dep, x_ND, x_ND_plus];

n0_NA=ni^2/NA; p0_ND=ni^2/ND;
n0_NA_plus=ni^2/NA_plus; p0_ND_plus=ni^2/ND_plus;

% Long-base assumption
% Lp=sqrt(Dp*tau_p);
% Ln=sqrt(Dn*tau_n);
% del_n_NA=n0_NA*(exp(q*V/(kB*T))-1)* exp(-abs(x_NA-xp)/Ln); % minority electron density
     in p(NA) region
% del_p_ND=p0_ND*(exp(q*V/(kB*T))-1)* exp(-abs(x_ND-xn)/Lp); % minority hole density in
     n(ND) region

% Short-base assumption
del_n_NA=n0_NA*(exp(q*V/(kB*T))-1)* (1-abs((x_NA-xp)/(xp-x_ppp))); % minority electron
     density in p(NA) region
del_p_ND=p0_ND*(exp(q*V/(kB*T))-1)* (1-abs((x_ND-xn)/(x_npp-xn))); % minority hole density
     in n(ND) region


n_NA=n0_NA+del_n_NA; p_ND=p0_ND+del_p_ND;
p_dep=zeros(1, length(x_dep)); n_dep=zeros(1, length(x_dep));
p_NA=ones(1, length(x_NA))*NA; % majority holes in p(NA) region
n_ND=ones(1, length(x_ND))*ND; % majority electrons in n(ND) region
n_NA_plus=ones(1, length(x_NA_plus))*n0_NA_plus;% assumption of uniform electrons in p++
p_ND_plus=ones(1, length(x_ND_plus))*p0_ND_plus;% assumption of uniform holes in n++
p_NA_plus=ones(1, length(x_NA_plus))*NA_plus; % majority holes in p++ region
n_ND_plus=ones(1, length(x_ND_plus))*ND_plus; % majority electrons in n++ region

n=[n_NA_plus, n_NA, n_dep, n_ND, n_ND_plus]; p=[p_NA_plus, p_NA, p_dep, p_ND, p_ND_plus];

Rj=(wg_width/2-xn)* Rs_rib_n+(wg_width/2+xp)* Rs_rib_p+(-wg_width/2-x_ppp)*
     Rs_slab_p+(x_npp-wg_width/2)* Rs_slab_n;
Cj=sqrt(q*epsilon0*epsilon_s/2/ (1/ND+1/NA)/(Vbi-V))*h_rib;
```

Listing 6.2 pn-Junction effective index and optical loss, MATLAB model, neff_V.m

```
% neff_V.m: effective index as a function of voltage for carrier-depletion phase modulation
% Wei Shi, UBC, 2012
% Usage, e.g.:
%    [del_neff alpha Rj Cj]=neff_V(1.55e-6, 220e-9, 500e-9, 90e-9, 3.47, 1.44, 1.44,
     50e-9, 1e-6, 1e-6, 500, 25, -1)
function [neff alpha Rj Cj]=neff_V(lambda, t, w, t_slab, n_core, n_clad, n_oxide,
     pn_offset, ds_n_plus, ds_p_plus, pts, T, V)
[n, p, xdoping, xn, xp, Rj, Cj]=pn_depletion(w, pn_offset, ds_n_plus, ds_p_plus, T, V,
     pts);

M=min(ds_n_plus-pn_offset+0.5e-6, ds_p_plus+pn_offset+0.5e-6)/w-0.5;

[xwg, TM_E_TEwg, neff0]=wg_TElike_1Dprofile_neff(lambda, t, w, t_slab, n_core, n_clad,
     n_oxide, pts, M);
Ewg=TM_E_TEwg(:,1)';

pts_x=length(xwg);
dxwg=zeros(1, pts_x);
dxwg(1)=xwg(2)-xwg(1); dxwg(pts_x)=xwg(pts_x)-xwg(pts_x-1);
for i=2:pts_x-1
   dxwg(i)=xwg(i+1)/2-xwg(i-1)/2;
end
```

```
n_wg=interp1(xdoping, n, xwg);
p_wg=interp1(xdoping, p, xwg);

del_ne=-3.64e-10*lambda^2*sum(conj(Ewg).*(n_wg*1e-6).*Ewg.*dxwg)/sum(conj(Ewg).*Ewg.*dxwg);
del_nh=-3.51e-6*lambda^2*sum(conj(Ewg).*(p_wg*1e-6).^0.8.*Ewg.*dxwg)/sum(conj(Ewg).
     *Ewg.*dxwg);
del_neff=del_ne+del_nh;
neff=neff0+del_neff;

del_alpha_e=3.52e-6*lambda^2*sum(conj(Ewg).*(n_wg*1e-6).*Ewg.*dxwg)/sum(conj(Ewg).
     *Ewg.*dxwg);
del_alpha_h=2.4e-6*lambda^2*sum(conj(Ewg).*(p_wg*1e-6).*Ewg.*dxwg)/sum(conj(Ewg).
     *Ewg.*dxwg);
alpha=del_alpha_e+del_alpha_h;
```

Listing 6.3 pn-Junction depletion, MATLAB model, neff_V_plot.m

```
% example:
% [neff alpha delta_neff delta_phi fc]=neff_V_plot(1.5e-6, 220e-9, 500e-9, 90e-9, 3.47,
     1.44, 1.44, 50e-9, 10e-6, 10e-6, 500, 25, -(1:5));

function [neff alpha delta_neff delta_phi fc] = neff_V_plot(lambda, t, w, t_slab, n_core,
     n_clad, n_oxide, pn_offset, ds_n_plus, ds_p_plus, pts, T, V);

neff=zeros(1, length(V)); alpha=zeros(1, length(V));
Rj=zeros(1, length(V)); Cj=zeros(1, length(V));
for i=1:length(V);
    [neff(i) alpha(i) Rj(i) Cj(i)]=neff_V(lambda, t, w, t_slab, n_core, n_clad, n_oxide,
         pn_offset, ds_n_plus, ds_p_plus, pts, T, V(i))
end

[neff_v0 alpha_v0]=neff_V(lambda, t, w, t_slab, n_core, n_clad, n_oxide, pn_offset,
     ds_n_plus, ds_p_plus, pts, T, 0);

delta_neff=neff-neff_v0;
alpha_dB=-10*log10(exp(-alpha));

figure; plot(-V, delta_neff)
figure; plot(-V, alpha_dB);

% Phase shift per cm
delta_phi=2*pi/lambda*delta_neff*1e-2/pi;% per cm
figure; plot(-V, delta_phi, 'linewidth', 2);

% Cut-off frequency
fc=1./(2*pi*Rj.*Cj)*1e-9;% in GHz
figure; plot(-V, fc, 'linewidth', 2);
```

Listing 6.4 Ring modulator, MATLAB model, RingMod.m

```
% RingMod.m: Ring modulator 1D model
% Usage, e.g.,
%   [Ethru Edrop Qi Qc Rj Cj]=RingMod(1.55e-6, 'all-pass', 10e-6, 0, 2*pi*10e-6, 500e-9,
     0, 1e-6, 1e-6, 25, 0);
%
% Wei Shi, UBC, 2012
% weis@ece.ubc.ca
%
function [Ethru Edrop Qi Qc tau_rt Rj Cj]=RingMod(lambda, Filter_type, r, Lc, L_pn, w,
     pn_offset, ds_n_plus, ds_p_plus, T, V);
%
% type: all-pass or add-drop
% r: radius
% Lc: coupler length
% Lh: heater length
%
% neff_pn, alpha_pn: effective index and free-carrier obsorption of the phase modulator
% Rj, Cj: junction resistance and capacitance of the phase modulator
```

```
%
% predetermined parameters
t=220e-9; t_slab=90e-9; n_core=3.47; n_clad=1.44; n_oxide=1.44; pts=200;
%
[neff_pn alpha_pn Rj Cj]=neff_V(lambda, t, w, t_slab, n_core, n_clad, n_oxide, pn_offset,
     ds_n_plus, ds_p_plus, pts, T, V);
%
% undoped waveguide mode and effective index
[xwg0 TM_E_TEwg0 neff0]=wg_TElike_1Dprofile_neff(lambda, t, w, t_slab, n_core, n_clad,
     n_oxide, pts, 2);
neff_exc=neff0;
del_lambda=0.1e-9;
[xwg1 TM_E_TEwg1 neff0_1]=wg_TElike_1Dprofile_neff(lambda+del_lambda, t, w, t_slab,
     n_core, n_clad, n_oxide, pts, 2);
ng=neff0-(neff0_1-neff0)/del_lambda*lambda;

alpha_wg_dB=5; % optical loss of intrinsic optical waveguide, in dB/cm
alpha_wg=-log(10^(-alpha_wg_dB/10));% converted to /cm
alpha_pn=alpha_wg+alpha_pn;
alpha_exc=alpha_wg; % optical loss of the ring cavity excluding the phase modulator

L_rt=Lc*2+2*pi*r;
L_exc=L_rt-L_pn;
phi_pn=(2*pi/lambda)*neff_pn*L_pn;
phi_exc=(2*pi/lambda)*neff_exc*L_exc;
phi_rt=phi_pn+phi_exc;

c=299792458;
vg=c/ng;
tau_rt=L_rt/vg;% round-trip time

A_pn=exp(-alpha_pn*100*L_pn); % attunation due to pn junciton
A_exc=exp(-alpha_exc*100*L_exc); % attunation over L_exc
A=A_pn*A_exc; % round-trip optical power attenuation

alpha_av=-log(A)/L_rt;% average loss of the cavity
Qi=2*pi*ng/lambda/alpha_av;

%coupling coefficients
k=0.2;
if (Filter_type=='all-pass')
   t=sqrt(1-k^2);
   Ethru=(-sqrt(A)+t*exp(-1i*phi_rt))/(-sqrt(A)*conj(t)+exp(-1i*phi_rt));
   Edrop=0;
   Qc=-(pi*L_rt*ng)/(lambda*log(abs(t)));
elseif (Filter_type=='add-drop')
   k1=k; k2=k1;
   t1=sqrt(1-k1^2); t2=sqrt(1-k2^2);
   Ethru=(t1-conj(t2)*sqrt(A)*exp(1i*phi_rt))/(1-sqrt(A)*conj(t1)*conj(t2)
        *exp(1i*phi_rt));
   Edrop=-conj(k1)*k2*sqrt(sqrt(A))*exp(1i*phi_rt/2)/(1-sqrt(A)*conj(t1)*conj(t2)
        *exp(1i*phi_rt));
   Qc1=-(pi*L_rt*ng)/(lambda*log(abs(t1)));
   Qc2=-(pi*L_rt*ng)/(lambda*log(abs(t2)));
   Qc=1/(1/Qc1+1/Qc2);
else
   error(1, 'The''Filter_type'' has to be ''all-pass'' or ''add-drop''.\n');
end
```

Listing 6.5 Waveguide mode profile and effective index calculation, MATLAB model, wg_TElike_1Dprofile_neff.m

```
% wg_TElike_1Dprofile.m - Effective Index Method - 1D mode profile
% Lukas Chrostowski, 2012
% modified by Wei Shi, 2012

% usage, e.g.:
%  [xwg, TM_E_TEwg]=wg_TElike_1Dprofile_neff (1.55e-6, 0.22e-6, 0.5e-6, 90e-9,3.47, 1,
     1.44, 100, 2);
```

```
% figure; plot(xwg, TM_E_TEwg(:,1))

function [xwg, TM_E_TEwg, neff_TEwg_1st]=wg_TElike_1Dprofile_neff (lambda, t, w, t_slab,
      n_core, n_clad, n_oxide, pts, M)

% TE (TM) modes of slab waveguide (core and slab portions):
[nTE,nTM]=wg_1D_analytic (lambda, t, n_oxide, n_core, n_clad);
if t_slab>0
   [nTE_slab,nTM_slab]=wg_1D_analytic (lambda, t_slab, n_oxide, n_core, n_clad);
else
   nTE_slab=n_clad; nTM_slab=n_clad;
end
[xslab, TE_Eslab, TE_Hslab, TM_Eslab, TM_Hslab]= wg_1D_mode_profile (lambda, t, n_oxide,
      n_core, n_clad, pts, M);

% TE-like modes of the etched waveguide (for fundamental slab mode):
[nTE,nTM]=wg_1D_analytic (lambda, w, nTE_slab(1), nTE(1), nTE_slab(1));
neff_TEwg_1st=nTM(1);
[xwg, TE_E_TEwg, TE_H_TEwg, TM_E_TEwg, TM_H_TEwg]= wg_1D_mode_profile (lambda, w,
      nTE_slab(1), nTE(1), nTE_slab(1), pts, M);
```

Listing 6.6 Ring modulator spectrum, MATLAB model, RingMod_spectrum.m

```
% calculate the ring modulator spectrum
% Wei Shi UBC, 2012
% weis@ece.ucb.ca

function [Ethru Edrop Qi Qc tau_rt Rj Cj]=RingMod_spectrum(lambda, Filter_type, r, Lc,
      L_pn, w, pn_offset, ds_n_plus, ds_p_plus, T, V);
%
Ethru=zeros(1, length(lambda));
Edrop=zeros(1, length(lambda));
Qi=zeros(1, length(lambda));
Qc=zeros(1, length(lambda));
tau_rt=zeros(1, length(lambda));
%
for i=1:length(lambda)
    [Ethru(i) Edrop(i) Qi(i) Qc(i) tau_rt(i) Rj Cj]=RingMod(lambda(i), Filter_type, r, Lc,
         L_pn, w, pn_offset, ds_n_plus, ds_p_plus, T, V);
end
```

Listing 6.7 Ring modulator spectrum plot, MATLAB, RingMod_spectrum_plot.m

```
% Plot the ring modulator spectrum
% Wei Shi, UBC, 2012
% weis@ece.ubc.ca
%
% RingMod_spectrum_plot;

c=299792458;
lambda=1e-9*(1530:0.1:1560);
Filter_type='add-drop';
pn_angle=2*pi; %
r=10e-6; Lc=0; L_pn=2*pi*r*pn_angle/(2*pi); w=500e-9; % WG parameters
pn_offset=0; ds_n_plus=1e-6; ds_p_plus=1e-6; % pn-junction design
T=25; V0=0; % temperature and voltage

[Ethru0 Edrop0 Qi0 Qc0 tau_rt0 Rj0 Cj0] = RingMod_spectrum (lambda, Filter_type, r, Lc,
      L_pn, w, pn_offset, ds_n_plus, ds_p_plus, T, V0);

figure;
plot(lambda*1e9, [10*log10(abs(Ethru0).^2); 10*log10(abs(Edrop0).^2)], 'linewidth', 2);
xlim([min(lambda) max(lambda)]*1e9);
set(gca, 'fontsize', 14);
xlabel({'\lambda (nm)'}, 'fontsize', 14);
ylabel({'Transmission (dB)'}, 'fontsize', 14);
legend('Through', 'Drop');
```

```
% zoom at one peak wavelength
lambda_zoom=1e-9*(1540.7:0.0025:1541);
V=-4:1:0;
lenV = length(V); lenLZ = length(lambda_zoom);
Ethru=zeros(lenV, lenLZ); Edrop=zeros(lenV, lenLZ);
A=zeros(lenV, lenLZ);
Qi=zeros(lenV, lenLZ); Qc=zeros(lenV, lenLZ);
Cj=zeros(lenV,1);      Rj=zeros(lenV,1);
for i=1:lenV
    [Ethru(i,:) Edrop(i,:) Qi(i,:) Qc(i,:) tau_rt(i,:) Rj(i,:)
         Cj(i,:)]=RingMod_spectrum(lambda_zoom, Filter_type, r, Lc, L_pn, w, pn_offset,
         ds_n_plus, ds_p_plus, T, V(i));
end

Qt=1./(1./Qi+1./Qc);% total Q
tp=Qt./(c/1541e-9*2*pi); % photon lifetime
tp_av=sum(tp, 2)/(length(lambda_zoom)); % average photon lifetime across over the spectrum
fcq=1./(2*pi*tp_av);
fcj=1./(2*pi*Rj.*Cj);
fc=1./(1./fcq+1./fcj);

figure; plot(lambda_zoom*1e9, 10*log10(abs(Ethru).^2), 'linewidth', 2);
set(gca, 'fontsize', 14);
xlabel({'\lambda (nm)'}, 'fontsize', 14);
ylabel({'Transmission (dB)'}, 'fontsize', 14);
legend({cat(2, num2str(-V'), char(ones(length(V),1)*'V'))}, 'Location', 'best',
      'fontsize', 14);

if strcmp(Filter_type,'add-drop')
    figure;
    plot(lambda_zoom*1e9, 10*log10(abs(Edrop).^2), 'linewidth', 2);
    set(gca, 'fontsize', 14);
    xlabel({'\lambda (nm)'}, 'fontsize', 14);
    ylabel({'Transmission (dB)'}, 'fontsize', 14);
    legend({cat(2, num2str(-V'), char(ones(length(V),1)*'V'))}, 'Location', 'best',
         'fontsize', 14);
end

figure;
plot(-V, [fcq fcj fc]*1e-9, 'linewidth', 2);
set(gca, 'fontsize', 14);
xlabel({'Voltage (V)'}, 'fontsize', 14);
ylabel({'Cutoff frequency (GHz)'}, 'fontsize', 14);
legend({'\tau_p determined','p-n junction determined','f_c'}, 'Location', 'NorthWest',
      'fontsize', 14);
```

Listing 6.8 Parameter definitions for the pn-junction phase modulator, used for electrical calculations in DEVICE, Listing 4.10, and optical calculations in MODE, Listing 6.10, modulator_setup_parameters.lsf

```
# Parameter definitions for the modulator, used for electrical calculations in DEVICE and
     optical calculations in MODE

# define wafer and waveguide structure
thick_rib = 0.13e-6;
width_rib = 0.5e-6;
thick_slab = 0.09e-6;
width_slab = 5e-6;
center_plateau = 3.75e-6;
width_plateau = 2.5e-6;

# define doping
center_pepi = 0.1e-6;  # pepi
thick_pepi = 0.3e-6;

x_center_p = -3.075e-6; # implant
x_span_p = 5.85e-6;
z_center_p = -0.105e-6;
```

```
z_span_p = 0.39e-6;
diff_dist_fcn = 1; # 0 for erfc, 1 for gaussian
face_p = 5; # upper z
width_junction_p = 0.1e-6;
surface_conc_p = 7e17*1e6;
reference_conc_p = 1e6*1e6;
x_center_n = 3.075e-6;
x_span_n = 5.85e-6;
z_center_n = -0.105e-6;
z_span_n = 0.39e-6;
face_n = 5; # upper z
width_junction_n = 0.1e-6;
surface_conc_n = 5e17*1e6;
reference_conc_n = 1e6*1e6;

x_center_p_contact = -4e-6; # contact
x_span_p_contact = 4e-6;
z_center_p_contact = -0.04e-6;
z_span_p_contact = 0.52e-6;
diff_dist_fcn_contact = 1; # 0 for erfc, 1 for gaussian
face_p_contact = 5; # upper z
width_junction_p_contact = 0.1e-6;
surface_conc_p_contact = 1e19*1e6;
reference_conc_p_contact = 1e6*1e6;
x_center_n_contact = 4e-6;
x_span_n_contact = 4e-6;
z_center_n_contact = -0.04e-6;
z_span_n_contact = 0.52e-6;
face_n_contact = 5; # upper z
width_junction_n_contact = 0.1e-6;
surface_conc_n_contact = 1e19*1e6;
reference_conc_n_contact = 1e6*1e6;

x_center_p_rib = -0.12e-6; # waveguide
x_span_p_rib = 0.36e-6;
z_center_p_rib = 0.1275e-6;
z_span_p_rib = 0.255e-6;
diff_dist_fcn_rib = 1; # 0 for erfc, 1 for gaussian
face_p_rib = 0; # lower x
width_junction_p_rib = 0.12e-6;
surface_conc_p_rib = 5e17*1e6;
reference_conc_p_rib = 1e6*1e6;
x_center_n_rib = 0.095e-6;
x_span_n_rib = 0.31e-6;
z_center_n_rib = 0.14e-6;
z_span_n_rib = 0.24e-6;
face_n_rib = 1; # upper x
width_junction_n_rib = 0.11e-6;
surface_conc_n_rib = 7e17*1e6;
reference_conc_n_rib = 1e6*1e6;

# define contacts
center_contact = 4.4e-6;
width_contact = 1.2e-6;
thick_contact = 0.5e-6;
voltage_start = -0.5;
voltage_stop = 4;
voltage_interval = 0.25;

# define simulation region
min_edge_length = 0.004e-6;
max_edge_length = 0.6e-6;
max_edge_length_override = 0.007e-6; # mesh override region
x_center = 0; x_span = 2*center_contact + width_contact;
y_center = 0; y_span = 1e-6; # irrelevant for 2D cross section
z_center = 0; z_span = 5e-6;

# define monitors
filename_mzi = 'mzi_carrier.mat';
```

Listing 6.9 Electrical simulation of the pn-junction phase modulator, in Lumerical DEVICE. This script simulates the DC characteristics of the junction, specifically the spatial charge density and junction capacitance versus voltage. Device and simulation parameters are defined in Listing 6.8; modulator_setup_device.lsf

```
# Electrical simulation of the pn-junction phase shifter
# Step 1: in Lumerical DEVICE; this script accomplishes:
# 1) Simulate the DC characteristics of the junction,
#   to export the spatial charge density, for different voltages
# 2) Calculates the junction capacitance versus voltage.
# 3) Calculates the resistance in each slab

newproject; redrawoff;

# modulator geometry variables defined:
modulator_setup_parameters;

# draw geometry
addrect;     # rib
set('name','rib');
set('material','Si (Silicon)');
set('x',x_center); set('x span', width_rib);
set('y',y_center); set('y span',y_span);
set('z min',thick_slab); set('z max',thick_slab+thick_rib);

addrect;     # slab
set('name','slab');
set('material','Si (Silicon)');
set('x',x_center); set('x span', width_slab);
set('y',y_center); set('y span',y_span);
set('z min',z_center); set('z max',thick_slab);

addrect;     # plateau
set('name','plateau_left');
set('material','Si (Silicon)');
set('x',-center_plateau); set('x span', width_plateau);
set('y',y_center); set('y span',y_span);
set('z min',z_center); set('z max',thick_slab+thick_rib);
copy;
set('name','plateau_right');
set('x',center_plateau);

addrect;     # contacts
set('name','anode');
set('material','Al (Aluminium) - CRC');
set('x',-center_contact); set('x span', width_contact);
set('y',y_center); set('y span',y_span);
set('z min',thick_slab+thick_rib); set('z max',thick_slab+thick_rib+thick_contact);
copy;
set('name','cathode');
set('x',center_contact);

addrect;     # oxide
set('name','oxide');
set('material','SiO2 (Glass) - Sze');
set('override mesh order from material database',1); set('mesh order',5);
set('override color opacity from material database',1); set('alpha',0.3);
set('x',x_center); set('x span',x_span);
set('y',y_center); set('y span',y_span);
set('z',z_center); set('z span',z_span);

# draw simulation region
adddevice;
set('min edge length',min_edge_length);
set('max edge length',max_edge_length);
set('x',x_center); set('x span',x_span-0.1e-6);
set('y',y_center); set('y span',y_span);
set('z',z_center); set('z span',z_span);
```

```
addmesh;
set('name','wg mesh');
set('max edge length',max_edge_length_override);
set('x',x_center); set('x span', width_rib);
set('y',y_center); set('y span',y_span);
set('z min',0); set('z max',thick_slab+thick_rib);
set('enabled',0);

# draw doping regions
adddope;
set('name','pepi');
set('dopant type','p'); # p type
set('x',x_center); set('x span',x_span);
set('y',y_center); set('y span',y_span);
set('z',center_pepi); set('z span',thick_pepi);

adddiffusion;
set('name','p implant');
set('x',x_center_p); set('x span',x_span_p);
set('y',y_center); set('y span',y_span);
set('z',z_center_p); set('z span',z_span_p);
set('dopant type','p');
set('face type',face_p);
set('junction width',width_junction_p);
set('distribution index',diff_dist_fcn);
set('concentration',surface_conc_p);
set('ref concentration',reference_conc_p);

adddiffusion;
set('name','n implant');
set('x',x_center_n); set('x span',x_span_n);
set('y',y_center); set('y span',y_span);
set('z',z_center_n); set('z span',z_span_n);
set('dopant type','n');
set('face type',face_n);
set('junction width',width_junction_n);
set('distribution index',diff_dist_fcn);
set('concentration',surface_conc_n);
set('ref concentration',reference_conc_n);

adddiffusion;
set('name','p++');
set('x',x_center_p_contact); set('x span',x_span_p_contact);
set('y',y_center); set('y span',y_span);
set('z',z_center_p_contact); set('z span',z_span_p_contact);
set('dopant type','p');
set('face type',face_p_contact);
set('junction width',width_junction_p_contact);
set('distribution index',diff_dist_fcn_contact);
set('concentration',surface_conc_p_contact);
set('ref concentration',reference_conc_p_contact);

adddiffusion;
set('name','n++');
set('x',x_center_n_contact); set('x span',x_span_n_contact);
set('y',y_center); set('y span',y_span);
set('z',z_center_n_contact); set('z span',z_span_n_contact);
set('dopant type','n');
set('face type',face_n_contact);
set('junction width',width_junction_n_contact);
set('distribution index',diff_dist_fcn_contact);
set('concentration',surface_conc_n_contact);
set('ref concentration',reference_conc_n_contact);

adddiffusion;
set('name','p wg implant');
set('x',x_center_p_rib); set('x span',x_span_p_rib);
set('y',y_center); set('y span',y_span);
set('z',z_center_p_rib); set('z span',z_span_p_rib);
set('dopant type','p');
```

```
set('face type',face_p_rib);
set('junction width',width_junction_p_rib);
set('distribution index',diff_dist_fcn_rib);
set('concentration',surface_conc_p_rib);
set('ref concentration',reference_conc_p_rib);

adddiffusion;
set('name','n wg implant');
set('x',x_center_n_rib); set('x span',x_span_n_rib);
set('y',y_center); set('y span',y_span);
set('z',z_center_n_rib); set('z span',z_span_n_rib);
set('dopant type','n');
set('face type',face_n_rib);
set('junction width',width_junction_n_rib);
set('distribution index',diff_dist_fcn_rib);
set('concentration',surface_conc_n_rib);
set('ref concentration',reference_conc_n_rib);

# draw monitors
addchargemonitor; # capacitance
set('monitor type',6); # 2D y-normal
set('integrate total charge',1);
set('x',x_center); set('x span',x_span);
set('z min',0); set('z max',thick_slab+thick_rib);
set('save data',1);
set('filename',filename_mzi);

# set contacts
addcontact;  # anode
setcontact('new_contact','name','anode');
setcontact('anode','geometry','anode');
addcontact;  # cathode
setcontact('new_contact','name','cathode');
setcontact('cathode','geometry','cathode');
setcontact('cathode','dc','fixed contact',0);
setcontact('cathode','dc','range start',voltage_start);
setcontact('cathode','dc','range stop',voltage_stop);
setcontact('cathode','dc','range interval',voltage_interval);

# 1) Simulate the DC characteristics of the junction,
#    export the spatial charge density, for different voltages
save('pn_wg_dcsweep.ldev');
run;

# 2) Calculates the junction capacitance versus voltage.
#    compare with experimental results
CV_baehrjones = [-0.4, 0.261; -0.25, 0.248; 0, 0.223; 0.25, 0.208; 0.5, 0.198; 0.75,
     0.190; 1.0, 0.184; 1.5, 0.175; 2.0, 0.168; 3.0, 0.157; 4.0, 0.150];

# perform two simulations, separated by 'dv', for each voltage step.
# this is used to determine the change in charge for the small voltage change,
# to find the capacitance.
vmin = -0.4; vmax = 4; N = 12;
dv = 0.025;
vdv = matrix(2*N,1);
vdv(1:2:(2*N)) = linspace(vmin,vmax,N);
vdv(2:2:(2*N)) = vdv(1:2:(2*N)) + sign(vmax)*dv;

switchtolayout;

# set contact bias
setcontact('cathode','dc','voltage table',vdv);
setcontact('cathode','dc','dc mode',2);
setcontact('cathode','dc','fixed contact',0);

# refine mesh for C calculation
setnamed('wg mesh','enabled',1);

# don't save this result to file
setnamed('monitor','save data',0);
```

```
save('pn_wg_cvanalysis.ldev');
run;

total_charge = getresult('monitor','total_charge');
Qn = e*pinch(total_charge.n);
Qp = e*pinch(total_charge.p);

Cn = abs(Qn(2:2:(2*N))-Qn(1:2:(2*N)))/dv;
Cp = abs(Qp(2:2:(2*N))-Qp(1:2:(2*N)))/dv;
V = vdv(1:2:(2*N));

# User should check for convergence; Cn and Cp should be equal:
#plotxy(V,Cn*1e15*1e-6,V,Cp*1e15*1e-6,"Voltage (V)","Capacitance (fF/um)");

# Final result:
plotxy(V,0.5*(Cn+Cp)*1e15*1e-6,CV_baehrjones(1:11,1),CV_baehrjones(1:11,2),"Voltage
      (V)","Capacitance (fF/um)");

#
# 3) Calculate the resistance in each slab
#    Add a contact in the middle then adjust simulation region
switchtolayout;

addrect;     # rib contact
set('name','r_contact');
set('material','Al (Aluminium) - CRC');
set('x',x_center); set('x span', 0.5*width_rib);
set('y',y_center); set('y span',y_span);
set('z min',z_center); set('z max',thick_slab+thick_rib+thick_contact);
set('override mesh order from material database',1);
set('mesh order',1);

vtest_max = 0.5;
addcontact;
setcontact('new_contact','name','r_test');
setcontact('r_test','geometry','r_contact');
setcontact('r_test','dc','fixed contact',0);
setcontact('r_test','dc','range start',0);
setcontact('r_test','dc','range stop',vtest_max);
setcontact('r_test','dc','range interval',0.1);

setcontact('anode','dc','fixed contact',1);
setcontact('cathode','dc','fixed contact',1);

setnamed('Device region','x',x_center);
setnamed('Device region','x span',x_span-0.1e-6);
setnamed('Device region','solver type','newton');
setnamed('Device region','x min',x_center);

save('pn_wg_R.ldev');
run;

test_result = getresult('Device region','r_test');
Itest = pinch(test_result.I);
Itest_max = abs(Itest(length(Itest)));
?"R_cathode = " + num2str(vtest_max/Itest_max * getnamed('Device region','norm
      length')/0.01) + " Ohm-cm";

switchtolayout;
setnamed('Device region','x',x_center);
setnamed('Device region','x span',x_span-0.1e-6);
setnamed('Device region','x max',x_center);

run;

test_result = getresult('Device region','r_test');
Itest = pinch(test_result.I);
Itest_max = abs(Itest(length(Itest)));
?"R_anode = " + num2str(vtest_max/Itest_max * getnamed('Device region','norm
      length')/0.01) + " Ohm-cm";
```

Listing 6.10 Optical simulation of the pn-junction phase modulator, in Lumerical MODE. This script loads the carrier density from electrical simulations in Listing 6.9, calculates the corresponding plasma dispersion effect, Equation (6.5), performs mode calculations, and determines the neff vs voltage, similar to results in Figure 6.5. The results are exported for compact modelling in INTERCONNECT such as the ring modulator in Listing 9.3. Device and simulation parameters are defined in Listing 6.8; modulator_setup_mode.lsf

```
# Optical simulation of the pn-junction phase modulator
# Step 2: in Lumerical MODE; this script accomplishes:
# 1) Loads the carrier density from electrical simulations, and
#    calculates the neff vs voltage
# 2) Exports the results for INTERCONNECT compact modelling.

newproject; redrawoff;

# modulator geometry variables defined:
modulator_setup_parameters;

# add material
np_material_name = 'silicon with carriers';
new_mat = addmaterial('np Density');
setmaterial(new_mat,'name',np_material_name);
setmaterial(np_material_name,'use soref and bennet model',1);
setmaterial(np_material_name,'Base Material','Si (Silicon) - Palik');

# add data source (np density grid attribute)
matlabload(filename_mzi); # read in charge dataset
addgridattribute('np Density');
importdataset(charge);  # attach to grid attribute
set('name',filename_mzi);

# define simulation region for MODE calculations
x_span = width_slab - 1e-6; # truncate the contact regions
z_span = 3e-6;
override_mesh_size = 0.01e-6;

# draw geometry
addrect;     # oxide
set('name','oxide');
set('material','SiO2 (Glass) - Palik');
set('override mesh order from material database',1); set('mesh order',5);
set('override color opacity from material database',1); set('alpha',0.3);
set('x',x_center); set('x span',width_slab);
set('y',y_center); set('y span',y_span);
set('z',0); set('z span',z_span);

addrect;     # rib
set('name','rib');
set('material',np_material_name);
set('x',x_center); set('x span', width_rib);
set('y',y_center); set('y span',y_span);
set('z min',thick_slab); set('z max',thick_slab+thick_rib);

addrect;     # slab
set('name','slab');
set('material',np_material_name);
set('x',x_center); set('x span', width_slab);
set('y',y_center); set('y span',y_span);
set('z min',z_center); set('z max',thick_slab);

# simulation region
addfde;
set('solver type','2D Y normal');
set('x',x_center); set('x span',x_span);
set('z',0);

addmesh;
set('name','wg mesh');
set('dx',override_mesh_size);
```

```
set('dz',override_mesh_size);
set('override y mesh',0);
set('x',x_center); set('x span', width_rib);
set('z min',0); set('z max',thick_slab+thick_rib);

# run simulation
V = voltage_start:voltage_interval:voltage_stop;
neff = matrix(length(V));

for (i=1:length(V)){
  switchtolayout;
  setnamed(filename_mzi,'V_cathode_index',i);
  findmodes;
  neff(i) = getdata('mode1','neff');
}

# write out
dneff = real(neff - neff(find(V>=0,1))); # relative change in index

la0 = getnamed("FDE","wavelength"); # central wavelength
rel_phase = 2*pi*dneff/la0*1e-2; # phase change /cm
alpha_dB_cm = -0.2*log10(exp(1))*(-2*pi*imag(neff)/la0);

plot(V,rel_phase, "Voltage (V)", "Relative phase (rad./cm)");
plot(V,alpha_dB_cm, "Voltage (V)", "loss (dB/cm)");

data = [V,dneff,imag(neff)];
write("modulator_neff_V.dat",num2str(data)); # for INTERCONNECT
```

Listing 6.11 Simulation of the doping offset on the excess optical loss of a PIN junction waveguide, in Lumerical MODE Solutions; wg_PIN.lsf

```
# wg_PIN.lsf - draw the PIN waveguide geometry in Lumerical MODE
new(1);
wg_2D_draw;

wavelength=1550e-9;

#N=5e20; P=1.9e20; # N++/ P++
#N=3e18; P=2e18; # N+ / P+
N=5e17; P=7e17; # N / P
alphaN_m = 3.52e-6*wavelength^2*N*100;
alphaP_m = 2.4e-6*wavelength^2*P*100;
k_P = wavelength * alphaP_m /4/pi;
k_N = wavelength * alphaN_m /4/pi;

matP = "P++";
temp = addmaterial("(n,k) Material");
setmaterial(temp, "name",matP);
setmaterial(matP, "Refractive Index", 3.47);
setmaterial(matP, "Imaginary Refractive Index", k_P);

matN = "N++";
temp = addmaterial("(n,k) Material");
setmaterial(temp, "name",matN);
setmaterial(matN, "Refractive Index", 3.47);
setmaterial(matN, "Imaginary Refractive Index", k_N);

# draw P++ doping
addrect; set("name", "P++"); set("material", matP);
set("y min", 0.8e-6);      set("y max", Ymax);
set("z min", 0);  set("z max", thick_Slab);
set("x min", Xmin); set("x max", Xmax);

# draw N++ doping
addrect; set("name", "N++"); set("material", matN);
set("y max", -0.8e-6);      set("y min", -Ymax);
set("z min", 0);  set("z max", thick_Slab);
set("x min", Xmin); set("x max", Xmax);
```

```
# define simulation parameters
meshsize     = 20e-9;  # maximum mesh size

# add 2D mode solver (waveguide cross-section)
addfde; set("solver type", "2D X normal");
set("x", 0);
set("y", 0);        set("y span", Y_span);
set("z max", Zmax); set("z min", Zmin);
set("wavelength", wavelength); set("solver type","2D X normal");
set("define y mesh by","maximum mesh step"); set("dy", meshsize);
set("define z mesh by","maximum mesh step"); set("dz", meshsize);
set("number of trial modes",1);

# define parameters to sweep
doping_offset_list=[0:.1:1.5]*1e-6; # sweep doping offset

select("FDE"); set("solver type","2D X normal"); # 2D mode solver
loss = matrix (length(doping_offset_list) );

for(ii=1:length(doping_offset_list)) {
  switchtolayout;
  setnamed("P++","y min", doping_offset_list(ii)+width_ridge/2);
  setnamed("N++","y max", -doping_offset_list(ii)-width_ridge/2);
  n=findmodes;
  loss(ii) =( getdata ("FDE::data::mode1","loss") ); # dB/m
}

plot (doping_offset_list, loss/100,"Doping offset","Excess Loss [dB/cm]", "","log10y");
matlabsave ('wg_PIN_low', doping_offset_list, loss);
```

Listing 6.12 Thermal modelling of a metallic heater above a strip waveguide, using MATLAB PDE ToolBox, Thermal_Waveguide.m

```
function pdemodel

% Geometry and parameter definitions
Width=50; % in unit microns
MetalWidth=1;
MetalThickness=0.5;
MetalLength=100;
MetalWGDistance=1;
WGWidth=0.5;
WGHeight=0.22;
kSi = 149e-6; % Thermal conductivity, W / (micron.K)
kSiO2 = 1.4e-6;
kAl = 250e-6;
Qsource=0.01; % Heat source in the metal [W]
Q = Qsource / MetalThickness / MetalWGDistance / MetalLength; % unit of W / micron^3;

%%%%%%%%%%%%%
% code generated by Matlab PDE ToolBox:
%%%%%%%%%%%%%
[pde_fig,ax]=pdeinit;
pdetool('appl_cb',9);
set(ax,'DataAspectRatio',[1 1 1]);
set(ax,'PlotBoxAspectRatio',[10 6 1]);
set(ax,'XLim',[-10 10]);
set(ax,'YLim',[-10 2]);
set(ax,'XTickMode','auto');
set(ax,'YTickMode','auto');

% Geometry description:
pderect([-50 50 2 -2],'Oxide');
pderect([-0.5 0.5 1.22 1.72],'Metal');
pderect([-0.25 0.25 0 0.22],'Si');
pderect([-50 50 -2 -100],'SiSubstrate');
set(findobj(get(pde_fig,'Children'),'Tag','PDEEval'),'String','Oxide+Metal+Si+SiSubstrate')
```

```
% Boundary conditions:
pdetool('changemode',0)
pdesetbd(14, 'neu', 1, '0','0')
pdesetbd(13, 'neu', 1, '0','0')
pdesetbd(10, 'neu', 1, '0','0')
pdesetbd(7, 'neu', 1, '0','0')
pdesetbd(6, 'neu', 1, '0','0')
pdesetbd(4, 'dir', 1, '1','0')

% Mesh generation:
setappdata(pde_fig,'Hgrad',1.3);
setappdata(pde_fig,'refinemethod','regular');
setappdata(pde_fig,'jiggle',char('on','mean',''));
setappdata(pde_fig,'MesherVersion','preR2013a');
pdetool('initmesh')
pdetool('refine')
pdetool('refine')

% PDE coefficients:
pdeseteq(1,...
  '1.4e-6!149e-6!250e-6!149e-6',...
  '0!0!0!0',...
  '(0)+(0).*(0.0)!(0)+(0).*(0.0)!(0.0002)+(0).*(0.0)!(0)+(0).*(0.0)',...
  '(1.0).*(1.0)!(1.0).*(1.0)!(1.0).*(1.0)!(1.0).*(1.0)',...
  '0:10',...
  '0.0',...
  '0.0',...
  '[0 100]')
setappdata(pde_fig,'currparam',...
  ['1.0!1.0!1.0!1.0          ';...
  '1.0!1.0!1.0!1.0          ';...
  '1.4e-6!149e-6!250e-6!149e-6';...
  '0!0!0.002!0              ';...
  '0!0!0!0                  ';...
  '0.0!0.0!0.0!0.0          '])

% Solve parameters:
setappdata(pde_fig,'solveparam',...
  char('0','22368','10','pdeadworst',...
  '0.5','longest','0','1E-4','','fixed','Inf'))

% Plotflags and user data strings:
setappdata(pde_fig,'plotflags',[1 1 1 1 1 1 7 1 0 0 0 1 1 1 0 0 0 1]);
setappdata(pde_fig,'colstring','');
setappdata(pde_fig,'arrowstring','');
setappdata(pde_fig,'deformstring','');
setappdata(pde_fig,'heightstring','');

% Solve PDE:
pdetool('solve')

%%% End of PDE ToolBox code.

%%%%%%%%%%%%%%%
% Extract data from PDE ToolBox and plot
%%%%%%%%%%%%%%%

pde_fig=findobj(allchild(0),'flat','Tag','PDETool');
u = get(findobj(pde_fig,'Tag','PDEPlotMenu'),'UserData');
h=findobj(get(pde_fig,'Children'),'flat','Tag','PDEMeshMenu');
hp=findobj(get(h,'Children'),'flat','Tag','PDEInitMesh');
he=findobj(get(h,'Children'),'flat','Tag','PDERefine');
ht=findobj(get(h,'Children'),'flat','Tag','PDEMeshParam');
p=get(hp,'UserData');
t=get(ht,'UserData');

% get geometry to overlay on plot
pdetool('export',2);
pause
```

```
g=evalin('base','g');
fid = wgeom(g, 'geom');

% Plot 2D data:
xlist =[-5:.05:5];
ylist =[-4:.05:2];
[X,Y] = meshgrid(xlist, ylist);
UXY=tri2grid(p,t,u,xlist,ylist);
figure; [c, h] = contourf(X,Y,UXY,20); colorbar;
c=colormap('hot'); c=c(end:-1:1,:); colormap(c)
hold all; h=pdegplot('geom');
set(h,'LineWidth',3,'Color','b');
global FONTSIZE
FONTSIZE=20;
xlabel ('x Position [\mum]','FontSize',FONTSIZE)
ylabel ('y Position [\mum]','FontSize',FONTSIZE)
pbaspect([ (xlist(end)-xlist(1))/(ylist(end)-ylist(1)) 1 1])
printfig ('thermal1');

% plot 1D
xlist =[0]; ylist =[-5:.01:2];
[X,Y] = meshgrid(xlist, ylist);
UXY=tri2grid(p,t,u,xlist,ylist);
figure; plot (Y, UXY,'LineWidth',3);
xlabel ('y Position [\mum]','FontSize',FONTSIZE)
ylabel ('Temperature','FontSize',FONTSIZE)
yl=ylim;ylim([0 yl(2)]); grid on
xlim([ylist(1) ylist(end)]);
printfig ('thermal2');

% Temperature inside waveguide
q=find(ylist>.09);
T_wg=UXY(q(1))

% Temperature inside metal
q=find(ylist>1.4);
T_metal=UXY(q(1))

% Thermal tuning efficiency calculations:
dndT = 1.87e-4;
FSR_fraction = dndT*T_wg*MetalLength / 1.55
mW_per_FSR = Qsource*1e3 / FSR_fraction


function printfig (file)
global FONTSIZE
set (gca, 'FontSize',FONTSIZE)
pdf = [ file '.pdf'];
print ('-dpdf', pdf); system([ 'pdfcrop ' pdf ' ' pdf ]);
```

References

[1] R. Soref and B. Bennett. "Electrooptical effects in silicon". *IEEE Journal of Quantum Electronics* **23**.1 (1987), pp. 123–129 (cit. on p. 217).

[2] G. T. Reed, G. Mashanovich, F. Y. Gardes, and D. J. Thomson. "Silicon optical modulators". *Nature Photonics* **4**.8 (2010), pp. 518–526 (cit. on p. 217).

[3] M. Nedeljkovic, R. Soref, and G. Z. Mashanovich. "Free-carrier electro-refraction and electroabsorption modulation predictions for silicon over the 1–14 micron infrared wavelength range". *IEEE Photonics Journal* **3**.6 (2011), pp. 1171–1180. DOI: 10.1109/JPHOT.2011.2171930 (cit. on pp. 217, 219).

[4] T. Baehr-Jones, R. Ding, Y. Liu, *et al.* "Ultralow drive voltage silicon traveling-wave modulator". *Optics Express* **20**.11 (2012), pp. 12 014–12 020 (cit. on pp. 224, 225).

[5] Xi Xiao, Hao Xu, Xianyao Li, *et al.* "25 Gbit/s silicon microring modulator based on misalignment-tolerant interleaved PN junctions". *Optics Express* **20**.3 (2012), pp. 2507–2515. DOI: 10.1364/OE.20.002507 (cit. on p. 226).

[6] Tom Baehr-Jones, Ran Ding, Ali Ayazi, *et al.* "A 25 Gb/s silicon photonics platform". *arXiv:1203.0767v1* (2012) (cit. on pp. 226, 234).

[7] W. Bogaerts, P. De Heyn, T. Van Vaerenbergh, *et al.* "Silicon microring resonators". *Laser & Photonics Reviews* **6**.1 (2012), pp. 43–73. (cit. on p. 226).

[8] Guoliang Li, Ashok V. Krishnamoorthy, Ivan Shubin, *et al.* "Ring resonator modulators in silicon for interchip photonic links". *IEEE Journal of Selected Topics in Quantum Electronics* **19**.6 (2013), p. 3401819(cit. on p. 226).

[9] Lukas Chrostowski, Samantha Grist, Jonas Flueckiger, *et al.* "Silicon photonic resonator sensors and devices". *Proceedings of SPIE Volume 8236; Laser Resonators, Microresonators, and Beam Control XIV* (Jan. 2012) (cit. on p. 230).

[10] Wei Shi, Xu Wang, Charlie Lin, *et al.* "Silicon photonic grating-assisted, contra-directional couplers". *Optics Express* **21**.3 (2013), pp. 3633–3650 (cit. on p. 232).

[11] Matthew Streshinsky, Ran Ding, Yang Liu, *et al.* "Low power 50 Gb/s silicon traveling wave Mach-Zehnder modulator near 1300 nm". *Optics Express* **21**.25 (2013), pp. 30 350–30 357 (cit. on p. 234).

[12] John E. Cunningham, Ivan Shubin, Xuezhe Zheng, *et al.* "Highly-efficient thermally-tuned resonant optical filters". *Optics Express* **18**.18 (2010), pp. 19055–19063 (cit. on p. 236).

[13] Tsung-Yang Liow, JunFeng Song, Xiaoguang Tu, *et al.* "Silicon optical interconnect device technologies for 40 Gb/s and beyond". *IEEE JSTQE* **19**.2 (2013), p. 8200312. DOI: 10.1109/JSTQE.2012.2218580 (cit. on pp. 236, 240).

[14] Qing Fang, Jun Feng Song, Tsung-Yang Liow, *et al.* "Ultralow power silicon photonics thermo-optic switch with suspended phase arms". *IEEE Photonics Technology Letters* **23**.8 (2011), pp. 525–527 (cit. on p. 236).

[15] *PDE – Partial Differential Equation Toolbox – MATLAB.* [Accessed 2014/04/14]. URL: http://www.mathworks.com/products/pde/ (cit. on p. 237).

[16] Michael R. Watts, Jie Sun, Christopher DeRose, *et al.* "Adiabatic thermo-optic Mach–Zehnder switch". *Optics Letters* **38**.5 (2013), pp. 733–735 (cit. on p. 240).

[17] T. Chu, H. Yamada, S. Ishida, and Y. Arakawa. "Compact 1 $\times$ N thermo-optic switches based on silicon photonic wire waveguides". *Optics Express* **13**.25 (2005), pp. 10 109–10 114 (cit. on p. 240).

7 Detectors

A critical component of monolithically integrated silicon photonics is the photodetector. The photodetector enables the conversion of an optical signal to an electrical signal. However, silicon is not an efficient light absorber at standard telecommunication wavelengths (1310 nm and 1550 nm). A number of techniques have been developed to build photodetectors on silicon including epitaxial germanium growth [1], III–V bonding [2], plasmonic absorption [3, 4], sub-bandgap silicon detection [5–10], and surface-state absorption [11]. The epitaxially grown germanium detector has emerged as the most practical detector technology due to its CMOS compatibility, relatively high responsivity, low size requirements and high speed [12–14]. In this chapter we will describe the design and characteristics of the germanium-on-silicon photodetector (Figure 7.1). The experimental results are based on detectors that were fabricated via OpSIS by IME in Singapore [12].

7.1 Performance parameters

There are three significant parameters that detail the performance of a photodetector. The first is the responsivity, a measure of how much electrical current is generated per unit optical power incident on the detector; the responsivity is wavelength dependent. The second parameter is the bandwidth, which denotes the speed at which the detector can respond to varying levels of optical signal. The final parameter is the dark current, the current from the detector when there is no incident optical power. A high dark current can contribute to the noise of the detector.

7.1.1 Responsivity

The responsivity is the key feature of a device which has the purpose of converting optical power to electrical current. The responsivity, R, is defined simply as the ratio of the incident optical power, P, to electrical current, I, and is usually given in amperes per watt (A/W):

$$R = \frac{I}{P}. \tag{7.1}$$

Another term similar to responsivity is the quantum efficiency, η. This parameter similarly denotes the conversion efficiency between optical and electrical power, but it

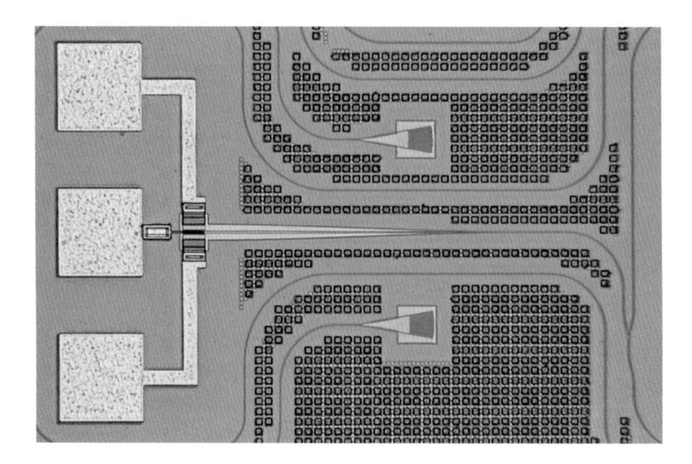

Figure 7.1 Optical image of a germanium-on-silicon detector.

is relative to the maximum possible current. Thus, a quantum efficiency of 1 denotes that every photon results in an electron hole pair that is captured by the detector. Converting from quantum efficiency to responsivity is done by the following relationship:

$$R = \eta \frac{q\lambda}{hc}, \tag{7.2}$$

where η is the quantum efficiency, q is the charge of an electron, λ is the wavelength of light, h is Planck's constant and c is the speed of light. The equation reveals that, for a quantum efficiency of 1, the detector has a maximum responsivity of 1.25 A/W at 1550 nm and 1.06 A/W at 1310 nm. The quantum efficiency of a detector can be quickly determined by dividing the measured responsivity by the maximum responsivity at a given wavelength. For example, a detector with a 1550 nm responsivity of 0.5 A/W has a quantum efficiency of 40%. It is possible to increase the quantum efficiency above unity by using techniques such as avalanche amplification.

7.1.2 Bandwidth

The bandwidth of a detector is the determining figure for how fast it responds to a modulated optical input. There are two main factors affecting the bandwidth of a detector: the transit time and the RC parasitic response.

Transit time

The transit time is simply the time it takes for carriers that are generated by an optical signal to be swept out of the detector's active region. However, not all the carriers arrive at the electrodes at once. Instead there is a range of times in which the carriers generated from an incident pulse will arrive at the electrodes of a detector. A detailed explanation of this effect can be found in Reference [15]. The velocity of the carriers is given by

$$v = \mu E, \tag{7.3}$$

Table 7.1 Silicon and germanium mobility and saturation velocity parameters [16].

	e^- (Si)	h^+ (Si)	e^- (Ge)	h^+ (Ge)
Mobility (cm^2/Vs)	1400	500	3900	1900
Sat. velocity (m/s)	1×10^5	0.7×10^5	0.7×10^5	0.63×10^5

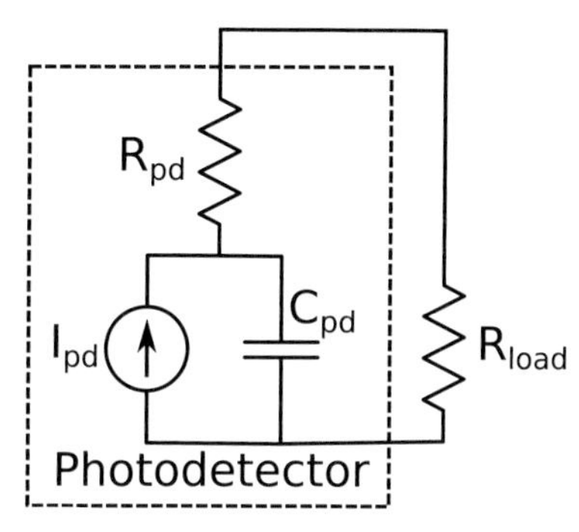

Figure 7.2 Equivalent circuit model for a photodetector.

where μ is the carrier mobility and E is the electric field. An important point to make is that the electrons and holes have different mobilities (values used in the model listed in Section 7.5.2). When biased with a high enough electric field, the carriers will reach a saturation velocity, v_{sat} (see Table 7.1). The mobility is thus modified to be

$$\mu_{\text{sat}} = \frac{\mu}{\sqrt{1 + \left(\frac{\mu E}{v_{\text{sat}}}\right)^2}}. \tag{7.4}$$

As an example, consider a 0.5 μm high germanium detector in which charges are generated uniformly. An instantaneous pulse of light hits the detector creating carriers. To collect the full charge, the holes and electrons must both travel the full height (h). The bandwidth is then given by [17]:

$$f_{\text{transit}} = 0.38\frac{v_{\text{sat}}}{h} = 60\,\text{GHz}. \tag{7.5}$$

RC response

The second factor in determining the bandwidth of a detector is the electrical impedance characteristic of the detector circuit. Parasitic electrical effects are often simplified to an RC circuit model as shown in Figure 7.2. The photodetector itself is shown in the dashed box as a current source with a parallel capacitor C_{pd} and single resistor, R_{pd}. The load on the detector is given by a second resistor, R_{load}. If we consider a circuit with 40 fF of junction capacitance, a detector resistance of 150 Ω and a load of 50 Ω, the RC time constant, τ_{RC}, is 2 ps. This is used to calculate the bandwidth due to RC limitations:

$$f_{RC} = \frac{1}{2\pi RC} = 20\,\text{GHz}. \tag{7.6}$$

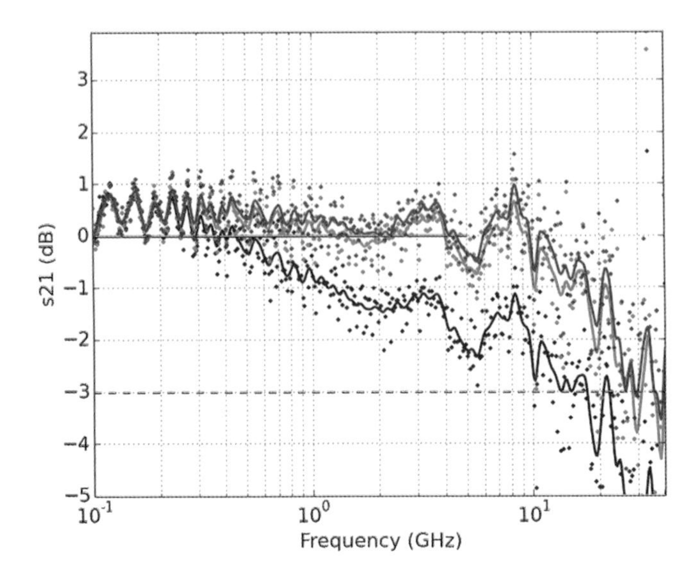

Figure 7.3 Detector small-signal modulation response for a detector with a width of 8 μm and length of 10 μm, for voltages 0, 2, and 4 V.

The total bandwidth which includes both *RC* and transit time effects is given by:

$$f = \left(\frac{1}{f_{RC}^2} + \frac{1}{f_{\text{transit}}^2} \right)^{-1/2} \tag{7.7}$$

Germanium detectors have been tested at very high bandwidths, e.g. 20–40 GHz, although the fastest detectors have achieved up to 120 GHz [18]. An example of a frequency response of a germanium detector is shown in Figure 7.3.

Dark current

As a DC current offset, the dark current can be subtracted from the total current to calculate the photocurrent signal. However, sufficient levels of dark current can contribute to the noise of a detector. There are two factors that contribute to the total dark current: bulk generation and surface generation.

Bulk generation is a volume-dependent mechanism that is shown to result predominantly from the Shockley–Read–Hall process [19]. The large lattice mismatch between silicon and germanium causes threading dislocations, which allow for the existence of mid bandgap states. At a low electric field, the bulk current density is relatively constant. However, as the electric field increases, band bending will result in a higher bulk current density that increases exponentially with the applied electric field; see Figure 7.4.

Surface generation is the second contributor to the dark current and is a result of surface defects such as dangling bonds. Surface passivation is more difficult with

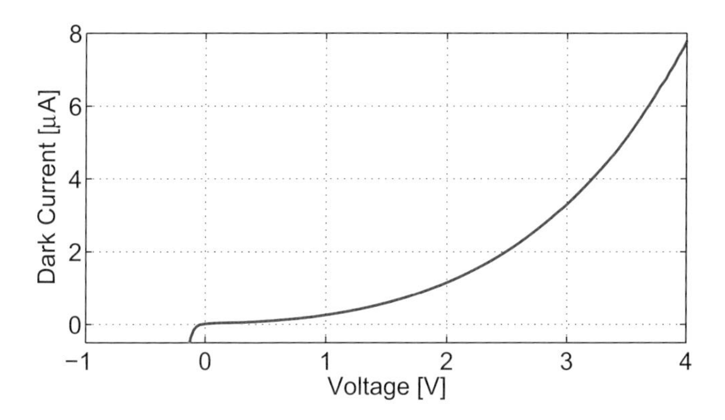

Figure 7.4 Dark current of a PIN photodetector. The dark current increases as a function of bias voltage.

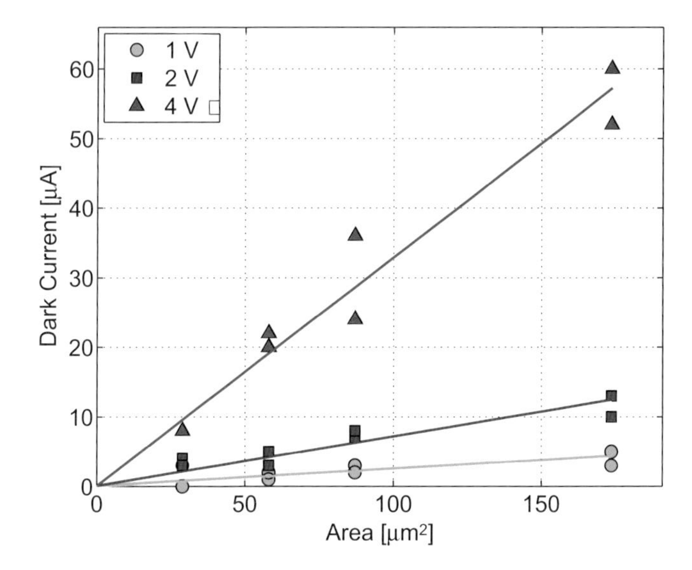

Figure 7.5 PIN detector, dark current versus area.

germanium than silicon as germanium is not fully passivated by silicon dioxide. Other materials such as germanium oxide have been used for passivation with some success [20].

The total dark current is given by [20]:

$$I_{dark} = J_{bulk} \cdot A + J_{surf}\sqrt{4\pi}\sqrt{A}. \tag{7.8}$$

If we assume that bulk current dominates, the dark current scales with detector area. Experimental results for Ge detectors are shown Figure 7.5, indicating an approximately linear increase of dark current with the area of the detector.

The shot noise is usually the dominant noise factor in germanium on silicon photodetectors. The shot noise due to the dark current is given by

$$I_n = \sqrt{2qI_{\text{dark}}BW}, \tag{7.9}$$

where I_{dark} is the dark current. For a dark current of 10 μA, and a bandwidth BW=10 GHz, this corresponds to a shot noise of 180 nA.

As seen in this example, faster bandwidths will place higher requirements on the dark current to provide the same signal-to-noise ratio. However, the electrical amplification in a receiver often contributes most of the noise in a receiver. Thus depending on the application, the dark current-induced noise can be negligible.

7.2 Fabrication

There are a number of different materials with which to build a detector. Here, we will focus on detectors that operate at the common telecommunications wavelengths of 1310 nm and 1550 nm. Owing to silicon's relatively large bandgap of 1.12 eV (1107 nm), it is an inefficient absorber in bulk. Three absorption material alternatives to bulk silicon are shown in Table 7.2. In particular, we focus on germanium, with the material absorption parameter plotted versus wavelength in Figure 7.6.

For simplicity of fabrication, it would be ideal if silicon could be used as the absorption material of a detector. While bulk silicon cannot be used as an absorption material at telecommunications wavelengths, a number of techniques have been developed to enable silicon-based detectors. These include damage-based detectors [5–10, 22], and plasmonic-Schottky detectors [3, 4]. However, these silicon-based detectors all suffer from some combination of low responsivity, high dark current, restrictive thermal budget and poor bandwidth. Hybrid III–V on silicon detectors solve these problems and show very high performance [23]. However, bonding III–V materials onto silicon greatly increases the process complexity.

While germanium can be grown in CMOS fabs using chemical vapour deposition (CVD) and is said to be "CMOS-compatible", efforts to grow germanium epitaxially on silicon have faced significant challenges (see Figure 7.7). The large lattice mismatch (over 4%) between silicon and germanium limits the quality of the germanium film [1]. Additionally, the low thermal budget of germanium creates significant limitations

Table 7.2 A selection of materials available for building detectors on a silicon platform for telecommunication wavelengths.

Material	Silicon	Germanium	III–V
Bandgap	Sub-bandgap	Indirect	Direct
Wavelength	Variable	Up to ~1570 nm	Variable
Processing	Easy	Medium	Hard
Performance	Low	High	High

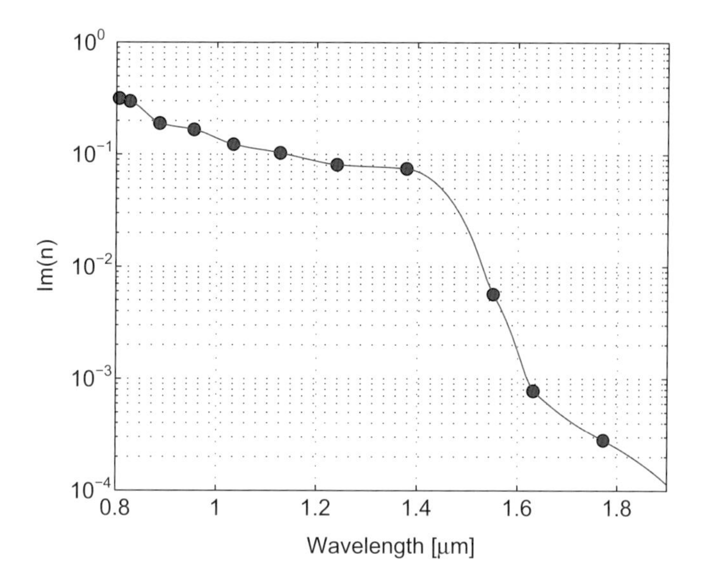

Figure 7.6 Germanium bulk material optical absorption constant. At 1550 nm, the imaginary refractive index is 0.005 67. Data from Reference [21].

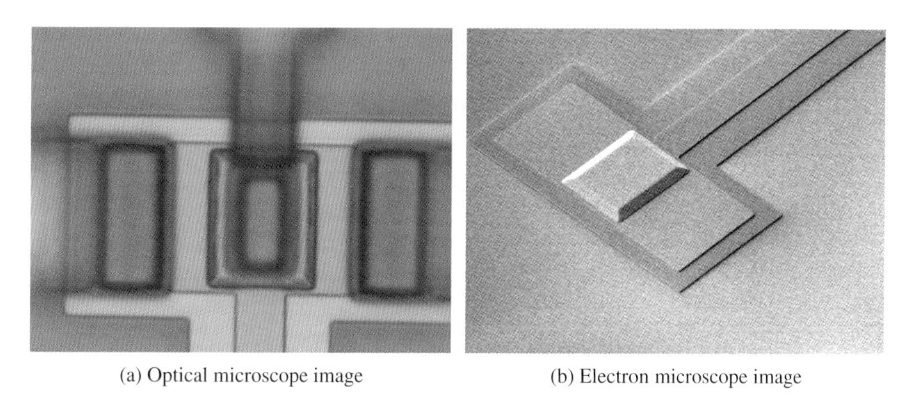

(a) Optical microscope image (b) Electron microscope image

Figure 7.7 Optical and SEM image of a germanium-on-silicon detector after epitaxy. Note the angled sidewalls and the plateau on top. A rib waveguide taper is used to enhance coupling to the germanium. Space is left on the sides of the detector for eventual contacts.

during the fabrication process. Defects during fabrication cause diminished detector performance including higher dark current. Significant effort has been devoted to dealing with these issues and the lattice mismatch in particular. One common technique is to grow a thin, low-temperature Ge or SiGe buffer layer and then a thicker layer of germanium on top [1]. A cross-section showing this buffer layer can be seen in Figure 7.8. The buffer layer will confine the defects to a small region and provide a mostly low-defect top layer. The responsivity versus wavelength for such a germanium detector is shown in Figure 7.9.

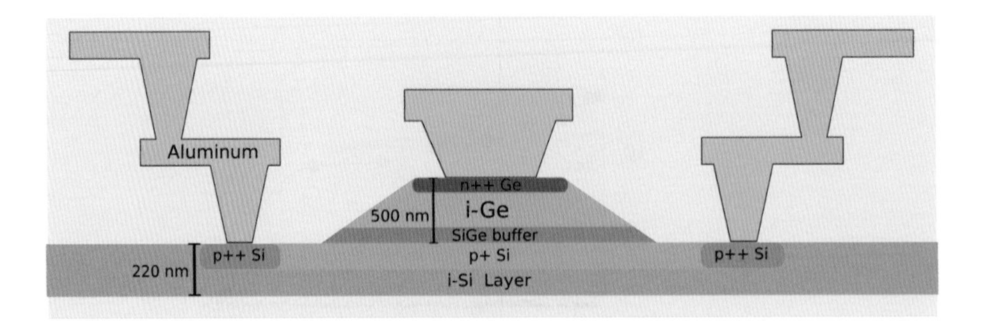

Figure 7.8 Cross-section of vertical PIN germanium-on-silicon detector.

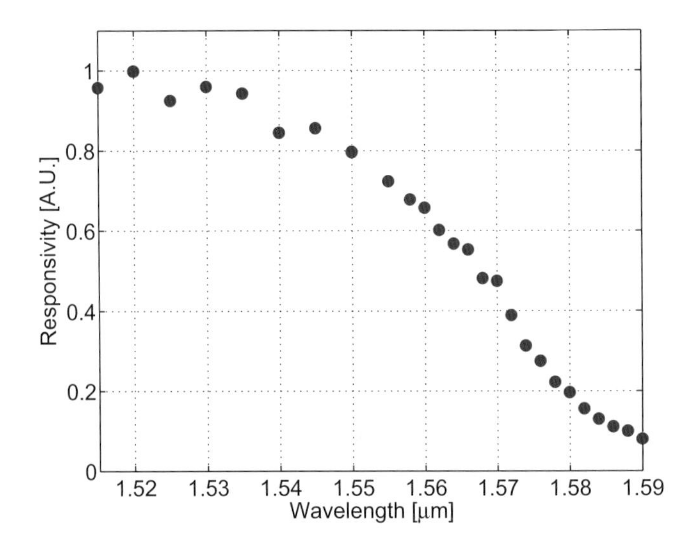

Figure 7.9 PIN detector, responsivity versus wavelength.

7.3 Types of detectors

A number of different types of detectors can be designed using the germanium-on-silicon material system. There are significant tradeoffs between the different detectors as will be discussed. Three common types of detectors with potential applications in silicon photonics will be covered. The discussion below will cover the photoconductive detector, PIN detector, and avalanche detector.

7.3.1 Photoconductive detector

Silicon photonic detectors are usually used in photoconductive mode as opposed to photovoltaic mode due to the need for linearity and speed. The photoconductive detector is perhaps the simplest detector design for this purpose as it is undoped except for the contacts. When a voltage is applied across the detector, incident photons generate charge carriers that are swept by the applied electric field to the terminals of the device.

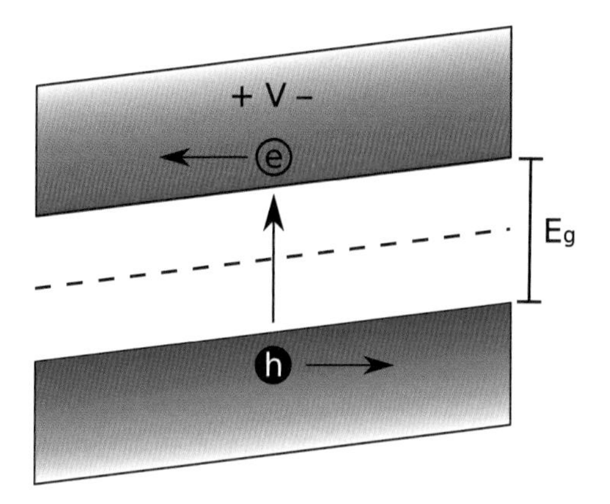

Figure 7.10 PIN detector, illustration of the band diagram.

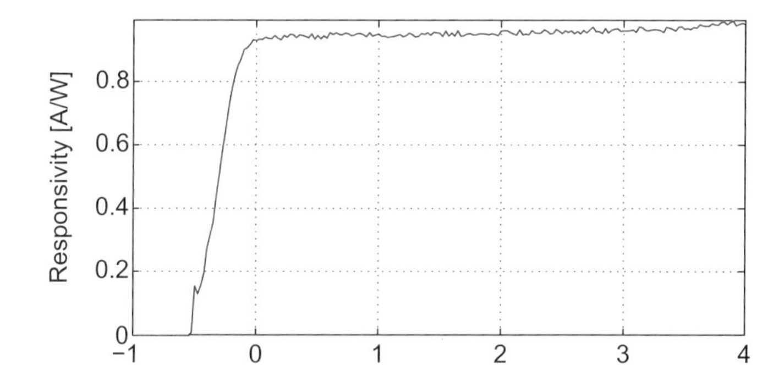

Figure 7.11 Responsivity of a PIN photodetector. In this case, the maximum responsivity is nearly achieved at 0 V reverse bias.

The metal–semiconductor–metal (MSM) detector is a variation on the photoconductive detector. This detector has two Schottky junctions on each metal–semiconductor contact, which limits the current flow at low bias. However, the low capacitance of this type of detectors allows for very high speeds, up to 300 GHz [24].

7.3.2 PIN detector

The PIN detector is the most common type of detector currently in use in silicon photonics [25–27]. One side of the detector is p-type and the other is n-type, which creates a built-in electric field in the centre intrinsic region, as shown in Figure 7.10. Light is absorbed in the intrinsic region and generated carriers are swept out by the electric field of the junction, which can be enhanced by an applied voltage. The result is a low or 0 V required bias voltage, a high responsivity and low capacitance, which results in high speed. The PIN detector has a responsivity at 0 V, unlike the photoconductive detector, but often has an applied reverse bias voltage to maximize the responsivity as shown in Figure 7.11.

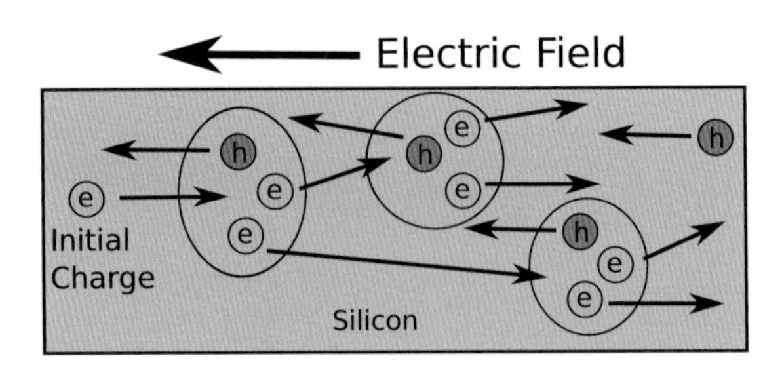

Figure 7.12 Diagram of avalanche multiplication in a system where only electrons cause impact ionization. A single electron from the left generates additional charge carriers through impact ionization. Note that the numbers of electrons and holes are always equal.

7.3.3 Avalanche detector

Both photoconductive detectors and PIN photodetectors are limited to a maximum responsivity defined by the bandgap of the absorption medium. Maximum responsivity, defined by Equation (7.1), or a quantum efficiency of 1, signify that each photon is converted into an electron–hole pair that is collected at the detector contacts. In practice, the quantum efficiency of that type of detectors is always less than 1 as finite carrier lifetime, metal absorption, and a variety of other factors work to degrade the detector performance.

One method of achieving quantum efficiency greater than unity is to employ avalanche gain. Carrier avalanche is a phenomenon that occurs when carriers travelling with high energy cause collisions that create additional electron–hole pairs in a process called impact ionization, as shown in Figure 7.12. In an avalanche photodetector, the current contribution of the photo-carriers can be multiplied many times by the addition of the avalanche generated carriers. The tendency for electrons and holes to undergo impact ionization is characterized by the ionization coefficients α_e for electrons and α_h for holes.

In a single carrier system, the ionization coefficient, α, is related to the carrier density in the x direction, $J(x)$, by [15]:

$$\frac{\partial J(x)}{\partial x} = \alpha J(x). \tag{7.10}$$

Thus, in a single carrier system, the current density will increase exponentially as a function of distance. Complications arise when the impact ionizations of both holes and electrons are considered. One parameter for characterizing the hole and electron avalanche is the ionization ratio, k, given by:

$$k = \frac{\alpha_e}{\alpha_h}, \tag{7.11}$$

where α_e and α_h are the ionization coefficients of electrons and holes, respectively.

The multiplication factor, M, gives the magnitude of increase of the current of the detector. For a two-carrier avalanche system, it is given by

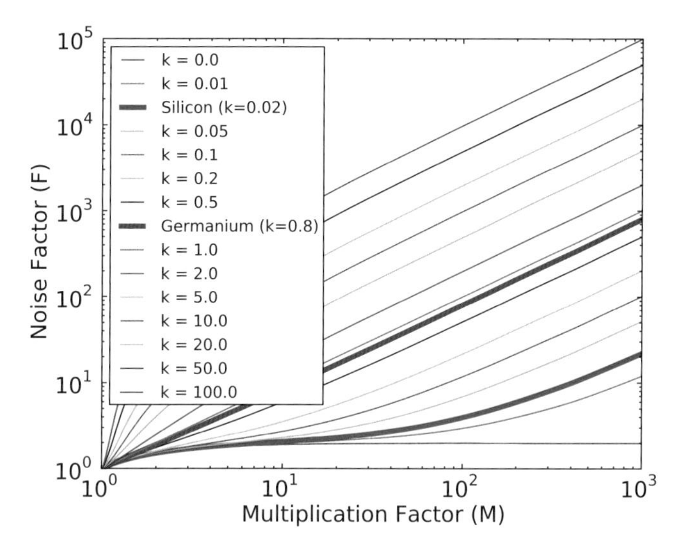

Figure 7.13 Noise factor as a function of the multiplication factor under electron injection. A number of k values are plotted, including the k values specific to silicon and germanium.

$$M = \frac{\alpha_e - \alpha_h}{\alpha_e e^{-(\alpha_e - \alpha_h) * w} - \alpha_h}, \tag{7.12}$$

where w is the width of the avalanche region.

It is preferable for only a single carrier to be the generator of the avalanche. When both carriers cause impact ionization concurrently, there are a number of undesirable consequences. Two-carrier avalanche multiplication will increase the noise of the detector as the avalanche generation proceeds in both directions. Similarly, the impact ionization effects will take more time and degrade the device bandwidth. The noise of an avalanche detector is usually given by the excess noise factor (F) (see also Figure 7.13). The general expression for the excess noise factor and the simplified version for electron injection is:

$$F(M) = \frac{\langle M^2 \rangle}{\langle M \rangle^2} \tag{7.13}$$

$$= M\left(1 - (1-k)\left(\frac{M-1}{M}\right)^2\right) \tag{7.14}$$

$$\approx kM + \left(2 - \frac{1}{M}\right)(1-k). \tag{7.15}$$

A number of geometries are possible for building avalanche photodetectors. A very simple geometry would include an absorption medium and a high electric field usually applied by reverse biasing a PIN junction. However, in this geometry the avalanche width (w) available for each carrier is different based on where they are absorbed. Thus a higher electric field must be applied to achieve a meaningful avalanche and breakdown is achieved relatively quickly.

Table 7.3 Breakdown electric field for electrons in silicon and germanium. Note that these values are approximate and will change with doping concentration, depletion width, temperature, and other factors.

	Electron (Si)	Electron (Ge)
Breakdown field (kV/cm)	300	100

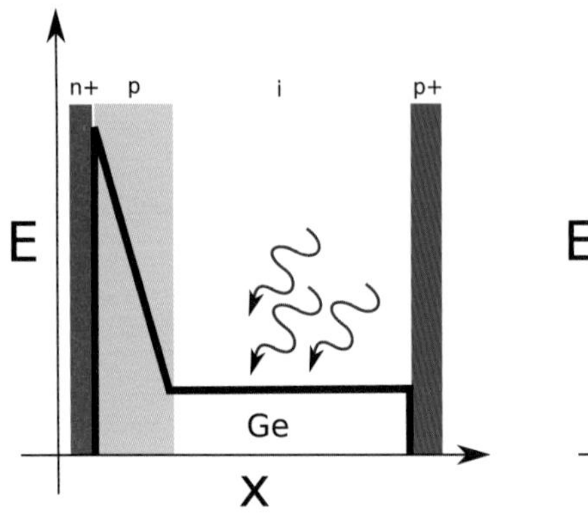

(a) Reach-through avalanche geometry with both absorption and avalanche occurring in the germanium.

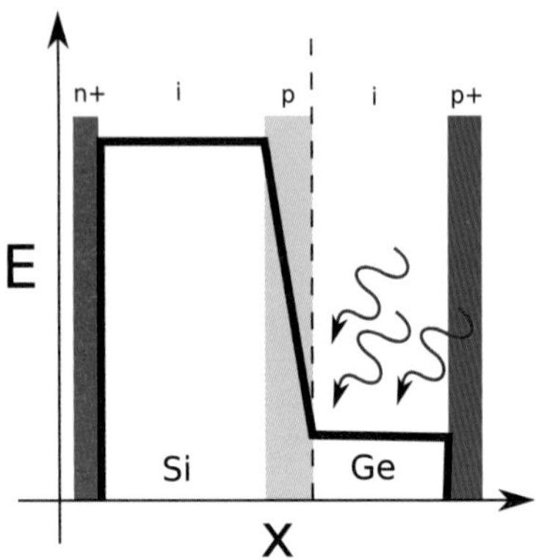

(b) Separate absorption charge multiplication (SACM) geometry with absorption in germanium and avalanche in silicon.

Figure 7.14 Avalanche detector geometries using reach-through and separate absorption (charge) multiplication configurations.

The reach-through detector is a more sophisticated avalanche detector design, illustrated in Figure 7.14a. This type of detector has a secondary doped region called the charge region that is used to create two electric field regions. The first region has a moderate electric field to encourage carrier drift. The second region has a high electric field to create impact ionization and gain.

As shown in Figure 7.13, the excess noise in germanium is significantly higher than in silicon. Many avalanche detectors attempt to use germanium for absorption and silicon for multiplication. This type of detector is aptly named a separate absorption (charge) multiplication detector, and illustrated in Figure 7.14b.

Charge region design

An important component of the avalanche detector design is the size and doping concentration in the charge region. The goal of this region is to reduce the electric field such that there is a low electric field in the absorption region and a high electric field in the avalanche region. A simple linear analysis can be used to approximate the width and concentration of the charge region (a more sophisticated design will require simulation with TCAD software), bearing mind the maximum fields before breakdown occurs as shown in Table 7.3.

For this example, we will consider the behaviour of a reach-through APD. The target peak electric field in the charge region is 300 kV/cm and the maximum electric field in the intrinsic region is 100 kV/cm. In practice, the maximum field in the intrinsic region must be lowered further due to edge effects. For this example, we assume the field in the germanium is constant at 50 kV/cm, and the average field in the charge region is 200 kV/cm. The total voltage is then computed by

$$V_{BD} = E_{\text{avg,charge}} \cdot w_{\text{charge}} + E_{\text{intrinsic}} \cdot w_{\text{intrinsic}}. \tag{7.16}$$

We can choose an intrinsic width of 10 μm for the detector and fix the charge width to 2 μm, which results in a required 90 V. The charge-doping concentration that will result in the correct electric field can then be calculated:

$$\frac{dE}{dx} = \frac{200\,\text{kV/cm}}{2\,\mu\text{m}} = 1\,\text{GV/cm}^2 = \frac{N \cdot q}{\epsilon_{ge}\epsilon_0}. \tag{7.17}$$

Solving for the doping concentration N, gives a value of about $9 \times 10^{17}/\text{cm}^2$. All components of the reach-through detector have been calculated at punch-through, but simulation is required to predict values for fabricated geometries.

7.4 Design considerations

There are a number of significant design choices to be made when building a germanium-on-silicon photodetector. A selection of the design parameters and their merits will be discussed in this section.

7.4.1 PIN junction orientation

The two commonly used orientations for the PIN junction are vertical (Figure 7.15a) and lateral (Figure 7.15b). In the vertical orientation, current must travel across the silicon/germanium interface which results in a higher dark current and lower responsivity due to the interfacial defects. The lateral junction avoids the interfacial region, but may require a higher voltage to sweep the carrier out of the intrinsic region in a timely manner as the contacts are both at the top of the germanium. In this section, we will focus on the vertical junction detector.

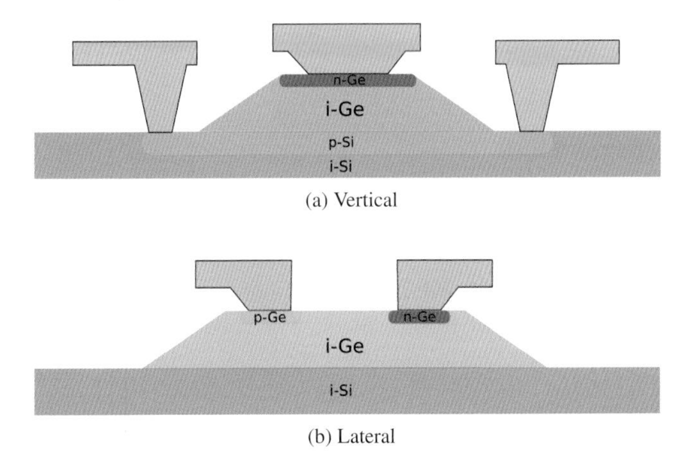

Figure 7.15 Germaniumon-on-silicon detectors with vertical (a) and lateral (b) junction geometries.

7.4.2 Detector geometry

The design decisions regarding the geometry of the detector have significant effects on the performance. Here, we will assume that we are dealing with a rectangle-like detector with a defined length, width, and height. While germanium detectors have been grown in etched silicon [18], there are a number of difficulties with fabrication. Thus we will focus on germanium grown epitaxially on un-etched silicon.

Detector length

Consider light incident on an absorptive medium that does not vary geometrically in the direction of incidence. This device will have a transmitted optical power of

$$P_{\text{tr}} = P_{\text{in}} e^{-\alpha L}. \tag{7.18}$$

Where P_{in} is the incident power, α is the coefficient of absorption, and L is the length of the device. The absorbed light is then given by

$$P_{\text{abs}} = P_{\text{in}} - P_{\text{tr}} = P_{\text{in}} \left(1 - e^{-\alpha L}\right). \tag{7.19}$$

From the above equation, the absorbed power and therefore the responsivity of the detector will increase as a function of length, but each increase in length will give exponentially diminishing returns. However, if we consider a PIN detector where the width is fixed and the capacitance per unit length is C_l, then the total capacitance of the detector is given by

$$C = C_l \cdot L. \tag{7.20}$$

Thus increasing the length of the detector will increase the capacitance and lower the bandwidth. The dark current will also increase linearly with length.

Experimental data for 5.8 μm wide Ge detectors with different lengths are shown in Figure 7.16. The fit for the exponential function, Equation (7.19), gives absorption constants of 0.22 /μm at 2 V reverse bias and 0.12 /μm at 4 V reverse bias. The approximate bandwidth due to RC limitations, as discussed in Section 7.1.2, is also shown in the figure. The bandwidth approximation is based on a value of 20 GHz for a 10 μm long device.

Detector width

Generally, a smaller width will lower the detector capacitance, increasing the bandwidth of the detector. Dark current will also drop as both bulk and surface rates are reduced. However, once the detector becomes small enough there are two effects that occur. First, the absorption length will increase as light spends less time confined to the germanium. Second, the contact to the silicon will need to be reduced in size, and the contact resistance may begin to counteract the increases in bandwidth.

Detector height

There are two concrete bounds on the height of germanium grown epitaxially on silicon using CVD. The lower bound to the height comes from the fact that the lattice mismatch

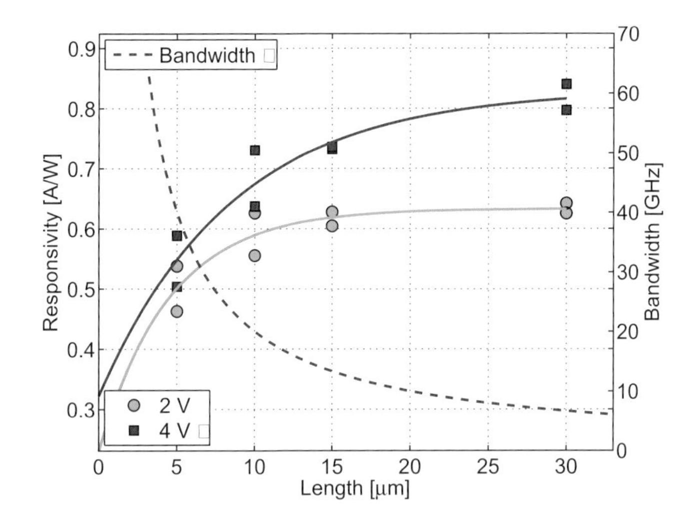

Figure 7.16 PIN detector, responsivity and bandwidth versus length. Germanium width is 5.8 μm.

between silicon and germanium necessitates a buffer region to confine the defects. This buffer region is usually 20–50 nm in height. Since the buffer has an extremely large number of defects, the buffer area should be kept to a small fraction of the total detector. If we limit the buffer to 20% of the detector volume, then a lower bound for the detector height can be set around 200 nm. The upper bound for the detector height is given by the thickness of germanium that can be reasonably grown. This will differ depending on the growth technique but, in practice, 2 μm is often given as an upper bound.

A further consideration in choosing the detector height is the effect that it will have on the bandwidth of the photodetector. A short detector will have a shorter transit time for the carriers, but a higher *RC* bandwidth, whereas a tall detector will have a long transit time and low *RC* delay. Assuming a simple model of a vertical PIN detector with thin P and N contacts, the tradeoff between transit time and *RC* delay can be estimated as shown in Figure 7.17.

7.4.3 Contacts

The placement and type of contacts is a critical piece of the detector design. The important considerations of detector contacts will be discussed qualitatively below. A more quantitative analysis will need a combination of simulation and experimental results.

Contact material

The material properties at the contact layer are of significant importance. Contact to silicon has been studied in-depth for a number of years with low-resistance, ohmic solutions available for both p-type and n-type silicon. Germanium contacts have had less development. In particular, the n-type germanium contact has caused a number of difficulties due to Fermi-level pinning. This pinning creates a large barrier for electron flow, as shown in Figure 7.18. The result is a Schottky contact that creates a high, non-Ohmic contact resistance.

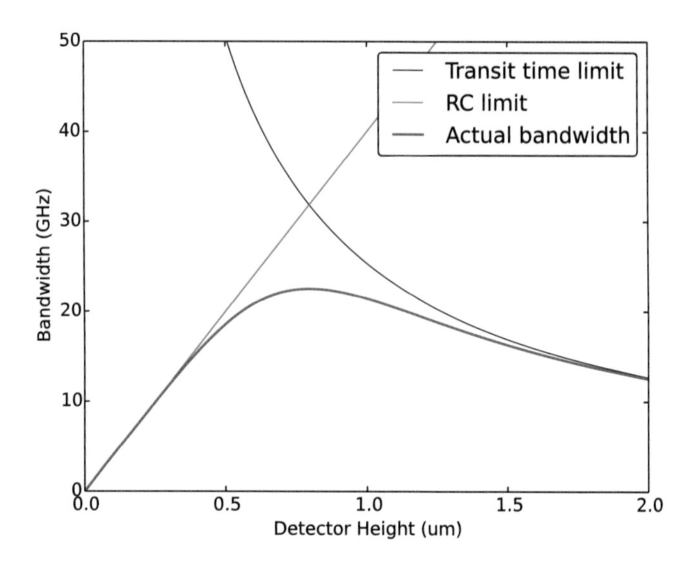

Figure 7.17 The approximate tradeoff between transit time and RC delay is shown as a function of detector height for a vertical PIN junction detector.

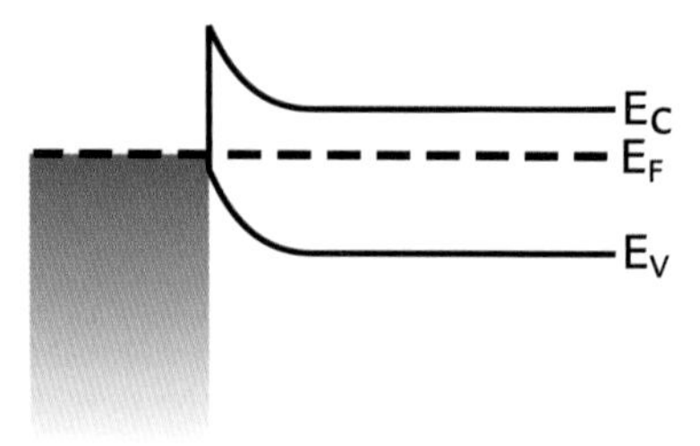

Figure 7.18 Fermi-level pinning of n-type germanium [28].

A number of techniques can be used to reduce the Fermi-level pinning. The choice of metal contact such as titanium or nickel can be used to offset the pinning [29]. Alternatively, a thin insulator can be inserted between the contact metal and the germanium [29]. With a very thin barrier, the tunnelling resistance is negligible compared to the reduction in resistance due to removal of the Fermi-level pinning. Creating a germanide – a germanium–metal hybrid – has also been shown to be effective in creating low-resistance contacts [30].

Contact geometry

There are a number of considerations when choosing the geometry of the detector concepts. The vertical PIN germanium detector has two important parameters with regards to the contact geometry, contact spacing, and contact area.

The spacing of the contacts refers to the distance that the silicon contacts will be placed from the germanium. There is a fundamental tradeoff between bandwidth and responsivity. Contacts that are too close to the germanium will absorb light and reduce

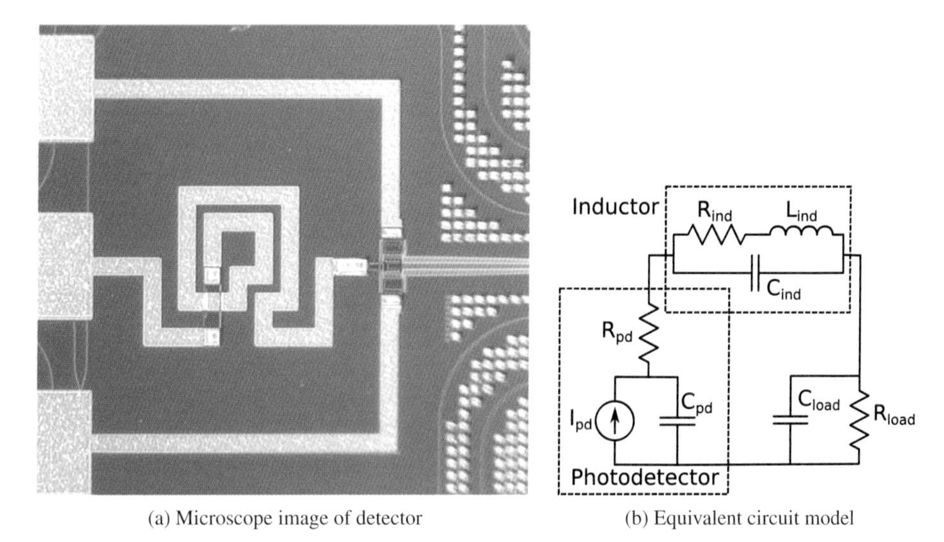

(a) Microscope image of detector (b) Equivalent circuit model

Figure 7.19 Optical image and equivalent circuit model of a germanium-on-silicon detector with inductive gain peaking [14].

bandwidth. However, if the silicon contacts are too far from the germanium, the resistance in the silicon will degrade the device bandwidth. A good design target would be to find the optical mode in the germanium detector and place the contacts far enough away that the electrical field magnitude of the mode is at least a factor of 100 less than the peak E-field. This should occur in just a few microns, but will depend on the width of the germanium.

The second important contact parameter is the contact area. This refers to size of the contact on top of the germanium. Larger contacts will result in higher metal absorption. If the contact is too small, the contact resistance may increase. The optimal contact sizing and placement is likely to be found by experiment.

Experimental results with different contact geometries have been reported in the literature, e.g. Reference [31].

7.4.4 External load on the detector

The load on the detector needs to be considered in the design and performance evaluation of the detector. It is possible to enhance the performance of a germanium photodetector using inductive gain peaking [14]. In this approach, an inductor is connected to the detector, as shown in Figure 7.19. This results in an increase in the frequency response, achieving a 60 GHz bandwidth.

7.5 Detector modelling

In this section, we model the germanium detector using TCAD numerical software. The detector considered is 50 μm long, 8 μm wide, and 500 nm thick germanium. The model is based on the detector fabricated at IME and described in Reference [12]. The simulation begins with 3D FDTD simulations of the silicon and germanium materials, in which light is incident on the detector via the taper waveguide. Owing to optical absorption in the germanium, the optical generation rate of charge carriers is calculated. This is exported into an electronic device solver. Electronic simulations are then performed to determine properties such as the dark current, responsivity, and modulation bandwidth.

7.5.1 3D FDTD optical simulations

The detector geometry is drawn and simulations are performed in 3D FDTD, via Listing 7.1, using the parameters defined in Listing 7.2. The simulation uses a mode source injected into the silicon. The simulation thus considers the coupling efficiency between the silicon waveguide and the germanium detector. The taper is long enough that it is adiabatic and has negligible insertion loss. Hence the mode source is injected at the end of the taper. The 3D simulations calculate the E and H fields. The absorbed power per unit volume is calculated at each point, at the central wavelength of the simulation, using the absorption coefficient in Figure 7.6:

$$L = -\frac{1}{2}\omega|E|^2 Im(\epsilon). \tag{7.21}$$

The cross-section images of the absorbed power (proportional to generation rate) are shown in Figures 7.20–7.23. In the simulation, y is the propagation direction (detector length), z is the wafer thickness direction, and x is in the detector width direction.

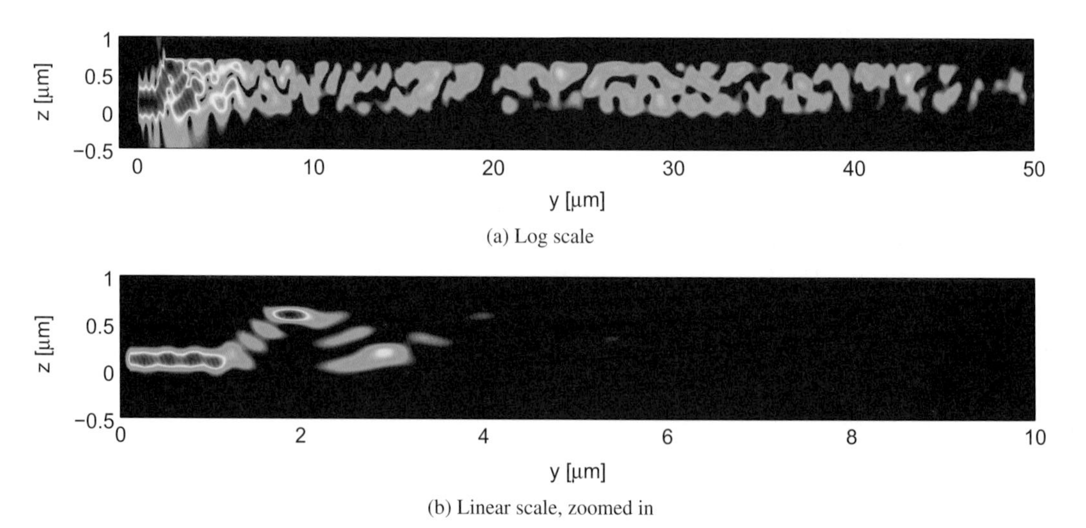

Figure 7.20 Simulated generation rate cross-section, viewed at the side of the detector: y is the optical propagation direction, and z is in the thickness direction.

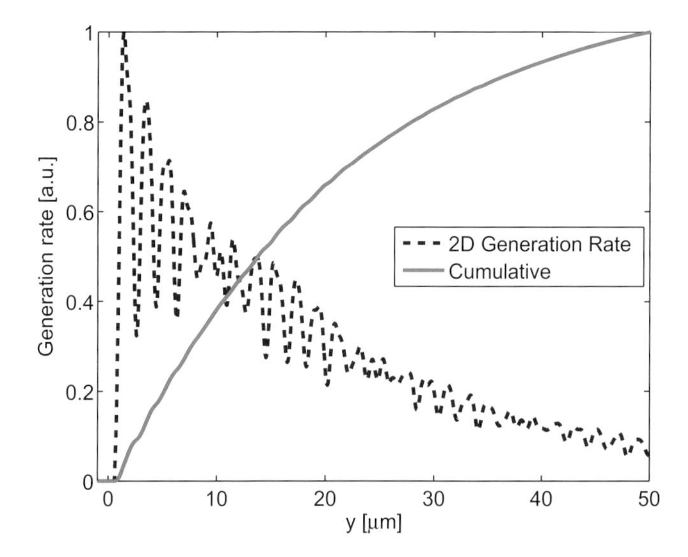

Figure 7.21 Generation rate along the length of the detector for each cross-section of the detector, versus position along the propagation direction, and the integrated generation rate (from 0 to y).

Figure 7.20 shows the cross-section of the generation rate vertically and along the length of the detector. Most of the light is absorbed in the first several microns of the germanium. Figure 7.21 show the generation rate summed over each detector cross-section, and plotted versus length. This shows an exponential decay of the field intensity as the light travels down the detector. When this plot is integrated from 0 to y, it results in an approximation for the total generation rate for a detector with length y. This is also shown in Figure 7.21. Using this graph, it is possible to identify approximately how long the detector should be made to absorb, for example, 90% of the light. This graph indicates that the detector should be 40 μm long. We should note that these simulations are all based on a germanium absorption model (Figure 7.6), which needs experimental validation particularly for thin Ge films. In light of the experimental data in Figure 7.16, it appears that the Ge model underestimates the absorption.

This detector is clearly highly multimode, owing to its large size, supporting hundreds of modes. The light therefore bounces up and down in the germanium and silicon regions. The field profile at the beginning of the detector, extracted from 3D FDTD simulations, can be decomposed into its modes. Such a simulation identifies that approximately 50% of the power is contained in the fundamental mode, 20% in the third mode, with the rest distributed among several high-order modes (e.g. modes 5, 39, 41, etc.). Note that only symmetric modes (in the x-axis) are excited since the light launched into the detector originates from the fundamental (symmetric) mode of the silicon waveguide. The modes experience a propagation loss of approximately 2000 dB/cm, dominated by the germanium absorption.

Figure 7.22 shows the top-view of the generation rate. This is shown at the bottom of the detector near the Ge–Si interface, in Figure 7.22a, and at just below the

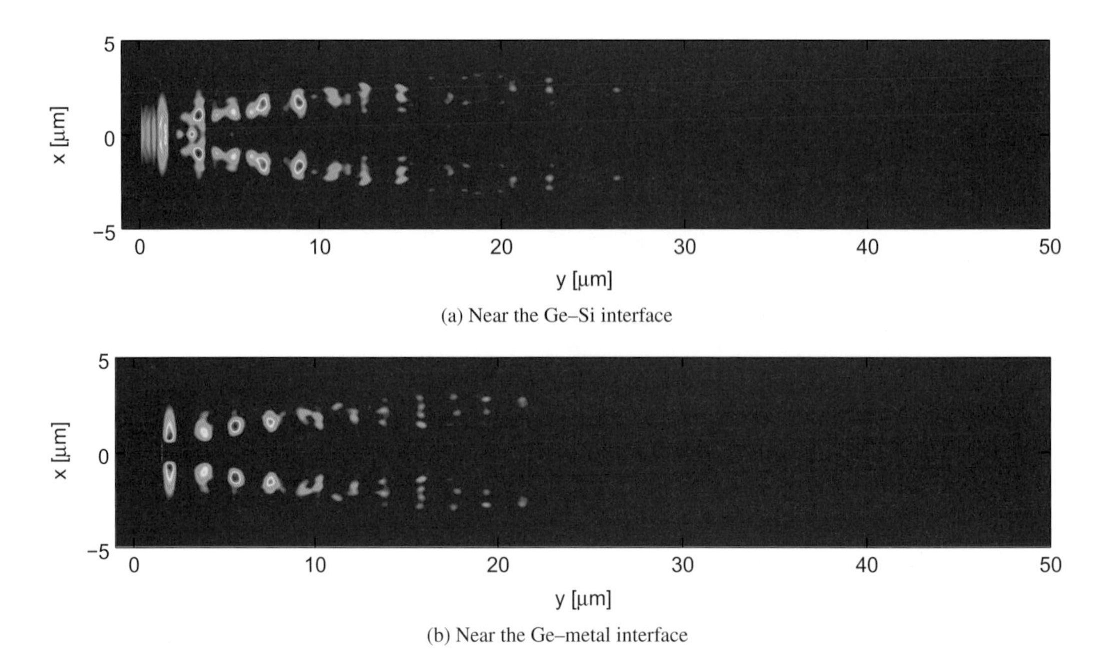

(a) Near the Ge–Si interface

(b) Near the Ge–metal interface

Figure 7.22 Top-view of the generation rate along the length of the detector.

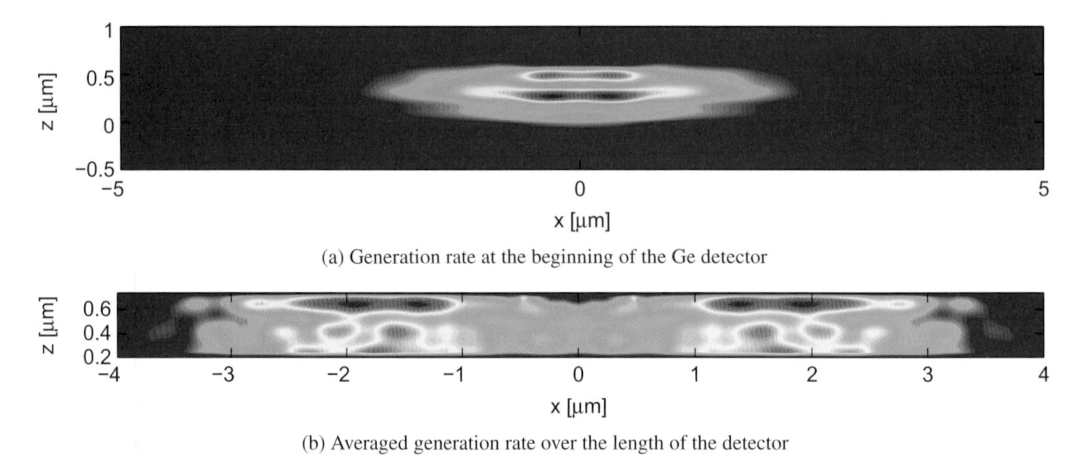

(a) Generation rate at the beginning of the Ge detector

(b) Averaged generation rate over the length of the detector

Figure 7.23 Simulated generation rate at the beginning of the detector, and averaged over the detector length.

metal–Ge interface, in Figure 7.22b. These plots identify a significant source of optical loss – namely the presence of light next to the metal contact. Note that the simulation can be repeated with and without the metal to determine the impact of the metal absorption on the responsivity.

Alternatively, eigenmode calculations can be performed for the detector, with and without the metal contact, to determine the impact of the metal contact. For the first two symmetric modes (TE polarized), the absorption increases on average by 220 dB/cm. This calculation indicates that the metal contact contributes to approximately 10% of the optical loss, hence a ~10% reduction in the responsivity due to the metal.

The three-dimensional (3D) generation-rate data are calculated assuming that every photon absorbed within the germanium contributes one electron–hole pair. The illumination is assumed to be at one optical wavelength then exported for the subsequent electrical simulations. Figure 7.23a shows the cross-section of the generation rate in the beginning of the detector (xy plane). Since the subsequent electrical simulations are performed in 2D, the generation rate is integrated over the length of the detector. That is, the 3D generation rates are averaged over the length dimension, resulting in a 2D generation rate as shown in Figure 7.23b; the electrical solver performs calculations keeping the length of the device in mind.

7.5.2 Electronic simulations

The electronic simulation is implemented as follows. First, the device and simulation parameters are defined, in Listing 7.2. The device is constructed as a 2D cross-section where the length is set to 50 µm, in Listing 7.3. This includes the doping profile.

The first part of the simulation is to define the material parameters. These parameters are primarily from literature, with adjustments made to match the experimental data. The assumption is that the primary source of dark current is the Ge–SiO_2 interface. Specifically, the surface recombination velocities at the Ge–SiO_2 interface are adjusted to fit the dark current response of the measured device at 1 V reverse bias.

Germanium model

- Type: Semiconductor
- Relative dielectric permittivity: 16
- Work function: 4.46 eV
- Effective mass (m^*/m_0): 0.56 for electrons; 0.39 for holes
- Bandgap model: $E_G(T) = E_G(0) - \frac{\alpha T^2}{T+\beta}$, where $\alpha = 0.000\,4774$, $\beta = 235$, $E_G(0) = 0.74$. This corresponds to 0.659 69 at 300 K.
- Intrinsic carrier concentration, n_i: $2.306\,05 \times 10^{13}$ cm^{-3}
- Mobility: electrons 3900 cm^2/Vs; holes 1900 cm^2/Vs
- Velocity saturation: electrons 6×10^6cm/s; holes 5.6×10^6cm/s
- Trap-assisted (Shockley–Read–Hall) recombination model: $A(T) = A(300) \left(\frac{T}{300}\right)^\eta$, where $A(300) = 1 \times 10^{-6}$, $\eta = -1.05$. The carrier lifetimes are assumed to be long, as in the bulk Ge material.
- Radiative recombination rate: 6.41×10^{-14}cm^3/s
- Auger recombination rate: 1×10^{-30}cm^6/s for both electrons and holes
- Surface recombination velocity (SRV) Ge–Si contact model: $A(T) = A(300) \left(\frac{T}{300}\right)^\eta$, where $A(300) = 100\,000$, $\eta = -2$; only for minority carriers.

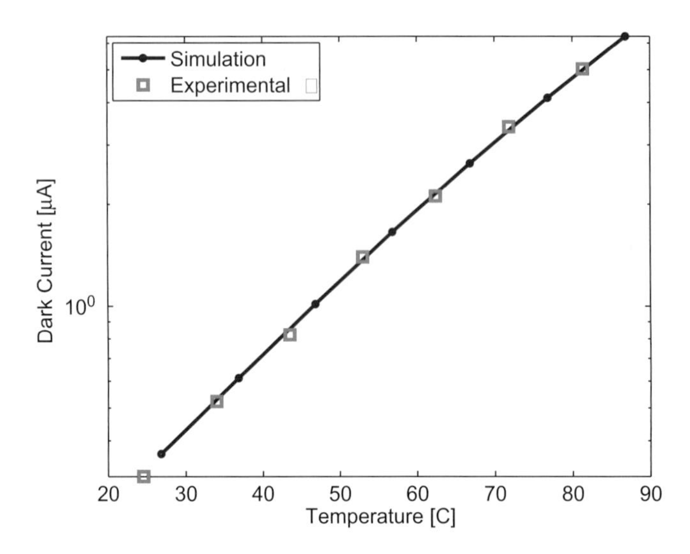

Figure 7.24 PIN detector, TCAD simulations for the dark current versus temperature, compared with experimental results in Reference [12].

- Surface recombination velocity (SRV) Ge–SiO_2: both minority and majority carriers; $A(300) = 200\,000$, $\eta = -5$.
- No band-to-band tunnelling/trap-assisted tunnelling modelled

Silicon contact model

- Type: Conductor
- Work function: 4.2 eV

Electronic simulation – configuration

- The Newton solver is used for all simulations.
- A half-space simulation reduces simulation time (the device is symmetric at $x = 0$).
- Two different illumination powers are used, as per the referenced publication: 0.9 mW for the responsivity simulation (giving an ideal photo-current of 1 mA), and 0.09 mW for the transient response simulation (steady-state photocurrent of 0.1 mA).
- The temperature for the responsivity simulation is set to 300 K.
- All calculations are performed for an unloaded device.
- The optical generation rate has been averaged in the direction of propagation.

Steady-state electrical simulations are performed in Listing 7.4. The simulation configures the contact voltages (1 V) and measures the current for temperatures ranging from 300 K to 360 K. The results are plotted in Figure 7.24, including a comparison with the experimental results in Reference [12]. Next the simulation is repeated for several voltages, to create a dark-current contour plot versus temperature and voltage, as shown in Figure 7.25.

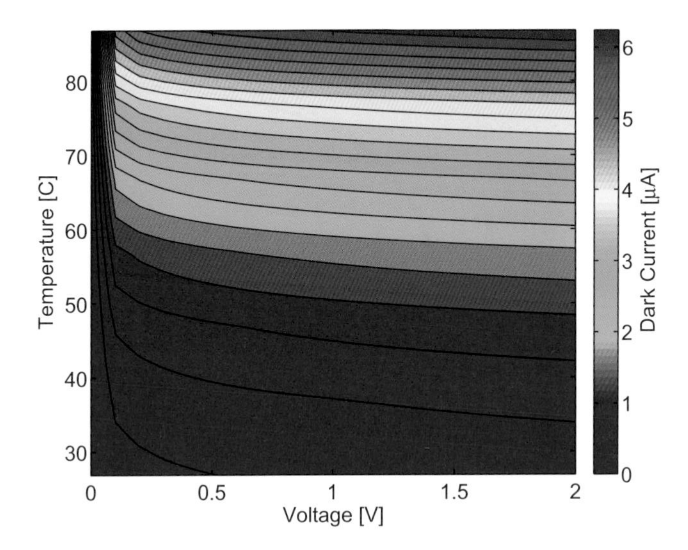

Figure 7.25 PIN detector, TCAD simulations for the dark current versus temperature and voltage.

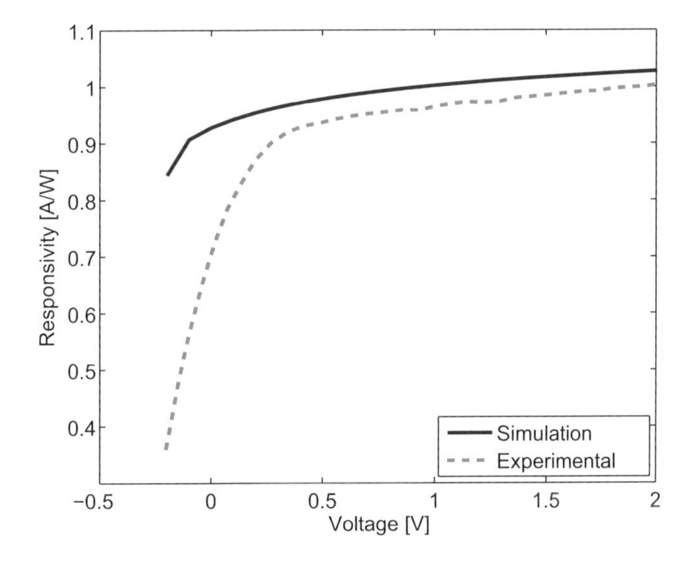

Figure 7.26 PIN detector, TCAD simulations for the responsivity versus voltage, compared with experimental results in Reference [12].

The responsivity is simulated next, by turning on the optical generation imported from FDTD. The responsivity is simulated for a range of voltages at 300 K, with results shown in Figure 7.26.

Transient simulations are performed next, to determine the step response when the light is turned on. This is implemented in Listings 7.5 and 7.6. The step response is plotted in Figure 7.27. Repeating these simulations for different voltages shows that the response changes very little, and the bandwidth is several GHz. Comparing these results

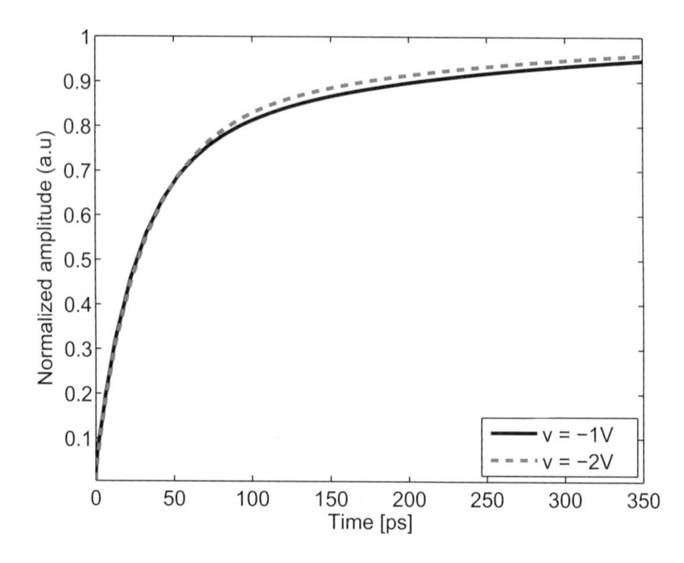

Figure 7.27 PIN detector, TCAD simulations for the transient step response for V = −1 V and −2 V.

with the calculated transit time from Equation (7.5), we can conclude that the device is not transit-time limited, but rather limited by the RC time constant.

7.6 Problems

7.1 What is the quantum efficiency of a germanium detector which has a responsivity of 0.7 A/W? If the absorption rate is constant with length and there are no losses due to scattering, metal, and all photo generated carriers contribute to the current, what would the quantum efficiency be after doubling the length?

7.2 The intrinsic region of a germanium detector is 1 μm wide. The detector is biased to saturation velocity. The detector series resistance is 300 Ω and the capacitance is 50 fF. Is the detector transit time limited or RC limited? What modifications could be done to the geometry to enhance the bandwidth?

7.3 Given a silicon avalanche region with 1 μm width, a charge region width of 100 nm and a germanium width of 2 μm calculate the approximate voltage and doping (in Atoms/cm^2) required for the SACM APD shown in Figure Figure 7.14b such that the electric field in the germanium is at saturation and the electric field in the silicon avalanche region is at breakdown.

7.4 Calculate the voltage and doping required for the SACM APD shown in Figure Figure 7.14b.

7.5 Perform simulations for the detector whereby the Ge metal electrode is split into two to reduce optical absorption. How much performance improvement is expected?

7.7 Code listings

Listing 7.1 Perform 3D FDTD simulations for the Ge detector, in Lumerical FDTD Solutions; detector_FDTD.lsf

```
# Optical simulation of detector
# Step 1: in Lumerical FDTD; this script accomplishes:
# 1) calculates the generation rate of eh pairs in Ge
# 2) Exports the results to DEVICE for the electrical modelling.

newproject; redrawoff;

# modulator geometry variables defined:
detector_setup_parameters;

# define simulation region for FDTD calculations

override_mesh_size = 0.025e-6;
override_mesh_size_smaller=0.02e-6;

# draw geometry
addrect;     # oxide
set('name','oxide');
set('material','SiO2 (Glass) - Palik');
set('override mesh order from material database',1); set('mesh order',5);
set('override color opacity from material database',1); set('alpha',0.3);
set('x',x_center); set('x span',oxide_width_x);
set('y',y_center); set('y span',oxide_width_y);
set('z',0); set('z span',oxide_thickness);

addrect;
set('name','silicon'); #silicon1
set('material','Si (Silicon) - Palik');
set('override mesh order from material database',1); set('mesh order',2);
set('x',x_center); set('x span',si_1_width_x);
set('y min',si_1_width_y_min); set('y max',si_1_width_y_max);
set('z',si_z); set('z span',si_1_thickness);

addrect;
set('name','silicon'); #silicon2
set('material','Si (Silicon) - Palik');
set('override mesh order from material database',1); set('mesh order',2);
set('x',x_center); set('x span',si_2_width_x);
set('y min',si_2_width_y_min); set('y max',si_2_width_y_max);
set('z',si_z); set('z span',si_2_thickness);


addpyramid;
set('name','Ge'); #germanium
set('material','Ge (Germanium) - Palik');
set('override mesh order from material database',1); set('mesh order',2);
set('x',x_center); set('x span bottom',ge_width_x_bottom); set('x span
     top',ge_width_x_top);
set('y',ge_y); set('y span bottom',ge_width_y_bottom); set('y span top',ge_width_y_top);
z_ge=(si_z+ge_thickness/2+si_1_thickness/2);
set('z span',ge_thickness); set('z',z_ge);


addpyramid;
set('name','cathode'); #PEC (
set('material','PEC (Perfect Electrical Conductor)');
set('override mesh order from material database',1); set('mesh order',2);
set('x',x_center); set('x span bottom',contact_width_x_bottom); set('x span
     top',contact_width_x_top);
set('y',contact_y); set('y span bottom',contact_width_y_bottom); set('y span
     top',contact_width_y_top);
z_pec=z_ge+ge_thickness/2+contact_thickness/2;
set('z',z_pec); set('z span',contact_thickness);
```

```
# simulation region
addfdtd;
set('x',x_center); set('x span',x_span);
set('y',y_center_region); set('y span',y_span);
set('z',z_center); set('z span',z_span);
set('x min bc', 'Anti-Symmetric');
set("mesh accuracy", 1);
set("simulation time", 4000e-15);

addmesh;
set('dz',override_mesh_size_smaller);
set('override y mesh',0);
set('override x mesh',0);
set('z',0.46e-6); set('z span',0.48e-6);

addmesh;
set('dz',override_mesh_size);
set('override y mesh',0);
set('override x mesh',0);
set('z',0.75e-6); set('z span',0.1e-6);



addmesh;
set('dz',override_mesh_size);
set('override y mesh',0);
set('override x mesh',0);
set('z',0.17e-6); set('z span',0.1e-6);


#source

addmode;
set("injection axis", "y-axis");
set('x',x_center); set('x span',x_span);
set('y',0);
set('z',z_center); set('z span',z_span);
set("wavelength start", lambda);
set("wavelength stop", lambda);
updatesourcemode;

setglobalmonitor("frequency points",1);

addprofile;
set('name','xy'); # top view
set("monitor type","2D Z-normal");
set('x',x_center); set('x span',x_span*1.5);
set('y',y_center_region); set('y span',y_span*1.5);
set('z',z_center);

addprofile;
set('name','xy2'); # top view, at the Ge-metal interface
set("monitor type","2D Z-normal");
set('x',x_center); set('x span',x_span*1.5);
set('y',y_center_region); set('y span',y_span*1.5);
set('z',si_1_thickness+ge_thickness-10e-9);

addprofile;
set('name','yz'); # cross-section along the direction of propagation
set("monitor type","2D X-normal");
set('x',x_center);
set('y',y_center_region); set('y span',y_span*1.5);
set('z',z_center); set('z span',z_span*1.5);

addprofile;
set('name','xz'); # cross-section perpendicular to the direction of propagation
set("monitor type","2D Y-normal");
set('x',x_center); set('x span',x_span*1.5);
set('y',ge_y-ge_width_y_top/2); # cross-section at the beginning of the Ge
set('z',z_center); set('z span',z_span*1.5);
```

```
addobject("CW_generation");
set('name','generation_rate');
set('x span',ge_width_x_bottom+0.04e-6); set("x",0);
set('y span',y_span+4e-6); set("y", y_center_region);
set('z span',ge_thickness+0.04e-6); # generation rate, slightly larger than Ge
set('z', z_ge);
set("export filename", generation_filename);
set("periods", 1);
set("average dimension", "y");
set("source intensity", 6e7); # power*sourceintensity(f)/sourcepower(f)
set("make plots", 1);
set("down sample x", 1);
set("down sample y", 1);
set("down sample z", 1);

# run simulation
save("vpd");
run;

runanalysis;

# Field profiles:

xy_x=pinch(getdata('xy','x'));
xy_y=pinch(getdata('xy','y'));
xy_E=pinch(abs(getdata('xy','Ex')));
image(xy_x,xy_y,xy_E);

xy2_x=pinch(getdata('xy2','x'));
xy2_y=pinch(getdata('xy2','y'));
xy2_E=pinch(abs(getdata('xy2','Ex')));
image(xy2_x,xy2_y,xy2_E);

yz_y=pinch(getdata('yz','y'));
yz_z=pinch(getdata('yz','z'));
yz_E=pinch(abs(getdata('yz','Ex')));
image(yz_y,yz_z,yz_E);

xz_x=pinch(getdata('xz','x'));
xz_z=pinch(getdata('xz','z'));
xz_E=pinch(abs(getdata('xz','Ex')));
image(xz_x,xz_z,xz_E);

matlabsave('vpd');
```

Listing 7.2 Set up the simulation parameters for the Ge detector, in Lumerical DEVICE and FDTD Solutions; detector_setup_parameters.lsf

```
# Parameter definitions for the detector, used for electrical calculations in DEVICE and
     optical calculations in FDTD
clear;

unfold_sim = 1;

# define the structure including the contacts
si_1_thickness = 0.22e-6;
si_1_width_y_min = 1e-6;
si_1_width_y_max = 70e-6;
si_1_width_x = 20e-6;
si_z=0.11e-6;

si_2_thickness = 0.22e-6;
si_2_width_y_min = -2e-6;
si_2_width_y_max = 1e-6;
si_2_width_x = 7e-6;
```

```
ge_thickness = 0.5e-6;
ge_width_y_top = 52.65e-6;
ge_width_y_bottom = 53.65e-6;
ge_width_x_top = 7e-6;
ge_width_x_bottom = 8e-6;
ge_y=27.825e-6;
oxide_thickness = 10e-6;
oxide_width_y = 75e-6;
oxide_width_x = 20e-6;
contact_thickness = 0.1e-6;
contact_width_y_top = 72e-6;
contact_width_y_bottom = 72e-6;
contact_width_x_top = 1.4e-6;
contact_width_x_bottom = 1e-6;
contact_y=37.625e-6;

# define doping
center_pepi = 0.35e-6;  # pepi doping
thick_pepi = 0.9e-6;
pepi_concentration= 1e+016;

x_center_p_contact = 0e-6; # contact doping
x_span_p_contact = 10e-6;
z_center_p_contact = 0.24495e-6;
z_span_p_contact = 0.0501e-6;
diff_dist_fcn_contact = 1; # 0 for erfc, 1 for gaussian
face_p_contact = 4; # lower z
width_junction_p_contact = 0.05e-6;
surface_conc_p_contact = 3.2e15*1e6;
reference_conc_p_contact = 1e8*1e6;

x_center_n_contact = 0e-6;
x_span_n_contact = 2e-6;
z_center_n_contact = 0.67e-6;
z_span_n_contact = 0.100001e-6;
face_n_contact = 5; # upper z
width_junction_n_contact = 0.1e-6;
surface_conc_n_contact = 1e18*1e6;
reference_conc_n_contact = 1e3*1e6;

# define simulation region
min_edge_length = 0.02e-6;
max_edge_length = 4e-6;
x_center = 0; x_span = 10e-6;
y_center = 35e-6; y_span = 51e-6; # irrelevant for 2D cross section
z_center = 0.25e-6; z_span = 1.5e-6;
y_center_region=24.5e-6;
x_center_region=x_center;
z_min_region=0.1e-6;
z_max_region=1e-6;

#simulation parameters
norm_length= 50e-6;
abs_lte_limit= 0.001;
rel_lte_limit= 0.001;
transient_max_time_step= 10e-12;
transient_min_time_step= 100e-15;
transient_sim_time_max=1e-9;
shutter_mode = "step on";
shutter_ton=transient_min_time_step;
shutter_tslew= 0;
solver_type = "newton";
generation_filename="Vpd_generation.mat";
generation_filename_transient="Vpd_generation_100uA.mat";
generation_filename_responsivity="Vpd_generation_1mA.mat";
lambda= 1.55e-6;
```

Listing 7.3 Configure the electrical steady-state simulation for the Ge detector, in Lumerical DEVICE; detector_setup_DEVICE_steadystate.lsf

```
# Steady State electrical simulation of the detector
#  setup script

newproject; redrawoff;

load ('vpd_materials2'); # thin film materials for Ge

redrawoff;

# modulator geometry variables defined:
detector_setup_parameters;
solver_mode=1; #DC

# draw geometry
addrect;     # oxide
set('name','oxide');
set('material', 'SiO2 (Glass) - Sze');
set('override mesh order from material database',1); set('mesh order',5);
set('override color opacity from material database',1); set('alpha',0.3);
set('x',x_center); set('x span',oxide_width_x);
set('y',y_center); set('y span',oxide_width_y);
set('z',0); set('z span',oxide_thickness);

addrect;
set('name','silicon'); #silicon1
set('material','Si (Silicon) [CONTACT]');
set('override mesh order from material database',1); set('mesh order',2);
set('x',x_center); set('x span',si_1_width_x);
set('y min',si_1_width_y_min); set('y max',si_1_width_y_max);
set('z',si_z); set('z span',si_1_thickness);

addrect;
set('name','silicon'); #silicon2
set('material','Si (Silicon) [CONTACT]');
set('override mesh order from material database',1); set('mesh order',2);
set('x',x_center); set('x span',si_2_width_x);
set('y min',si_2_width_y_min); set('y max',si_2_width_y_max);
set('z',si_z); set('z span',si_2_thickness);


addpyramid;
set('name','Ge'); #germanium
set('material','Ge (Germanium) thin film 2um');
set('override mesh order from material database',1); set('mesh order',2);
set('x',x_center); set('x span bottom',ge_width_x_bottom); set('x span
     top',ge_width_x_top);
set('y',ge_y); set('y span bottom',ge_width_y_bottom); set('y span top',ge_width_y_top);
z_ge=(si_z+ge_thickness/2+si_1_thickness/2);
set('z span',ge_thickness); set('z',z_ge);


addpyramid;
set('name','cathode'); #Al
set('material','Al (Aluminium) - CRC');
set('override mesh order from material database',1); set('mesh order',2);
set('x',x_center); set('x span bottom',contact_width_x_bottom); set('x span
     top',contact_width_x_top);
set('y',contact_y); set('y span bottom',contact_width_y_bottom); set('y span
     top',contact_width_y_top);
z_pec=z_ge+ge_thickness/2+(4*contact_thickness)/2;
set('z',z_pec); set('z span',4*contact_thickness);

# draw simulation region
adddevice;
set('min edge length',min_edge_length);
set('max edge length',max_edge_length);
if (unfold_sim) {
```

```
  set('x min',x_center);
} else {
  set('x min',x_center - 0.5*x_span+0.1e-6);
}
set('x max',x_center + 0.5*x_span-0.1e-6);
set('y',y_center);
set('z min',z_min_region); set('z max',z_max_region);
set('norm length', norm_length);
set('solver mode', solver_mode);
set('solver type', solver_type);
set('rel lte limit',rel_lte_limit);
set('abs lte limit',abs_lte_limit);
set('transient min time step',transient_min_time_step);
set('transient max time step',transient_max_time_step);

addimportgen;
matlabload(generation_filename_responsivity);
igen = rectilineardataset(x,y,z);
igen.addparameter('v',0); #dummy
igen.addattribute('G',G);
importdataset(igen);
set ("enabled", 0);
set("name", "pulse");

# draw doping regions
adddope;
set('name','pepi');
set('dopant type','p'); # p type
set('x',x_center); set('x span',x_span);
set('y',y_center); set('y span',y_span);
set('z',center_pepi); set('z span',thick_pepi);
set('concentration',pepi_concentration );

adddiffusion;
set('name','p-');
set('x',x_center_p_contact); set('x span',x_span_p_contact);
set('y',y_center); set('y span',y_span);
set('z',z_center_p_contact); set('z span',z_span_p_contact);
set('dopant type','p');
set('face type',face_p_contact);
set('junction width',width_junction_p_contact);
set('distribution index',diff_dist_fcn_contact);
set('concentration',surface_conc_p_contact);
set('ref concentration',reference_conc_p_contact);

adddiffusion;
set('name','n++');
set('x',x_center_n_contact); set('x span',x_span_n_contact);
set('y',y_center); set('y span',y_span);
set('z',z_center_n_contact); set('z span',z_span_n_contact);
set('dopant type',"n");
set('face type',face_n_contact);
set('junction width',width_junction_n_contact);
set('distribution index',diff_dist_fcn_contact);
set('concentration',surface_conc_n_contact);
set('ref concentration',reference_conc_n_contact);

# set contacts

addcontact;  # anode
setcontact('new_contact','dc','name','anode');
setcontact('anode','geometry','silicon');
addcontact;  # cathode
setcontact('new_contact','dc','name','cathode');
setcontact('cathode','geometry','cathode');
setcontact('cathode','dc','fixed contact',0);

save('vpd');
```

Listing 7.4 Run the electrical steady-state simulation and analyze the result, for the Ge detector, in Lumerical DEVICE; detector_run_DEVICE_steadystate.lsf

```
# Steady State electrical simulation of the detector
# Step 2a: in Lumerical DEVICE; this script accomplishes:
# 1) import generation rate data from FDTD
# 2) Calculate the dark current and temperature dependance of it
# 3) Calculate responsivity of the steady state simulation

# First run this to configure:
# detector_setup_DEVICE_steadystate;
# then check parameters


# 1) Simulate the dark current temperature dependance of the device at v=1 volt

if (0) {
setcontact('anode','dc','fixed contact',1);
setcontact('anode','dc','voltage',-1);
setcontact('cathode','dc','fixed contact',1);
setcontact('cathode','dc','voltage',0);
save("vpd");

Tmin=300;
Tmax=360;
T=linspace(Tmin, Tmax,7);
cathode_I= matrix(length(T));
for (i=1:length(T)){
  switchtolayout;
  setnamed('Device region','simulation temperature',T(i));
  run;
  cathode_I(i) = getdata("Device region", "cathode.I");
}
T=T-273.15;

liow_fig13a_uA = [
24.599056603773583,0.3131313106003993;
34.009433962264154,0.5223345074266843;
43.490566037735846,0.8213531263060052;
52.971698113207545,1.3973305213983964;
62.382075471698116,2.1124494476604796;
71.86320754716981,3.387774774440254;
81.34433962264151,5.021745520652797];
liow_fig13a_uA = [ 24.599,0.31313; 34.009,0.52233; 43.490,0.82135;
52.971,1.39733; 62.382,2.11244; 71.863,3.38777; 81.344,5.02174];
nfig13a = size(liow_fig13a_uA); nfig13a = nfig13a(1);

refdataT = liow_fig13a_uA(1:nfig13a,1);
refdataIuA = liow_fig13a_uA(1:nfig13a,2);
refdataIuA_interp=interp(refdataIuA,refdataT,T);

if (unfold_sim) {
  cathode_I = 2*cathode_I;
}

plot(T,cathode_I*1e6,(refdataIuA_interp), "Temperature in degrees Celsius", "Dark current
     in uA at V=-1(v)");
legend("DEVICE simulation","Reference, p+ Si w/ anneal");

matlabsave ('vpd_darkcurrent_1');
}

# 2) Dark current versus temperature and versus bias voltage
if (0) {
vmin=0;
vmax=2;
vnum=21;
switchtolayout;
setcontact('anode','dc','fixed contact',0);
setcontact('anode','dc','range start',vmin);
```

```
setcontact('anode','dc','range stop',-vmax);
setcontact('anode','dc','range num points',vnum);
setcontact('cathode','dc','fixed contact',1);
setcontact('cathode','dc','voltage',0);
save("vpd");

Tmin=300;
Tmax=360;
T=linspace(Tmin, Tmax,7);
cathode_I_image= matrix(vnum,length(T));
for (i=1:length(T)){
  switchtolayout;
  setnamed('Device region','simulation temperature',T(i));
  run;
  cathode_I_image(1:vnum,i) = getdata("Device region", "cathode.I");
}

if (unfold_sim) {
  cathode_I_image = 2*cathode_I_image;
}

T=T-273.15;
V=linspace(vmin,vmax,vnum);

T2=linspace(min(T),max(T),100);
V2=linspace(min(V),max(V),200);
Ik2=interp(cathode_I_image,V,T,V2,T2);
image(V2,T2,1e6*Ik2,"photodetector bias voltage (V)","Temperature in degrees Celsius",
      "Dark Current (uA)");

matlabsave ('vpd_darkcurrent_2');
}


# 3) Simulate the responsivity under illumination
if (1) {
switchtolayout;
select("pulse");
set("enabled", 1);
vmin=-0.2;
vmax=2;
vnum=23;

setcontact('anode','dc','fixed contact',0);
setcontact('anode','dc','range start',-vmin);
setcontact('anode','dc','range stop',-vmax);
setcontact('anode','dc','range num points',vnum);
setcontact('cathode','dc','fixed contact',1);
setcontact('cathode','dc','voltage',0);
save("vpd");

Pin=0.9e-3; # Watts
switchtolayout;
setnamed('Device region','simulation temperature',300);
run;

liow_fig15_pvpd = [ -0.2027,0.3591; -0.1748,0.4163;
-0.1503,0.4701; -0.1258,0.5157; -0.09895,0.5621;
-0.04545,0.6428; -0.005244,0.6965; 0.03146,0.7407;
0.07342,0.7845; 0.1282,0.8245; 0.1958,0.8687;
0.2692,0.9002; 0.3426,0.9195; 0.4160,0.9307;
0.4895,0.9357; 0.5629,0.9429; 0.6363,0.9482;
0.7097,0.9522; 0.7832,0.9550; 0.8566,0.9586;
0.9300,0.9585; 1.0034,0.9654; 1.0769,0.9700;
1.1503,0.9732; 1.2237,0.9719; 1.2944,0.9736;
1.3706,0.9805; 1.4440,0.9828; 1.5174,0.9856;
1.5909,0.9887; 1.6643,0.9914; 1.7377,0.9924;
1.8111,0.9972; 1.8846,0.9987; 1.9580,1.0005;
1.9982,1.0030];
nliowR = size(liow_fig15_pvpd); nliowR = nliowR(1);
```

```
I=-getdata("Device region", "anode.I");
if (unfold_sim) {
  I = 2*I;
}
I_norm=I/max(I);
V=linspace(vmin,vmax,vnum);
Resp=I/Pin;
plotxy(V,I_norm,
      liow_fig15_pvpd(1:nliowR,1),liow_fig15_pvpd(1:nliowR,2),
      "photo detector bias voltage(v)","I_(photo,norm)=I_(photo)/I_(norm) a.u ");
plot(V,Resp,"photo detector bias voltage(v)","Responsivity (A/W) ");
matlabsave('vpd_responsivity');
}
```

Listing 7.5 Configure the electrical transient simulation for the Ge detector, in Lumerical DEVICE; detector_setup_DEVICE_transient.lsf

```
# Trransient electrical simulation of the detector
#  setup script

detector_setup_DEVICE_steadystate;

solver_mode=2; #Transient

select('Device region');
set('solver mode', solver_mode);
set('solver type', solver_type);
set("abs lte limit", abs_lte_limit);
set("rel lte limit", rel_lte_limit);
set("shutter mode",shutter_mode);
set("shutter ton", shutter_ton);
set("shutter tslew", shutter_tslew);
set("transient max time step", transient_max_time_step);
set("transient min time step", transient_min_time_step);

select('pulse');
matlabload(generation_filename_transient);
igen = rectilineardataset(x,y,z);
igen.addparameter('v',0); #dummy
igen.addattribute('G',G);
importdataset(igen);
set ("enabled", 1);

voltage_table = [0,0];
time_table= [0, transient_sim_time_max];

setcontact('anode','transient','fixed contact',1);
setcontact('anode','transient','voltage',-0);
setcontact('cathode','transient','fixed contact',0);
setcontact("cathode","transient", "voltage table", voltage_table );
setcontact("cathode","transient", "voltage time steps", time_table);

save("vpd_transient");
```

Listing 7.6 Run the electrical transient simulation and analyze the result, for the Ge detector, in Lumerical DEVICE; detector_run_DEVICE_transient.lsf

```
# Trransient electrical simulation of the detector
# Step 2b: in Lumerical DEVICE; this script accomplishes:
# 1) Set up and calculate normalized response of the transient simulation

# First run this to configure:
# detector_setup_DEVICE_transient;
# then check parameters

v_list = [0, -0.5, -1, -2];
```

```
for (i=1:length(v_list)) {

switchtolayout;
setcontact('anode','transient','voltage', v_list(i));
?V=getcontact('anode','transient','voltage');

run;

I=getdata("Device region", "cathode.I");
t=getdata("Device region", "cathode.t");

if (unfold_sim) {
  I = 2*I;
}

# Take the central derivative of the step to get the impulse response
N= length(t);
th = t(2:(N-1));
Nh = N-2;
dI = I(3:N) - I(1:Nh);
dt = t(3:N) - t(1:Nh);
dIdt = dI/dt;


# Interpolate the impulse response to plot it on the same figure as the step
t_interp = th(1):0.1e-12:th(Nh); #uniform time grid
dIdt_interp = interp(dIdt, th,t_interp);
plotxy(t_interp*1e12,dIdt_interp/max(dIdt_interp),t*1e12,I/max(I),"time (ps)","Normalized
     amplitude (a.u) ");

# take fft of the original impulse response to get the frequency response
H = fft(dIdt_interp,2,1);
w = fftw(t_interp - t_interp(1),2,1);
Nw = length(w);
three_dB=w>0;
three_dB=three_dB*0.001;
plot(1e-9*w(2:Nw)/2/pi,20*log10(abs(H(2:Nw))/max(abs(H))) , log10(three_dB(2:Nw)),
     "Frequency (GHz)", " Normalized response (dB)");
legend(num2str(V), "3dB line ");

matlabsave('vpd_transient' + num2str(V));
}
```

References

[1] L. Colace, G. Masini, F. Galluzzi, *et al.* "Metal–semiconductor–metal near-infrared light detector based on epitaxial Ge/Si". *Applied Physics Letters* **72**.24 (1998), pp. 3175–3177 (cit. on pp. 259, 264, 265).

[2] Hsu-Hao Chang, Ying-hao Kuo, Richard Jones, Assia Barkai, and John E. Bowers. "Integrated hybrid silicon triplexer". *Optics Express* **18**.23 (2010), pp. 23 891–23 899 (cit. on p. 259).

[3] Ilya Goykhman, Boris Desiatov, Jacob Khurgin, Joseph Shappir, and Uriel Levy. "Locally oxidized silicon surface-plasmon Schottky detector for telecom regime". *Nano Letters* **11**.6 (2011), pp. 2219–2224 (cit. on pp. 259, 264).

[4] Ilya Goykhman, Boris Desiatov, Jacob Khurgin, Joseph Shappir, and Uriel Levy. "Waveguide based compact silicon Schottky photodetector with enhanced responsivity in the telecom spectral band". *Optics Express* **20**.27 (2012), pp. 28 594–28 602 (cit. on pp. 259, 264).

[5] J. D. B Bradley, P. E. Jessop, and A. P. Knights. "Silicon waveguide-integrated optical power monitor with enhanced sensitivity at 1550 nm". *Applied Physics Letters* **86**.24 (2005), pp. 241 103–241 103 (cit. on pp. 259, 264).

[6] A. P. Knights, J. D. B. Bradley, S. H. Gou, and P. E. Jessop. "Silicon-on-insulator waveguide photodetector with self-ion-implantation-engineered-enhanced infrared response". *Journal of Vacuum Science & Technology A* **24**.3 (2006), pp. 783–786 (cit. on pp. 259, 264).

[7] M. W. Geis, S. J. Spector, M. E. Grein, *et al.* "Silicon waveguide infrared photodiodes with >35 GHz bandwidth and phototransistors with 50 AW-1 response". *Optics Express* **17**.7 (2009), pp. 5193–5204. DOI: 10.1364/OE.17.005193 (cit. on pp. 259, 264).

[8] J. K. Doylend, P. E. Jessop, and A. P. Knights. "Silicon photonic resonator-enhanced defect-mediated photodiode for sub-bandgap detection". *Optics Express* **18**.14 (2010), pp. 14 671–14 678 (cit. on pp. 259, 264).

[9] Jason J. Ackert, Abdullah S. Karar, Dixon J. Paez, *et al.* "10 Gbps silicon waveguide-integrated infrared avalanche photodiode". *Optics Express* **21**.17 (2013), pp. 19 530–19 537. DOI: 10.1364/OE.21.019530 (cit. on pp. 259, 264).

[10] Richard R. Grote, Kishore Padmaraju, Brian Souhan, *et al.* "10 Gb/s Error-free operation of all-silicon ion-implanted-waveguide photodiodes at 1.55". *IEEE Photonics Technology Letters* **25**.1 (2013), pp. 67–70 (cit. on pp. 259, 264).

[11] Jason J. Ackert, Abdullah S. Karar, John C. Cartledge, Paul E. Jessop, and Andrew P. Knights. "Monolithic silicon waveguide photodiode utilizing surface-state absorption and operating at 10 Gb/s". *Optics Express* **22**.9 (2014), pp. 10 710–10 715 (cit. on p. 259).

[12] Tsung-Yang Liow, Kah-Wee Ang, Qing Fang, *et al.* "Silicon modulators and germanium photodetectors on SOI: monolithic integration, compatibility, and performance optimization". *IEEE Journal of Selected Topics in Quantum Electronics* **16**.1 (2010), pp. 307–315 (cit. on pp. 259, 276, 280, 281).

[13] R. Ichikawa, S. Takita, Y. Ishikawa, and K. Wada. "Germanium as a material to enable silicon photonics". *Silicon Photonics II*. Ed. by David J. Lockwood and Lorenzo Pavesi. Vol. 119. Topics in Applied Physics. Springer Berlin Heidelberg, 2011, pp. 131–141. ISBN: 978-3-642-10505-0. DOI: 10.1007/978-3-642-10506-7_5 (cit. on p. 259).

[14] Ari Novack, Mike Gould, Yisu Yang, *et al.* "Germanium photodetector with 60 GHz bandwidth using inductive gain peaking". *Optics Express* **21**.23 (2013), pp. 28 387–28 393. DOI: 10.1364/OE.21.028387 (cit. on pp. 259, 275).

[15] M. C. Teich and B. E. A. Saleh. *Fundamentals of Photonics*. Canada, Wiley Interscience (1991), p. 3 (cit. on pp. 260, 268).

[16] S. M. Sze and K. K. Ng. *Physics of Semiconductor Devices*. Wiley Interscience, 2006 (cit. on p. 261).

[17] Sheila Prasad, Hermann Schumacher, and Anand Gopinath. *High-speed Electronics and Optoelectronics: Devices and Circuits*. Cambridge University Press, 2009 (cit. on p. 261).

[18] L. Vivien, A. Polzer, D. Marris-Morini, *et al.* "Zero-bias 40Gbit/s germanium waveguide photodetector on silicon". *Optics Express* **20** (2012), pp. 1096–1101. DOI: 10.1364/OE.20.001096 (cit. on pp. 262, 272).

[19] Kah-Wee Ang, Joseph Weisheng Ng, Guo-Qiang Lo, and Dim-Lee Kwong. "Impact of field-enhanced band-traps-band tunneling on the dark current generation in germanium pin photodetector". *Applied Physics Letters* **94**.22 (2009), p. 223 515 (cit. on p. 262).

[20] Mitsuru Takenaka, Kiyohito Morii, Masakazu Sugiyama, Yoshiaki Nakano, and Shinichi Takagi. "Dark current reduction of Ge photodetector by GeO_2 surface passivation and gas-phase doping". *Optics Express* **20**.8 (2012), pp. 8718–8725 (cit. on p. 263).

[21] Edward Palik. *Handbook of Optical Constants of Solids*. Elsevier, 1998 (cit. on p. 265).

[22] A. P. Knights, D. F. Logan, P. E. Jessop, *et al.* "Deep-levels in silicon waveguides: a route to monolithic integration". *Photonics Conference (PHO), 2011 IEEE*. IEEE. 2011, pp. 461–462 (cit. on p. 264).

[23] Alexander W. Fang, Richard Jones, Hyundai Park, *et al.* "Integrated AlGaInAs-silicon evanescent racetrack laser and photodetector". *Optics East 2007*. International Society for Optics and Photonics. 2007, 67750P (cit. on p. 264).

[24] Govind P Agrawal. *Fiber-optic Communication Systems*. Vol. 1. 1997 (cit. on p. 267).

[25] Tom Baehr-Jones, Ran Ding, Ali Ayazi, *et al.* "A 25 Gb/s silicon photonics platform". *arXiv:1203.0767v1* (2012) (cit. on p. 267).

[26] Amit Khanna, Youssef Drissi, Pieter Dumon, *et al.* "ePIX-fab: the silicon photonics platform". *SPIE Microtechnologies*. International Society for Optics and Photonics. 2013, 87670H (cit. on p. 267).

[27] A. Mekis, S. Abdalla, D. Foltz, *et al.* "A CMOS photonics platform for high-speed optical interconnects". *Photonics Conference (IPC)*. IEEE. 2012, pp. 356–357 (cit. on p. 267).

[28] Yi Zhou, Masaaki Ogawa, Xinhai Han, and Kang L. Wang. "Alleviation of Fermi-level pinning effect on metal/germanium interface by insertion of an ultrathin aluminum oxide". *Applied Physics Letters* **93**.20 (2008), p. 202105 (cit. on p. 274).

[29] A. Dimoulas, P. Tsipas, A. Sotiropoulos, and E. K. Evangelou. "Fermi-level pinning and charge neutrality level in germanium". *Applied Physics Letters* **89**.25 (2006), p. 252110 (cit. on p. 274).

[30] R. R. Lieten, V. V. Afanasev, N. H. Thoan, *et al.* "Mechanisms of Schottky barrier control on n-type germanium using Ge_3N_4 interlayers". *Journal of the Electrochemical Society* **158**.4 (2011), H358–H362 (cit. on p. 274).

[31] Y. Painchaud, M. Poulin, F. Pelletier, *et al.* "Silicon-based products and solutions". *Proc. SPIE*. 2014 (cit. on p. 275).

8 Lasers

This chapter discusses one of the most challenging aspects of silicon photonics, namely the laser. It is highly desirable to have a silicon-compatible material that can provide optical emission and optical gain, for light sources (lasers, LEDs) and for on-chip optical amplifiers. Silicon is an indirect-band semiconductor, hence very inefficient at light generation. The common semiconductors used to make lasers at wavelengths where silicon is transparent ($> 1.1\ \mu m$), such as InP-based compounds, have a crystal-lattice constant that is much larger than silicon, hence are difficult to grow on silicon. Germanium is the closest-matched material and has a smaller bandgap than silicon; however, it is also an indirect-band semiconductor.

Present-day multi-project wafer (MPW) foundries (see Section 1.5.5) do not provide monolithic or hybrid-integrated lasers, and users rely on external lasers. While edge and grating couplers have both seen improvements in coupling efficiency, the lack of an on-chip source limits the potential applications of these chips. Laser integration is not widely available, and in turn the design of lasers for silicon photonics is an evolving research area. While there are design methodologies for the various laser-integration approaches described below, these depend on the type of approach, hence silicon photonic laser design is not a typical "text-book" topic.

This chapter describes the challenges associated with integrating lasers and optical amplifiers on the silicon photonics platform. It begins with the easiest method of getting light on the chip – namely using external lasers. We then discuss approaches with increasing level of difficulty: co-packaging, epitaxial bonding (hybrid lasers), monolithic growth, and germanium lasers.

8.1 External lasers

One approach for silicon photonic systems is to consider the laser as an optical power supply [1], similar to electrical power supplies, both delivering a constant amplitude. There are several advantages for keeping the laser off-chip.

(1) Thermal management is simplified by keeping the laser off-chip. Lasers are typically 10%–30% efficient, hence an on-chip laser would be a contributing heat source; the heat output is 3–10 times higher than the optical power of the laser itself. The laser may require temperature stabilization, particularly if a specific

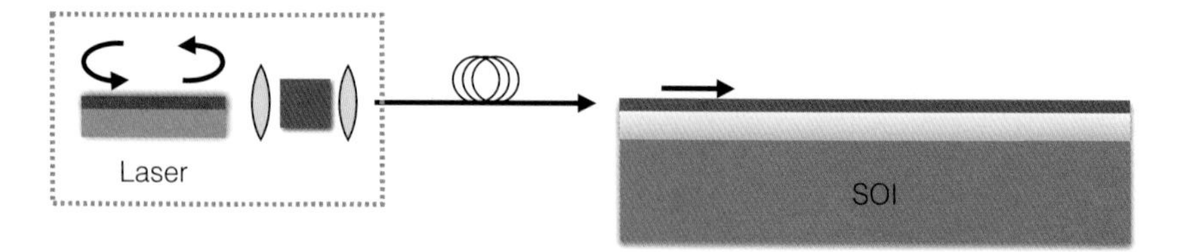

Figure 8.1 Externally packaged semiconductor laser. Laser includes optical isolator and lenses to couple the light into a polarization-maintaining single-mode optical fibre. The laser is coupled to the silicon photonic chip using edge or grating couplers.

wavelength is required; this would require cooling the entire silicon photonic chip, which further adds to the thermal and power budget.

(2) Heat-sinking can be more effective when implemented off-chip, leading to improved laser performance. The specific challenge in silicon photonics is the buried oxide, which serves as a thermal isolator.

(3) Mature III–V laser fabrication offers highly efficient lasers, with high output power, and a high yield, that are commercially available.

(4) Concerns about the manufacturability and compatibility of III–V materials and lasers in a CMOS foundry, as well as yield and reliability concerns.

(5) Multi-wavelength laser sources, available as external components, can be used to directly feed a WDM silicon photonic system such as that described in Section 13.1.2.

(6) Optical isolation can be built-in to the laser package to ensure stable operation. For on-chip lasers, there is the challenge of either building on-chip optical isolators or making the laser feedback insensitive and minimizing optical feedback.

(7) The laser light can be delivered to the silicon photonic chip by using a polarization-maintaining fibre.

When investigating which option to pursue, a cost analysis should be performed to compare (1) manufacturing the laser and silicon photonic chips separately in their own optimized facilities, and integrating them after; or (2) having a facility that can manufacture both parts simultaneously.

The integration of the laser into the system can be either via optical fibre, as illustrated in Figure 8.1, or by using a co-packaging approach (Section 8.3). The fibre-connected approach has been used to develop commercial products. For example, TeraXion's coherent receiver uses an external laser as the local oscillator to mix with the received signals [2].

8.2 Laser modelling

Next, we describe a common method of modelling semiconductor lasers – via the phenomenological laser rate equation model. We also discuss key performance parameters.

To build a laser requires a resonator (cavity) such as a Fabry–Perot cavity constructed with Bragg gratings (Section 4.5.6) or a ring resonator (Section 4.4). The cavity can be described by its quality factor, or its photon lifetime. The photon lifetime of the cavity is related to the total cavity losses, α_{tot}, and to the cavity quality factor, Q, as follows:

$$\tau_p = \frac{n_g}{\alpha_{\text{tot}} c} = \frac{Q}{\omega}, \tag{8.1}$$

where ω is the optical frequency.

We require a mechanism for the light to escape (e.g. mirror reflectivity less than 100%). This necessarily contributes an additional source of loss for the optical gain material. Thus, the total cavity loss has two contributions, $\alpha_{\text{tot}} = \alpha_m + \alpha_i$, where α_m are the output coupling losses (transmission through the mirrors), and α_i are the internal losses.

A primary design consideration is to determine the optical gain required to obtain lasing, and is called the material threshold gain, g_{th}. It is found by the condition that the round-trip modal optical gain in a cavity (here a Fabry–Perot cavity) is equal to unity, or simply that gain = loss:

$$r_1 r_2 e^{(\Gamma g_{th} - \alpha_{\text{tot}})L} = 1, \tag{8.2}$$

where $r_{1,2}$ are the mirror field reflectivities, L is the cavity length, and Γ is the confinement factor, which describes the fraction of the light inside the optical gain material. A typical semiconductor laser cavity has a modal threshold gain requirement of $\Gamma g_{th} =$ 10–100 cm^{-1}.

The laser rate equations describe the number of photons inside the laser, S, as a function of the electrical current injection, I, via differential equations that couple the photons and carriers (usually electrons, N). These equations can be solved analytically or numerically, for the steady-state optical output power, or in the time domain when the current is directly modulated:

$$\frac{dN}{dt} = \frac{I}{q} - \frac{N}{\tau_s} - G \cdot S \tag{8.3}$$

$$\frac{dS}{dt} = \left(G - \frac{1}{\tau_p}\right) \cdot S + \beta \frac{N}{\tau_s}. \tag{8.4}$$

The laser cavity is described by a photon lifetime τ_p; the carriers have a lifetime in the laser's active region τ_s; the optical gain is G; a small portion, β, of the spontaneous emission is coupled into the lasing optical mode; η_i is the carrier injection efficiency; and q is the charge of the electron. These equations are for S and N as quantities (e.g. 10^5 photons in the cavity). Alternatively, they can be expressed as densities (see e.g. Reference [3]).

The simple phenomenological model for the optical gain for the single mode laser, G, is

$$G = \frac{G_o(N - N_o)}{1 + \epsilon S} = \frac{B\Gamma c}{V n_g} \frac{N - N_o}{1 + \epsilon S}, \tag{8.5}$$

where the parameters are the material differential gain, B, the carrier density at transparency N_o, the optical confinement factor, $\Gamma = V_{\text{active}}/V_{\text{cavity}}$, the laser's active volume, $V = V_{\text{active}}$, the optical mode volume, V_{cavity}, the group velocity, n_g, and a gain compression factor, ϵ (the optical gain reduces for high optical powers). The gain parameters for different semiconductors are commonly available in textbooks and journal publications.

The light output power from a laser is:

$$P_{out} = \frac{SV_{\text{cavity}}h\nu}{\tau_p}\frac{\alpha_m}{\alpha + \alpha_m} = \eta_i \frac{\alpha_m}{\alpha + \alpha_m}\frac{h\nu}{q}(I - I_{th}). \tag{8.6}$$

The external quantum slope efficiency is:

$$\eta_{ext} = \eta_i \frac{\alpha_m}{\alpha + \alpha_m}. \tag{8.7}$$

The efficiency in converting electric power to optical power is known as the wall-plug efficiency:

$$\eta_{\text{wall-plug}} = \frac{\text{optical power output}}{\text{electrical power input}} = \frac{P_{out}}{I \cdot V}. \tag{8.8}$$

The external quantum efficiency, namely the photon output rate versus the carrier injection rate, is

$$\eta = \frac{P}{I}\frac{q}{h\nu}. \tag{8.9}$$

Typical III–V semiconductor lasers today have efficiencies of 10%–30%.

Advanced models

The simple model described above assumes that the laser is an optically isolated zero-dimensional object, namely that it can be described by single quantities for the carrier and photon densities. There are several limitations to this model, and more sophisticated models will generally be required to model the laser behaviour for integrated silicon photonic systems. The following are a few important modelling considerations, most of which are addressed by commercial optoelectronic TCAD software packages (see Section 2.3).

- Thermal: lasers operate at very high carrier densities and thus need to be operated at low-enough temperatures to ensure a long lifetime. This requires thermal modelling to ensure adequate heat sinking. The laser behaviour is also highly temperature dependent, affecting the optical gain (gain is generally reduced at high temperatures) and the index of refraction (leading to a temperature-dependent lasing wavelength).
- Optical feedback: lasers are sensitive to optical feedback, and thus are generally manufactured with optical isolators. Typically, it is necessary to maintain less than −30 dB of optical feedback in order to achieve laser power and wavelength stability [4–6]. An active area of research is designing silicon photonic-specific lasers that are less sensitive to optical feedback [7], in order to reduce the feedback sensitivity, and thus enable isolator-free operation. Modelling thus

needs to predict the optical feedback from the photonic circuit, as well as predict the laser behaviour in the presence of this feedback.

- Multiple optical interfaces in the cavity: these will lead to non-uniform spatial field profiles, which will affect the carrier densities and gain profiles.
- Long cavity: particularly for external cavity lasers where the reflectors are in the silicon, dynamics (time-domain) need to be considered as these may impact laser stability and noise.
- Laser linewidth and noise: certain applications such as coherent communications require narrow linewidth lasers. The model needs to simulate this, particularly in the presence of optical feedback.
- Multiple optical modes: external cavities are used to stabilize the laser wavelength, and this needs to be included in the model. This is necessary to ensure that the laser is operating at a single optical wavelength (rather than multi-mode).
- Optical injection locking can also be included [8].

8.3 Co-packaging

8.3.1 Pre-made laser

The approach developed and commercialized by Luxtera involves a laser micro-package that is actively aligned and attached to the silicon photonic chip using grating couplers [9]. The micro-packaged lasers are fabricated in a MEMS process on a silicon wafer and contain numerous lasers on one wafer. This allows the lasers to be wafer-level tested prior to dicing. The laser micro-package contains a laser diode, ball lens, optical isolator, corner reflector, and electrical contacts. The light is coupled to the silicon photonic chip using grating couplers, as illustrated in Figure 8.2a.

Using self-alignment structures, passive alignment of a laser ceramic sub-mount micro-package onto silicon photonic grating couplers was recently demonstrated [10].

Another approach is to edge-couple the lasers to the silicon chip. This approach is illustrated in Figure 8.3a.

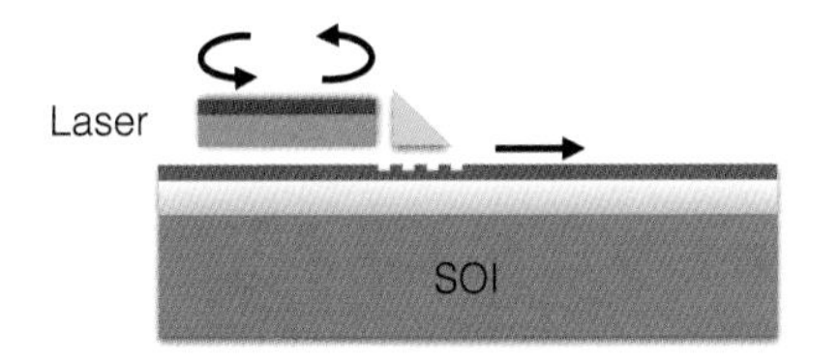

(a) Laser assembly includes a corner reflector, a lens, and possibly an isolator.

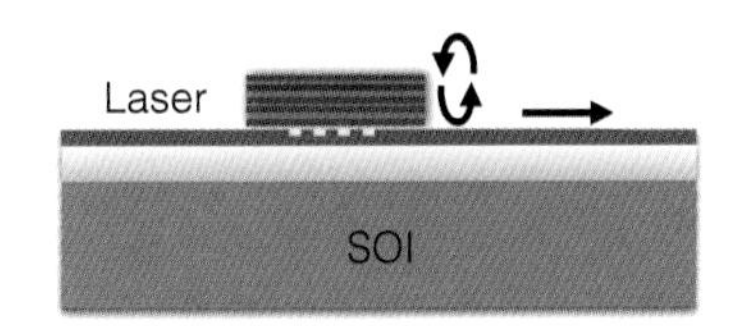

(b) Co-packaged vertical cavity surface emitting laser (VCSEL) array.

Figure 8.2 Co-packaged lasers for silicon photonics, using grating couplers. Optical interfaces between the lasers and the silicon photonic chip are either grating couplers or edge couplers.

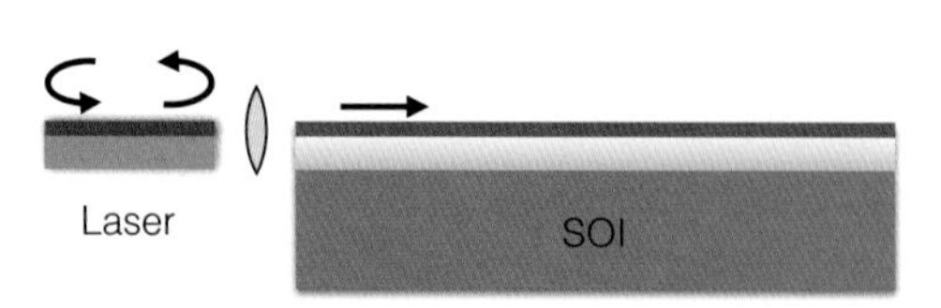

(a) Interfaced using a lens assembly, a waveguide, or directly butt-coupled.

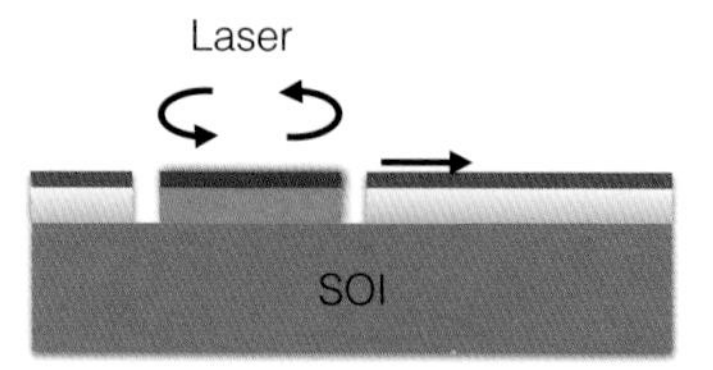

(b) Recessed in an etched pit on the silicon photonic chip; using either a waveguide or directly butt-coupled.

Figure 8.3 Co-packaged lasers for silicon photonics, using edge couplers. Optical interfaces between the lasers and the silicon photonic chip are either grating couplers or edge couplers.

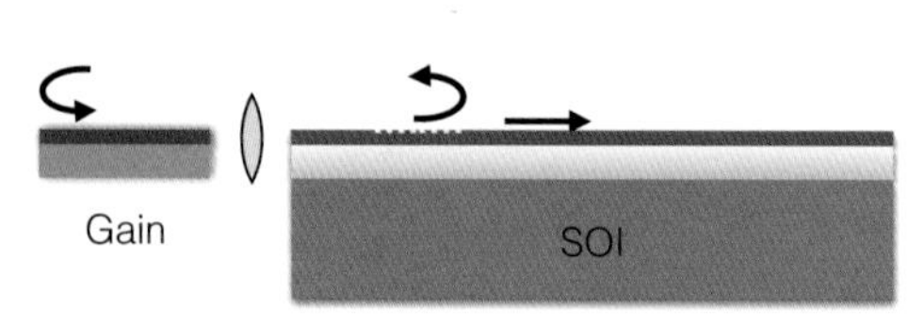

(a) Interfaced using an edge coupler or grating coupler.

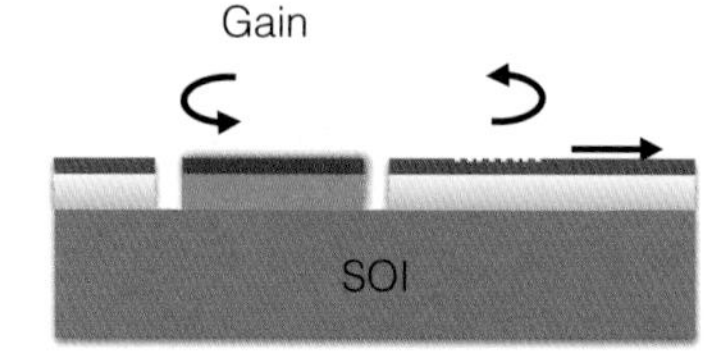

(b) Gain material recessed in an etched pit on the silicon photonic chip. Interfaced using an edge coupler, with a waveguide or directly butt-coupled.

Figure 8.4 External cavity lasers, co-packaged with a silicon photonic chip. The laser cavity can be implemented using one reflector in the III–V material, while the other is implemented in the silicon on insulator wafer. The reflector in the SOI can be based on reflective ring resonators [11, 12], reflective disk resonators [13], Bragg gratings (Section 4.5), or other approaches.

8.3.2 External cavity lasers

In this approach, a III–V gain chip is optically coupled to an on-chip silicon photonic reflector [11, 12, 14–17]. This creates an external cavity configuration, with the gain on one chip (reflective semiconductor optical amplifier, RSOA), and the optical feedback on the silicon photonic chip. The advantage of this approach is that the silicon photonic chip can control the laser properties (e.g. wavelength) using on-chip electrical controls. Since the optical feedback is provided on-chip, the light is already present on the chip and can be used for useful functionality. The insertion loss from the gain medium to the silicon chip needs to be low; since this loss is part of the cavity, a large loss will increase the laser threshold and decrease the efficiency. Edge-coupled insertion losses of 1 dB would be comparable to what has been achieved in the tapers used in hybrid lasers, except that only one coupler is needed per laser, whereas two tapers are needed in the hybrid laser. The optical feedback can be defined by using ring resonators [11, 12] or Bragg gratings, for example. This is illustrated in Figure 8.4a.

The RSOA can also be flip-chip bonded onto the silicon photonic chip to form an external cavity defined in the silicon [16]. In Reference [16], the oxide was removed and metal contacts were deposited on the silicon which improves the heat sinking capability. The laser was edge coupled to the silicon photonic waveguides.

8.3.3 Etched-pit embedded epitaxy

An interesting approach developed by Skorpios takes unprocessed III–V epitaxy and incorporates it in etched sites on an SOI wafer [18]. The end result is a planar silicon wafer with III–V material embedded within, and provides the designer with the flexibility of defining the optical cavities and waveguides in either or both the InP and SOI materials, as illustrated in Figure 8.3b and Figure 8.4b. The fabrication process involves a standard silicon photonics front-end fabrication flow to define the waveguides and modulators. Pits for the III–V epitaxy are etched down to the silicon substrate. The oxide is removed in order to improve the thermal conductivity and to match the height of the III–V gain region with the silicon waveguide. The III–V materials are metal bonded and the InP substrate is removed to leave a planar wafer. Metal bonding offers the following advantages: higher thermal conductivity; provides a bottom electrical contact; and alleviates lattice and thermal expansion coefficient mismatch issues between the silicon and the III–V material. Several fabrication steps are performed to fill the regions between the III–V and the SOI wafer, to define the waveguides in the III–V material the III–V-to-SOI coupling region, and to hermetically seal the wafer using silicon dioxide. Standard back-end processing is used for the electrical interconnects. The laser achieved 8 mW of output power at 250 mA; this represents an efficiency of 4% using the definition in Equation (8.9).

An etched pit in the silicon photonic chip to hold another chip was also pursued for CMOS electronics integration [19]. A planar surface was beneficial for subsequent lithography of the metal interconnects, to connect the optical circuits with the electronics on separate chips.

8.4 Hybrid silicon lasers

The hybrid laser is a heterogeneous integration approach of bonding III–V materials such as InP onto the silicon photonic platform [19–21]. This approach enables the design of lasers, semiconductor optical amplifiers, detectors, and amplitude and phase modulators, using III–V materials. Hybrid integration is presently the most successful method for making lasers integrated on silicon. The combination of the two materials brings to the silicon platform the advantage of the high optical gain and efficient light emission available in III–V materials.

The fabrication process begins with the typical SOI fabrication, including the etches for waveguides, and optionally the doping for modulators [22, 23] and growth of the Ge detector. This is followed by bonding unprocessed III–V materials (die or wafer) onto the silicon photonic die or wafer. This is done in the back-end process of the CMOS foundry, namely, the high-temperature processing (e.g. anneal) of the silicon photonics is performed prior to bonding. Finally, the back-end metal interconnects are added, both

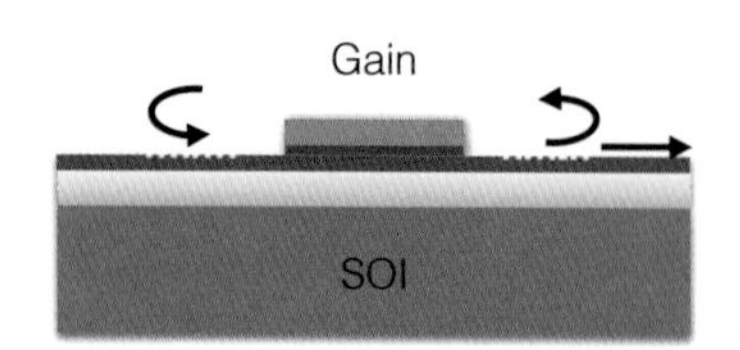

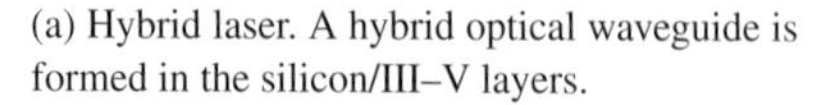

(a) Hybrid laser. A hybrid optical waveguide is formed in the silicon/III–V layers.

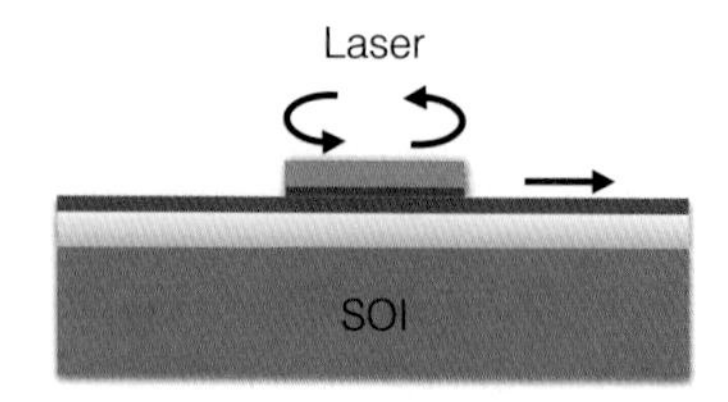

(b) Bonded laser. The laser cavity is implemented in the III–V material. The light is coupled to the silicon evanescently [29].

Figure 8.5 Hybrid silicon lasers, where the gain material is bonded to the silicon waveguide. The laser cavity can be implemented either in the III–V material or in the SOI material. Light is evanescently coupled between the two materials.

for the III–V and silicon contacts. Bonding can be achieved by several means including molecular wafer bonding [19, 24], or adhesive bonding [25–28].

There are two main approaches for the design of the hybrid waveguide, regarding the coupling between the silicon and the III–V material. The first, shown in Figure 8.5a, is based on evanescent coupling, where most of the light is guided by the silicon waveguide, and a small amount of light is evanescently coupled to the III–V material. In this configuration, the optical cavity is defined in the silicon using a variety of reflector structures such as Bragg gratings and ring resonators. The second approach is to build the cavity in the III–V material and to evanescently couple the laser to the silicon waveguides, as illustrated in Figure 8.5b. It is possible to have the light mainly confined in the III–V material and coupling is performed using tapered waveguides to a cavity defined in silicon [27]. Another method is to mainly confine the light in the III–V material, but evanescently couple to a silicon Bragg grating waveguide thereby resulting in a distributed feedback (DFB) laser [25].

A major advantage of the evanescent coupling approach is that the laser cavity is defined using the same masks and lithography that are used for the rest of the photonic circuit. Thus, the alignment between the laser cavity and the circuit is ensured. Furthermore, the III–V material is an unprocessed wafer and does not need to be carefully aligned to the silicon. The challenge, however, is in patterning the III–V material to define the tapers for the transition region between the hybrid waveguide and the silicon waveguide, to ensure low optical loss. The design of the transition from the hybrid InP–Si waveguide to the silicon waveguide can be modelled by using 3D FDTD. Achieved taper losses range from 0.3 dB per taper (passive only) to 1–2 dB (with active material). One challenge is that part of the taper is built using III–V material that possibly may be under-pumped or may have surface recombination. Another challenge is regarding the III–V etch resolution and alignment, in part since the III–V material is relatively thick. This results in the taper tips being relatively large (e.g., 400 nm in Reference [27]) and thus contributing to insertion loss between the III–V–silicon and silicon-only regions. Precise alignment is also required to minimize insertion loss.

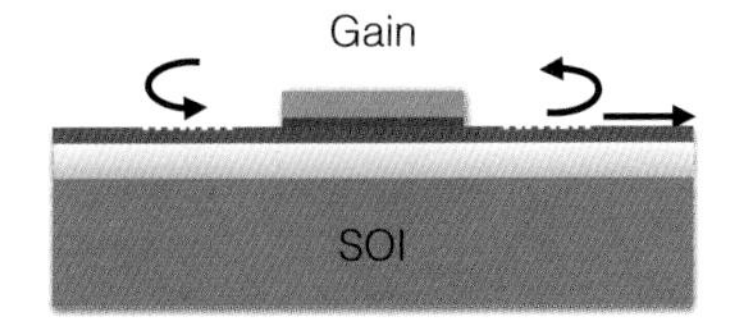

(a) Gain material on top of the silicon waveguide.

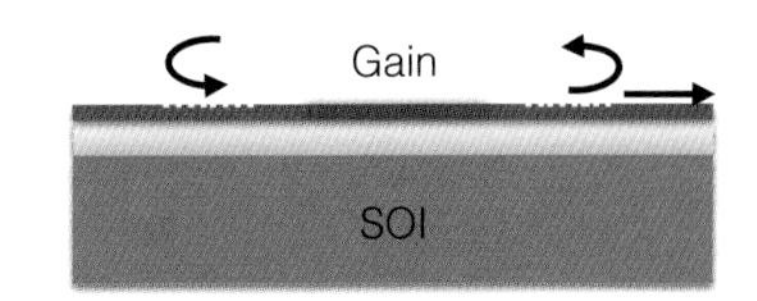

(b) Gain material embedded within the silicon waveguide.

Figure 8.6 Monolithic integrated laser, where the gain material (e.g. InP, Ge, etc.) is epitaxially grown directly on the silicon on insulator wafer. A hybrid optical waveguide is formed in the silicon–III–V layers. The laser cavity can be implemented using a variety of reflectors implemented in the silicon on insulator wafer.

8.5 Monolithic lasers

Compared to the integration using bonding approaches, monolithic integration of an optical gain material, such as III–V (InP, GaAs) or Ge, would be highly advantageous in terms of high density, system reliability, increased functionality, and the use of well-established low-cost silicon fabrication techniques (see Figure 8.6). We discuss several approaches here, which have not reached the same level of maturity as the more established approaches described above.

It is useful to review the history in the progress of III–V semiconductor lasers [30], specifically in terms of the threshold current density. Prior to the advent of the double heterostructure laser, the threshold densities of semiconductor lasers in the 1960s were in the region of 10–100 kA/cm^2. The double heterostructure provided a means to create a population inversion purely by electron and hole injection from two wider-bandgap materials. Thus the optical-gain material does not need to be doped, hence reducing the optical absorption. The volume is also significantly reduced by the confinement of carriers in the thin gain region, hence a lower current is required to maintain the carrier density necessary for optical gain. After many years of development, this led to a threshold current density of 4300 A/cm^2 at room temperature in the late 1960s. In the 1980s, quantum wells provided further reduction down to 160 A/cm^2 [31], and lower. This was made possible by advances in semiconductor growth using molecular beam epitaxy. Finally, even smaller gain materials were fabricated, quantum dots, and this reduced the threshold current density even further, e.g. 16 A/cm^2 in 2000 [32]. Decades of research were required to develop the semiconductor laser to a practical and efficient device, where the threshold current density was steadily reduced down from $>10^5$ A/cm^2 and corresponding currents in the amperes, to $<10^3$ A/cm^2 and very small devices with milliampere currents.

8.5.1 III–V Monolithic growth

The direct epitaxial growth of III–V materials on silicon has been a topic of interest for over 20 years [33], and there have been numerous challenges. There have been several demonstrations of III–V materials epitaxially grown on silicon wafers, including demonstrations of Si-based lasers [34–37].

The challenges lie in material property mismatches among different materials, such as lattice constant and thermal expansion, which lead to high-density defects, cracking, and delamination. The most challenging has been the introduction of threading dislocations in the III–V material due to the large lattice-constant mismatch between the silicon and the III–V material. There have been several approaches at reducing dislocations and improving material quality. The first is a body of work using germanium. Germanium is an ideal intermediary material between GaAs (an important III–V material) and silicon because of its complete miscibility with silicon and the close lattice match between bulk Ge and GaAs at room temperature. Germanium is currently routinely used in CMOS processors and silicon photonics, hence already CMOS compatible. The first step is to create a Ge virtual substrate on Si, and methods have included direct Ge layer deposition [38]; a graded buffer layer [34, 39]; Ge condensation [40]; and the aspect ratio trapping (ART) technique [36]. In the ART technique, the threading dislocations are trapped at the vertical sidewalls of a thin layer of 400 nm-wide trenches; a laser with 3 kA/cm^2 threshold current density was achieved [36].

Another approach at reducing dislocations is to use quantum dots as the active material [35], in which a laser with 900 A/cm^2 threshold current density was achieved. Quantum dots, owing to their small size, are less sensitive to material defects [41]. Furthermore, AlAs can be used as a nucleation layer to further improve the growth quality of the GaAs buffer layer on silicon substrates for subsequent quantum-dot growth. A 650 A/cm^2 threshold current density was demonstrated for an InAs/GaAs quantum-dot laser epitaxially grown in silicon operating at 1.3 μm [37].

These efforts are very impressive and provide a possible means of integrating III–V materials in silicon. However, a challenge with this approach is the compatibility with the CMOS process. First, the epitaxy is grown using MBE or MOCVD techniques, which typically require high temperatures, 400–600 °C in Reference [37]. The epitaxy could in principle be done on the SOI wafer prior to the CMOS front-end processing, if the materials are deemed compatible. Or, selective-area growth could be performed on processed wafers, most likely after the front-end processing, if growth temperatures can be kept low enough.

8.5.2 Germanium lasers

Germanium has been proposed as a CMOS compatible gain medium though its photoemission efficiency is hindered by its indirect band gap. Recently, the first electrically driven laser using a germanium gain medium on silicon was demonstrated [42]. This section details the challenges of this approach.

Although germanium is an indirect-band semiconductor (Figure 8.7a), it can be made to operate in direct-band transitions, and thus provide a means to achieve stimulated emission. This is accomplished under tensile strain [43], which causes a reduction of the conduction band minimum (Γ point in the band-structure) relative to the indirect-band conduction minimum (L point); see Figure 8.7b. Although it is still indirect under

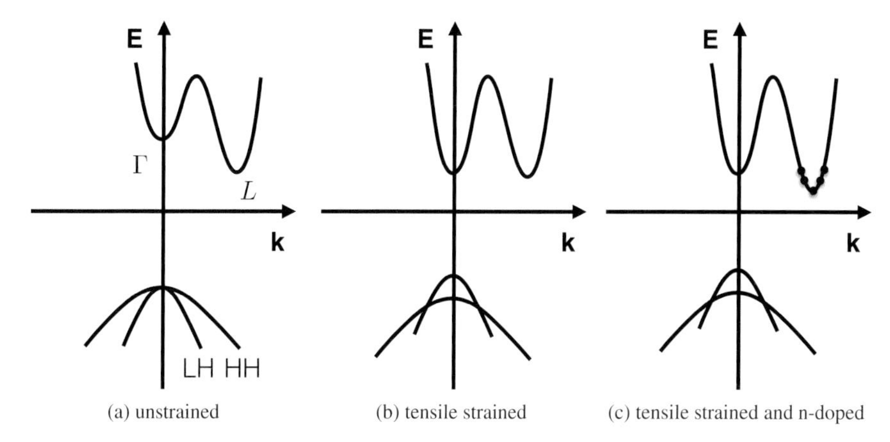

(a) unstrained (b) tensile strained (c) tensile strained and n-doped

Figure 8.7 Band structure diagram for germanium, after Reference [44]. Band structure electron and hole energy (E), and crystal momentum (k). Band edges supporting direct (Γ) and in-direct (L) transitions. The valence band has heavy-hole (HH) and light-hole (LH) bands.

typically achievable strains (e.g. $<2\%$), doping with n-type material causes a filling of the lowest conduction band (L), leaving the direct band (Γ) available for direct stimulated emission transitions, which are necessary to have a population inversion for lasing (Figure 8.7c).

Understanding the conditions necessary to obtain optical gain in germanium are critical to guide experiments and for laser design [43–45]. W. W. Chow presented a comprehensive model for optical gain in germanium [44]. The model is semiclassical, and includes carrier kinetic energy, carrier–light interaction, Coulomb interaction between carriers, many-body effects, and optical absorption from free carriers. The simulations predict the optical gain spectra for different strain and doping densities, for a given carrier density. For example, a gain of $300\,\mathrm{cm}^{-1}$ is predicted for a 0.2% tensile strain and doping density of 2.5×10^{19} cm^{-3}, and a carrier density of 8×10^{18} cm^{-3}. For comparison, the gain in un-doped un-strained bulk GaAs reaches similar values of optical material gain at carrier densities 2–3 times lower. This number is also small compared to a typical semiconductor quantum-well optical gain under similar conditions, e.g. $3000\,\mathrm{cm}^{-1}$ in GaAs. Next, the net optical gain is found by subtracting the free-carrier optical absorption losses. Finally, for a particular carrier density, the current density is found, which is due to spontaneous emission, Schockley–Read–Hall, and Auger scattering. This provides relationships between the net optical gain versus the current density. The results predict that a net gain of several hundred per centimeter is possible for various scenarios. For a 0.2% tensile strain and doping density of 4×10^{19} cm^{-3}, the carrier density for positive gain is about 1000 A/cm^2, which is similar to a typical semiconductor laser. The simulations also predict that the carrier density can be further reduced by a factor of ten for increased strain, e.g. 1.8% strain, down to <100 A/cm^2. These simulations provide guidance that under a large-enough strain, it should be possible to obtain significant optical amplification at reasonable current densities. Recently, a 3.1% strain was demonstrated [46].

A typical semiconductor laser cavity has a threshold gain requirement of 10–100 cm^{-1}. This suggests that the optical confinement factor should be large when considering a bulk germanium gain material, so that the material gain requirement is kept small, as predicted by the model described above. However, the germanium optical gain is quite different than in typical III–V semiconductors, and a recent study has shown that low-loss cavities are not a viable path towards reducing the the laser threshold current and efficiency and that larger cavity losses are desirable for a high slope efficiency [45]. Laser design, and optimized geometries, will thus evolve as the materials, and the optical gain achievable, improve over time.

Given the required material gain for a particular optical resonator, and knowing the material gain versus current density, it is possible to predict the current density at the threshold of the laser. This important figure of merit for semiconductor lasers is the threshold current density, J_{th}, expressed in A/cm^2. As an example, the electrically pumped Ge laser reported in Reference [42] showed a threshold current density of 300 000 A/cm^2, several orders of magnitude higher than that for typical semiconductor lasers. For the laser dimensions given, this corresponds to a current of 0.81 A. The output power reached 1 mW, which corresponds to a slope efficiency of 1 mW/0.81 A = 0.0012 W/A. The external quantum efficiency, calculated using Equation (8.9), is:

$$\eta = \frac{P}{I}\frac{q}{h\nu} = \frac{1\,\mathrm{mW}}{0.81\,\mathrm{A}}\frac{1}{0.8\,\mathrm{eV}} = 0.0015 = 0.15\%. \tag{8.10}$$

Typical III–V semiconductor lasers today have efficiencies of 10%–30%, two orders of magnitude higher.

The comparison between the simulations and published experimental results to date suggests that there is significant room for improvement in the performance of the germanium laser. It will be interesting to see the future evolution of the germanium laser, and see whether similar improvements as in the III–V laser history will be possible, perhaps with similar approaches, such as exploring different materials for heterostructures, larger strain [45, 46], quantum-confinement effects such as in quantum wells [47] and dots [48], etc. It should also be noted that the higher-strain devices are expected to be much more efficient, though also operate at much longer wavelengths [45], e.g. 2–3 μm. Such longer wavelengths may be well suited for short-reach optical interconnects.

8.6 Alternative light sources

A few other options have been researched for light source integration: (1) optically pumped Raman lasers [49]; (2) silicon nano-crystals as a gain material imbedded in the silicon waveguide [50]; and (3) black-body radiation [51]. Namely, a hot piece of silicon waveguide will emit light, which for typical on-chip temperatures has peak emissions at wavelengths longer than 2 μm. This can be used for spectroscopy applications, particularly for long wavelength integrated photonics [52]. Another related topic is surface plasmon-polaritons [53], which may be useful in developing compact lasers integrated with silicon.

8.7 Problem

8.1 Simulate the mode profile of a hybrid silicon laser. Calculate the optical confinement factor.

References

[1] Yurii A Vlasov. "Silicon CMOS-integrated nano-photonics for computer and data communications beyond 100G". *IEEE Communications Magazine, IEEE* **50**.2 (2012), s67–s72 (cit. on p. 295).

[2] Y. Painchaud, M. Poulin, F. Pelletier, *et al.* "Silicon-based products and solutions". *Proc. SPIE.* 2014 (cit. on p. 296).

[3] L. A. Coldren, S. W. Corzine, and M. L. Mashanovitch. *Diode Lasers and Photonic Integrated Circuits.* Wiley Series in Microwave and Optical Engineering. John Wiley & Sons, 2012. ISBN: 9781118148181 (cit. on p. 297).

[4] Roy Lang and Kohroh Kobayashi. "External optical feedback effects on semiconductor injection laser properties". *IEEE Journal of Quantum Electronics* **16**.3 (1980), pp. 347–355 (cit. on p. 298).

[5] R. W. Tkach and Andrew R. Chraplyvy. "Regimes of feedback effects in 1.5-μm distributed feedback lasers". *Journal of Lightwave Technology* **4**.11 (1986), pp. 1655–1661 (cit. on p. 298).

[6] P. Bala Subrahmanyam, Y. Zhou, L. Chrostowski, and C. J. Chang-Hasnain. "VCSEL tolerance to optical feedback". *Electronics Letters* **41**.21 (2005), pp. 1178–1179 (cit. on p. 298).

[7] Laurent Schares, Yoon H. Lee, Daniel Kuchta, Uzi Koren, and Len Ketelsen. "An 8-wavelength laser array with high back reflection tolerance for high-speed silicon photonic transmitters". *Optical Fiber Communication Conference.* Optical Society of America. 2014, Th1C–3 (cit. on p. 298).

[8] Chih-Hao Chang, Lukas Chrostowski, and Connie J. Chang-Hasnain. "Injection locking of VCSELs". *IEEE Journal of Selected Topics in Quantum Electronics* **9**.5 (2003), pp. 1386–1393 (cit. on p. 299).

[9] Peter De Dobbelaere, Ali Ayazi, Yuemeng Chi, *et al.* "Packaging of silicon photonics systems". *Optical Fiber Communication Conference.* Optical Society of America. 2014, W3I–2 (cit. on p. 299).

[10] Bradley Snyder, Brian Corbett, and Peter OBrien. "Hybrid integration of the wavelength-tunable laser with a silicon photonic integrated circuit". *Journal of Lightwave Technology* **31**.24 (2013), pp. 3934–3942 (cit. on p. 299).

[11] Tao Chu, Nobuhide Fujioka, and Masashige Ishizaka. "Compact, lower-power-consumption wavelength tunable laser fabricated with silicon photonic-wire waveguide micro-ring resonators". *Optics Express* **17**.16 (2009), pp. 14 063–14 068 (cit. on p. 300).

[12] Shuyu Yang, Yi Zhang, David W. Grund, *et al.* "A single adiabatic microring-based laser in 220 nm silicon-on-insulator". *Optics Express* **22**.1 (2014), pp. 1172–1180 (cit. on p. 300).

[13] Wei Shi, Han Yun, Wen Zhang, *et al.* "Ultra-compact, high-Q silicon microdisk reflectors". *Optics Express* **20**.20 (2012), pp. 21 840–21 846. DOI: 10.1364/OE.20.021840 (cit. on p. 300).

[14] Nobuhide Fujioka, Tao Chu, and Masashige Ishizaka. "Compact and low power consumption hybrid integrated wavelength tunable laser module using silicon waveguide resonators". *Journal of Lightwave Technology* **28**.21 (2010), pp. 3115–3120 (cit. on p. 300).

[15] Keita Nemoto, Tomohiro Kita, and Hirohito Yamada. "Narrow-spectral-linewidth wavelength-tunable laser diode with Si wire waveguide ring resonators". *Applied Physics Express* **5**.8 (2012), p. 082701 (cit. on p. 300).

[16] Shinsuke Tanaka, Seok-Hwan Jeong, Shigeaki Sekiguchi, *et al.* "High-output-power, single-wavelength silicon hybrid laser using precise flip-chip bonding technology". *Optics Express* **20**.27 (2012), pp. 28 057–28 069 (cit. on p. 300).

[17] A. J. Zilkie, P. Seddighian, B. J. Bijlani, *et al.* "Power-efficient III-V/silicon external cavity DBR lasers". *Optics Express* **20**.21 (2012), pp. 23 456–23 462 (cit. on p. 300).

[18] Timothy Creazzo, Elton Marchena, Stephen B. Krasulick, *et al.* "Integrated tunable CMOS laser". *Optics Express* **21**.23 (2013), pp. 28 048–28 053 (cit. on p. 301).

[19] M. J. R. Heck, J. F. Bauters, M. L. Davenport, *et al.* "Hybrid silicon photonic integrated circuit technology". *IEEE Journal of Selected Topics in Quantum Electronics,* **19**.4 (2013), p. 6100117. DOI: 10.1109/JSTQE.2012.2235413 (cit. on pp. 301, 302).

[20] A. Fang, H. Park, O. Cohen, *et al.* "Electrically pumped hybrid AlGaInAs–silicon evanescent laser". *Optics Express* **14** (2006), pp. 9203–9210 (cit. on p. 301).

[21] B. Ben Bakir, A. Descos, N. Olivier, *et al.* "Electrically driven hybrid Si/III-V Fabry-Perot lasers based on adiabatic mode transformers". *Optics Express* **19** (2011), pp. 10 317–10 325 (cit. on p. 301).

[22] Guang-Hua Duan, Jean-Marc Fedeli, Shahram Keyvaninia, and Dave Thomson. "10 Gb/s integrated tunable hybrid III-V/si laser and silicon mach-zehnder modulator". *European Conference and Exhibition on Optical Communication.* Optical Society of America. 2012, Tu–4 (cit. on p. 301).

[23] Andrew Alduino, Ling Liao, Richard Jones, *et al.* "Demonstration of a high speed 4-channel integrated silicon photonics WDM link with hybrid silicon lasers". *Integrated Photonics Research, Silicon and Nanophotonics.* Optical Society of America. 2010, PDIWI5 (cit. on p. 301).

[24] Di Liang, Gunther Roelkens, Roel Baets, and John E. Bowers. "Hybrid integrated platforms for silicon photonics". *Materials* **3**.3 (2010), pp. 1782–1802. DOI: 10.3390/ma3031782 (cit. on p. 302).

[25] S. Keyvaninia, S. Verstuyft, L. Van Landschoot, *et al.* "Heterogeneously integrated III-V/silicon distributed feedback lasers". *Optics Letters* **38**.24 (2013), pp. 5434–5437. DOI: 10.1364/OL.38.005434. URL: http://ol.osa.org/abstract.cfm?URI=ol-38-24-5434 (cit. on p. 302).

[26] Stevan Stankovic, Richard Jones, Matthew N. Sysak, *et al.* "1310-nm hybrid III–V/Si Fabry–Perot laser based on adhesive bonding". *IEEE Photonics Technology Letters* **23**.23 (2011), pp. 1781–1783 (cit. on p. 302).

[27] Shahram Keyvaninia, Gunther Roelkens, Dries Van Thourhout, *et al.* "Demonstration of a heterogeneously integrated III-V/SOI single wavelength tunable laser". *Optics Express* **21**.3 (2013), pp. 3784–3792 (cit. on p. 302).

[28] S. Keyvaninia, S. Verstuyft, S. Pathak, *et al.* "III-V-on-silicon multi-frequency lasers". *Optics Express* **21**.11 (2013), pp. 13 675–13 683. DOI: 10.1364/OE.21.013675. URL: http://www.opticsexpress.org/abstract.cfm?URI=oe-21-11-13675 (cit. on p. 302).

[29] Joris Van Campenhout, Pedro Rojo Romeo, Philippe Regreny, *et al.* "Electrically pumped InP-based microdisk lasers integrated with a nanophotonic silicon-on-insulator waveguide circuit". *Optics Express* **15**.11 (2007), pp. 6744–6749 (cit. on p. 302).

[30] Zhores Alferov. "Double heterostructure lasers: early days and future perspectives". *IEEE Journal of Selected Topics in Quantum Electronics* **6**.6 (2000), pp. 832–840 (cit. on p. 303).

[31] W. T. Tsang. "Extremely low threshold (AlGa) As graded-index waveguide separate-confinement heterostructure lasers grown by molecular beam epitaxy". *Applied Physics Letters* **40**.3 (1982), pp. 217–219 (cit. on p. 303).

[32] Gyoungwon Park, Oleg B. Shchekin, Diana L. Huffaker, and Dennis G. Deppe. "Low-threshold oxide-confined 1.3-μm quantum-dot laser". *IEEE Photonics Technology Letters* **12**.3 (2000), pp. 230–232 (cit. on p. 303).

[33] R. Fischer, H. Morkoc, D. A. Neumann, *et al.* "Material properties of high-quality GaAs epitaxial layers grown on Si substrates". *Journal of Applied Physics* **60**.5 (1986), pp. 1640–1647 (cit. on p. 303).

[34] O. Kwon, J. J. Boeckl, M. L. Lee, *et al.* "Monolithic integration of AlGaInP laser diodes on SiGe/Si substrates by molecular beam epitaxy". *Journal of Applied Physics* **100** (2006), p. 013103 (cit. on p. 304).

[35] Z. Mi, J. Yang, P. Bhattacharya, and D. L. Huffaker. "Self-organised quantum dots as dislocation filters: the case of GaAs-based lasers on silicon". *Electronics Letters* **42**.2 (2006), pp. 121–123 (cit. on p. 304).

[36] J. Z. Li, J. M. Hydrick, J. S. Park, *et al.* "Monolithic integration of GaAs/InGaAs lasers on virtual Ge substrates via aspect-ratio trapping". *Journal of the Electrochemical Society* **156**.7 (2009), H574–H578 (cit. on p. 304).

[37] A. D. Lee, Qi Jiang, Mingchu Tang, *et al.* "InAs/GaAs quantum-dot lasers monolithically grown on Si, Ge, and Ge-on-Si substrates". *IEEE Journal of Selected Topics in Quantum Electronics* **19**.4 (2013), p. 1901107. DOI: 10.1109/JSTQE.2013.2247979 (cit. on p. 304).

[38] D. Choi, E. Kim, P. C. McIntyre, and J. S. Harris. "Molecular-beam epitaxial growth of III–V semiconductors on Ge/Si for metal-oxide-semiconductor device fabrication". *Applied Physics Letters* **92** (2008), p. 203502 (cit. on p. 304).

[39] M. T. Currie, S. B. Samavedam, T. A. Langdo, C. W. Leitz, and E. A. Fitzgerald. "Controlling threading dislocation densities in Ge on Si using graded SiGe layers and chemical-mechanical polishing". *Applied Physics Letters* **72** (1998), p. 1718 (cit. on p. 304).

[40] H. J. Oh, K. J. Choi, W. Y. Loh, *et al.* "Integration of GaAs epitaxial layer to Si-based substrate using Ge condensation and low-temperature migration enhanced epitaxy techniques". *Journal of Applied Physics* **102** (2007), p. 054306 (cit. on p. 304).

[41] Zetian Mi, Jun Yang, Pallab Bhattacharya, Guoxuan Qin, and Zhenqiang Ma. "High-performance quantum dot lasers and integrated optoelectronics on Si". *Proceedings of the IEEE* **97**.7 (2009), pp. 1239–1249 (cit. on p. 304).

[42] R. Camacho-Aguilera, Y. Cai, N. Patel, *et al.* "An electrically pumped germanium laser". *Optics Express* **20** (2012), pp. 11 316–11 320 (cit. on pp. 304, 306).

[43] J. Liu, X. Sun, D. Pan, *et al.* "Tensile-strained, n-type Ge as a gain medium for monolithic laser integration on Si". *Optics Express* **15** (2007), pp. 11 272–11 277 (cit. on pp. 304, 305).

[44] Weng W. Chow. "Model for direct-transition gain in a Ge-on-Si laser". *Applied Physics Letters* **100**.19 (2012), 191113. DOI: http://dx.doi.org/10.1063/1.4714540 (cit. on p. 305).

[45] Birendra Dutt, Devanand S. Sukhdeo, Donguk Nam, *et al.* "Roadmap to an efficient germanium-on-silicon laser: strain vs. n-type doping". *IEEE Photonics Journal, IEEE* **4**.5 (2012), pp. 2002–2009 (cit. on pp. 305, 306).

[46] M. J. Sess, R. Geiger, R. A. Minamisawa, *et al.* "Analysis of enhanced light emission from highly strained germanium microbridges". *Nature Photonics* **7**.6 (2013), pp. 466–472 (cit. on pp. 305, 306).

[47] Yan Cai, Zhaohong Han, Xiaoxin Wang, *et al.* "Analysis of threshold current behavior for bulk and quantum well germanium laser structures". *IEEE Journal of Selected Topics in Quantum Electronics* **19**.4 (2013), 1901009 (cit. on p. 306).

[48] Xuejun Xu, Sho Narusawa, Taichi Chiba, *et al.* "Silicon-based light emitting devices based on Ge self-assembled quantum dots embedded in optical cavities". *IEEE Journal of Selected Topics in Quantum Electronics* **18**.6 (2012), pp. 1830–1838 (cit. on p. 306).

[49] Ozdal Boyraz and Bahram Jalali. "Demonstration of a silicon Raman laser". *Optics Express* **12**.21 (2004), pp. 5269–5273 (cit. on p. 306).

[50] L. Pavesi, L. Dal Negro, Ca. Mazzoleni, G. Franzo, and F. Priolo. "Optical gain in silicon nanocrystals". *Nature* **408**.6811 (2000), pp. 440–444 (cit. on p. 306).

[51] M. U. Pralle, N. Moelders, M. P. McNeal, *et al.* "Photonic crystal enhanced narrow-band infrared emitters". *Applied Physics Letters* **81**.25 (2002), pp. 4685–4687 (cit. on p. 306).

[52] Richard Soref. "Toward silicon-based longwave integrated optoelectronics (LIO)". *Integrated Optoelectronic Devices 2008.* International Society for Optics and Photonics. 2008, p. 689809 (cit. on p. 306).

[53] Pierre Berini and Israel De Leon. "Surface plasmon-polariton amplifiers and lasers". *Nature Photonics* **6**.1 (2012), pp. 16–24 (cit. on p. 306).

Part IV

System design

9 Photonic circuit modelling

This chapter describes the process of optoelectronic parameter extraction starting from the geometric parameters in components, followed by their use in circuit simulations. The techniques described can be used for components such as waveguides, couplers, Y-junctions, grating couplers, edge couplers, waveguide couplers, ring resonators, modulators, and filters and photodetectors, as described in Parts II and III. For fixed-cell designs, the components can be simulated once using the appropriate physical solvers and the optoelectronic parameters can be extracted for later use. For parameterized-cell designs, it is necessary to perform parameter sweeps over a range of possible geometries to create lookup tables or validated phenomenological models. We then show examples of how these extracted optoelectronic parameters can be used in photonic circuit simulations.

9.1 Need for photonic circuit modelling

Silicon photonics is a technology enabling large-scale integration of photonic components into photonic circuits. There is increasing interest in photonic integrated circuits in silicon systems for a variety of applications including on-chip and inter-chip communication systems. Broad adoption of silicon photonics technology for circuits and systems requires standardization in the design flow that is similar to what is available for electrical circuit design. This chapter describes methods for modelling optical circuits. The methods presented here are integrated as part of a design methodology that takes advantage of commercial electronic-design automation (EDA) tools for circuit design and layout, as described in Chapter 10.

Numerical methods such as finite-difference time domain (FDTD) and eigen-mode solvers coupled with solutions to optoelectronic equations are workhorses in silicon photonic component-level design. These methods unfortunately do not scale well as the number of components in the photonic circuit increases. Thus, modelling approaches that are computationally efficient yet accurately represent complex nanophotonic devices need to be employed in circuit simulations. In addition, the physical layout can affect the circuit response and needs to be considered. These issues have been addressed by the CMOS electronics industry, and are beginning to be incorporated into silicon photonic design. Several photonic circuit modelling and design approaches are discussed in Section 2.6.

Circuit modelling tools utilize physical-level optoelectronic simulations, and/or experimental data, to build compact models for the optical elements. This integrated approach allows designers to study, for example, the influence of optical feedback from components such as grating couplers on the optical response of the optical circuit, taking into account the physical layout. Such a unified methodology is essential for understanding the performance and designing for future complex silicon photonic systems.

One of the key challenges in the process is the translation from the largely geometric parameters that can be easily extracted from EDA tools and the optoelectronic parameters that are necessary for the simulation of a photonic circuit. For example, after design and layout, properties such as waveguide width, bend radius, gap distance in waveguide couplers, electrical contact positions, and so on, can be extracted easily from EDA tools. However, photonic circuit simulation requires optoelectronic parameters such as effective/group index, dispersion, S parameters, and information about the dependence of the effective index on applied voltage or temperature. These quantities cannot easily be determined from the geometric parameters of the layout, but require a combination of physics based solvers such as eigenmode solvers, FDTD, and electrical device solvers.

9.2 Components for system design

There are two primary design flows for the components available to the system designer.

(1) The *top-down approach* is aimed at system designers. The designer specifies the high-level circuit parameters, known as *design-intent-based parameters*, illustrated in Figure 9.1. In optical circuits, this includes parameters such as the target operating wavelength, the modulation bandwidth, etc. The design methodology enables a high-level design approach by providing appropriate compact models as well as physical implementations (layouts) of components based on these design-intent parameters. The available components can either be discrete cells, or parameterized.

(2) The *bottom-up approach* is used by the component developers. These components are designed by component-level designers using a variety of physical-level parameters (e.g. width, length), and include performance parameters (operating wavelength, bandwidth), which the system designer can choose from. A design of experiment (DOE) approach can be used [1, 2], with the component development flow illustrated in Figure 9.2. These components have both compact models and physical layouts.

These design flows can co-exist, with some components being fixed cells from a library, and others parameterized with design-intent-based parameters.

9.3 Compact models

There are several ways to represent a component in a circuit simulation. Desirable properties for a compact model include the following.

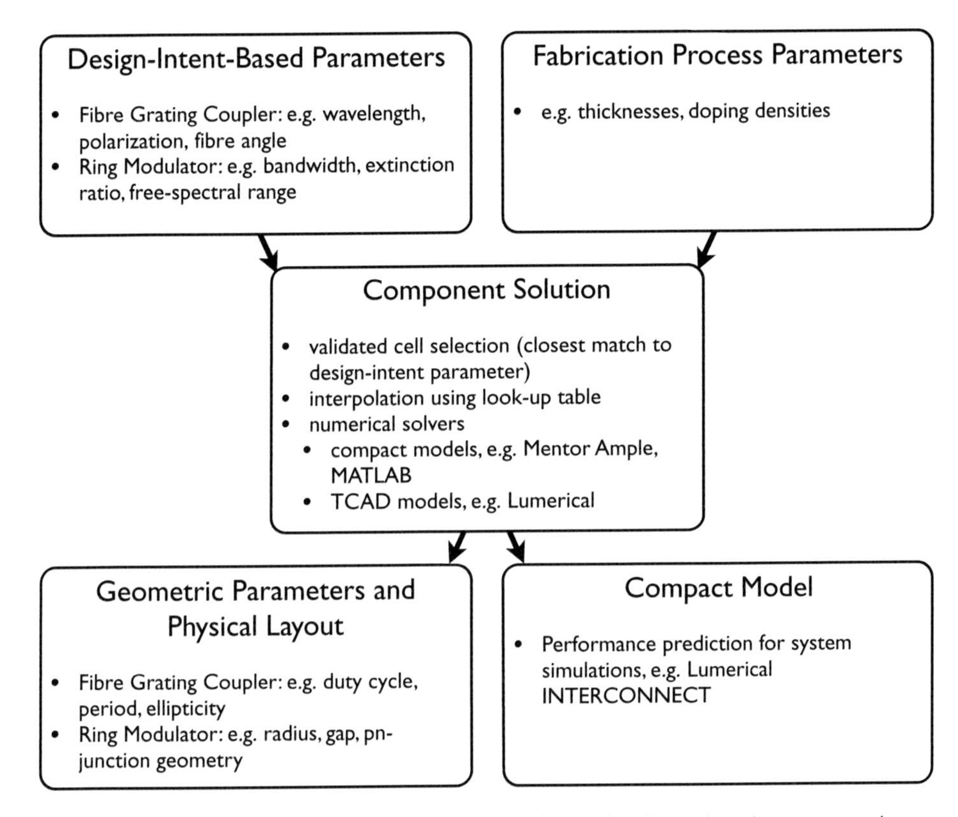

Figure 9.1 Diagram illustrating the methodology for using design-intent-based components in system design.

(1) Accuracy: while a compact model provides enormous computation complexity savings, the circuit model needs to be sufficiently accurate.
(2) Wavelength dependence: this is necessary in order to be able to model a circuit over the wavelength range of interest. In many cases, the wavelength dependence cannot be ignored, such as for a waveguide in a ring resonator, since it is the group index that determines the free-spectral range, see Equations (3.5) and (4.27).
(3) Parameterized versus the geometry: a model that covers a range of physical parameters is necessary to study the circuit dependence on physical parameters, such as waveguide thickness, width, etc. This can be implemented by a collection of single-element compact models accessed via a look-up table (and possibly interpolation), or a multi-dimensional parameterized model.

It should be noted that different levels of model complexity may be required for different applications. For example, for a simple 10/90% optical power splitter, perhaps only the $|t|$ and $|\kappa|$ coefficients of the directional coupler may be required (magnitude only), as found in Section 4.1. For a ring resonator, the phase needs to be included, as described in Section 4.1.2, and the directional coupler can be considered as a "point-coupler" for the ring model in Section 4.4.1. These parameters can be either constants,

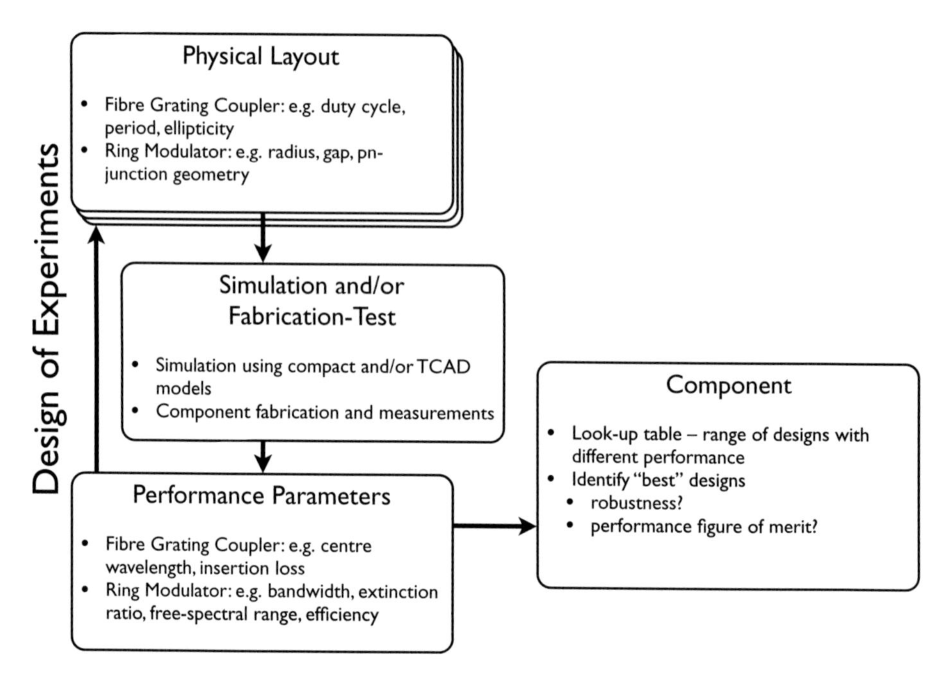

Figure 9.2 Diagram illustrating the methodology for component development using a design of experiment (DOE) approach.

or wavelength dependent, as per simulation requirements. Finally, reflections may also need to be considered.

In the following, we describe two common methods for constructing compact models, with an emphasis on capturing all necessary information for complete linear optical system analysis at multiple wavelengths.

9.3.1 Empirical or equivalent circuit models

Empirical models are the standard approach used in electronics design. For example, a SPICE model for a transistor can have over 100 parameters in a suitable function that is used to fit experimental (or simulated) data. It is common in the electronics industry to create compact models extracted from experimental data. Numerous tools can build an equivalent electrical circuit model for a set of measurements, e.g. a frequency response (e.g. agilent advanced design system – ADS). Other tools can can curve-fit the frequency response to empirical models such as polynomial or rational functions [3, 4] (e.g. available in MATLAB RF Toolbox). The advantage of such compact models is that they are just that – compact – particularly when compared to the original measured data. By choosing an appropriate function, the curve-fitting effectively smooths over the measurement noise, thus leading to well-behaved system simulations. These compact models are thus highly suitable for large-scale system simulations. The main challenge is identifying appropriate model equations and keeping track of the range of validity (e.g. valid frequency range).

One reason why some empirical models are inappropriate is that they have too many parameters; this leads to curve-fit capturing measurement errors (e.g. systematic calibration errors, measurement noise). In general, it is beneficial to have an understanding of the expected variation and to use as simple a model as possible. For example, a (linear) resistor is typically modelled by a single parameter, R. If we fit experimental data with an N-order polynomial, then some measurement error would be captured by the fit, possibly resulting in an incorrect model.

In Section 9.4, we will use both a polynomial function (Equation (9.2)) and a rational function (Equation (9.4)) to build empirical compact models.

9.3.2 S-parameters

Scattering parameters (S-parameters) are used to describe the behaviour of a linear time-invariant network. Traditionally, this concept is used to describe the response of electrical devices as a function of frequency. S-parameters are particularly useful for experimental characterization of electrical devices, using a vector network analyzer (VNA). They can also be used to describe optical devices, with experiments conducted using an optical vector network analyzer (ONA). In both cases, measurements are performed over a range of frequencies – typically GHz frequencies in the electrical RF domain and THz frequency in the optical domain. The S-parameters are generally complex, meaning they include an amplitude response and a phase response. In most examples considered in Parts II and III, the main focus was on the amplitude response, while ignoring the phase. When building circuits (e.g. ring resonators, interferometers), the phase response needs to be included in the compact model.

A two-optical-port device (e.g. section of a waveguide, a waveguide taper, a waveguide bend, a fibre grating coupler), can be described as a matrix (illustrated in Figure 4.32a):

$$\begin{pmatrix} b_1 \\ b_2 \end{pmatrix} = \begin{pmatrix} S_{11} & S_{12} \\ S_{21} & S_{22} \end{pmatrix} \begin{pmatrix} a_1 \\ a_2 \end{pmatrix} = S \begin{pmatrix} a_1 \\ a_2 \end{pmatrix}, \tag{9.1}$$

where a_1 describes the light incident on the device on port 1, b_1 describes the reflected light, b_2 describes the transmitted light, S_{11} is the reflection coefficient, and S_{21} is the forward gain. The other parameters are similarly defined for light incident on port 2 and, in particular, there can be light incident on both ports simultaneously. Common parameters for describing optical devices are the return loss, $RL = -20\log_{10}|S_{11}|$ [dB], and the insertion loss, $IL = -20\log_{10}|S_{21}|$ [dB]. Since passive optical devices are reciprocal, the same transmission result should be obtained by having the incident light on port 2, or $S_{12} = S_{21}$.

S-parameters are commonly used in analog electronics design, whereby a designer wishes to incorporate the component in a circuit design. For example, the experimental S-parameters for components such as wirebonds and optical modulators can be used to design electrical drivers, and trans-impedance amplifiers for detectors. These parameters can be used directly by the EDA tool. S-parameters can also be used for system-level simulations, where multiple components can be cascaded and the overall response of the system can be very conveniently determined.

S-parameters are excellent for experimental data; however, the measurement data include noise and other sources of error that are then included in subsequent system analysis, and the parameters are measured at many frequency points (e.g. 10 000). The use of these parameters in system modelling thus has two main challenges: the incorporation of measurement error, and computational complexity. The computation inefficiency is particularly troublesome for very large system modelling (e.g. thousands of optical components), particularly when executing time-domain simulations, which require the frequency domain S-parameters to be converted into an impulse response filter (e.g. finite or infinite impulse response filters).

Just as with experimental data, S-parameters can be extracted from FDTD simulations. Constructing compact models from numerical simulations has the same benefits as experimentally derived compact models – as an effective data reduction technique that leads to more stable and fast system-level simulations. However, similar to experimental results, FDTD simulations are not 100% accurate. There is always a residual numerical error due to the finite mesh size, spurious reflections from the simulation boundaries, etc.

9.4 Directional coupler – compact model

In the following, we describe the process of starting with an FDTD simulation, determining the S-parameters, and curve-fitting these to build a very simple empirical model. This example is concerned with the directional coupler, with an emphasis on a simple model for t and κ parameters (both amplitude and phase). This empirical model could then be used to construct a compact model with a circuit modelling tool, e.g. Lumerical INTERCONNECT.

Next, we use S-parameters directly within circuit simulations. We address some of the challenges associated with using S-parameters for passive devices, namely, that we need to ensure that the parameters are passive. Next, we use these S-parameters to construct a ring modulator.

9.4.1 FDTD simulations

To simulate the S-parameters by using FDTD, first the geometry is constructed, as shown in Figure 9.3. This example considers the directional coupler used in a ring modulator, with rib waveguides. The top waveguide can eventually be used to form a ring resonator, and the bottom waveguide is the coupler.

Since this is a four-port device, in general four simulations would need to be conducted: input on port 1, with measurements on ports 1, 2, 3, and 4. This is followed by a second simulation with input on port 2, then a third with input on port 3, and finally a fourth with port 4 as input. However, we can often make use of symmetry to simplify the measurements or simulations. In this case, the device is symmetric about the vertical axis, hence the results for light input on port 3 would be the same as when input on port 1.

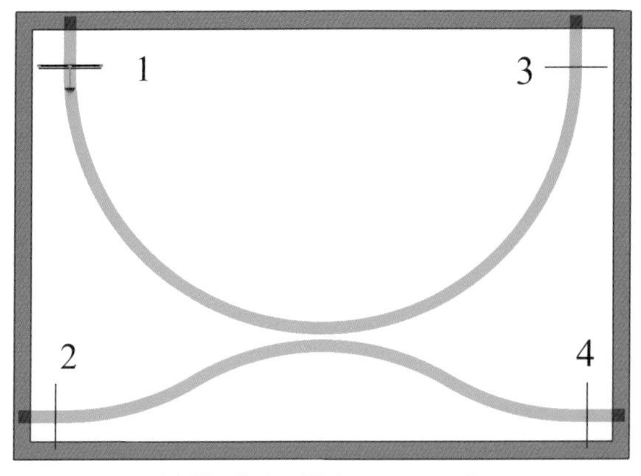

(a) Simulation #1, input on port 1.

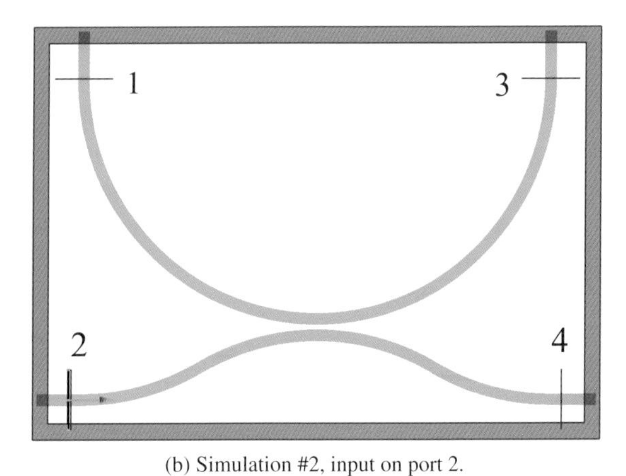

(b) Simulation #2, input on port 2.

Figure 9.3 Directional coupler device and simulation configuration. The radius is 10 μm, and the gap is 180 nm. Reprinted with permission [5].

The first simulation is shown in Figure 9.3a. Light is incident on port 1 using a mode source (pointing down). Four field monitors are positioned at the four waveguides. In addition, mode expansion monitors are used to isolate the component of the field travelling in a particular direction, and also to ensure the measurement takes into account only the light in the fundamental mode of the waveguide. For example, to find the S_{11} parameters, the field monitor is below the source and measures the reflected light (travelling up in the figure). The data in each of the monitors are then normalized to the power in the input waveguide, which results in the complex S-parameters. These can be written as magnitude and phase. This first simulation gives the four S-parameters: S_{11}, S_{21}, S_{31}, and S_{41}.

The second simulation is shown in Figure 9.3b. Light is now incident on port 2 (pointing right). Similar measurements and calculations are conducted. This second simulation gives the four S-parameters: S_{12}, S_{22}, S_{32}, and S_{42}.

Since the device is symmetric, the remaining eight parameters do not need to be simulated.

9.4.2 FDTD S-parameters

There are important considerations when extracting S-parameters from FDTD simulations of waveguide-based devices: (1) the waveguide modes are not necessarily power orthogonal; (2) the vectorial nature of the EM modal fields introduces additional complexity compared to a scalar quantity like voltage; and (3) small numerical errors in the S-matrix may cause it to violate passivity, particularly for devices with extremely low insertion loss such as directional couplers. This can be fixed by using a variety of methods, and is discussed in Reference [5] and Section 9.4.4.

Waveguide power orthogonality

Power orthogonality means that the total power in the waveguide is the sum of the power in each individual mode present. Fortunately, in non-absorbing waveguides, the waveguide modes are power orthogonal, which greatly simplifies the analysis. While there is always some absorption in real waveguides, in most practical circuits it is sufficiently low that we can make the approximation that the waveguide modes are power orthogonal. The implication of orthogonality is that S-parameters can be constructed on a mode-by-mode basis, and the system can be excited one mode at a time (or with any linear combination). Care should be exercised when this approximation is invalid, for example, in photodetectors, where the absorption is strong. In this case, a modal analysis is inaccurate and a full-wave simulation must be considered, as done in Section 7.5.

Vectorial nature of waveguide modes

In general, a standard sign and phase convention for the E and H fields of forward- and backward-propagating modes should be used for all ports of a component. Typically, in a non-absorbing waveguide, the transverse E and H fields are chosen to be real-valued. The backward-propagating mode has the same field profile but the tangential H field and normal E field components are reversed in sign. The S-parameters are the complex coefficients of the forward- and backward-propagating modes, once all modes have been normalized to carry the same power. This means that the S-parameters can have additional negative signs that may not be expected. For example, the S_{21} coefficient in a 180° U-bend will have an additional negative sign for TE-like modes but not for TM-like modes, as can be seen Figure 9.4. This is simply due to the fact that the physical fields in the waveguide are the S-parameter coefficients multiplied by the (fully vectorial) modal fields of the waveguide and a 180° rotation will reverse the direction of some field components. Furthermore, we must be careful when using symmetry considerations to reduce the number of simulations or measurements that need to be made since the symmetry of the modal fields must be considered.

For the directional coupler, symmetry, reciprocity and the above considerations allow us to obtain the eight additional S-parameters, as follows: $S_{13} = S_{31}$, $S_{23} = -S_{41}$, $S_{33} = S_{11}$, $S_{43} = -S_{21}$, $S_{14} = S_{41}$, $S_{24} = S_{42}$, $S_{34} = -S_{12}$, and $S_{44} = S_{22}$.

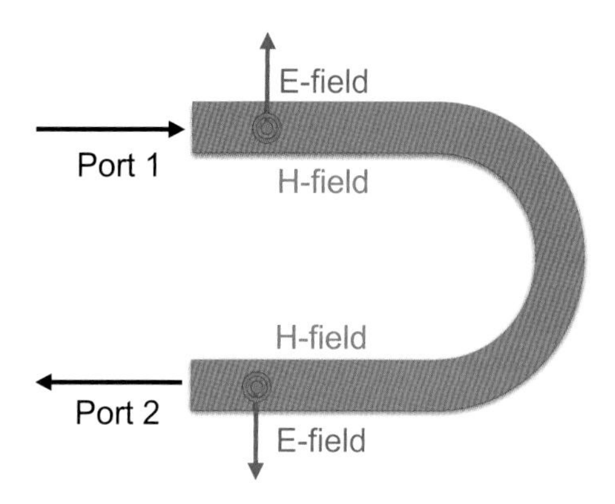

Figure 9.4 Simulation EM field considerations for defining S-parameters.

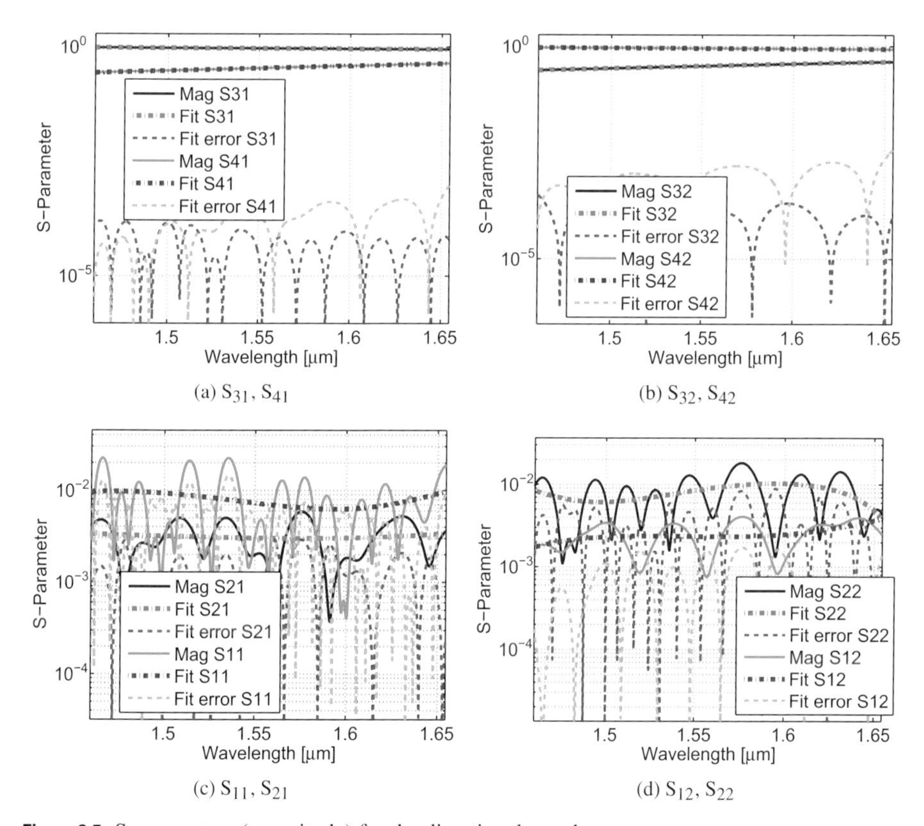

(a) S_{31}, S_{41} (b) S_{32}, S_{42}

(c) S_{11}, S_{21} (d) S_{12}, S_{22}

Figure 9.5 S-parameters (magnitude) for the directional coupler.

Directional coupler S-parameters

The simulated S-parameters for the device are shown in Figure 9.5 (magnitude response) and Figure 9.6 (phase response), for light incident on ports 1 and 2. The parameters typically used to describe the coupler, namely the through field coupling coefficient, t, and the cross-field coupling coefficient, κ, as in Equations (4.1) and (4.2), are equivalent to the magnitude S-parameters shown in Figure 9.5a and Figure 9.5b.

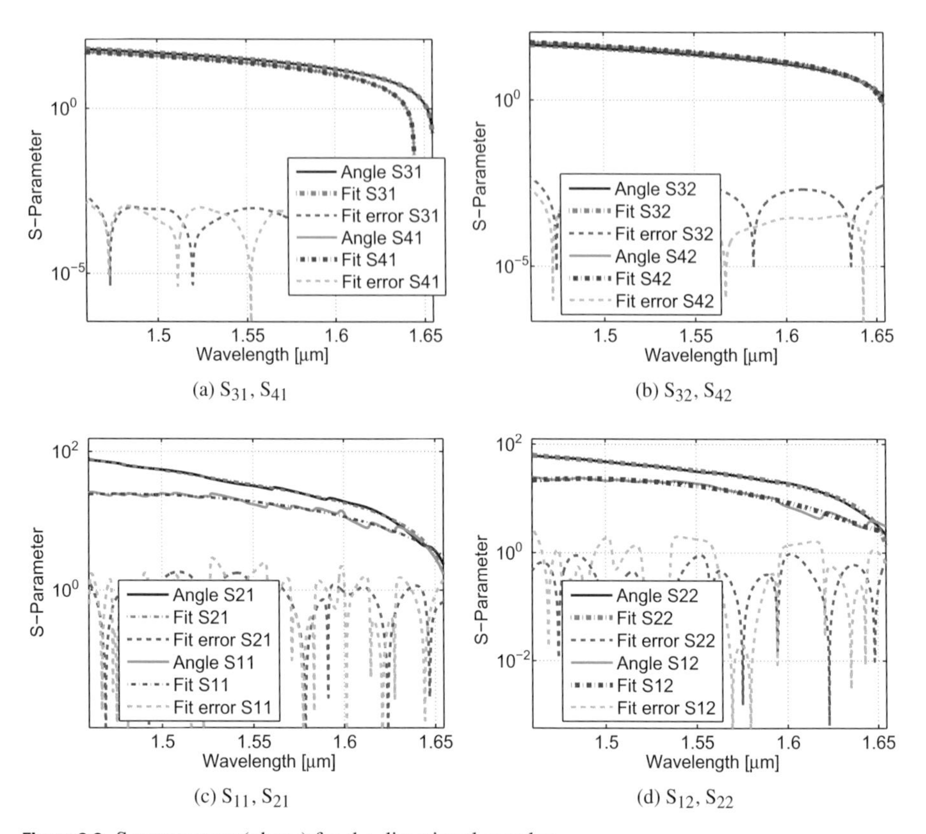

(a) S_{31}, S_{41}

(b) S_{32}, S_{42}

(c) S_{11}, S_{21}

(d) S_{12}, S_{22}

Figure 9.6 S-parameters (phase) for the directional coupler.

In this example, $t = 0.933\,079$ and $\kappa = 0.347\,666$, at the center wavelength, and these parameters are wavelength dependent, as found in Figure 4.8. Note that $t^2+\kappa^2 = 0.9915$, representing a 0.85% power loss in the coupler.

In contrast to Section 4.1, where we only considered t and κ, here we also consider the reflection (S_{11}, S_{22}; return loss) of the device, as well as the parasitic coupling from the input to its adjacent waveguide (S_{21}, S_{12}; unintentional contra-directional coupling). The results are plotted in Figure 9.5c and Figure 9.5d. We can see that the S-parameters have a strong and oscillatory wavelength dependence, suggesting that there are multiple reflection and scattering sites present in the device leading to Fabry–Perot modes and standing waves. However, these are very small. The field coupling coefficients are approximately in the range of 0.001 to 0.01, implying that the return loss is typically better than 40 dB.

The S-parameters are complex, and their phases are plotted in Figure 9.6. Note that, in this example, the phase includes the propagation of light in the waveguides, and hence accounts for the time delay and dispersion. This phase information is critical in system modelling, for example if this coupler is used in an interferometer or resonator. The phase for the through and cross-coupling S-parameters is well-behaved, as expected, similar to light travelling through a waveguide. However, the phase of the reflected light shows numerous phase transitions typical of resonant structures.

Table 9.1 S-parameters for a four port device. The S-parameters are indexed in the table as S_{RC}, where R is the row (output), C is the column (input). e.g., for S_{21}, port 2 is the output and port 1 is the input.

S	1	2	3	4
1	S_{11}	S_{12}	S_{13}	S_{14}
2	S_{21}	S_{22}	S_{23}	S_{24}
3	S_{31}	S_{32}	S_{33}	S_{34}
4	S_{41}	S_{42}	S_{43}	S_{44}

In total, there are 16 S-parameters (see Table 9.1), each being a matrix of complex values with one point per wavelength. In this example the spectra contain 1001 points, hence there are $4 \times 4 \times 1001$ complex values, for a total of 32 032 real numbers.

9.4.3 Empirical model – polynomial

In order to reduce the amount of data and reduce computational complexity, we curve-fit each S-parameter using a fourth-order polynomial

$$y = a_0 + a_1(x - \lambda_0) + a_2(x - \lambda_0)^2 + a_3(x - \lambda_0)^3 \tag{9.2}$$

as a Taylor expansion about the mid-point of the spectrum, λ_0. For curve-fitting, it is best to have all the values in a similar range, which was achieved by using microns for the wavelength rather than meters. The choice of fitting function is important for minimizing the error and, in this example, the fit had the lowest error when the data were plotted as a function of wavelength rather than frequency, and also as field-coupling coefficients rather than power-coupling coefficients. This was convenient since the field-coupling coefficients are equal in value to the S-parameters. For the through and cross S-parameters, the fit error was typically between 10^{-3} and 10^{-4}.

However, for the reflection and parasitic coupling, where the S-parameters are oscillatory, the low-order polynomial does not fit the data well. The best this function can provide in this case is an average S-parameter, neglecting the oscillations versus wavelength for both the magnitude and phase response. Hence, the fitting eliminates the ripple in the spectrum due to reflections within the device. On one hand this simplifies the model, while on the other hand we are losing subtle information about the device. It is possible to use more parameters for the fitting function in order to capture these oscillations, and also we can use a Fourier series. In this example, in order to obtain a reasonable fit, >100 parameters are required. Whether these small oscillations are important in the system performance can be ascertained by comparing a system in which the directional coupler is modelled either by the compact circuit model, or by the S-parameters. If the difference is not important for the system simulation, the simulation can be simplified and speeded up by using the compact model found with this approach. We note that this compact model contains a total of 65 real numbers, in contrast to the original 32 032, and hence represents a significant saving.

Table 9.2 Compact model for a directional coupler (magnitude).

S	1	2	3	4
1	0.007 27, −0.0393, 0.217, 3.93	0.002 36, −0.001 52, 0.0234, 1.28	S_{31}	S_{32}
2	0.00 301, −0.001 93, 0.0356, 0.216	0.008 54, 0.0634, −0.139, −8.89	S_{41}	S_{42}
3	0.933, −0.341, −0.729, −0.804	0.347, 0.862, 0.522, −1.56	S_{11}	S_{12}
4	0.348, 0.857, 0.495, −1.18	0.900, −0.379, −0.286, 6.24	S_{21}	S_{22}

Table 9.3 Compact model for a directional coupler (phase).

S	1	2	3	4
1	19.1, −132, −539, 3168	17.2, −168, −553, 7643	S_{31}	S_{32}
2	33.1, −333, 1011, −6975	32.6, −286, 200, −3440	S_{41}	S_{42}
3	31.4, −325, 238, −155	23.1, −228, 167, −100	S_{11}	S_{12}
4	24.1, −276, 202, −137	27.3, −276, 202, −132	S_{21}	S_{22}

The compact model coefficients for the directional coupler shown in 9.3, based on the model in Equation (9.2) are listed in Table 9.2 (for the magnitude) and Table 9.3 (for the phase). In the tables, the convention for which parameter is which is given in Table 9.1. Referring to Equation (9.2), the values in the table are for a_0, a_1, a_2, a_3, and $\lambda_0 = 1.5511\ \mu\text{m}$. By symmetry, the parameters for light injected in ports 3 and 4 are taken from the results from ports 1 and 2.

9.4.4 S-parameter model passivity

Efficient circuit simulations can be by achieved by using S-parameters directly. This is particularly simple for the optical transmission response of an optical circuit, built using components with known S-parameters. For time-domain simulations, the circuit modelling tool internally needs to convert the S-parameters into a time-domain representation. In this section, we use the results of the FDTD simulations in a circuit simulation, using Lumerical INTERCONNECT.

Passivity assessment

In order to use S-parameters of passive optical components in circuit simulations, we first need to make sure that the parameters are valid, namely that they meet the passivity requirements. A passive model needs to be causal (impulse response is 0 for $t < 0$), stable (impulse response decays in time), passive (no amplification), and reciprocal. One test that can be applied to the S-parameters is the norm-2 test [6], which verifies if the system is passive, namely

$$||S(j\omega)||_2 \leq 1. \tag{9.3}$$

For the S-parameters shown in Figures 9.5 and 9.6, the passivity test is performed for the 4×4 complex S-parameter matrix for each wavelength. The result is plotted in Figure 9.7. As can be seen, the result is greater than 1 for some wavelengths. Thus, the

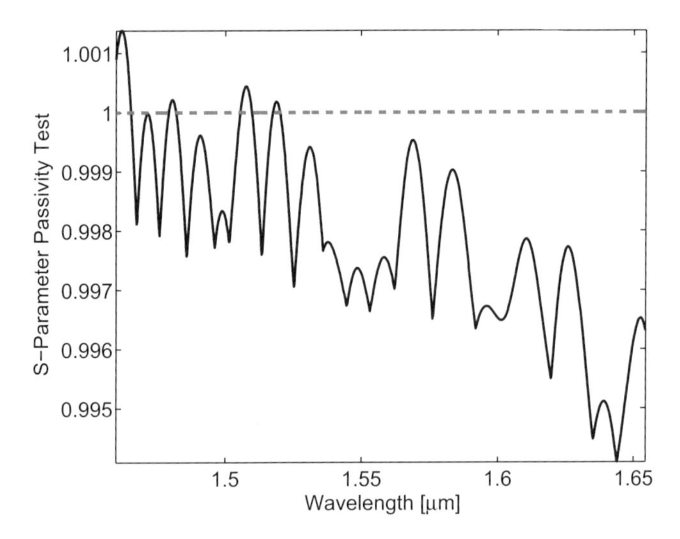

Figure 9.7 Passivity test on the S-parameters. This example reveals that the S-parameters are not passive for all frequencies.

model is not passive, hence, these parameters should be used with caution, particularly if used to construct resonant structures such as ring resonators. S-parameters are often not passive as a result of error, either measurement, or in this case, numerical, error.

Passivity enforcement

When importing S-parameters for system simulations, it is necessary to ensure that the parameters are passive, namely that the model does not generate energy. This would clearly be problematic for structures such as ring resonators and would give non-physical results where gain is observed in the simulations.

In this section, we employ the rational modelling techniques provided open-source by B. Gustavsen [7] to condition the S-parameters. The approach is as follows:

(1) Assess the passivity of the S-parameters [8]. If it is not passive, then proceed to step (2).

(2) Curve-fit the S-parameters to a rational function with a series of poles, instead of a polynomial, using the vector fitting technique [3, 9, 10]:

$$\mathbf{S}(s) = \sum_{m=1}^{N} \frac{\mathbf{R}_m}{s - a_m}, \tag{9.4}$$

where $\mathbf{S}(s)$ are the S-parameters; $s = j2\pi f$ is the complex frequency, with f being the optical frequency; a_m are the poles; and $\mathbf{R}_m$ are the residues matrices.

(3) Obtain a passive version of the pole-residue function, by perturbing the model until it becomes passive [11].

(4) The pole-residue compact model can be used directly in circuit simulations, can be converted into a lumped equivalent circuit model [12], or the S-parameters can be generated and exported. The choice depends on the tool preference. In this section, we export the S-parameters for circuit modelling.

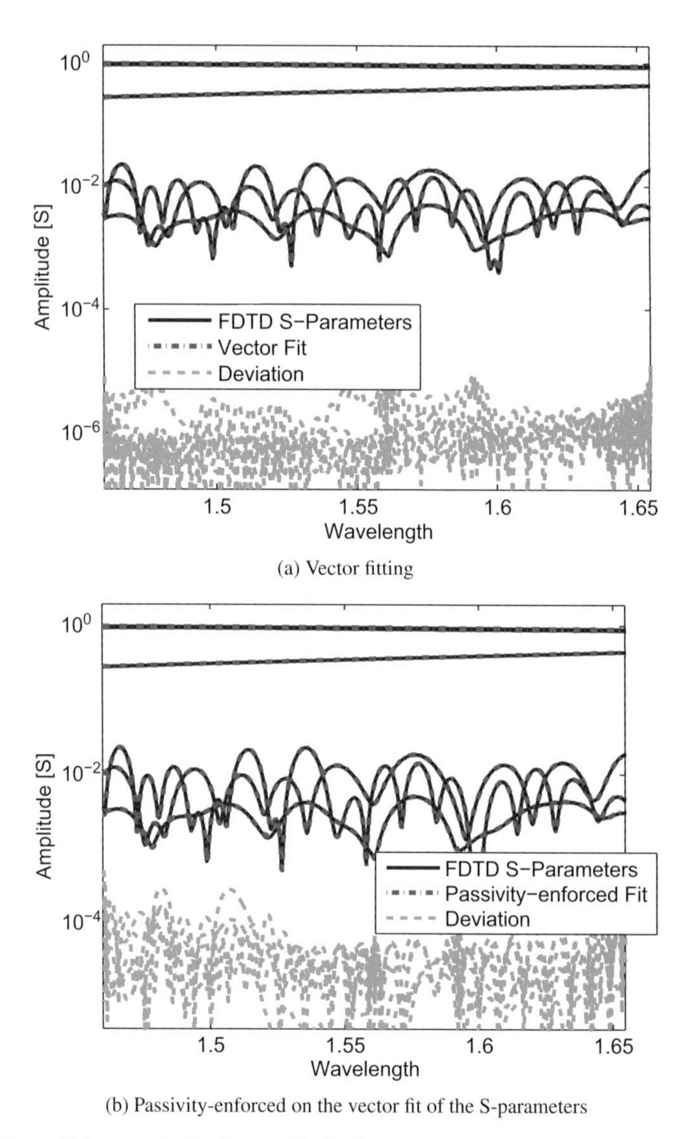

(a) Vector fitting

(b) Passivity-enforced on the vector fit of the S-parameters

Figure 9.8 Vector fitting results for the amplitude S-parameter response of the directional coupler. Fitting error is plotted for each S-parameter.

The code listed in Listing 9.2 uses the open-source functions [7] as follows.

(1) Construct an S-parameter matrix from the two FDTD simulations. The S-parameter matrix should satisfy reciprocity, i.e. $\mathbf{S} = \mathbf{S}'$. Symmetry of the device means that $S_{14} = S_{32}$ and $S_{23} = S_{41}$. Together, reciprocity and symmetry imply that $S_{41} = S_{32}$, and $S_{14} = S_{23}$. This reduces the S-parameters to only six unique S-parameters for this four-port optical device (instead of the total 16 S-parameters in the 4×4 matrix). Since the simulations resulted in eight S-parameters, we average the two redundant ones, namely S_{12} with S_{21}, and S_{41} and S_{32}.

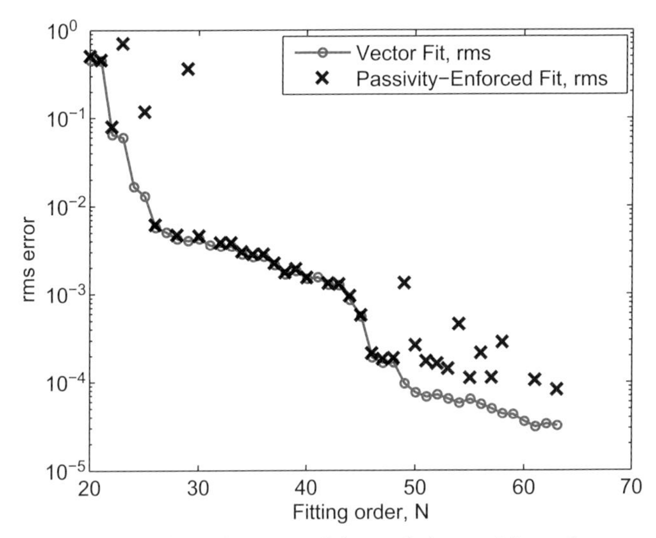

(a) Convergence from the vector fitting and the passivity enforcement algorithms

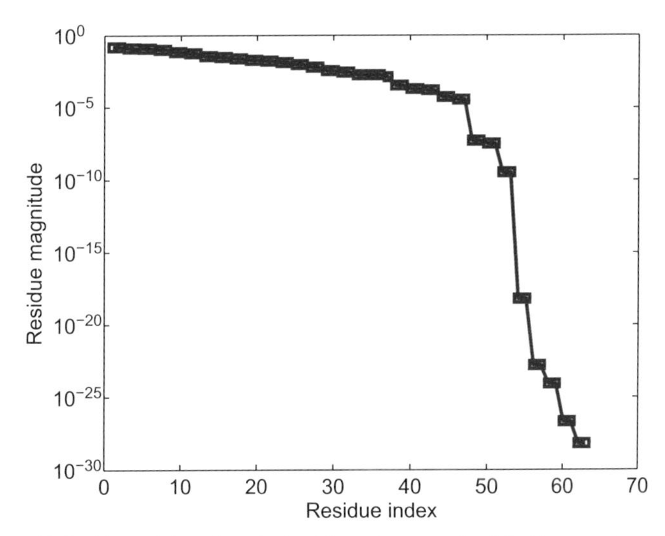

(b) Residues of the poles, obtained from the vector fit of the S-parameters

Figure 9.9 Convergence and residues for the vector fitting.

(2) Assess the passivity of the S-parameters. If they are passive, stop and export the S-parameter matrix.

(3) Perform a loop over a range of N values, as in Equation (9.4).
- Perform the vector fit.
- Perform the passivity enforcement, and check if it was successful.
- Calculate the root mean square (rms) error between the original S-parameters, and the two results above.
- Terminate when the rms is below a threshold and the S-parameters are passive.

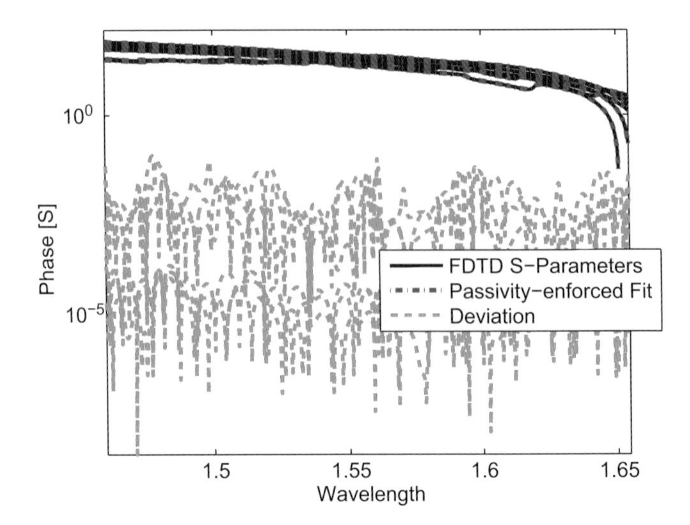

Figure 9.10 Passivity-enforced vector fitting results for the phase S-parameter response of the directional coupler.

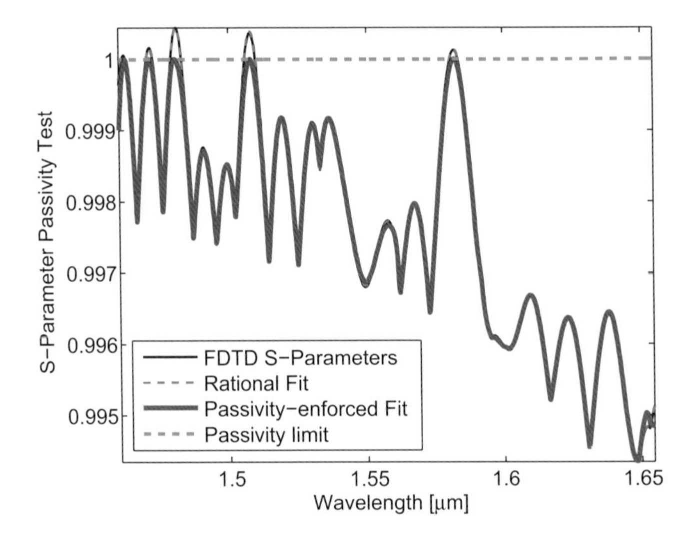

Figure 9.11 Passivity test on the S-parameters after vector fitting and passivity enforcement. The S-parameters have been made passive for all frequencies. Reprinted with permission [5].

(4) Plot the result of the scan of N values, to visualize the convergence, Figure 9.9a, and residues $\mathbf{R}_m$, Figure 9.9b.
(5) Plot the S-parameters and the passivity-enforced vector fit functions, Figures 9.8–9.10.
(6) Plot the passivity of the S-parameters, Figure 9.11.
(7) Export the new S-parameters.

By visual inspection, the number of peaks in the S-parameter spectra total approximately 36. Assuming each peak requires a pair of complex conjugate poles, this provides an estimate for the number of parameters necessary to obtain a good fit to be 72. Looking at the convergence results (Figure 9.9a), an error of less than 10^{-4} was obtained with 63 parameters. The plot of the residues $\mathbf{R}_m$ (Figure 9.9b) shows that it is the first 50 or so parameters that play the most important role in fitting the function, in ensuring that all the peaks are represented in the fit. Beyond these 50 parameters, the additional poles have very small amplitude coefficients, $\mathbf{R}_m$, thus make little impact in the shape of the fitting – only the rms error decreases.

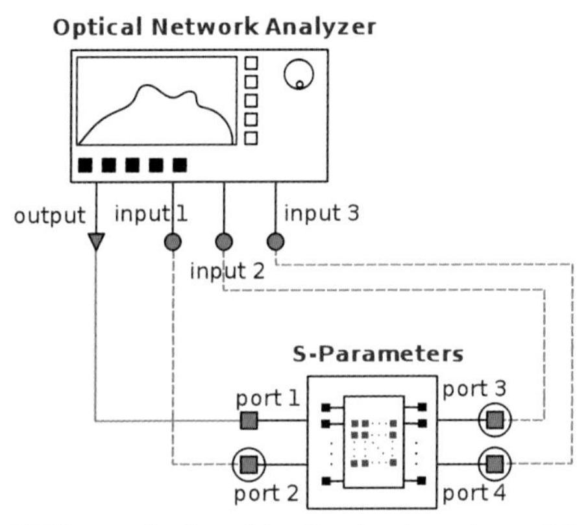

(a) Characterization of the directional coupler model.

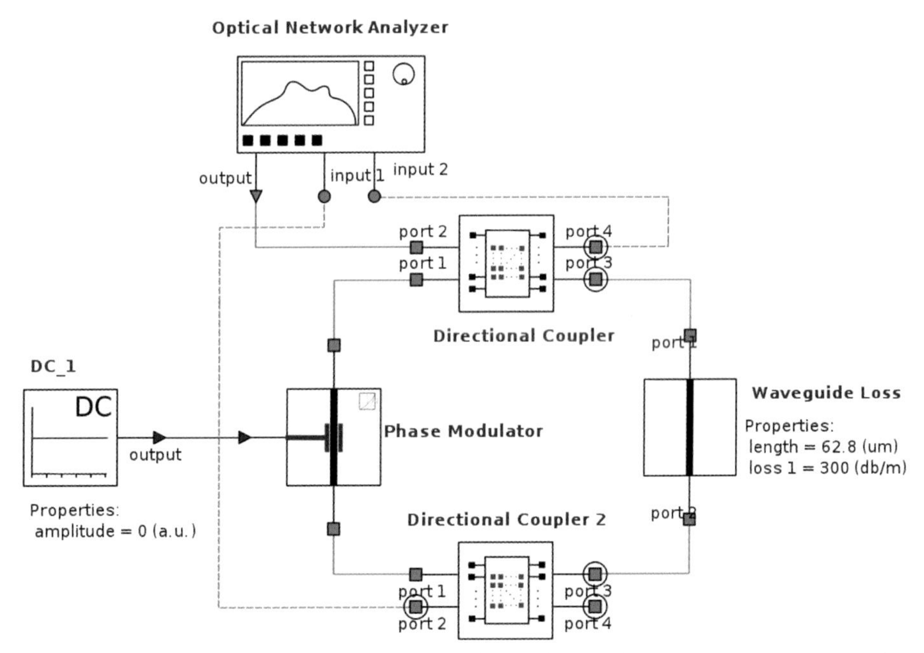

(b) Characterization of a ring modulator constructed with two directional couplers, phase modulator, and an additional waveguide loss. Reprinted with permission [5].

Figure 9.12 Circuit modelling schematic diagrams, in Lumerical INTERCONNECT. The S-parameters for the directional coupler are used to create the model used by the simulator. The Optical Network Analyzer is used to measure the optical transmission spectra.

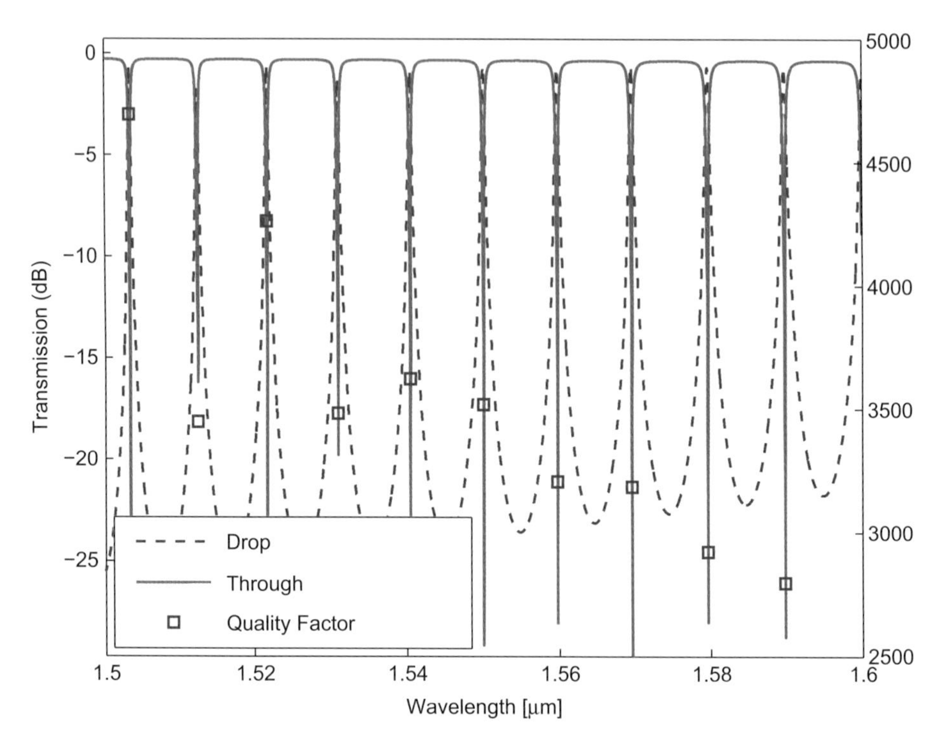

Figure 9.13 Optical spectrum of the circuit in Figure 9.12, in Lumerical INTERCONNECT. The S-parameters shown in Figures 9.8–9.10 are used for the directional coupler. Reprinted with permission [5].

9.5 Ring modulator – circuit model

As shown in Figure 9.12, the compact models can be used to build sub-circuits in Lumerical INTERCONNECT. In this section, we use the compact model for the directional coupler found above (passivity-enforced S-parameters) to build a ring resonator, as illustrated in Figure 9.13. This is accomplished by loading the S-parameters into an optical N-port S-parameter import object, adding the waveguide loss (3 dB/cm), adding a phase modulator, and connecting the circuit as illustrated to make a ring modulator. This is implemented in Listing 9.3. The optical transmission spectrum of the through and drop ports is shown in Figure 9.13. On this plot, the quality factors for each resonance are also indicated.

9.6 Grating coupler – S-parameters

In this section, we describe the method of simulating the S-parameters for a fibre grating coupler. This is presented in Listing 9.4. Two simulations are performed: (1) an output grating coupler where the light is injected from the waveguide (port 1), Figure 5.7b, and (2) an input grating coupler where the light is injected from the optical fibre (port 2),

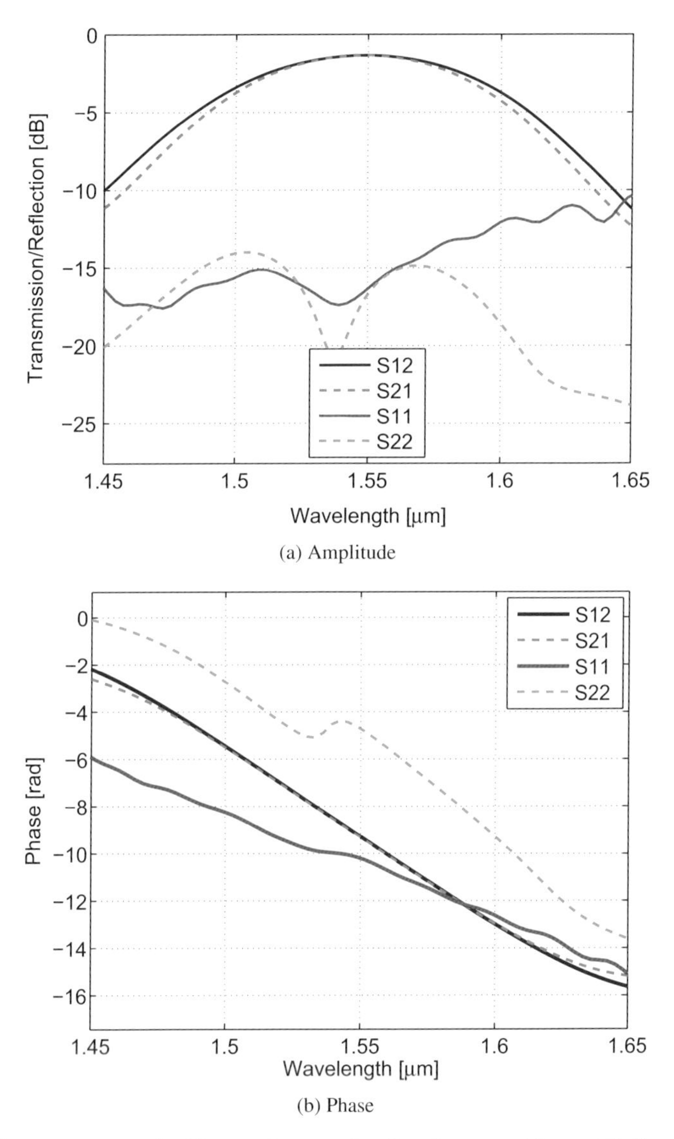

(a) Amplitude

(b) Phase

Figure 9.14 S-parameters for the fibre grating coupler. Port 1: waveguide; port 2: fibre. S_{21}: transmission from the waveguide to the fibre; S_{12}: transmission from the fibre to the waveguide; S_{11}: reflection back to the waveguide; S_{22}: reflection back to the fibre.

Figure 5.7a. Mode-expansion monitors are used to isolate the component of the field travelling in a particular direction, and also to ensure the measurement is taking into account only the light in the fundamental mode of the waveguide. The data in each of the monitors are then normalized to the power in the input port, which results in the complex S-parameters (note normalization is critical especially in this case where the waveguides have a different geometry). The four S-parameters are then calculated, plotted as shown in Figure 9.14, and recorded. Note that due to numerical simulation error, $S_{12} \neq S_{21}$ for wavelengths away from the central simulation wavelength.

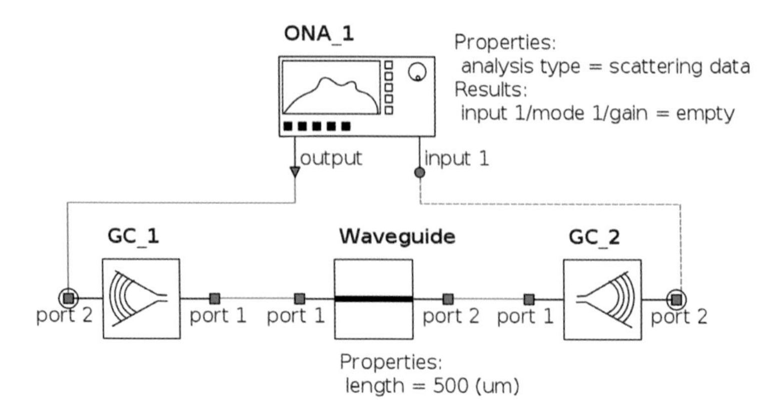

Figure 9.15 Circuit modelling schematic diagram, in Lumerical INTERCONNECT. The S-parameters for the fibre grating coupler are used to create a model of two couplers on a chip connected by a 500 μm long strip waveguide. The Optical Network Analyzer is used to measure the optical transmission spectra.

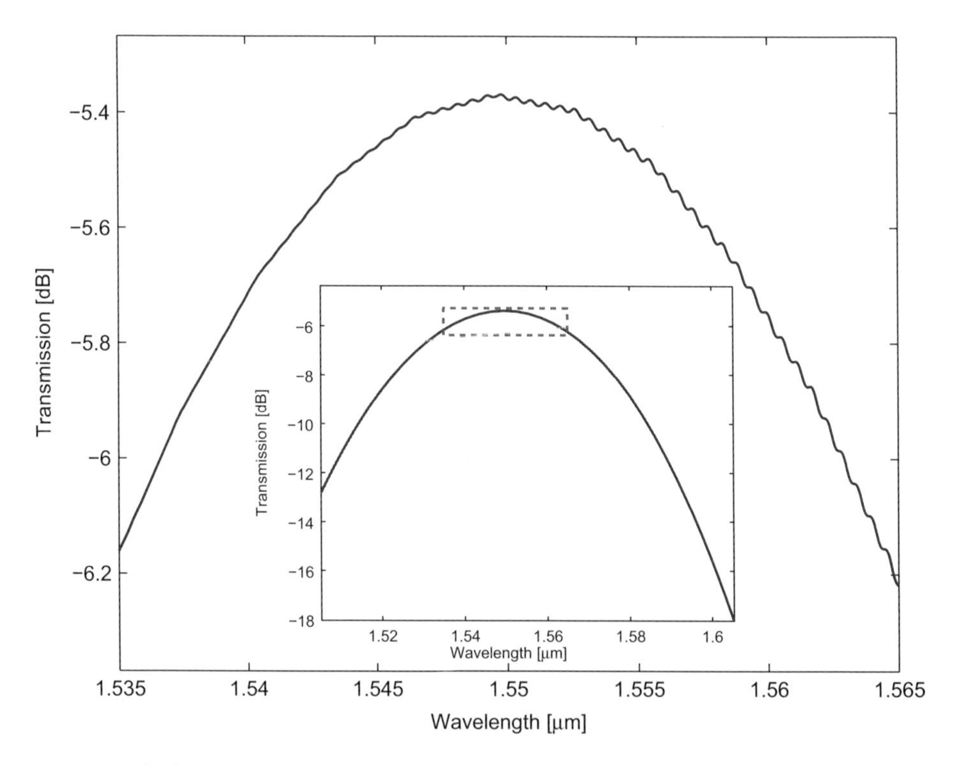

Figure 9.16 Optical spectrum of the circuit in Figure 9.15, in Lumerical INTERCONNECT. The S-parameters shown in Figure 9.14 are used for the grating couplers. The main plot is for a zoomed-in wavelength span of 30 nm. Inset: same plot but with a wavelength span of 100 nm. The dotted box identifies the zoom range for the main figure.

9.6.1 Grating coupler circuits

As shown in Figure 9.15, the S-parameters are used to create a component that can then be used in photonic circuits. The simple circuit illustrated consists of two grating couplers connected by a 500 μm long strip waveguide. The optical transmission spectrum of the two couplers is shown in Figure 9.16. Owing to the back-reflections from the grating coupler in the waveguide (S_{11} parameter), a weak Fabry–Perot cavity is set up in this circuit. This is evident in the small ripples visible in the transmission spectrum.

9.7 Code listings

Listing 9.1 FDTD Simulation for a directional coupler, with S-parameter export to MATLAB

```
# Directional coupler simulations
# Lumerical, Jonas Flueckiger, 2014

#based on script provided by Lukas Chrostowski, Miguel Angel Guillen Torres
# inputs:
#   radius:  radius, centre of waveguide
# DC_angle: angle for the bottom coupler (typical 30 degrees)
# gap:   directional coupler gap
# Lc:  coupler length, 0=point coupling
# wavelength: centre wavelength, e.g., 1550e-9
# wg_width: width of waveguide, e.g,. 500e-9;

figs = 1; # 1: plots s params, Norm, and abs(S-S*); 0: no plots

#Define data export
xml_filename="directional_coupler_map.xml";
table = "directional_coupler";
#open file to write table
lookupopen(xml_filename,table);
#prepare data structure
design = cell(6);
extracted = cell(4);
design{1} = struct; design{1}.name = "gap";
design{2} = struct; design{2}.name = "radius";
design{3} = struct; design{3}.name = "wg_width";
design{4} = struct; design{4}.name = "wg_height";
design{5} = struct; design{5}.name = "Lc";
design{6} = struct; design{6}.name = "SiEtch2";
extracted{1} = struct; extracted{1}.name = "s-param";
extracted{2} = struct; extracted{2}.name = "transmission_coeff";
extracted{3} = struct; extracted{3}.name = "coupling_coeff";
extracted{4} = struct; extracted{4}.name = "insertion_loss";

# Variables definition
num_wg_width = 1; # numer of data points for wg width;
num_gap = 1;   # number of data points for gap.
num_Lc = 1;     # number of data points coupling length.
num_radius = 1; # number of data points radius
num_wg_height = 1;
num_SiEtch2 = 1; # number of Etch thicknesses
lambda_min = 1.5e-6; # [m]
lambda_max = 1.6e-6; # [m]

# Design parameters to sweep
if (num_wg_width ==1){wg_width = [0.5e-6];}
else{wg_width = linspace(0.40e-6, 0.6e-6,num_wg_width);}

if (num_radius==1){radius = [5e-6];}
else {radius = linspace(5e-6,50e-6,num_radius);}

if (num_gap==1){gap =[200e-9];}
else{gap = linspace(180e-9,220e-9,num_gap);}
```

```
if (num_Lc==1){Lc=[0];}
else {Lc = linspace(0,15,num_Lc);}

if (num_wg_height==1){si_thickness = [220e-9];}
else {si_thickness = linspace(210e-9,230e-9,num_wg_heigth);}

if (num_SiEtch2==1){Etch2=[130e-9];}
# Etch2=[130e-9]; #i.e., the slab thickness is 220-130 = 90 nm

dc_angle=30;

newproject; #Creates new layout environment

#Define materials
materials; # creates a dispersive material model.
Material_Si = "Si (Silicon) - Dispersive & Lossless";
Material_Ox ="SiO2 (Glass) - Const";

MESH = 1;  # test,
MESH = 3;
#MESH = 4;  # final
SIM_TIME = 3000e-15; #Simulation time.

#Set global source parameters.
FREQ_PTS = 1001;
FREQ_CENTRE = c/1.55e-6;
setglobalsource("set frequency",1);
setglobalsource("center wavelength", 1.55e-6);
setglobalsource("center frequency",FREQ_CENTRE);
#setglobalsource("frequency span",c/(lambda_max-lambda_min));
setglobalsource("frequency span",10000e9);

#Set global monitor parameters
setglobalmonitor("frequency center", FREQ_CENTRE);
setglobalmonitor("frequency span", 10000e9);
setglobalmonitor("frequency points",FREQ_PTS); # Must be odd, so that the centre frequency
      is actually amongst the selected values.


FDTD_above=800e-9; # Extra simulation volume added, 0.8 um on top and bottom
FDTD_below=800e-9;

maxvzWAFER=si_thickness(1)+FDTD_above; minvzWAFER=-FDTD_below;

redrawoff;

# ************************************************************************************
## WAVEGUIDE FOR COUPLER:
# ************************************************************************************
for (rr=1:length(radius))
{
for (gg=1:length(gap)){
switchtolayout;
groupscope("::model");
deleteall;
addgroup; set("name","wg");
set("x",0);set("y",0);
groupscope("wg");

# straight part of DC:
addrect;
set("y span",wg_width(1));
set("x span",Lc(1));
set("y",-radius(rr));

copy(0,-wg_width(1)-gap(gg)); #copy of selected object and move in dy
#set("y",-radius-wg_width-gap);

# extra waveguide above ring
```

```
addrect;
set("x span", wg_width(1));
set("x",-radius(rr)-Lc(1)/2);
set("y min",0);
set("y max", 2e-6+4e-6);

copy;
set("x",-get("x"));

# Ring
addring;
set("theta start",-90);
set("theta stop",0);
set("inner radius",radius(rr)-wg_width(1)/2);
set("outer radius",radius(rr)+wg_width(1)/2);
set("x",Lc(1)/2);
set("y", 0);

copy;
set("theta start",180);
set("theta stop",270);
set("x",-Lc(1)/2);

# bottom Directional coupler
copy;
set("theta start", 90-dc_angle);
set("theta stop", 90);
set("x",Lc(1)/2);
set("y",-2*radius(rr)-wg_width(1)-gap(gg));

copy;
set("theta start", 90);
set("theta stop", 90+dc_angle);
set("x",-Lc(1)/2);

copy;
set("theta start", -90);
set("theta stop", -90+dc_angle);
set("x",-Lc(1)/2-2*sin(dc_angle/360*2*pi)*radius(rr));
set("y",-2*(1-cos(dc_angle/360*2*pi))*radius(rr)-wg_width(1)-gap(gg));

copy;
set("theta start", -90-dc_angle);
set("theta stop", -90);
set("x",-get("x"));

addrect;
set("y span",wg_width(1));
wg_bottom_input_y=-2*(1-cos(dc_angle/360*2*pi))*radius(rr)-radius(rr)-wg_width(1)-gap(gg);
set("y",wg_bottom_input_y);
wg_bottom_input_x=-Lc(1)/2-2*sin(dc_angle/360*2*pi)*radius(rr);
set("x max",wg_bottom_input_x);
set("x min",get("x max")-2e-6-3e-6);

copy;
set("x",-get("x"));

selectall;
set("material",Material_Si);
set("z min",0);
set("z max",si_thickness(1));

# ********************************************************************************
#Waveguide slab Si partial etch 2 (130 nm deep) 3;0
# ********************************************************************************
 maxvxWAFER = Lc(1)/2+2*sin(dc_angle/360*2*pi)*radius(rr)+2e-6;
 minvxWAFER = -maxvxWAFER;
 minvy=-2*(1-cos(dc_angle/360*2*pi))*radius(rr)-radius(rr)-4*wg_width(1)-gap(gg);
 maxvy=0.5e-6;
```

```
 addrect; set("name","slab");
 set("y max",maxvy+4e-6);
 set("y min",minvy-4e-6);
 set("x span",2*maxvxWAFER+8e-6);set('x', 0);
 set('material', Material_Si);
 set('z max',si_thickness(1)-Etch2(1));
 set('z min',0);
 set("alpha",0.5); #transparency of the object.

# buried Oxide
addrect; set("name", "Oxide");
set("x min", minvxWAFER-3e-6); set("y min", minvy-4e-6);
set("x max", maxvxWAFER+3e-6); set("y max", maxvy+4e-6);
set("z min", -2e-6); set("z max", 0);
set("material", Material_Ox);
set("alpha",0.52);

# Cladding oxide
addrect; set("name", "Cladding");
set("x min", minvxWAFER-3e-6); set("y min", minvy-4e-6);
set("x max", maxvxWAFER+3e-6); set("y max", maxvy+4e-6);
set("z min", 0); set("z max", 2.3e-6);
set("material", Material_Ox);
set("alpha",0.5);
set("override mesh order from material database", 1);
set("mesh order", 4); # make the cladding the background, i.e., "send to back".


# ********************************************************************************
# Add the simulation area
# ********************************************************************************
groupscope("::model"); #f you want to delete all objects in the simulation, set the group
      scope the root level (i.e. ::model).

MonitorSpanY=3e-6+wg_width(1);
SimMarginX=0.0e-6;
SimMarginY=0.5e-6;
SourceMarginX=2e-6;
MonitorMarginX=2e-6;

addfdtd; #Add simulation area
set("x min", minvxWAFER-SimMarginX); set("x max", maxvxWAFER+SimMarginX);
set("y min", minvy-SimMarginY); set("y max", maxvy);
set("z min", minvzWAFER); set("z max", maxvzWAFER);
set("mesh accuracy", MESH);
set("simulation time",SIM_TIME);
if (MESH<3) {
    set("z min bc","Metal"); set("z max bc","Metal");
    set("y min bc","Metal");
}

addmode; set("name","simulation_port2");
# Add source to the simulation environment.
#set("override global source settings",0);
set("injection axis","x-axis"); set("direction","Forward");
set("x", wg_bottom_input_x-1000e-9); # simulate up to the layout port (wg_width x wg_width
      box)
set("y span",MonitorSpanY);
set("y", wg_bottom_input_y);
set("z min", minvzWAFER); set("z max", maxvzWAFER);
set('override global source settings', 0);
set('mode selection', 'fundamental TE mode');
set("enabled",0);

#set("center frequency",FREQ_CENTRE);
#set("frequency span",c/(lambda_max-lambda_min));

addmode; set("name","simulation_port1");
#Adds source to the simulation environment.
#set("override global source settings",0);
```

```
set("injection axis","y-axis"); set("direction","Backward");
set("x", -radius(rr)-Lc(1)/2); set("x span",MonitorSpanY);
set("y", 200e-9);
set("z min", minvzWAFER); set("z max", maxvzWAFER);
set('override global source settings', 0);
set('mode selection', 'fundamental TE mode');

#set("center frequency",FREQ_CENTRE);
#set("frequency span",c/(lambda_max-lambda_min));

#**********************************************************
#Monitors
#**********************************************************

XcoordPort2=wg_bottom_input_x;  # bottom left input

addpower; set("name", "Port2");# Add power monitor for Port 2
set('monitor type', '2D X-normal');
set('y', wg_bottom_input_y);
set("y span",MonitorSpanY);
set("z min", minvzWAFER); set("z max", maxvzWAFER);
set("x", XcoordPort2);
set('override global monitor settings', 0);
addmesh; #To avoid phase error. force monitor on mesh
set('y', wg_bottom_input_y);
set("z min", minvzWAFER); set("z max", maxvzWAFER);
set('x', XcoordPort2);
set('y span', 0);
set('x span', 0);
set('y span',0);
set("set maximum mesh step", 1); set("override y mesh", 1); set("dy", 20e-9);
set("override x mesh", 1); set("dx", 20e-9);
set("override z mesh", 1); set("dz", 20e-9);


addpower; set("name", "Port1");# Add power monitor for Port 1
set('monitor type', '2D Y-normal');
set('y', 0); # unfortunately, monitor cannot overlap source
set("x",-radius(rr)-Lc(1)/2);
set("x span",MonitorSpanY);
set("z min", minvzWAFER); set("z max", maxvzWAFER);
set('override global monitor settings', 0);
addmesh; #To avoid phase error. force monitor on mesh
set('y', 0);
set('y span', 0);
set('x span', 0);
set('y span',0);
set("z min", minvzWAFER); set("z max", maxvzWAFER);
set('x',-radius(rr)-Lc(1)/2);
set("set maximum mesh step", 1); set("override x mesh", 1); set("dx", 20e-9);
set("override y mesh", 1); set("dy", 20e-9);
set("override z mesh", 1); set("dz", 40e-9);


addpower; set("name", "Port4");
set('monitor type', '2D X-normal');
set('y', wg_bottom_input_y);
set("y span",MonitorSpanY);
set("z min", minvzWAFER); set("z max", maxvzWAFER);
set("x", -XcoordPort2);
set('override global monitor settings', 0);
addmesh; #To avoid phase error. force monitor on mesh
set('y', wg_bottom_input_y);
set("z min", minvzWAFER); set("z max", maxvzWAFER);
set('x', -XcoordPort2);
set('y span', 0);
set('x span', 0);
set('y span',0);
set("set maximum mesh step", 1); set("override y mesh", 1); set("dy", 20e-9);
set("override x mesh", 1); set("dx", 20e-9);
```

```
set("override z mesh", 1); set("dz", 40e-9);


# Add power monitor for Port 3
addpower; set("name", "Port3");
set('monitor type', '2D Y-normal');
set('y', 0);
set("x",radius(rr)+Lc(1)/2);
set("x span",MonitorSpanY);
set("z min", minvzWAFER); set("z max", maxvzWAFER);
set('override global monitor settings', 0);
addmesh; #To avoid phase error. force monitor on mesh
set('y', 0);
set("z min", minvzWAFER); set("z max", maxvzWAFER);
set('x',radius(rr)+Lc(1)/2);
set('y span', 0);
set('x span', 0);
set('y span',0);
set("set maximum mesh step", 1); set("override x mesh", 1); set("dx", 20e-9);
set("override y mesh", 1); set("dy", 20e-9);
set("override z mesh", 1); set("dz", 40e-9);


addmodeexpansion; #the fraction of power transmitted into any modes
set('name', 'expansion_v');
set('monitor type', '2D X-normal');
set('y', wg_bottom_input_y);
set("y span",MonitorSpanY);
set("z min", minvzWAFER); set("z max", maxvzWAFER);
set("x", -XcoordPort2);
set('frequency points',1);
Mode_Selection = 'fundamental TE mode';
set('mode selection', Mode_Selection);
setexpansion('Port2expa','Port2');
setexpansion('Port4expa','Port4');

addmodeexpansion;
set('name', 'expansion_h');
set('monitor type', '2D Y-normal');
set('y', 0);
set("x",radius(rr)+Lc(1)/2);
set("x span",MonitorSpanY);
set("z min", minvzWAFER); set("z max", maxvzWAFER);
set('frequency points',1);
set('mode selection', Mode_Selection);
setexpansion('Port1expa','Port1');
setexpansion('Port3expa','Port3');

select("FDTD"); setview("extent"); # zoom to extent

# refine mesh in coupling region
if (MESH>2) {
  mesh_span = 1e-6;
  addmesh;
  set('y', -radius(rr)-wg_width(1)/2-gap(gg)/2);
  set('y span', mesh_span);
  set('x', 0);
  set('x span', 5e-6);
  set('z',si_thickness(1)/2);
  set('z span', 500e-9);
  set("set maximum mesh step", 1);
  set("override x mesh", 1); set("dx", 30e-9);
  set("override y mesh", 1); set("dy", 30e-9);
  set("override z mesh", 1); set("dz", 40e-9);
}

######
# RUN
######
```

```
save('DC_Sparam');
run;

Port1=getresult("expansion_h","expansion for Port1expa");
Port2=getresult("expansion_v","expansion for Port2expa");
Port3=getresult("expansion_h","expansion for Port3expa");
Port4=getresult("expansion_v","expansion for Port4expa");
#for substracting phase of wag
neff = getresult("expansion_h", "neff");
f=neff.f;
neff = pinch(neff.neff,2,1);
k = neff*2*pi*f/c;
L1=radius(rr)*pi/2;
L2=radius(rr)*pi*2*dc_angle/360*2;
f=Port1.f;

S11=Port1.a/Port1.b; S21=Port2.b/Port1.b;
S31=Port3.a/Port1.b; S41=Port4.a/Port1.b;

if (figs==1){
  plot (c/f*1e6, 10*log10(abs(S11)^2), 10*log10(abs(S21)^2), 10*log10(abs(S31)^2),
       10*log10(abs(S41)^2), 'Wavelength (um)', 'Transmission (dB)', 'Input Port1; g=' +
       num2str(gap(gg)*1e9) + 'nm r=' + num2str(radius(rr)*1e6) + ' um');
  legend ('S11 (backreflection)','S21 (cross backreflection)', 'S31 (through)', 'S41
       (cross)');
}
?"Center wavelength: " + num2str( c/f(length(f)/2+0.5) );
?"Through power at center wavelength (Input1): " +
     num2str(abs(S31(length(f)/2+0.5))^2*100) + "%.";
?"  Cross power at center wavelength (Input1): " +
     num2str(abs(S41(length(f)/2+0.5))^2*100) + "%.";
?"                            Loss (Input1): " +
     num2str(100-abs(S31(length(f)/2+0.5))^2*100-abs(S41(length(f)/2+0.5))^2*100) + "%.";

############################################################

# Input on Port 2:
switchtolayout;
select("simulation_port1"); set("enabled",0);
select("simulation_port2"); set("enabled",1);
run;

Port1=getresult("expansion_h","expansion for Port1expa");
Port2=getresult("expansion_v","expansion for Port2expa");
Port3=getresult("expansion_h","expansion for Port3expa");
Port4=getresult("expansion_v","expansion for Port4expa");
#for substracting phase of wg
neff = getresult("expansion_v", "neff");
f=neff.f;
neff = pinch(neff.neff,2,1);
k = neff*2*pi*f/c;
L1=radius(rr)*pi/2;
L2=radius(rr)*pi*2*dc_angle/360*2;

f=Port2.f;
S12=Port1.a/Port2.a; S22=Port2.b/Port2.a;
S32=Port3.a/Port2.a; S42=Port4.a/Port2.a;


if (figs==1){
  plot (c/f*1e6, 10*log10(abs(S12)^2), 10*log10(abs(S22)^2), 10*log10(abs(S32)^2),
       10*log10(abs(S42)^2), 'Wavelength (um)', 'Transmission (dB)', 'Input Port2; g=' +
       num2str(gap(gg)*1e9) +'nm r=' + num2str(radius(rr)*1e6) + ' um');
  legend ('S12 (cross backreflection)','S22 (backreflection)', 'S32 (cross)', 'S42
       (through)');
}
?"Through power at center wavelength (Input2): " +
     num2str(abs(S42(length(f)/2+0.5))^2*100) + "%.";
?"  Cross power at center wavelength (Input2): " +
     num2str(abs(S32(length(f)/2+0.5))^2*100) + "%.";
```

```
?"                              Loss (Input2): " +
     num2str(100-abs(S42(length(f)/2+0.5))^2*100-abs(S32(length(f)/2+0.5))^2*100) + "%.";


############################################################
#Export data
###########################################################

#export .xml and .sparam (individual file for s-params)
n_ports = 4;
mode_label = "TE";
mode_ID = "1";
input_type = "transmission";
filename = "dc_R=" + num2str(radius(rr)*1e6) + ",gap=" + num2str(gap(gg)*1e9) + ",Lc="
     +num2str(Lc(1)*1e6) + ",wg="+num2str(wg_width(1)*1e9) ;

if(fileexists(filename)) {rm(filename);} # old files with same name get deleted.
system('mkdir -p sparam_files'); file1="sparam_files/" + filename+".sparam";
system('mkdir -p txt_files'); file2="txt_files/" + filename+".txt";
if (fileexists(file1)) { rm(file1);} # delete Sparam file if it already exists
if (fileexists(file2)) { rm(file2);} # delete text file if it already exists

write(file2,"Center wavelength: " + num2str( c/f(length(f)/2+0.5) ));
write(file2,"Mesh accuracy: " + num2str(MESH));
write(file2,"Through power at center wavelength (Input1): " +
     num2str(abs(S31(length(f)/2+0.5))^2*100) + "%.");
write(file2," Cross power at center wavelength (Input1): " +
     num2str(abs(S41(length(f)/2+0.5))^2*100) + "%.");
write(file2,"                              Loss (Input1): " +
     num2str(100-abs(S31(length(f)/2+0.5))^2*100-abs(S41(length(f)/2+0.5))^2*100) + "%.");
write(file2,"Through power at center wavelength (Input2): " +
     num2str(abs(S42(length(f)/2+0.5))^2*100) + "%.");
write(file2," Cross power at center wavelength (Input2): " +
     num2str(abs(S32(length(f)/2+0.5))^2*100) + "%.");
write(file2,"                              Loss (Input2): " +
     num2str(100-abs(S42(length(f)/2+0.5))^2*100-abs(S32(length(f)/2+0.5))^2*100) + "%.");


#Symmetry note: due to sign convention (in FDTD) and vectorial nature of E there is a sign
     change
S13=S31; S23=-S41; S33=S11; S43=-S21;
S14=-S32; S24=S42; S34=-S12; S44=S22;

# Sx1 Input on port 1
S11_data=[f,abs(S11),unwrap(angle(S11))];
write(file1,"('port 1','TE',1,'port 1',1,'transmission')"
      +endl+"("
      +num2str(length(f))
      +",3)"
      +endl+num2str(S11_data)
      );
S21_data=[f,abs(S21),unwrap(angle(S21))];
write(file1,"('port 2','TE',1,'port 1',1,'transmission')"
      +endl+"("
      +num2str(length(f))
      +",3)"
      +endl+num2str(S21_data)
      );
S31_data=[f,abs(S31),unwrap(angle(S31))];
write(file1,"('port 3','TE',1,'port 1',1,'transmission')"
      +endl+"("
      +num2str(length(f))
      +",3)"
      +endl+num2str(S31_data)
      );
S41_data=[f,abs(S41),unwrap(angle(S41))];
write(file1,"('port 4','TE',1,'port 1',1,'transmission')"
      +endl+"("
      +num2str(length(f))
      +",3)"
```

```
        +endl+num2str(S41_data)
        );

 # Sx2 Input on port 2
 S12_data=[f,abs(S12),unwrap(angle(S12))];
 write(file1,"('port 1','TE',1,'port 2',1,'transmission')"
        +endl+"("
        +num2str(length(f))
        +",3)"
        +endl+num2str(S12_data)
        );
 S22_data=[f,abs(S22),unwrap(angle(S22))];
 write(file1,"('port 2','TE',1,'port 2',1,'transmission')"
        +endl+"("
        +num2str(length(f))
        +",3)"
        +endl+num2str(S22_data)
        );
 S32_data=[f,abs(S32),unwrap(angle(S32))];
 write(file1,"('port 3','TE',1,'port 2',1,'transmission')"
        +endl+"("
        +num2str(length(f))
        +",3)"
        +endl+num2str(S32_data)
        );
 S42_data=[f,abs(S42),unwrap(angle(S42))];
 write(file1,"('port 4','TE',1,'port 2',1,'transmission')"
        +endl+"("
        +num2str(length(f))
        +",3)"
        +endl+num2str(S42_data)
        );

 # Sx3 Input on port 1
 S13_data=[f,abs(S13),unwrap(angle(S13))];
 write(file1,"('port 1','TE',1,'port 3',1,'transmission')"
        +endl+"("
        +num2str(length(f))
        +",3)"
        +endl+num2str(S13_data)
        );
 S23_data=[f,abs(S23),unwrap(angle(S23))];
 write(file1,"('port 2','TE',1,'port 3',1,'transmission')"
        +endl+"("
        +num2str(length(f))
        +",3)"
        +endl+num2str(S23_data)
        );
 S33_data=[f,abs(S33),unwrap(angle(S33))];
 write(file1,"('port 3','TE',1,'port 3',1,'transmission')"
        +endl+"("
        +num2str(length(f))
        +",3)"
        +endl+num2str(S33_data)
        );
 S43_data=[f,abs(S43),unwrap(angle(S43))];
 write(file1,"('port 4','TE',1,'port 3',1,'transmission')"
        +endl+"("
        +num2str(length(f))
        +",3)"
        +endl+num2str(S43_data)
        );

 # Sx4 Input on port 4
 S14_data=[f,abs(S14),unwrap(angle(S14))];
 write(file1,"('port 1','TE',1,'port 4',1,'transmission')"
        +endl+"("
        +num2str(length(f))
        +",3)"
        +endl+num2str(S14_data)
```

```
       );
S24_data=[f,abs(S24),unwrap(angle(S24))];
write(file1,"('port 2','TE',1,'port 4',1,'transmission')"
       +endl+"("
       +num2str(length(f))
       +",3)"
       +endl+num2str(S24_data)
       );
S34_data=[f,abs(S34),unwrap(angle(S34))];
write(file1,"('port 3','TE',1,'port 4',1,'transmission')"
       +endl+"("
       +num2str(length(f))
       +",3)"
       +endl+num2str(S34_data)
       );
S44_data=[f,abs(S44),unwrap(angle(S44))];
write(file1,"('port 4','TE',1,'port 4',1,'transmission')"
       +endl+"("
       +num2str(length(f))
       +",3)"
       +endl+num2str(S44_data)
       );


#test passivity of Smatrix
S_norm = matrix(1,FREQ_PTS);
S_err = S_norm;
for (ff=1:FREQ_PTS)
{
  S=[S11(ff),S12(ff),S13(ff),S14(ff);
     S21(ff),S22(ff),S23(ff),S24(ff);
     S31(ff),S32(ff),S33(ff),S34(ff);
     S41(ff),S42(ff),S43(ff),S44(ff)];

  S_norm(ff)=norm(S);
          S_err(ff) = max(abs( S-transpose(S)));
}

if (max(S_err) > 0.05){
  ? '******* Warning: S parameters violate reciprocity by more than 5% *********';
}
if (max(S_norm) > 1.01){
  ? '******* Warning: S parameters not passive *********';
}
else {if (max(S_norm) <1){
  ? '******* S parameters are passive ********';
}}
if (figs ==1 ){
  plot (f, S_norm, 'Wavelength (um)', 'Norm |S|', 'g=' + num2str(gap(gg)*1e9) + 'nm r=' +
        num2str(radius(rr)*1e6) + ' um');
  plot(c/f*1e6, S_err, 'Wavelength (um)', 'abs(s-transpose(S))','Reciprocity');
}


#XML export
#if coupling coeff frequency then loop over frequency. for now it is at 1550e-9;

design{1}.value = gap(gg);
design{2}.value = radius(rr);
design{3}.value = wg_width(1);
design{4}.value = si_thickness(1);
design{5}.value = Lc(1);
design{6}.value = Etch2(1);
extracted{1}.value = file1;
extracted{2}.value = num2str(abs(S42(length(f)/2+0.5))^2);
extracted{3}.value = num2str(abs(S32(length(f)/2+0.5))^2);
extracted{4}.value = num2str(1-abs(S42(length(f)/2+0.5))^2 - abs(S32(length(f)/2+0.5))^2);

#write design/extracted pair
lookupwrite( xml_filename, design, extracted );
```

```
?"radius loop: " + num2str(rr/length(radius)*100) + "% complete";
}
?"gap loop: " + num2str(gg/length(gap)*100) + "% complete";
}
?"Simulation - complete ";
lookupclose( filename );
```

Listing 9.2 Fit of S-parameters using a rational model with pole-residue form, using the vector fitting (FV) technique, and enforcement of passivity constraint.

```
function VectorFitting_Sparam ()
clear; close all; FONTSIZE=20;

file='../DC_ringmod_type1_R=10,gap=180,Lc=0,wg=500,lambda=1550,mesh=2,angle=30.mat';
load (file)

c=3e8; wavelength=c./f*1e6;
fSCALING = 1e14; f=f/fSCALING; s=1i*2*pi*f; Np=length(f);
hwait = waitbar(0,'Please wait...');

% average the S parameters for the symmetric parameters that were simulated twice,
% by considering amplitude and phase separately
S1221 = (abs(S12)+abs(S21))/2 .* exp ( 1i * (unwrap(angle(S12)) + unwrap(angle(S21)) ) /2
    );
S12=S1221; S21 = S1221;
S4132 = (abs(S41)+abs(S32))/2 .* exp ( 1i * (unwrap(angle(S41)) + unwrap(angle(S32)) ) /2
    );
S41=S4132; S32 = S4132;
S13=S31; S23=S41; S33=S11; S43=S21; S14=S32; S24=S42; S34=S12; S44=S22;


for i=1:Np
  Sparam = [ [ S11(i),S12(i),S13(i),S14(i)];
    [ S21(i),S22(i),S23(i),S24(i)];
    [ S31(i),S32(i),S33(i),S34(i)];
    [ S41(i),S42(i),S43(i),S44(i)] ] ;
  Test1(i) = norm(Sparam);
  Sparam_w (:,:,i)=Sparam;
end

% Check if S-Parameters are already passive; if not, perform Vector Fit,
% otherwise, export.
if ~isempty(find(Test1>1))

  % rational fit, sweep number of parameters, N:
  optsN=20:1:100;
  opts.poletype='lincmplx'; opts.parametertype='S'; opts.stable=0; opts.Niter_out=10;
  Npassive=[]; rms3passive=[];
  for i=1:length(optsN)
    %%%%%%%%%%%%%%%%%%%%%%%%%%%%%%%%%%%%%%%%%%%%%%%%%%%%%%%%%%%%%
    opts.N=optsN(i);  % Vector Fit:
    [SER,rmserr,Hfit,opts2]=VFdriver(Sparam_w,s,[],opts);

    %%%%%%%%%%%%%%%%%%%%%% Enforce Passivity:
    [SER,H_passive,opts3,wintervals]=RPdriver(SER,s,opts);
    % Note: added output parameter wintervals to RPdriver.

    % Find rms error:
    tell=0; Nc=length(SER.D);
    for col=1:Nc
      for row=col:Nc % makes assumption that S = S'
        tell=tell+1; % make a single vector:
        Sparam10(tell,:)= squeeze(Sparam_w(row,col,:)).';
        fit10(tell,:)= squeeze(Hfit(row,col,:)).';
        fitP(tell,:)= squeeze(H_passive(row,col,:)).';
      end
    end
    rms2(i) = sqrt(sum(sum(abs((Sparam10-fit10).^2))))/sqrt(4*Np);
    rms3(i) = sqrt(sum(sum(abs((Sparam10-fitP).^2))))/sqrt(4*Np);
```

```
  % Determine if the Passivity Enforcement was successful.
  if (rms3(i) < 1) && isempty(wintervals)
    Npassive=[Npassive optsN(i)]; rms3passive = [rms3passive rms3(i)];
  end

  waitbar (i/length(optsN), hwait);
  if (rms3(i) < 1e-4) && isempty(wintervals) ; break; end
end
close (hwait);

% Plot error versus number of fitting parameters:
figure;
semilogy(optsN(1:i),rms2,'ro-', 'LineWidth',2, 'MarkerSize',7); hold all;
labels={}; labels{end+1} = 'Vector Fit, rms';
plot (Npassive, rms3passive, 'kx', 'MarkerSize',14, 'LineWidth',3);
labels{end+1} = 'Passivity-Enforced Fit, rms';
xlabel('Fitting order, N'); ylabel('rms error');
for ii=1:length(labels); labels{ii}=[labels{ii} ' ' char(31) ]; end;
legend (labels,'Location','NorthEast');
printfig(file,'convergence');

% Plot residue values:
figure;
residues=sort(squeeze(prod(prod(abs(SER.R),1),2).^(1/4)), 1, 'descend');
semilogy(residues,'-s', 'LineWidth',3, 'MarkerSize',8);
xlabel('Residue index'); ylabel('Residue magnitude');
printfig(file,'residues');

%%%%%%%%%%%%%%%%%%%%%%%%%%%%%%%%%%%%%%%%%%%%%%
% Plot S-Parameters + fit functions:
figure;
Nc=length(SER.D);
for row=1:Nc
  for col= row:Nc
    dum1=squeeze(Sparam_w(row,col,:));
    dum2=squeeze(H_passive(row,col,:));
    h1=semilogy(wavelength,abs(dum1),'b','LineWidth',3); hold on
    h2=semilogy(wavelength,abs(dum2),'r-.','LineWidth',4);
    h3=semilogy(wavelength,abs(dum2-dum1),'g--','LineWidth',3);
  end
end
hold off
xlabel('Wavelength'); ylabel('Amplitude [S]');
axis tight; Yl =ylim; ylim ([Yl(1)*10 Yl(2)*2]);
labels={}; labels{end+1} = 'FDTD S-Parameters';
labels{end+1} = 'Passivity-enforced Fit'; labels{end+1}='Deviation';
for i=1:length(labels); labels{i}=[labels{i} ' ' char(31)]; end;
legend (labels,'Location','Best');
printfig(file,'VF2a');

figure;
Nc=length(SER.D);
for row= 1:Nc %1
  for col= row:Nc %3
    dum1=squeeze(Sparam_w(row,col,:));
    dum2=squeeze(H_passive(row,col,:));
    h1=semilogy(wavelength,unwrap(angle(dum1)),'b','LineWidth',3); hold on
    h2=plot(wavelength,unwrap(angle(dum2)),'r-.','LineWidth',4);
    h3=plot(wavelength,abs(unwrap(angle(dum2))-unwrap(angle(dum1))),'g--','LineWidth',3);
  end
end
hold off
xlabel('Wavelength'); ylabel('Phase [S]');
axis tight; Yl =ylim; ylim ([Yl(1)*10 Yl(2)*2]);
labels={}; labels{end+1} = 'FDTD S-Parameters';
labels{end+1} = 'Passivity-enforced Fit'; labels{end+1}='Deviation';
for i=1:length(labels); labels{i}=[labels{i} ' ' char(31)]; end;
legend (labels,'Location','Best');
printfig(file,'VF2p');
```

```
  %%%%%%%%%%%%%%%%%%%%%%%%%%%%%%
  % Passivity test results:
  for i=1:length(f)
    Test0(i) = norm(Sparam_w(:,:,i));
    Test1(i) = norm(Hfit(:,:,i));
    Test2(i) = norm(H_passive(:,:,i));
  end
  figure;
  plot (wavelength,Test0,'LineWidth',2); hold all;
  plot (wavelength,Test1,'--','LineWidth',2)
  plot (wavelength,Test2,'LineWidth',4)
  plot(wavelength,ones(length(f),1),'--','LineWidth',3);
  labels={}; labels{1} = 'FDTD S-Parameters'; labels{2} = 'Rational Fit'; labels{3} =
       'Passivity-enforced Fit'; labels{4}='Passivity limit';
  for i=1:length(labels); labels{i}=[labels{i} ' ' char(31)]; end; legend
       (labels,'Location','Best');
  axis tight
  xlabel ('Wavelength [\mum]');
  ylabel ('S-Parameter Passivity Test');
  printfig(file,'passivitytest3');
end

%%%%%%%%%%%%%%%%%%%%%%%%%%%%%%%%%%%%%%%
% export S parameters to INTERCONNECT
fid = fopen([file '.sparam'],'w'); Nc=4;
for row= 1:Nc
  for col= 1:Nc
    fprintf(fid,'%s\n',[ '(''port ' num2str(col) ''',''TE'',1,''port ' num2str(row)
         ''',1,''transmission'')' ] );
    fprintf(fid,'%s\n',[ '(' num2str(Np) ',3)' ] );
    dum2=squeeze(H_passive(row,col,:));
    mag = abs(dum2);
    phase = unwrap (angle(dum2)); % figure; plot (wavelength, phase);
    for i=1:Np
      fprintf(fid,'%g %g %g\n', f(i)*1e14, mag(i), phase(i));
    end
  end
end
fclose(fid);


function printfig (file, b)
global PRINT_titles;
PRINT_titles=0;
FONTSIZE=20;
set(get(gca,'xlabel'),'FontSize',FONTSIZE);
set(get(gca,'ylabel'),'FontSize',FONTSIZE);
set(get(gca,'title'),'FontSize',FONTSIZE-5);
set(gca,'FontSize',FONTSIZE-2);
if PRINT_titles==0
  delete(get(gca,'title'))
end
%a=strfind(file,'.'); file(a)=',';
pdf = [file(1:end-4) '_' b '.pdf'];
print ('-dpdf','-r300', pdf);
system([ 'pdfcrop ' pdf ' ' pdf ' &' ]);
% system(['acroread ' pdf '.pdf &']);
```

Listing 9.3 Script to build a ring modulator compact model in INTERCONNECT, using the S-parameters for the directional coupler, and the phase vs. voltage data from the pn-junction

```
# Script to build a ring modulator compact model in INTERCONNECT
# Uses:
#      S-Parameters for Directional Coupler
#      Phase vs. voltage data from PN junction
# by Jonas Flueckiger
```

```
switchtolayout;
deleteall;

R=15e-6;

#Add Optical Network Analyser
elementName = addelement('Optical Network Analyzer');
setnamed(elementName, 'x position', 200);
setnamed(elementName,'y position',100);
setnamed(elementName, 'input parameter', 'center and range');
setnamed(elementName, 'center frequency', 193.1e12);
setnamed(elementName, 'frequency range', 10000e9);
setnamed(elementName, 'plot kind', 'wavelength');
setnamed(elementName, 'relative to center', false);
setnamed(elementName, 'number of input ports', 2);
setnamed(elementName, 'name', 'Optical Network Analyzer');

#Add directional couplers
#Add N-port S-parameter element
?elementName = addelement('Optical N Port S-Parameter');
setnamed(elementName,'x position',300);
setnamed(elementName,'y position',400);
setnamed(elementName, 'passivity','test'); # make sure s-param file gets tested
setnamed(elementName, 'reciprocity','test');
setnamed(elementName, 'load from file','true');
setnamed(elementName, 's parameters filename', 'dc_R=5,gap=200,Lc=0,wg=500.sparam');
setnamed(elementName, 'name', 'Directional Coupler 1');

copy(0,200);
set('name','Directional Coupler 2');

#Add waveguide to make ring
elementName = addelement('Straight Waveguide');
setnamed(elementName, 'x position', 500);
setnamed(elementName, 'y position',500);
rotateelement(elementName);
setnamed(elementName, 'length', pi*R);
setnamed(elementName, 'loss 1', 300);
# Waveguide is here only to provide loss; propagation is taken into account by the
      directional coupler
setnamed(elementName, 'effective index 1', 0);
setnamed(elementName, 'group index 1', 0);
setnamed(elementName, 'name','WG');

#Add phase modulator
elementName = addelement('Optical Modulator Measured');
setnamed(elementName, 'x position', 150);
setnamed(elementName, 'y position',500);
flipelement(elementName);
rotateelement(elementName);
setnamed(elementName, 'operating frequency', 'user defined');
setnamed(elementName, 'frequency', 193.1e12);
setnamed(elementName, 'length', pi*30e-6);
setnamed(elementName, 'load from file', false);
setnamed(elementName, 'measurement type','effective index');
setnamed(elementName, 'name','Phase Modulator');

#Add DC source
elementName = addelement('DC Source');
setnamed(elementName, 'x position', 5);
setnamed(elementName, 'y position',500);
setnamed(elementName, 'amplitude', 0);
setnamed(elementName, 'name', 'DC Source');

connect('WG', 0, 'Directional Coupler 1', 3);
connect('WG', 1, 'Directional Coupler 2', 2);
connect('Phase Modulator', 0, 'Directional Coupler 1', 1);
connect('Phase Modulator', 2, 'Directional Coupler 2', 0);
connect('DC Source', 0, 'Phase Modulator', 1);
connect('Optical Network Analyzer', 1, 'Directional Coupler 1', 2);
```

```
connect('Optical Network Analyzer', 0, 'Directional Coupler 1', 0);
connect('Optical Network Analyzer', 2, 'Directional Coupler 2', 1);

run;

t1=getresult("Optical Network Analyzer", "input 1/mode 1/gain");
q2=getresult("Optical Network Analyzer", "input 2/mode 1/peak/quality factor");
t2=getresult("Optical Network Analyzer", "input 2/mode 1/gain");
wvl1= t1.getparameter('wavelength');
t1= t1.getattribute("'TE' gain (dB)");
t2= t2.getattribute("'TE' gain (dB)");
q2wvl = q2.getparameter('wavelength');
q2 = q2.getattribute("'TE' quality factor");
angle1=getresult("Optical Network Analyzer", 'input 1/mode 1/angle');
angle1=angle1.getattribute("'TE' angle (rad)");


plot(wvl1*1e6,t1,t2,'wavelength [micron]','Amplitude [dB]','Ring Modulator Spectrum');
legend('through','drop');

plot(wvl1*1e6,unwrap(angle1),'wavelength [micron]','Phase [rad]','Ring Modulator
     Spectrum');
legend('through');

#switchtolayout;
```

Listing 9.4 Grating coupler 2D FDTD simulations to extract the S-parameters; http://siepic.ubc.ca/files/GC_S_extraction.lsf

```
# S Parameter extraction for the grating coupler
# Port 1 = fibre
# Port 2 = waveguide

newproject;
redrawoff;

GC_init;
GC_setup_fibre;

#from waveguide to fibre
select("waveguide_source"); set("enabled",1);
select("fibre::fibre_mode"); set("enabled",0);
run;

T = getresult("waveguide","expansion for T");
fibre = getresult("fibre::fibre_modeExpansion","expansion for fibre_top");
f = T.f;
S22 = T.b / T.a;
S12 = fibre.a*sqrt(fibre.N) / (T.a*sqrt(T.N));

#from fibre to waveguide
switchtolayout;
select("waveguide_source"); set("enabled",0);
select("fibre::fibre_mode"); set("enabled",1);
run;

T = getresult("waveguide","expansion for T");
fibre = getresult("fibre::fibre_modeExpansion","expansion for fibre_top");
S11 = fibre.a / fibre.b;
S21 = T.b*sqrt(T.N) / (fibre.b*sqrt(fibre.N));

plot(c/f*1e6,10*log10(abs([S12, S21,S11,S22])),'Wavelength
     (micron)','Transmission/Reflection (dB)');
legend('S12','S21','S11','S22');
plot(c/f*1e6,unwrap(angle([S12, S21,S11,S22])),'Wavelength (micron)','Phase (rad)');
legend('S12','S21','S11','S22');

# export S parameters for INTERCONNECT
```

```
Sdata = [ f, abs(S11), unwrap(angle(S11)), abs(S21), unwrap(angle(S21)), abs(S12),
     unwrap(angle(S12)), abs(S22), unwrap(angle(S22)) ];

filename = "GC_Sparam.dat";
rm(filename);
format long;
write(filename,num2str(Sdata));
format short;

matlabsave ("GC_Sparam");
```

References

[1] Gyung-Jin Park. "Design of experiments". *Analytic Methods for Design Practice.* Springer, 2007, pp. 309–391 (cit. on p. 314).

[2] *Design of Experiments (DOE) with JMP.* [Accessed 2014/04/14]. URL: http://www.jmp.com/applications/doe (cit. on p. 314).

[3] Bjorn Gustavsen and Adam Semlyen. "Rational approximation of frequency domain responses by vector fitting". *IEEE Transactions on Power Delivery* **14**.3 (1999), pp. 1052–1061 (cit. on pp. 316, 325).

[4] Robert X. Zeng and Jeffery H. Sinsky. "Modified rational function modeling technique for high speed circuits". *Microwave Symposium Digest, 2006. IEEE MTT-S International* (2006), pp. 1951–1954 (cit. on p. 316).

[5] Lukas Chrostowski, Jonas Flueckiger, Charlie Lin, *et al.* "Design methodologies for silicon photonic integrated circuits". *Proc. SPIE, Smart Photonic and Optoelectronic Integrated Circuits XVI* 8989 (2014), 15 pages (cit. on pp. 319, 320, 328, 329, 330).

[6] *Enforce passivity of S-parameters – MATLAB makepassive.* [Accessed 2014/04/14]. URL: http://www.mathworks.com/help/rf/makepassive.html (cit. on p. 324).

[7] *The Vector Fitting Web Site.* [Accessed 2014/04/14]. URL: http://www.sintef.no/Projectweb/VECTFIT (cit. on pp. 325, 326).

[8] Bjorn Gustavsen and Adam Semlyen. "Fast passivity assessment for S-parameter rational models via a half-size test matrix". *IEEE Transactions on Microwave Theory and Techniques* **56**.12 (2008), p. 2701 (cit. on p. 325).

[9] Bjorn Gustavsen. "Improving the pole relocating properties of vector fitting". *IEEE Transactions on Power Delivery* **21**.3 (2006), pp. 1587–1592 (cit. on p. 325).

[10] Dirk Deschrijver, Michal Mrozowski, Tom Dhaene, and Daniel De Zutter. "Macromodeling of multiport systems using a fast implementation of the vector fitting method". *IEEE Microwave and Wireless Components Letters* **18**.6 (2008), pp. 383–385 (cit. on p. 325).

[11] Bjorn Gustavsen. "Fast passivity enforcement for S-parameter models by perturbation of residue matrix eigenvalues". *IEEE Transactions on Advanced Packaging* **33**.1 (2010), p. 257 (cit. on p. 325).

[12] Bjorn Gustavsen. "Computer code for rational approximation of frequency dependent admittance matrices". *IEEE Transactions on Power Delivery* **17**.4 (2002), pp. 1093–1098 (cit. on p. 325).

10 Tools and techniques

This chapter describes the tools used by photonic integrated circuit designers, in particular for those focusing on the physical implementation of the design. We begin with a discussion of process design kits (PDKs) typically provided by the fabrication foundry. This is followed by a discussion of what electronic design automation (EDA) tools offer, including a library of components, schematic capture, (schematic-driven) layout, and design rule checking. We also provide suggestions for space-efficient photonic mask layout.

10.1 Process design kit (PDK)

A process design kit is a set of documentation and data files that describe a fabrication process at a semiconductor foundry and enable the user to complete a design. A typical PDK contains: documentation including technology details, mask layout instructions, and design rules; a library of cells such as modulators, detectors, etc.; component models and/or experimental data; and design verification tools. PDKs usually contain the foundry's proprietary information and trade secrets, thus are not always openly available.

In this section, we describe a Generic Silicon Photonics (GSiP) PDK, which is implemented in Mentor Graphics (Pyxis and Calibre) and Lumerical INTERCONNECT tools and is available for download. The purpose of this kit is to demonstrate the functionality of a silicon photonics design flow implementation, with no restrictions to its distribution. This kit can be adapted for different fabrication processes, and also provides insight into what PDK and libraries available today offer [1, 2].

The components of the GSiP PDK include the following.

- Fabrication process parameters, mask layer table.
- Library: a small example library of components, including fibre grating couplers; waveguides, waveguide bends, and a splitter; a ring modulator; and an electrical bond pad.
  - Component symbols for schematic capture: using the provided components and/or user-provided components, circuits can be designed at the schematic level.

  - Component models: the library components include circuit models implemented in Lumerical INTERCONNECT.
  - Component physical layout: mask layout for components is implemented in fixed-layout cells (i.e. GDS, e.g. Y-branch splitter) and parameterized (i.e. PCells, e.g. ring modulator).
- Schematic capture: this functionality allows the designer to create a schematic of their system. This stage of the design includes defining the connectivity between components (netlist), labelling components and ports, and choosing parameters for the PCells.
- Circuit simulations: the schematic is exported as a netlist, and loaded in the circuit simulation tool, Lumerical INTERCONNECT. Component models are loaded from the PDK. The connectivity is imported from the netlist. A simulation testbench can be defined to perform specific simulations, such as optical transmission spectrum, time domain characterization, etc. Functionality of the system can be simulated and verified.
- Schematic-driven layout (SDL): prior to the advent of modern EDA tools, the design focus was on drawing polygons in the physical layout. In this approach, the components and connections are already defined, and the task is to place the components and route using the already-defined connectivity (place and route). The components and connectivity are imported from the schematic and (automatically) instantiated. The connectivity is graphically represented. The interactive (or automated) routing tool is used to complete the metal and optical routing.
- Waveguide routing: electrical routing has been largely automated in the EDA tools. In photonics designs, the tool needs additional considerations for smooth waveguide bends, different types of waveguides, wide low-loss waveguides for long-distance routing, and waveguide crossings.
- Design rule checking (DRC): foundry supplied rules include minimum feature size, minimum spacing, inclusion and exclusion rules. Basic rules are included in this PDK for typical minimum feature sizes. These rules are primarily concerned with what is permitted by the foundry to yield a manufacturable design.
- Layout versus schematic (LVS): while DRC identifies errors that violate manufacturing rules, it does not identify circuit or construction errors. LVS plays this role, by comparing the schematic and the physical layout. It operates on the layout and identifies components and connections, creates a netlist, and compares the netlist with the original design schematic. It identifies circuit differences, to isolate errors such as net errors (broken waveguides or metallic interconnects, disconnected optical and electrical ports, accidental crossings of interconnects), and component errors (missing components, incorrect component placed, wrong PCell parameters such as incorrect ring resonator radius).
- Tiling: one of the manufacturability design rules concerns the density of the patterns. In order to meet the minimum density rules, tiles are added to the layout, typically to the silicon and metal layers. This is necessary to ensure planarity during the chemical mechanical polish (CMP) fabrication step, and also to have an

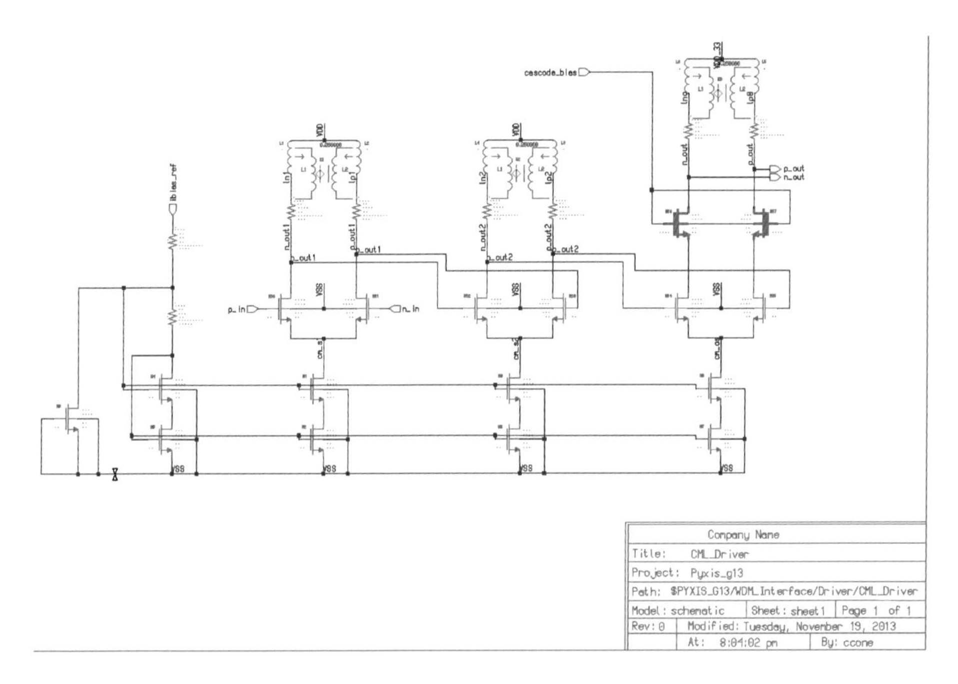

Figure 10.1 Mentor Graphics – electrical design.

etch density that is as uniform as possible leading to reduced process variability. This GSiP does not include the tiling script as it uses a proprietary language.

- Sub-system design example and tutorial: a two-channel wavelength division multiplexed (WDM) optical transmitter using ring modulators. Example includes schematic, circuit simulations including optical spectra and eye diagrams, schematic-driven layout, design rule checking, layout versus schematic, and post-layout extraction.
- Electronic/photonic co-design: the GSiP PDK contains both optical and electronic generic technology, which enables two separate chips to be co-designed. The tools provide for data exchange between the optical and the electrical simulations. Namely, one can design a CMOS modulator driver, and the resulting waveform can be used to drive the modulator. Similarly, the light received by the detector is converted to a photo-current, which drives the trans-impedance amplifier. With this approach, it is possible to design a complete CMOS-to-photonic-to-CMOS transmitter/receiver. In this flow, both the photonic and electronic generic PDKs can be replaced with actual foundry-provided PDKs, which offers the designer flexibility in choosing which foundries are used to fabricate each of the two chips. This design flow is presently based on data exchange, hence does not offer co-simulation of electronics and optics, and would not be suitable for the simulations that require lock-step self-consistent simulations, such as microwave photonic electro-optical oscillators. An example of an electrical design is shown in Figure 10.1.

Table 10.1 Generic Silicon Photonic PDK – fabrication process parameters

Parameter	Target value	Reference
Silicon thickness	220 nm	[4–7]
Silicon etch 1	60 nm	[1]
Silicon etch 2	130 nm	[1]
Buried oxide thickness	2000 nm	[4–7]
Germanium thickness	500 nm	[1]
Doping (N)	$5 \times 10^{17} cm^{-3}$	[1]
Doping (P)	$7 \times 10^{17} cm^{-3}$	[1]
Doping (N++)	$5 \times 10^{20} cm^{-3}$	[1]
Doping (P++)	$1.9 \times 10^{20} cm^{-3}$	[1]

10.1.1 Fabrication process parameters

Silicon thickness and etch

There are several silicon thicknesses used by researchers and industry, including 3 μm (e.g., Kotura [3]), 300 nm (e.g. Luxtera), 260 nm (e.g. NRC), and 220 (e.g. IMEC [4], LETI [5], OpSIS [1], IME [6]). The GSiP PDK uses the 220 nm standard, with parameters listed in Table 10.1. With this thickness, several etch depths are useful.

(1) Grating coupler etch: a shallow etch of 60 nm [1] to 70 nm [4] is optimal for grating couplers.

(2) Rib waveguide etch: a separate etch can be used for the rib waveguides [1]. In this case, the etch depth is chosen as a compromise between (a) reducing rib waveguide bend loss, which requires a deeper etch, and (b) reducing the electrical resistance in pn-junction modulators, which requires a shallower etch.

GDS layer map

Table 10.2 lists the mask layers used in the Generic Silicon Photonics technology. The masks included etch and deposition (for silicon and germanium), implantation for electrically active devices, and metallization. Miscellaneous layers are included to aid in the design flow.

Design rules

This describes the rules between geometries in the layout. This includes minimum feature sizes, etc., and is described further in Section 10.1.6.

10.1.2 Library

Figure 10.2 shows examples of schematic symbols in the library. As an example, the library includes a parameterized double-bus racetrack modulator (can be a point-coupled ring modulator), with a physical mask layout as shown in Figure 6.10. The parameters for this device include: radius, directional coupler gap and length, waveguide width, parameters controlling the dimensions of the doped regions, and a pn-junction offset relative to the centre of the waveguide.

Table 10.2 Generic Silicon Photonic PDK – GDS layer map.

Name	Description	GDS #
	Materials and etches	
Si	Full thickness silicon; for waveguides and active devices	1
SiEtch1	Silicon shallow etch; used to define grating couplers	2
SiEtch2	Silicon shallow etch; used to define rib waveguides	3
SiEtch3	Silicon deep etch; trench for edge coupling or cleaving	4
Ge	Ge growth for detector	5
OxEtch	Oxide etch; access to the silicon waveguides	6
	Implants	
N	Doping (N)	20
P	Doping (P)	21
N+	Doping (N+)	22
P+	Doping (P+)	23
Npp	Doping (N++)	24
Ppp	Doping (P++)	25
GeN	Germanium N doping; for Ge detectors	26
GeP	Germanium P doping; for Ge detectors	27
Defect1	Si defects 1; for ion implanted defect-mediated detectors	28
	Metals	
VC	Metal contact via between N++/P++ layers and Metal1	40
M1	Metal 1 for interconnects	41
V1	Via 1, above Metal 1	42
M2	Metal 2 for interconnects	43
VL	Last Via, connecting M1 (or M2) with last metal	44
ML	Last Metal, for interconnects and for electrical probe/bond pads	45
MLOpen	Opening for the Last Metal; for electrical probing	46
	Miscellaneous	
M1KO	Tiling keep-out for M1	60
M2KO	Tiling keep-out for M2	61
MLKO	Tiling keep-out for ML	62
SiKO	Tiling keep-out for Si	63
fp	Design outline, used for error checking and floorplanning	63
Dicing	Dicing lanes	65
Text	Text comments, will not be printed; used for automated measurements	66
DRCex	Exclusion layer for DRC checking	67
devrec	LVS: device recognition layer	68
pinrec	LVS: pin recognition layer	69
fbrtgt	LVS: fibre target pin layer	81
bndtgt	LVS: electrical bond target pin layer	82

10.1.3 Schematic capture

The schematic diagram for a two-channel optical transmitter is shown in Figure 10.3. Components are instantiated using a symbol selector or searching through the library.

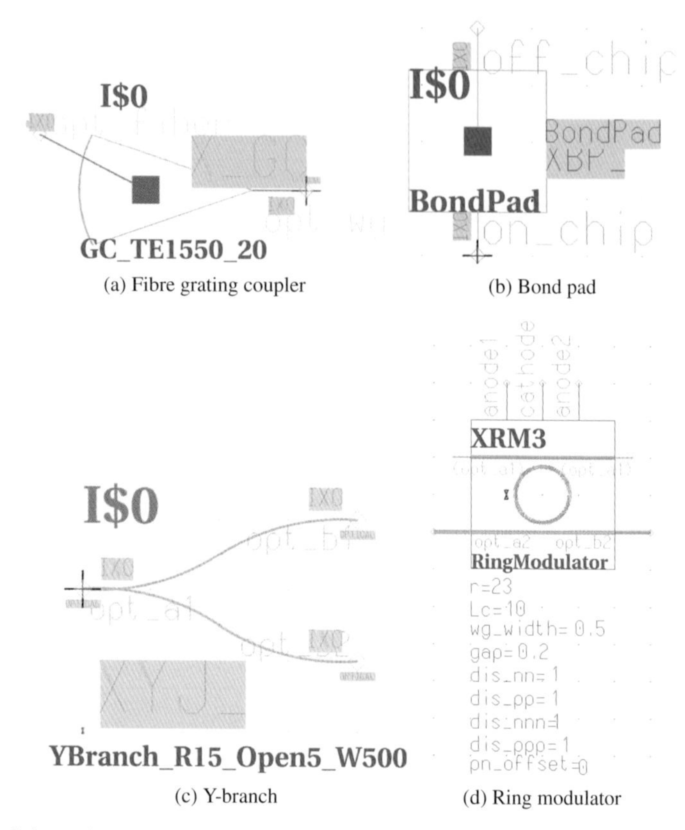

(a) Fibre grating coupler (b) Bond pad (c) Y-branch (d) Ring modulator

Figure 10.2 Schematic symbols for library components. Reprinted with permission [8].

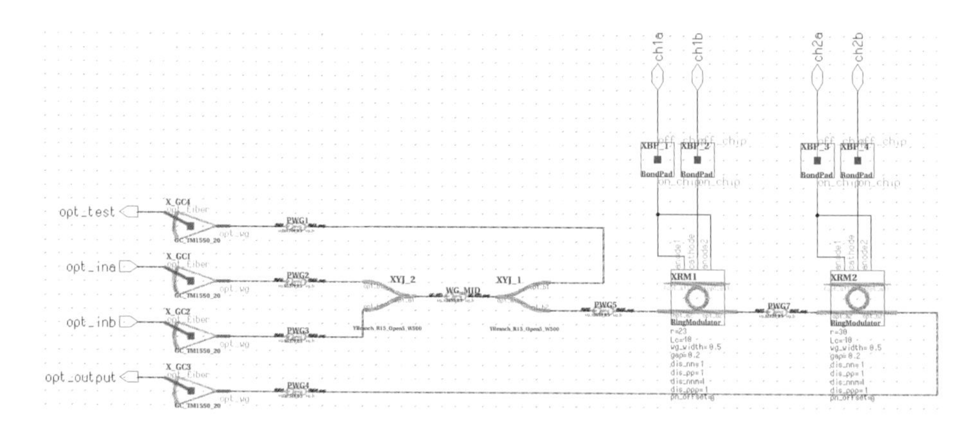

Figure 10.3 Schematic for the example system. Reprinted with permission [8].

Wires are added to connect the components (both optical and electrical connections). The interface signals to the chip (bond pads, grating couplers) are established by labelling the pin net, and adding input/output ports.

The schematic is further updated to insert "pwg" waveguide devices between components to help track the total routed waveguide lengths in the layout. The "pwg" device is used in the layout for waveguide routing and post-layout extraction. This functionality can be used to impose routing constraints or initial waveguide length estimates.

The schematic is checked for errors and saved. The optical signals are annotated with a different colour to distinguish from the electrical signals. Additionally, rule checks ensure that optical signals are represented point to point and run only between optical pins.

10.1.4 Circuit export

Pyxis exports a netlist of the schematic for circuit simulations (Listing 10.1). The output netlist appears as follows. The format for the netlist is chosen to be based on SPICE. The schematic is represented as a sub-circuit (.subckt command), with the input/output terminals defined. Next, each component is instanced, using a format consisting of a unique label (e.g. XYJ_2); a series of nets this component is connected to; and the library component name (e.g. YBranch_R15_Open5_W500). The library component name is common throughout the library (symbol, simulation compact model, layout). Parameterized cells (e.g. RingModulator) have additional parameters and values listed (e.g. $r = 30$).

Listing 10.1 Netlist export of circuit schematic.

```
.CONNECT GROUND 0
*
* MAIN CELL: Component pathname : $PYXIS_GSIP/FbrTx/wdm2
*
.subckt WDM2 OPT_OUTPUT OPT_INB OPT_INA OPT_TEST CH2B CH2A CH1B CH1A

  XYJ_1 WG_MID PWG1 PWG5 YBranch_R15_Open5_W500
  XBP_4 CH2B N$30 BondPad
  XBP_3 CH2A N$28 BondPad
  XBP_2 CH1B N$20 BondPad
  XBP_1 CH1A N$22 BondPad
  X_GC3 OPT_OUTPUT PWG4 GC_TM1550_20
  X_GC4 OPT_TEST PWG1 GC_TM1550_20
  X_GC2 OPT_INB PWG3 GC_TM1550_20
  X_GC1 OPT_INA PWG2 GC_TM1550_20
  XYJ_2 WG_MID PWG2 PWG3 YBranch_R15_Open5_W500
  XRM2 PWG7 PWG4 N$28 N$28 N$30 RingModulator r=30 w=0.5 gap=0.2 Lc=10
  XRM1 PWG5 PWG7 N$22 N$22 N$20 RingModulator r=23 w=0.5 gap=0.2 Lc=10
.ends WDM2
```

For circuit simulations, the schematic is imported into Lumerical INTERCONNECT. In order to have the same visual representation of the schematic, the positions of the instances are also exported (Listing 10.2).

Listing 10.2 Instance positions for the circuit schematic.

```
###Instance positions for : $PYXIS_GSIP/FbrTx/wdm2
portin opt_ina -11.0 1.0 0 @false
portin opt_inb -11.0 0.0 0 @false
portout opt_test -11.0 2.0 0 @true
portout opt_output -11.0 -1.0 0 @true
portbi ch1a -1.25 4.25 90 @false
portbi ch1b -0.5 4.25 90 @false
portbi ch2a 1.75 4.25 90 @false
```

```
portbi ch2b 2.5 4.25 90 @false
RingModulator XRM1 -1.25 0.25 0 @false
RingModulator XRM2 1.75 0.25 0 @false
YBranch_R15_Open5_W500 XYJ_2 -5.75 0.5 0 @true
GC_TM1550_20 X_GC1 -9.25 0.75 0 @false
GC_TM1550_20 X_GC2 -9.25 -0.25 0 @false
GC_TM1550_20 X_GC3 -9.25 -1.25 0 @false
GC_TM1550_20 X_GC4 -9.25 1.75 0 @false
BondPad XBP_1 -1.25 2.5 0 @false
BondPad XBP_2 -0.5 2.5 0 @false
BondPad XBP_3 1.75 2.5 0 @false
BondPad XBP_4 2.5 2.5 0 @false
YBranch_R15_Open5_W500 XYJ_1 -4.25 0.5 0 @false
```

Note that the above netlist does not include the "pwg" waveguide devices. These can be included during the export if they are desired for the simulation.

10.1.5 Schematic-driven layout

Next, a new layout is created using the schematic component connectivity. In this mode, both the layout and schematic are visible simultaneously, and the schematic components are semi-automatically ("AutoInst") instantiated in the layout. The instantiation is typically done in groups (e.g. first all the modulators, then all the pads, then all the grating couplers). The "AutoInst" function tries to preserve the relative positions and orientation of all cells in the schematic. During the layout, alignment tools are used to ensure that waveguide ports line up horizontally and vertically, to minimize the number of S-bends and keep the waveguides straight. The layout is shown in Figure 10.4.

It is at this stage in the layout that the placement of components is established, with consideration for testing (Design for test; see Section 12.3). Especially critical is the location of the optical input/outputs, and their relative position with respect to the electrical ones. See also Listing 10.3.

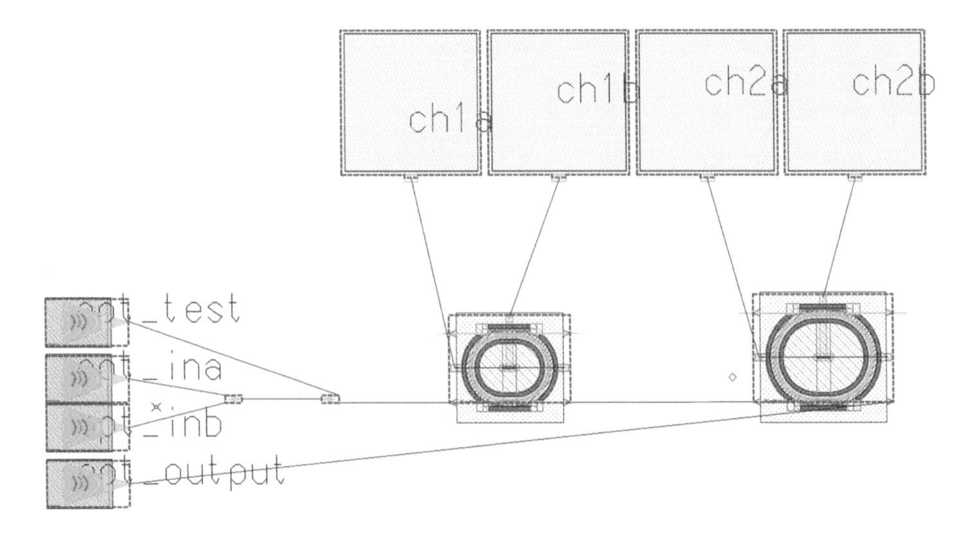

Figure 10.4 Layout for the example system, prior to routing. Reprinted with permission [8].

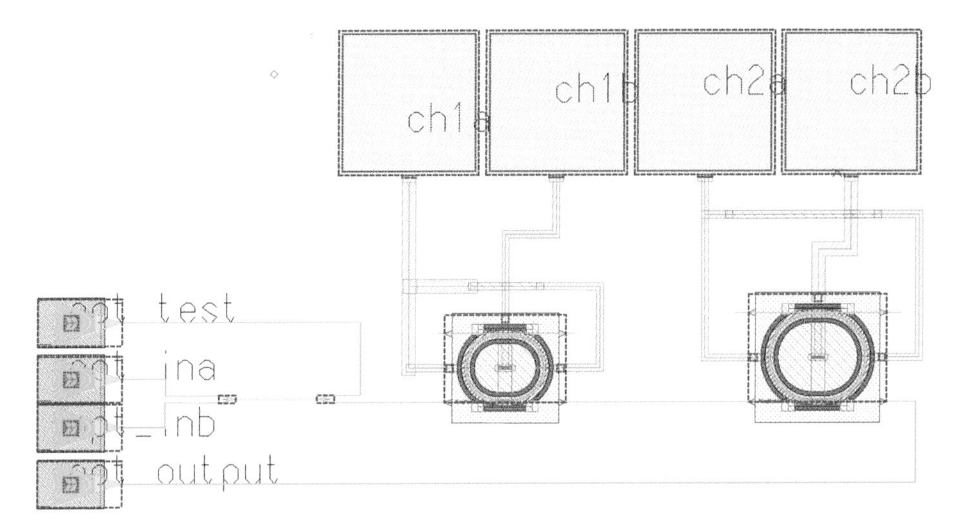

Figure 10.5 Layout for the example system, after routing. Reprinted with permission [8].

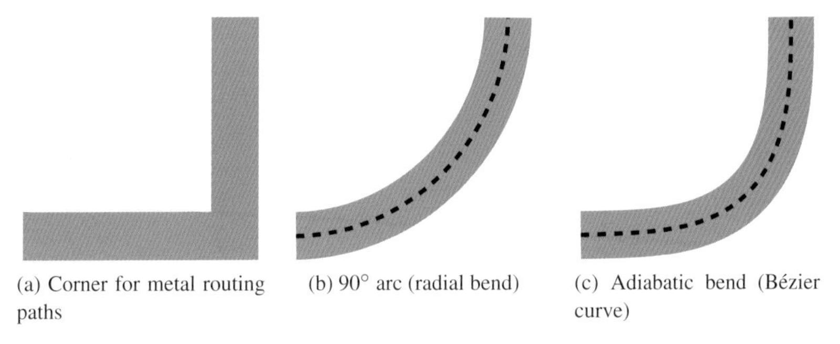

(a) Corner for metal routing paths

(b) 90° arc (radial bend)

(c) Adiabatic bend (Bézier curve)

Figure 10.6 Possible 90° bends for metallic and optical interconnects. Reprinted with permission [8].

Next, electrical and optical routing is performed by using the Pyxis "IRoute" interactive routing tool. This creates path objects for both electrical and optical routes, and the optical routes have sharp 90° corners, as shown in Figure 10.5.

The optical routes are then converted to optical waveguides. This is done using the "Make PWGs" function. This function implements the following features.

- Waveguide bends: instead of a 90° perpendicular corner like in metal interconnects (Figure 10.6a), optical waveguides require smooth bends, which can be in the form of radial bends (circles, Figure 10.6b), or other adiabatic curves [9, 10] (Figure 10.6c). The choice in the bend parameters is dependent on the performance requirements (insertion loss, back-reflections), the type of waveguide, the wavelength, and the polarization (e.g. TM modes require larger bend radii). This PDK gives the designer the choice of radius and bend type (radial versus adiabatic Bézier), and is defined for different types of waveguides (strip, rib).

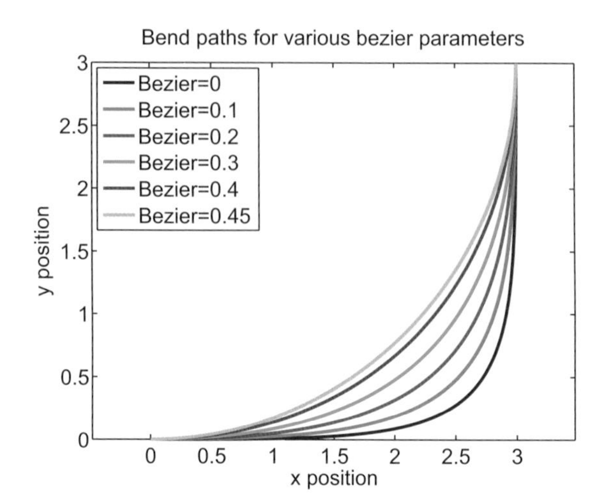

Figure 10.7 Bézier bend with length L =3 μm: waveguide path.

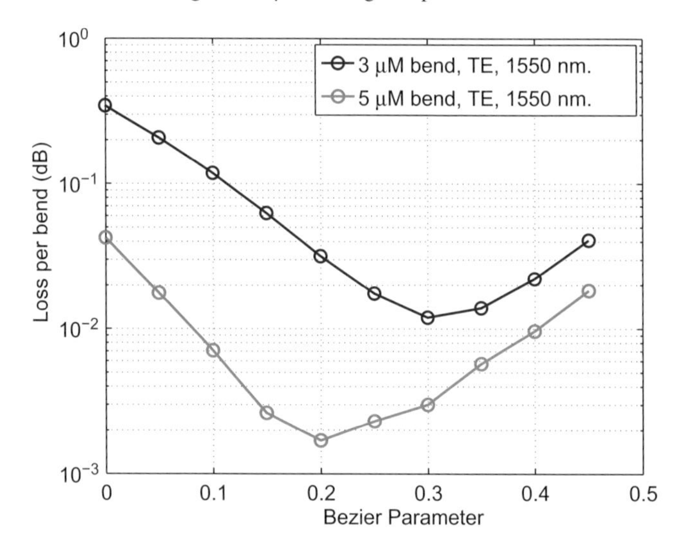

Figure 10.8 3D FDTD simulation results for the Bézier bends for L = 3 μm, and L = 5 μm. Note that when the Bézier parameter is 0.45, this is approximately a conventional arc with constant radius. Reprinted with permission [8].

As described in Section 3.3, the dominant loss mechanism in strip waveguides (for TE polarization) is mode-mismatch loss, which can be reduced by varying the curvature continuously. An adiabatic 90° bend based on a Bézier curve can have a lower optical insertion loss than a traditional arc with a constant radius of curvature. This structure has been modeled for TE polarization at 1550 nm in strip waveguides, using the method described in Section 3.3.1. The Bézier parameter describes the variation from a constant radius (Bézier = 0.45) to a natural Bézier curve (Bézier = 0.45), as shown in Figure 10.7. As shown in Figure 10.8, much lower losses are achieved for the adiabatic bends based on Bézier curves.

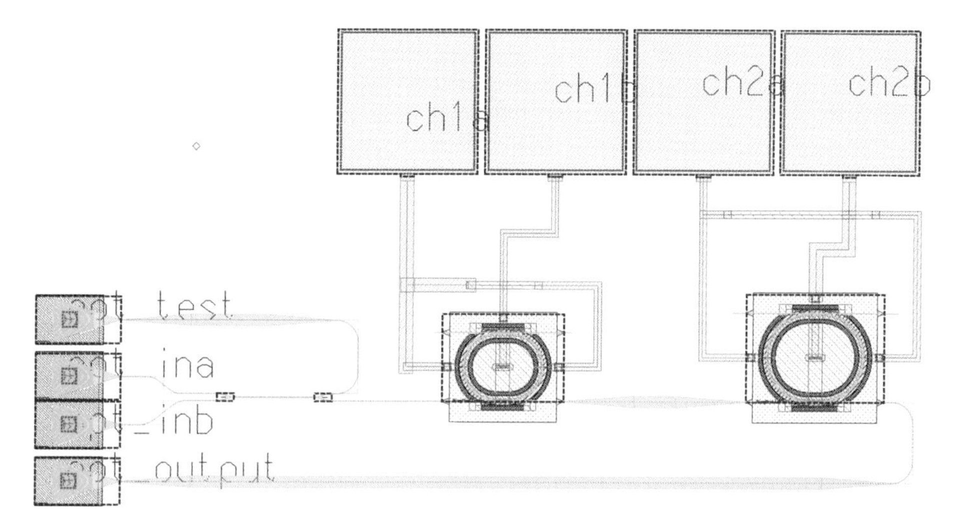

Figure 10.9 Layout for the example system, with optical paths converted into waveguides. Reprinted with permission [8].

The best 5 μm bend has the same performance as a radial 20 μm bend, whereas the best 3 μm bend has the same performance as a radial 6 μm bend (compare with Figure 3.28a). Hence, these bends are more compact for the same allowable insertion loss. The designer must choose the appropriate parameter based on the waveguide operation. It should be noted that there is little improvement in using adiabatic bends for TM polarization since the dominant loss mechanism is radiative loss, as opposed to mode-mismatch loss.

- S-bends: this structure is used to smoothly connect two disjointed and offset straight waveguides. Optionally, it is automatically inserted when necessary.
- Automated augmented waveguides: for long-distance optical routing, it is desirable to use lower-loss waveguides. This is done by converting from a strip waveguide in the bend regions to a wide rib waveguide [9, 11]. The wide waveguides can be either single mode (700 nm wide with 0.27 dB/cm loss in Reference [9]) or multi-mode (3 μm wide with 0.026 dB/cm loss in a passive process, and 0.75 dB/cm loss in a full-flow, Reference [11]). Experimental results for devices fabricated by IME in a full-flow process show that the propagation loss is <0.06 dB/cm; these waveguides had 3 μm wide ribs, with a 5 μm slab.

  This PDK implements an optional automated waveguide augmentation. The parameters include a length threshold for when to use the augmentation, a taper length, and the waveguide width. (See also Figure 10.9.)
- Waveguide crossings: unlike electrons, photons can cross each other without interacting. This simplifies the fabrication process since it does require multi-layer routing (as in electronics) – and creating multi-layer photonic circuits is challenging. Low-loss and low-cross-talk waveguide crossings have been designed and fabricated [12, 13], and have been used in routing, and incorporated into devices such as ring resonators [12]. The crossings are added by the designer, similar to the other components.

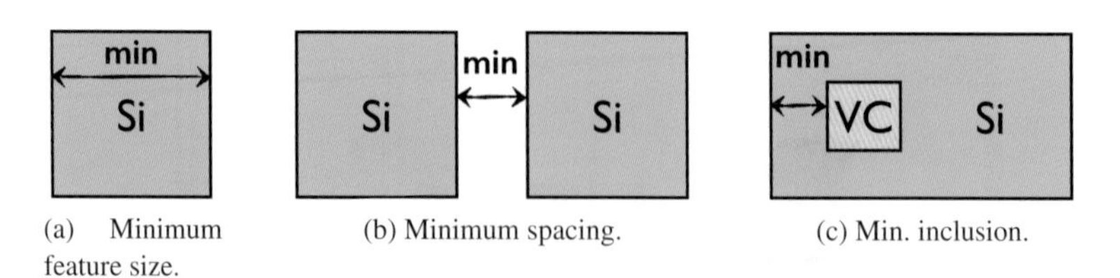

(a) Minimum feature size. (b) Minimum spacing. (c) Min. inclusion.

Figure 10.10 Types of design rule checking (DRC) rules.

Listing 10.3 Netlist export of circuit schematic, after post-layout extraction to add waveguide lengths.

```
.CONNECT GROUND 0
*
* MAIN CELL: Component pathname : $PYXIS_GSIP/FbrTx/wdm2
*
.subckt WDM2 OPT_OUTPUT OPT_INB OPT_INA OPT_TEST CH2B CH2A CH1B CH1A

    X_PWG5 PWG5 PWG5_PWG pwg wg_length=36.95 wg_width=0.5
    XYJ_1 WG_MID_PWG PWG1_PWG PWG5 YBranch_R15_Open5_W500
    X_WG_MID WG_MID WG_MID_PWG pwg wg_length=36.95 wg_width=0.5
    X_PWG4 PWG4 PWG4_PWG pwg wg_length=727.516 wg_width=0.5
    X_PWG3 PWG3 PWG3_PWG pwg wg_length=96.279 wg_width=0.5
    X_PWG2 PWG2 PWG2_PWG pwg wg_length=96.279 wg_width=0.5
    X_PWG1 PWG1 PWG1_PWG pwg wg_length=314.716 wg_width=0.5
    XBP_4 CH2B N$30 BondPad
    XBP_3 CH2A N$28 BondPad
    XBP_2 CH1B N$20 BondPad
    XBP_1 CH1A N$22 BondPad
    X_GC3 OPT_OUTPUT PWG4 GC_TM1550_20
    X_GC4 OPT_TEST PWG1 GC_TM1550_20
    X_GC2 OPT_INB PWG3 GC_TM1550_20
    X_GC1 OPT_INA PWG2 GC_TM1550_20
    XYJ_2 WG_MID PWG2_PWG PWG3_PWG YBranch_R15_Open5_W500
    X_PWG7 PWG7 PWG7_PWG pwg wg_length=121.25 wg_width=0.5
    XRM2 PWG7_PWG PWG4_PWG N$28 N$28 N$30 RingModulator r=30 w=0.5 gap=0.2 Lc=10
    XRM1 PWG5_PWG PWG7 N$22 N$22 N$20 RingModulator r=23 w=0.5 gap=0.2 Lc=10
.ends WDM2
```

10.1.6 Design rule checking

There are three primary rule types: minimum feature size, minimum spacing, and minimum inclusion. These are illustrated in Figure 10.10. Additional rules to be considered include density rules,

Two example rules are listed in Listing 10.4. In this example, the minimum feature size and minimum spacing are defined for the silicon layer to be 200 nm. A DRC exclusion allows the designer to skip these rules in the regions defined by the "DRCex" layer. A DRC rule file for a foundry may consist of hundreds of rules.

Listing 10.4 Example Mentor Graphics Calibre DRC rules.

```
Si.Width {
 @ Si layer: Width, minimum: 0.2
 OUTSIDE ( INT Si < 0.2 REGION) DRCex
}

Si.Space {
 @ Silicon: Space, minimum: 0.2
 OUTSIDE (EXT Si < 0.2 REGION) DRCex
}
```

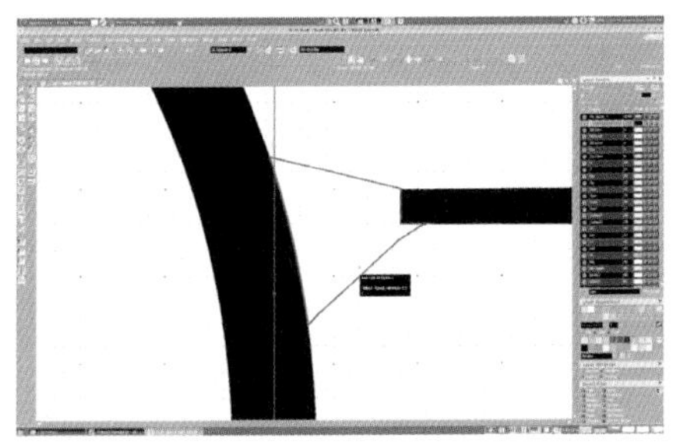

(a) Minimum space error flagged.

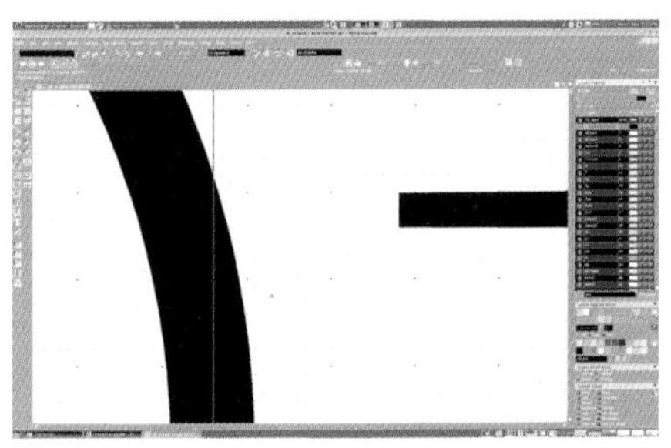

(b) User fixes error; flag is removed.

Figure 10.11 Interactive DRC checking, flagging errors while user is editing. Reprinted with permission [8].

The rule checker operates in two modes. (1) Interactive – as the layout is constructed, the tool graphically reports the errors; this allows the designer to correct the errors immediately during layout. This is shown in Figure 10.11. Interactive checking operates on a small fraction of the layout, namely the portion of the layout that is being edited and in view. (2) Sign-off verification – the full layout is exported and checked for errors. An error report is provided graphically and as a list.

10.1.7 Layout versus schematic

The first task in layout versus schematic (LVS), is for the software to parse the mask layout file and identify structures from the drawn shapes. The software first performs an extraction to finds known devices (e.g. ring modulator, grating coupler, etc.) and then determines their connectivity. It then creates a netlist based on the layout, and compares it with the original schematic netlist. See Listing 10.5.

The extraction makes use of device recognition layers to simplify the process. The first layer ("devrec") is used to mark the extent of a particular device with a polygon, as well as to mark the area with a text label identifying the device name (e.g. "Ring-Modulator"). The second layer ("pinrec") is used to mark the location and name of the pins; both electrical (e.g. anode1, anode2, cathode), and optical (e.g. opt_a2, opt_b2). A series of logical operations is used to find the geometries within the device recognition layers. Measurements are then performed to extract the device parameters (e.g. gap in a directional coupler). Additional operations are performed to find the connectivity between components.

Listing 10.5 LVS extracted netlist from mask layout.

```
* SPICE NETLIST
***************************************

.SUBCKT RingModulator opt_a2 opt_b2 anode1 anode2 cathode
.ENDS
***************************************
.SUBCKT YBranch_R15_Open5_W500 opt_a1 opt_b1 opt_b2
.ENDS
***************************************
.SUBCKT BondPad off_chip on_chip
.ENDS
```

```
**************************************
.SUBCKT GC_TM1550_20 opt_fiber opt_wg
.ENDS
**************************************
.SUBCKT GC_TE1550_20 opt_fiber opt_wg
.ENDS
**************************************
.SUBCKT wdm2tx_routed ch1a ch1b ch2a ch2b opt_output opt_inb opt_ina opt_test
** N=71 EP=8 IP=0 FDC=12
X0 4 7 56 56 57 RingModulator $X=288250 $Y=79700 $D=0
X1 7 10 58 58 59 RingModulator $X=504000 $Y=79700 $D=0
X2 3 2 1 YBranch_R15_Open5_W500 $X=127800 $Y=79700 $D=1
X3 3 5 4 YBranch_R15_Open5_W500 $X=197150 $Y=79700 $D=1
X4 ch1a 56 BondPad $X=212300 $Y=237000 $D=2
X5 ch1b 57 BondPad $X=317400 $Y=237000 $D=2
X6 ch2a 58 BondPad $X=422500 $Y=237000 $D=2
X7 ch2b 59 BondPad $X=527600 $Y=237000 $D=2
X8 opt_output 10 GC_TM1550_20 $X=-1700 $Y=7800 $D=3
X9 opt_inb 1 GC_TM1550_20 $X=-1700 $Y=46900 $D=3
X10 opt_ina 2 GC_TM1550_20 $X=-1700 $Y=80400 $D=3
X11 opt_test 5 GC_TM1550_20 $X=-1700 $Y=119500 $D=3
.ENDS
**************************************
```

10.2 Mask layout

In this section, we provide further discussion on generating mask layouts.

10.2.1 Components

In order to create photonic circuit layouts, as described in Section 10.1.5, we require a library of component layouts. These can be created in a variety of software packages using different approaches. Layouts can be manually drawn using primitive elements such as polygons, or created using elaborate scripts to create parameterized layouts. Manual layout is very well suited for design components with simple geometries, such as the edge couplers presented in Figure 5.6. Scripted layout is necessary for complex structures such as focusing gratings couplers, ring resonators, etc.

Using the Mentor Graphics mask layout Pyxis tool, and the AMPLE scripting language, parameterized components are created. An example of a grating coupler script is provided in the scripts described in Section 5.2.3. The designer instantiates these components and is presented with choices for the parameters, as shown in Figure 10.12.

10.2.2 Layout for electrical and optical testing

A sample device layout is shown in Figure 10.13a. The design contains GS probe pads for electrical testing. The device is connected to a distant pair of grating couplers for optical input and optical output via a fibre array, Figure 10.13b. The complete layout of the test device including electrical and optical pads is shown in Figure 10.14. Total size of the largest device considered is approximately 2.8 mm $\times$ 0.4 mm = 1.1 mm^2, limited by (1) the size of the device (or fibre array), height in this case, and (2) the minimum separation of the optical and electrical probes and associated mechanics, here the width.

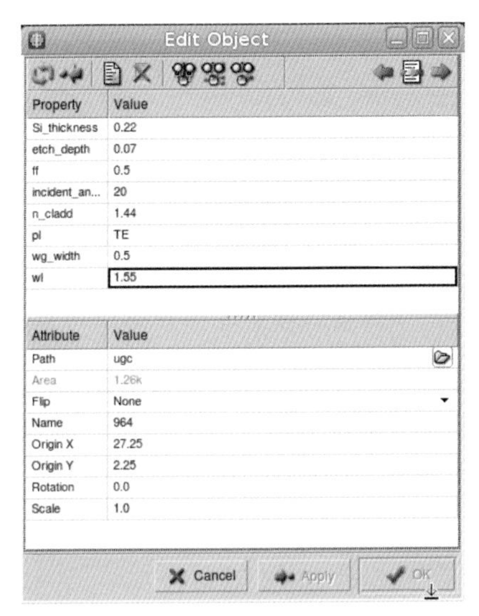

(a) Grating coupler [14]

(b) Adiabatic 2 × 2 splitter [15]

Figure 10.12 Parameterized cell layouts.

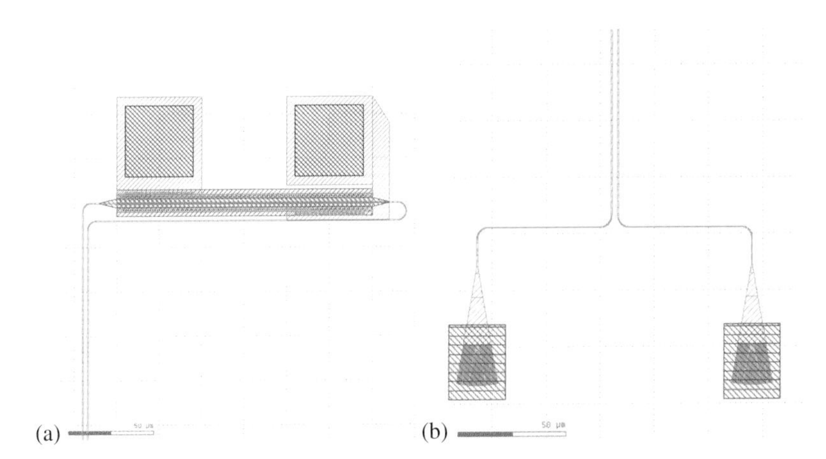

Figure 10.13 (a) Single device to be tested. The design contains GS probe pads for electrical testing. (b) A pair of grating couplers (input, output).

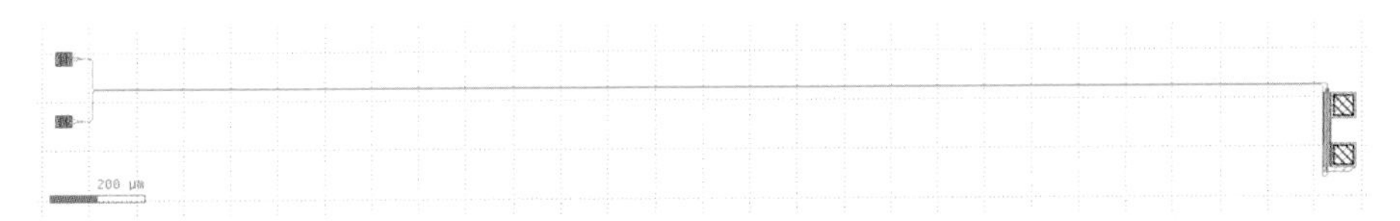

Figure 10.14 Layout of the test device including electrical and optical pads. Total size of the largest device is approximately 2.8 mm × 0.4 mm = 1.1 mm^2, limited by the width of the fibre array and minimum separation of the optical and electrical probes and associated mechanics.

The key consideration is the spacing between the electrical and optical probes. This will depend on the physical setup implemented. For experimental setups described in Section 12.2, 1 mm is a sufficient distance. Further "Design for test" considerations are presented in Section 12.3.

10.2.3 Approaches for fast GDS layout

The fastest method of laying out devices is to tile them, putting them side-by-side. In this method, each device is drawn separately and multiple devices are merged into a single GDS design. This method is particularly useful when the designer is constrained in time and requires a large number of devices in a large area to be designed.

An improvement in space efficiency can be obtained by varying the dimensions (e.g. overall length of the device) and by arranging them in L-shapes.

10.2.4 Approaches for space-efficient GDS layout

In order to save space, the designer can find methods of laying out the devices and routing the waveguides so as to minimize the amount of "white" or unused space. This, however, requires additional waveguide routing complexity. This can be implemented by writing a script to create the routing waveguides.

Space-efficient layout has clear financial benefits, but it comes with potential risks that need to be considered.

- Optical cross-talk: cross-talk can occur between optical waveguides; see Section 4.1.7. This requires that the waveguides be separated far enough. For strip waveguides, 3 μm is considered safe enough to ensure negligible cross-talk overscale distances. The designer should consider what is the necessary cross-talk level tolerance and perform directional coupler simulations. Another source of cross-talk is between the fibre grating couplers. They need to be spaced far enough apart so that light to/from the optical fibres does not enter adjacent couplers. For single-mode fibres, this is rarely a concern, but for multi-mode fibres, the couplers spacing should be larger than the core size (e.g. $>50\,\mu m$).
- Electrical cross-talk: careful microwave design needs to be considered to ensure that there is no cross-talk between adjacent devices and to suppress microwave resonances that may introduce oscillations in the frequency response.
- Measurement challenges: unanticipated measurement challenges may arise. For example, opto-mechanical setup constraints may limit how close the optical fibres can be placed to the electrical probes.

An approach for the dense packing of devices is presented in the following example. In a typical silicon photonic device design cycle, numerous variations of the device are usually considered with suitable parameter variation. In this example, 26 devices were optimally assembled. The method of increasing the device packing density consists of an array of fibre array optical probe pads with routing waveguides between the couplers as shown in Figure 10.15. This routing method can be extended up to 16 pairs of fibre

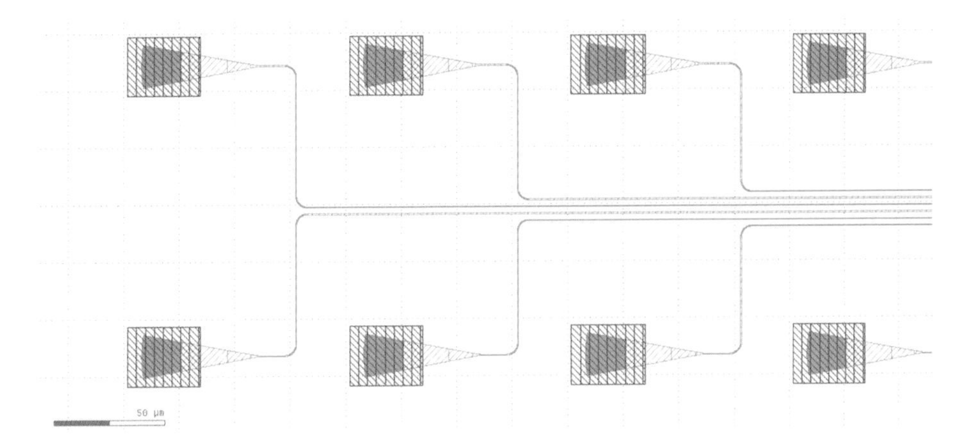

Figure 10.15 Method of increasing the device packing density. This consists of an array of fibre array optical probe pads with routing waveguides between the couplers.

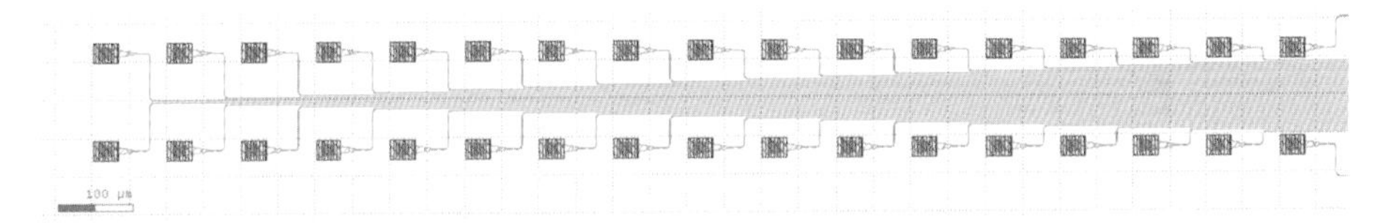

Figure 10.16 Method of increasing the device packing density. Sixteen pairs of fibre array optical probe pads can be routed such that the waveguides run between the couplers. This creates a waveguide bundle to which the devices will be connected.

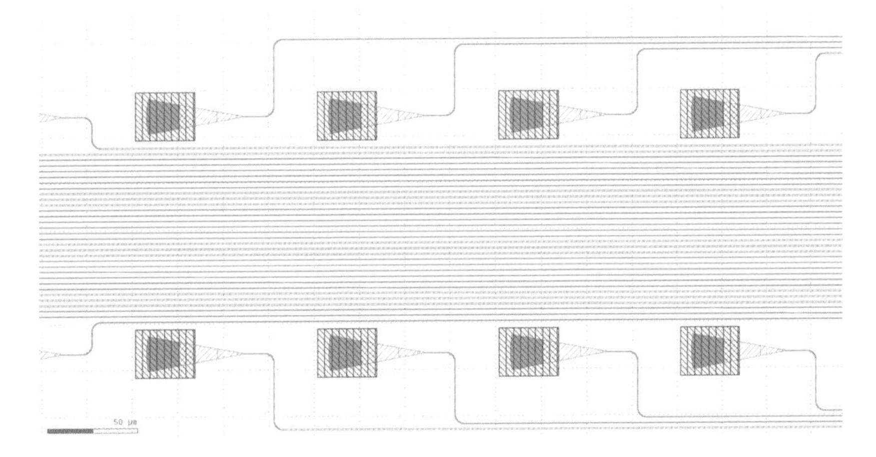

Figure 10.17 Method of increasing the device packing density. If additional devices are required, the waveguides can be run on the outside of the couplers.

array optical probe pads, routed such that the waveguides run between the couplers (Figure 10.16). This design used a 3 µm centre-to-centre waveguide spacing, sufficient to avoid cross-talk between adjacent waveguides. This creates a waveguide bundle to which the devices will be connected. If more than 16 devices are required, additional waveguides can be routed on the outside of the grating couplers (Figure 10.17).

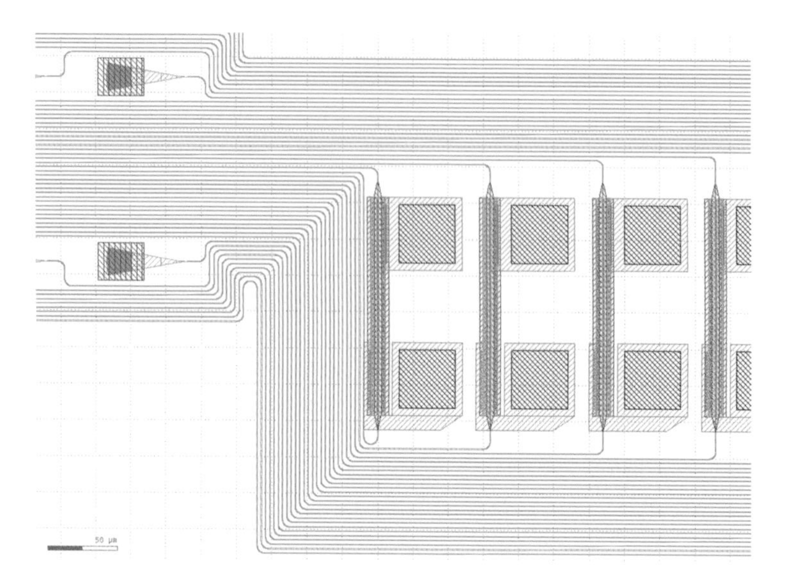

Figure 10.18 Method of increasing the device packing density. The devices are connected to the waveguide bundle. The left-most device is connected to the left-most fibre grating couplers, in order to ensure a sufficient distance between optical and electrical probes.

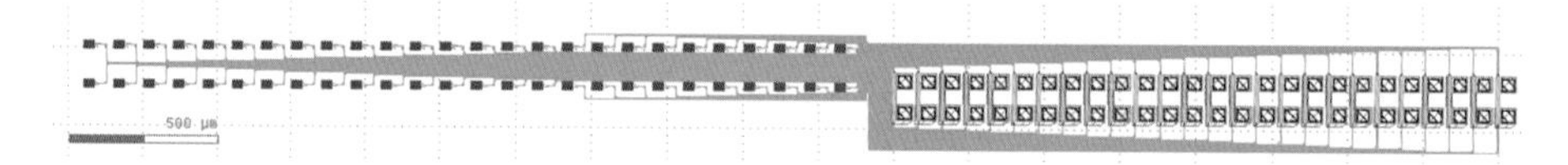

Figure 10.19 Method of increasing the device packing density. Twenty-six devices (right side) are connected to a waveguide bundle with 26 pairs of grating couplers (left side). The entire area is 4.7 mm $\times$ 0.4 mm = 1.8 mm^2. If the devices had been laid out side-by-side without this routing approach, the total area would have been 20.7 mm^2. This packaging arrangement represents a compression of 11.5 $\times$ over the simple device-tiling approach.

Next, devices are connected to the waveguide bundle (Figure 10.18). The left-most device is connected to the left-most fibre grating couplers, in order to ensure a sufficient distance between optical and electrical probes. In the overall layout, in Figure 10.19, 26 devices (on the right side) are connected to a waveguide bundle with 26 pairs of grating couplers (on the left side). The entire area is 4.7 mm $\times$ 0.4 mm = 1.8 mm^2. If the devices were laid out side-by-side without this routing approach, the total area would have been 20.7 mm^2. This packaging arrangement represents a compression of a factor of 11.5 over the simple device-tiling approach.

References

[1] Tom Baehr-Jones, Ran Ding, Ali Ayazi, *et al.* "A 25 Gb/s silicon photonics platform". *arXiv:1203.0767v1* (2012) (cit. on pp. 349, 352).

[2] *NSERC CREATE Silicon Electronic Photonic Integrated Circuits (Si-EPIC) program.* [Accessed 2014/04/14]. URL: http://www.siepic.ubc.ca (cit. on p. 349).
[3] D. Feng, S. Liao, P. Dong, *et al.* "High-speed Ge photodetector monolithically integrated with large cross-section silicon-on-insulator waveguide". *Applied Physics Letters* **95** (2009), p. 261105 (cit. on p. 352).
[4] *ePIXfab – The silicon photonics platform – IMEC Standard Passives.* [Accessed 2014/04/14]. URL: http://www.epixfab.eu/technologies/49-imecpassive-general (cit. on p. 352).
[5] *ePIXfab – The silicon photonics platform – LETI Full Platform.* [Accessed 2014/04/14]. URL: http://www.epixfab.eu/technologies/fullplatformleti (cit. on p. 352).
[6] *Agency for Science, Technology and Research (A *STAR) Institute of Microelectronics (IME).* [Accessed 2014/07/21]. URL: http://www.a-star.edu.sg/ime/ (cit. on p. 352).
[7] *Europractice Imec-ePIXfab SiPhotonics Passives technology.* [Accessed 2014/04/14]. URL: http://www.europractice-ic.com/SiPhotonics_technology_passives.php (cit. on p. 352).
[8] Lukas Chrostowski, Jonas Flueckiger, Charlie Lin, *et al.* "Design methodologies for silicon photonic integrated circuits". *Proc. SPIE, Smart Photonic and Optoelectronic Integrated Circuits XVI* 8989 (2014), pp. 8989–9015 (cit. on pp. 354, 356, 357, 358, 359, 361).
[9] Wim Bogaerts and S. K. Selvaraja. "Compact single-mode silicon hybrid rib/strip waveguide with adiabatic bends". *IEEE Photonics Journal* **3**.3 (2011), pp. 422–432 (cit. on pp. 357, 359).
[10] Matteo Cherchi, Sami Ylinen, Mikko Harjanne, Markku Kapulainen, and Timo Aalto. "Dramatic size reduction of waveguide bends on a micron-scale silicon photonic platform". *Optics Express* **21**.15 (2013), pp. 17 814–17 823 (cit. on p. 357).
[11] Guoliang Li, Jin Yao, Hiren Thacker, *et al.* "Ultralow-loss, high-density SOI optical waveguide routing for macrochip interconnects". *Optics Express* **20**.11 (May 2012), pp. 12035–12039. DOI: 10.1364/OE.20.012035 (cit. on p. 359).
[12] W. Bogaerts, P. Dumon, D. Thourhout, and R. Baets. "Low-loss, low-crosstalk crossings for silicon-on-insulator nanophotonic waveguides". *Optics Letters* **32** (2007), pp. 2801–2803 (cit. on p. 359).
[13] Yi Zhang, Shuyu Yang, Andy Eu-Jin Lim, *et al.* "A CMOS-compatible, low-loss, and low-crosstalk silicon waveguide crossing". *IEEE Photonics Technology Letters* **25** (2013), pp. 422–425 (cit. on p. 359).
[14] Yun Wang, Jonas Flueckiger, Charlie Lin, and Lukas Chrostowski. "Universal grating coupler design". *Proc. SPIE* **8915** (2013), 89150Y. DOI: 10.1117/12.2042185 (cit. on p. 363).
[15] Han Yun, Wei Shi, Yun Wang, Lukas Chrostowski, and Nicolas A. F. Jaeger. "2×2 adiabatic 3-dB coupler on silicon-on-insulator rib waveguides". *Proc. SPIE, Photonics North 2013* **8915** (2013), p. 89150V (cit. on p. 363).

11 Fabrication

In this chapter, we discuss the impact of manufacturing variability of silicon photonic integrated circuits. The dominant variations in silicon photonics are silicon thickness and feature size. These variations appear from wafer to wafer, as well as within a single photonic integrated circuit. Smoothing due to lithography is also important to consider. Methods of including these variations in the design process are discussed. Finally, some experimental results from on-chip test structures are presented to illustrate the manufacturability and non-uniformity challenge of silicon photonics.

11.1 Fabrication non-uniformity

Photonic integrated circuits (PICs) often require precise matching of the central wavelength and the waveguide propagation constants between components on a chip (e.g. ring modulators, optical filters), particularly for wavelength division multiplexing. Understanding the fabrication variability is critical to developing strategies (e.g. thermal tuning) for system implementation, and for determining the cost implications for such compensation strategies (e.g. power consumption).

There have been several studies on the fabrication non-uniformity including intra-device uniformity (e.g. CROWs [1]), within-wafer, wafer-to-wafer, and batch-to-batch variations [2–5]. The dominant fabrication parameter that results in device variation has been identified to be the silicon thickness variation, followed by lithography (e.g. waveguide width) variations.

Zortman *et al.* [2] found that the thickness variation across 10 cm led to ± 1000 GHz variation in TE resonator wavelength, whereas the width variation contributed to ± 200 GHz. From measurements of TE and TM resonators, they extracted the dimensional variations to be ± 5 nm in both thickness and waveguide width (or diameter). The reader is also referred to References [3, 5]. In [5], TE-polarization Bragg gratings, using both strip and rib waveguides, were used to extract thickness variations of approximately ± 5 nm, which led to resonance shifts up to 10 nm; these results are consistent with Reference [2].

An example of fabrication non-uniformity, and its impact on device performance, is shown in Figure 11.1. The devices are Bragg gratings, with lengths ranging from 325 μm to 4.9 mm. Theoretically, as the length is increased, the bandwidth should remain

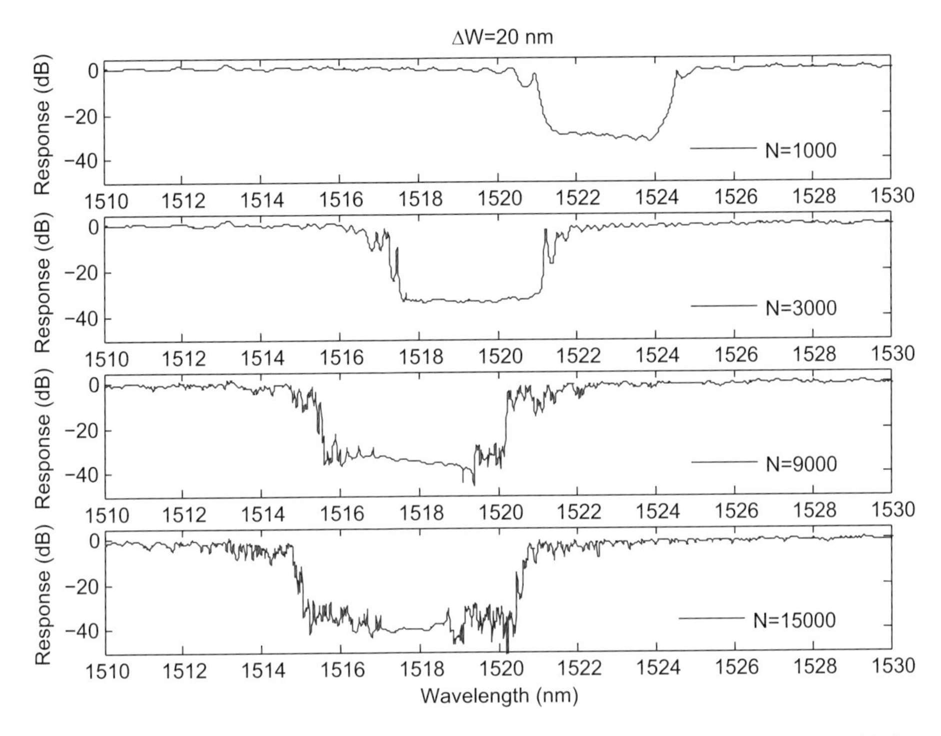

Figure 11.1 Measured transmission spectra of Bragg gratings with a 20 nm corrugation width for various lengths, showing bandwidth broadening effect with increasing length, and wavelength variations due to fabrication variations. Devices: strip waveguides with air cladding, $W = 500$ nm, $\Lambda = 325$ nm.

constant, or decrease if the coupling is weak, as per Equation (4.33). On the contrary, we experimentally observe that the stop band becomes broader when the grating gets longer, as shown in Figure 11.1. This is due to the waveguide geometry variations along the fabricated chip. It is also seen that the central wavelength varies for the four devices shown.

11.1.1 Lithography process contours

The computation lithography models described in Section 4.5.4 not only produce a simulation of the expected structure, but they also can generate several outputs corresponding to the anticipated range for feature sizes, e.g. variations in the waveguide width. As shown in Figure 11.2, the three lines correspond to the nominal device geometry, and two minimum and maximum process contours (e.g. corresponding to over- and under-exposure). These simulated geometries provide insight into the anticipated variability to identify designs that will be challenging to manufacture (e.g. directional couplers with very small gaps, slot waveguides), or the lithography-simulated structures can be exported for optical simulations (as in Section 4.5.4).

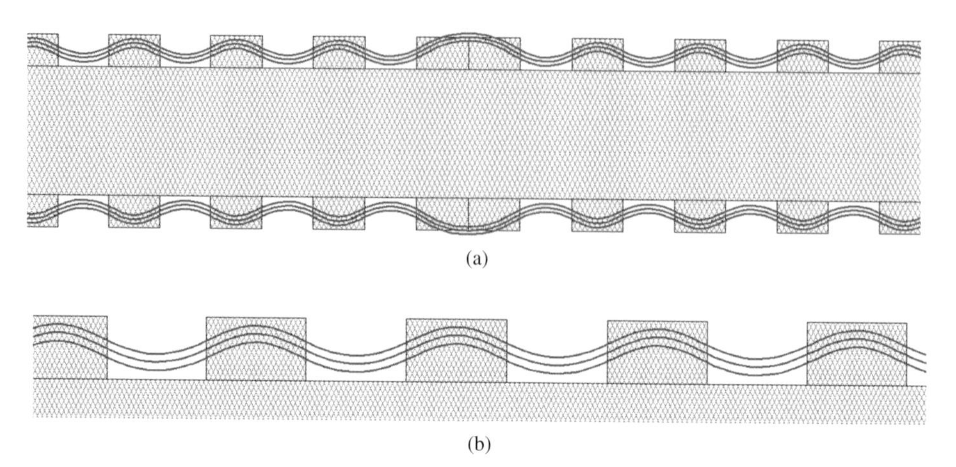

Figure 11.2 Computational lithography model showing the fabrication process contours for Bragg gratings considered in Section 4.5.6.

11.1.2 Corner analysis

Fabrication process corners is a simple design of experiment (DoE) technique that considers typical process variations for each parameter. Typically a ± 3-σ variation is considered. For example, the thickness of the SOI layer may be specified as 220 nm, with a ± 10 nm 3σ variation. Similarly, other fabrication parameters will have variations, such as the lithography line-width, etch depth, mask overlay (alignment error between two masks), doping concentration, and thickness of deposited material. Additionally, stimulus and environmental parameters can be included in this analysis, including temperature, voltage, etc.

In electronic design, each process parameter is assigned a name: typical (T), slow (S), and fast (F). For two process parameters, we can thus consider the nominal design (TT), four process corners (SS, FF, SF, FS), and four edges (TS, TF, ST, FT), for a total of nine possibilities. For each point, a simulation is conducted; all simulations are collated to understand the sensitivity to fabrication. This can be extended to numerous process dimensions.

First, we consider the two-dimension process corners for a strip waveguide. As shown in Figure 11.3a, we consider the nominal design as a 500×220 nm waveguide. The process variations include thickness, 220 ± 10 nm, and waveguide width, 500 ± 10 nm. The corners are indicated by the points on the square.

Next, we consider a three-dimension process corners analysis for a rib waveguide. This includes the waveguide height (nominal is 220 nm), waveguide width (nominal is 500 nm), and slab height (nominal is 90 nm). Each process parameter is assumed to have a ± 10 nm 3σ variation. The process corners to be simulated are shown in Figure 11.3b. Thus, the following parameters are varied: waveguide width (490, 500, 510 nm), total height (210, 220, 230 nm), and slab height (80, 90, 100 nm). The results of the simulation for the waveguide effective index and group index are shown in Figure 11.4a and 11.4b, respectively. Note that, in the rib waveguide, the slab height

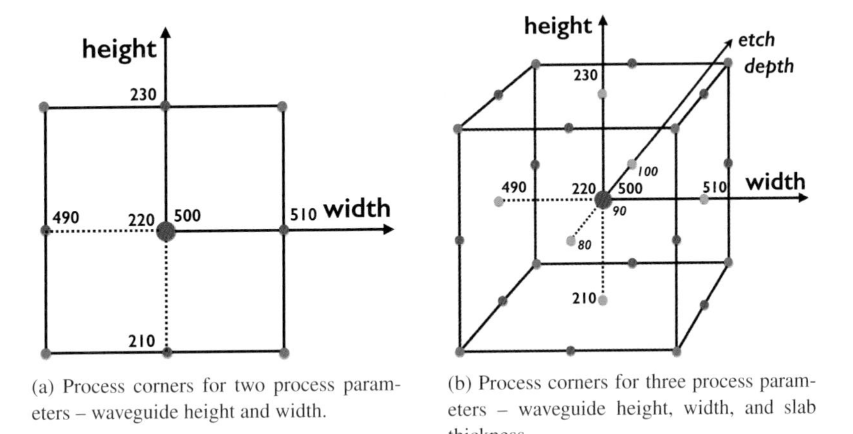

(a) Process corners for two process parameters – waveguide height and width.

(b) Process corners for three process parameters – waveguide height, width, and slab thickness.

Figure 11.3 Process corners. The box indicates the range of devices resulting from the process. Points indicate the process conditions that will be simulated.

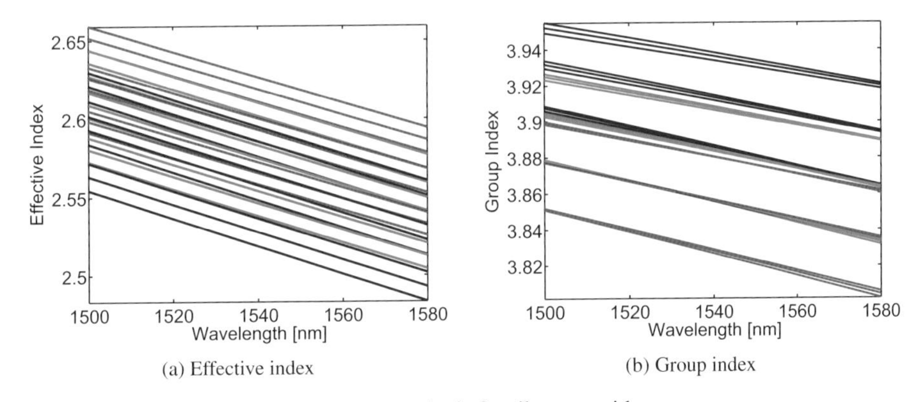

(a) Effective index

(b) Group index

Figure 11.4 Optical spectra from a corner analysis for rib waveguide.

variation is not only due to the etch variation, but also due to the SOI thickness. Thus, the slab height and waveguide thickness are partially correlated since the original SOI thickness affects both the waveguide height and the slab height. In such cases, correlation between process effects could be included. However, the analysis presented here will yield a worst-case prediction. The results of Figure 11.4b should be compared to the experimental results in Figure 3.21b; it is seen that the experimental results lie within the predicted range due to fabrication variation.

For multi-dimensional process corners, the number of process corner simulations is 3^N (i.e. 27 simulations in Figure 11.4a and 11.4b). To simplify the analysis and reduce simulation time, we can reduce the corner analysis to only the nominal and corners, for a total of $2^N + 1$ simulations, where N is the number of dimensions (i.e. for three dimensions; TTT, SSS, SSF, SFS, SFF, FSS, FSF, FFS, FFF).

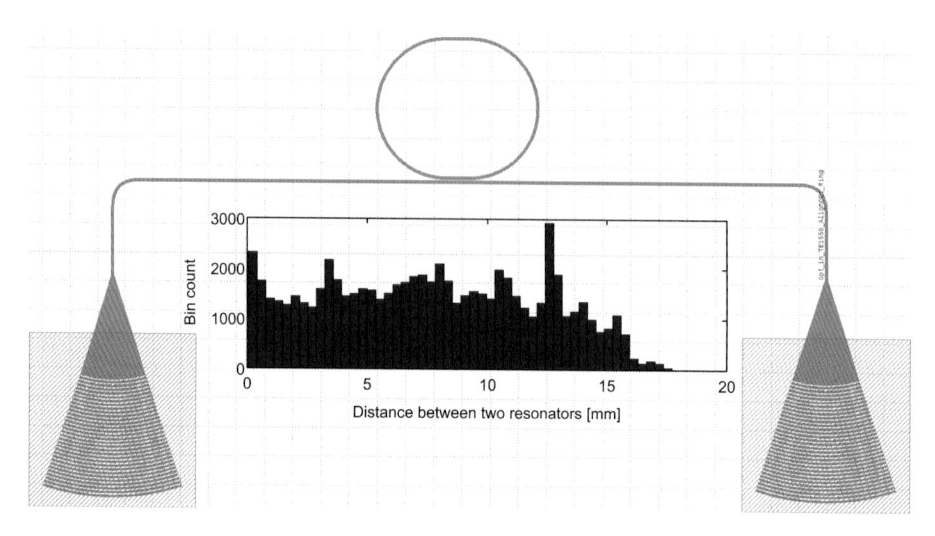

Figure 11.5 Mask layout of the racetrack resonator with a pair of fibre grating couplers on a 127 μm pitch for automated optical testing. Inset: histogram for the distribution of the distances between devices, for the 68 635 combinations. From Reference [6].

11.1.3 On-chip non-uniformity, experimental results[1]

In this section, we present experimental results illustrating the variation between devices within a chip. First, we map the variations of the device across the chip. Then, we analyze the variations as a function of the distance between components typically seen in PICs, namely ranging from hundreds of microns to millimetres. In contrast to Section 11.1.2, where process variations were considered for one isolated device at a time, with no consideration for neighbouring devices, here we find that the variation in multiple devices is strongly correlated to the distance. Thus, process corner analysis provides the variations observed on an absolute scale (i.e. wafer-to-wafer, batch-to-batch), but neglects on-chip correlation of process parameters, which leads to reduced on-chip device variation. The implication is the importance of very compact layouts when components need to be matched, leading to reduced (but not eliminated) trimming cost.

The results presented in this section are based on the fabrication, test, and analysis of 371 identical racetrack resonators on a 16 × 9 mm chip fabricated by the IME A*STAR silicon photonics foundry. The measured die was located close to the centre of the wafer. The unit cell for testing is shown in Figure 11.5, and consists of two fibre grating couplers (GCs) designed for 1550 nm quasi-TE operation [7], 220 nm thick silicon on insulator (SOI) strip waveguides with a 500 nm width, connected to a TE polarization racetrack resonator [8] with a 12 μm radius, and a directional coupler with a 4.5 μm length and 200 nm gap. Devices were placed between 60 μm and 18 mm apart. To obtain insight into the statistics of the wavelength variation, the 371 resonators were compared with each other resonator on the chip, for a total of (371 choose 2) = 68 635

[1] A version of this section has been published [6]: Chrostowski, L., Wang, X., Flueckiger, J., Wu, Y., Wang, Y., Fard, S. Talebi. "Impact of fabrication non-uniformity on chip-scale silicon photonic integrated circuits", *OSA Optical Fiber Communication Conference*, p. Th2A–37, 2014. Reprinted by permission.

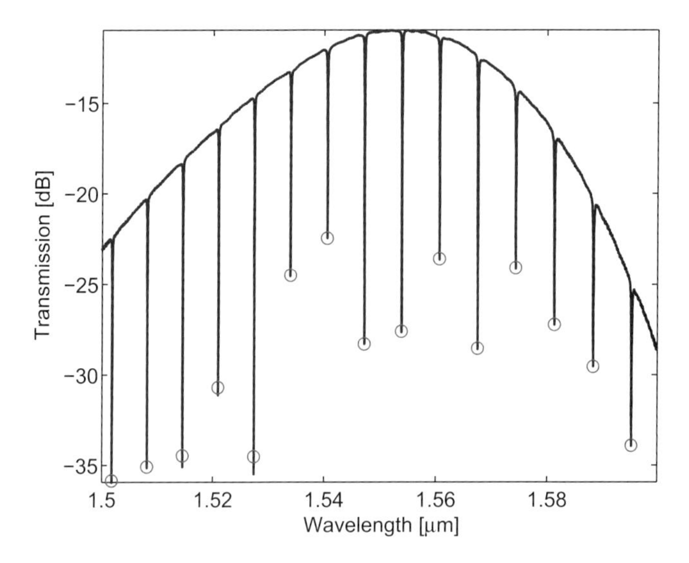

Figure 11.6 Optical spectrum of a typical device with resonance wavelengths identified. From Reference [6].

combinations; a histogram of these distances is shown inset in Figure 11.5. Note that >1300 comparison were made for resonators less than 200 μm aparts.

Automated measurements were performed on all devices. The spectra were sampled at a 10 pm resolution for rapid sweeping. This limits the precision of this study to 10 pm and, of less concern, the extinction ratios in the spectra were limited due to under-sampling. The typical quality factors for the resonators were 10 000–30 000. The fibre–fibre insertion loss was typically 11 dB, due to a large distance between the chip and the grating couplers for safer automated probing. A peak finding algorithm was used for data analysis. A typical spectrum is shown in Figure 11.6, where the free-spectral range (FSR) is ~7 nm.

Ring resonators

It was found that the wavelength variability across the chip was greater than the FSR of the device. Hence it was not possible to directly determine the variation of the resonance wavelength since it was not known which mode was which from the spectrum by inspection. The following method was applied.

(1) From the peak position information, and knowing the resonator length, the group index (n_g) of the ring waveguide was calculated by

$$FSR = \frac{c}{n_g L}. \tag{11.1}$$

(2) The n_g and wavelength for all resonators and all peaks were plotted; see Figure 11.7. A clear relationship is evident, where each downward diagonal line corresponds to a resonator azimuthal mode, m. This relationship originates from (a) the resonator wavelength being dependent on the effective index of the

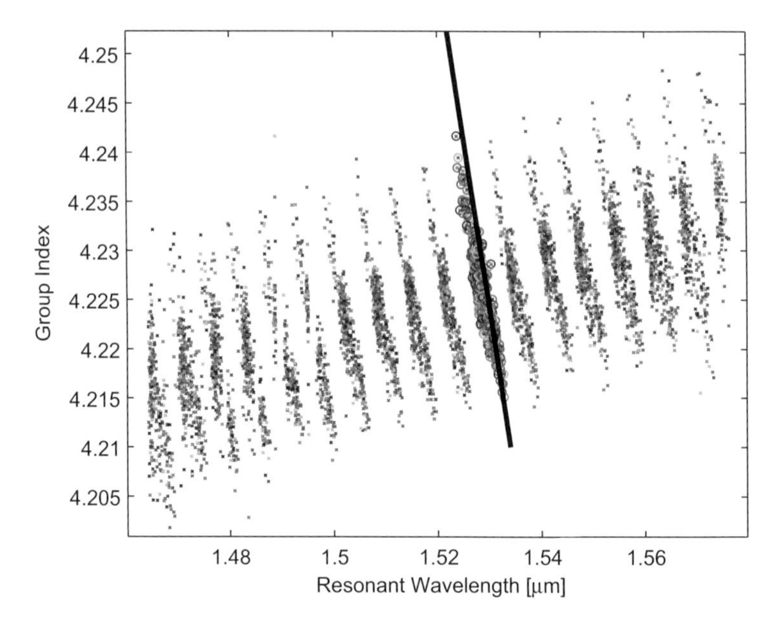

Figure 11.7 Extracted group index versus resonance wavelength for the 15 modes measured, for the 371 resonators. From Reference [6].

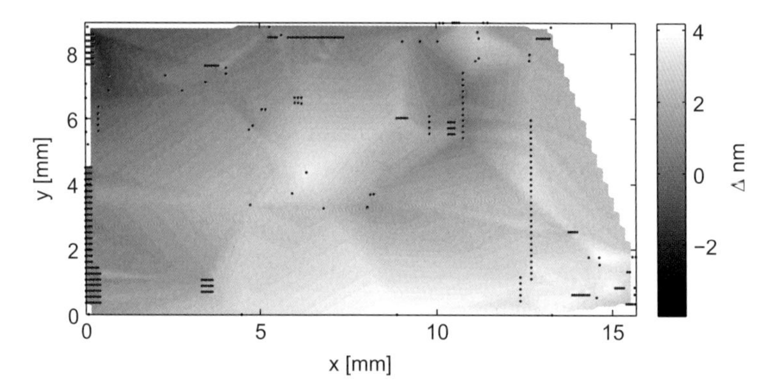

Figure 11.8 Resonance wavelength deviation contours (in nm) versus physical position on the chip. Resonance is chosen from the mode selected from the line in Figure 11.7. The "0" contour corresponds to the mean wavelength. From Reference [6].

waveguide via $m\lambda_m = n_{\text{eff}}(\lambda_m)L$, and (b) the physical variations (e.g. thickness, width) that change the effective index of the waveguide, n_{eff}, and simultaneously affect the group index.

(3) Using this correlation, it is possible to select all the data points that belong to one mode by finding the points closest to a chosen line, as drawn in Figure 11.7.

It is seen that the maximum variation of the resonator wavelength across the chip is approximately 10 nm. Figure 11.8 shows a map of the resonator wavelength versus position.

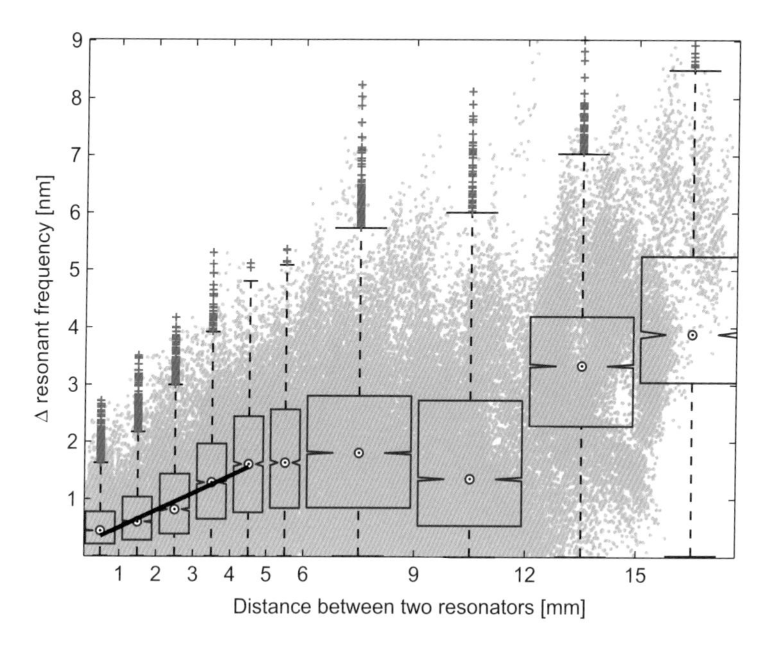

Figure 11.9 Resonator wavelength mismatch, related to distance between devices across the chip. Scatter plot of the difference in resonator wavelength, versus distance between them (68 635 data points in purple). Overlaid boxplot for statistics with bins for distances ranging 0–1 mm, 1–2, 2–3, 4–5, 5–6, 6–9, 9–12, 12–15, and 15–18 mm. The box represents the central 50% of the data. The lower and upper boundary lines are at the 25%/75% quantiles. The middle circles indicate the median. The two vertical lines extend maximally to 1.5 times the height of the box but not past the range of the data. Data points outside this range are considered outliers and are marked as red ("+") markers. Notches are the 95% confidence intervals. From Reference [6].

To obtain insight into the statistics of the wavelength variations, each of the 371 resonators was compared with each other resonator on the chip, for a total of 371 choose 2 = 68 635 comparisons. We calculate the difference in resonator wavelength (y-axis), versus the distance between each pair of devices (x-axis), as shown in the scatter plot in Figure 11.9. This clearly shows that the worst-case deviation between two resonators is approximately linearly proportional to the distance between them. For example, two resonators that are 1 mm apart show at most a 2–3 nm difference in wavelength; however, when they are 4 mm apart, the difference increases to 5 nm. These points were statistically analyzed, and represented using box plots in Figure 11.9. This was done by binning the points in distance (1 mm bins for the first few mm, larger for longer distances).

The probability distribution functions for the wavelength mismatch are shown in Figure 11.10. The probability distribution functions (p.d.f.) for the wavelength mismatch do not follow a Gaussian distribution. This is clear from the box plots, which show a significant number of outliers, as well as from the p.d.f. graphs. Curiously the distributions become nearly uniform with a sharp maximum cutoff for large enough distance ranges (e.g. 4–5 mm).

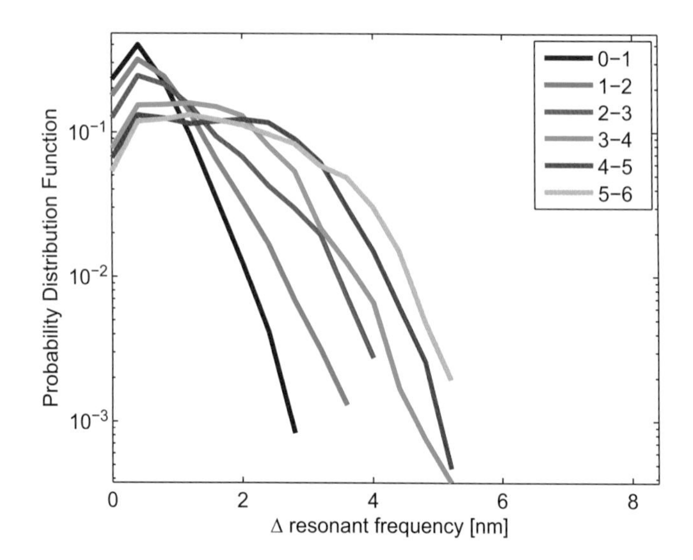

Figure 11.10 Resonator wavelength mismatch, related to distance between devices across the chip. Probability distribution functions for the expected wavelength difference between resonators 0–1 mm apart, etc., with data from Figure 11.9. From Reference [6].

For distances $\geq$ 6 mm, the variations show trends of "bunching", which is likely due to the arrangement of the resonators on the layout and the limited sample size. The resonators were not distributed with positions following a random uniform distribution function, but rather these devices were placed space-permitting between other components on a dense PIC layout.

The data also show that the wavelength variation is less than 9 nm for any two devices on the chip, but only for resonators separated by at least 6 mm and, importantly, this variation saturates for longer separate distances to the worst-case uncorrelated statistics.

Further insight can be obtained by looking at the data in a smaller range of distances, namely between 0 and 3 mm, as shown in Figure 11.11. Important information from this analysis includes the median, which increases linearly up to a distance of 5 mm. This indicates that the fabrication variations are correlated over short-distance scales. At short distances, a line of best fit for the median wavelength deviation versus distance is

$$\bar{\lambda}_{\text{ring}} = 0.20\,\text{nm/mm} \cdot d + 0.37\,\text{nm}, \tag{11.2}$$

as shown in Figure 11.10. This gives the expected (average) value for the deviation between two resonant devices on a chip as a function of distance.

The y-intercept in Equation (11.2) also predicts that two resonators in very close proximity will have an average wavelength mismatch of 0.37 nm. Since the FSR is 7 nm, this corresponds to an intrinsic variation of $\sim 0.1\pi$ in this resonator. Assuming a heater efficiency of 0.8 mW/FSR [9], this suggests that we will need on average 0.37 nm/7 nm FSR$\cdot$0.8 mW/FSR = 0.04 mW per pair of resonators that need matching. This value is useful for predicting the typical tuning power required, for example, in a

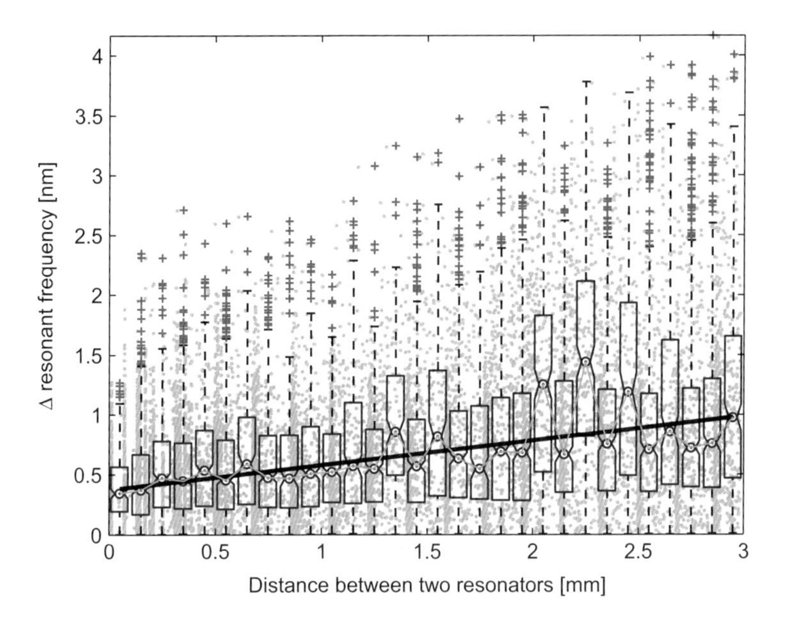

Figure 11.11 Zoom of Figure 11.9 – scatter and box plot of the difference in resonator wavelength, versus distance between them, for distances ranging from 0 to 3 mm, with boxes in 100 μm steps. From Reference [6].

two-ring Vernier filter [10]. Note that, for resonators that are on opposite sides of the chip, the tuning requirement in the worst case will be the full FSR.

Grating couplers

Using the same methodology, an analysis was done on the grating couplers' peak wavelength (which is $\lambda_{FGC} = 1.555\,\mu m$ in Figure 11.6) and insertion loss (which is $\sim$ 11 dB for two couplers in Figure 11.6). For the grating coupler central wavelength, the contour map and deviation statistics are shown in Figures 11.12 and 11.13. The wavelength variation similarly increases linearly from 0 to 5 mm, with an expected average variation of:

$$\bar{\lambda}_{FGC} = 0.10\,\text{nm/mm} \cdot d + 0.80\,\text{nm}. \tag{11.3}$$

We note that the variations in resonator and grating coupler wavelengths have different spatial and statistical distributions, suggesting they have different physical origins. In addition to the SOI thickness, the grating couplers are primarily sensitive to etch depth. Based on FDTD simulations, these grating couplers have a central wavelength that is sensitive to the geometric parameters as follows: 1.82 nm/nm for SOI thickness, 1.9 nm/nm for etch depth, and 0.215 nm/nm for the width of the grating fingers.

The variation of the insertion loss of the grating couplers is shown in Figure 11.14.

The results of this study are useful for photonic integrated circuit system optimization. Specifically, these results can be used to estimate the power consumption required for trimming large arrays of micro-ring resonator or Mach–Zehnder switches. These

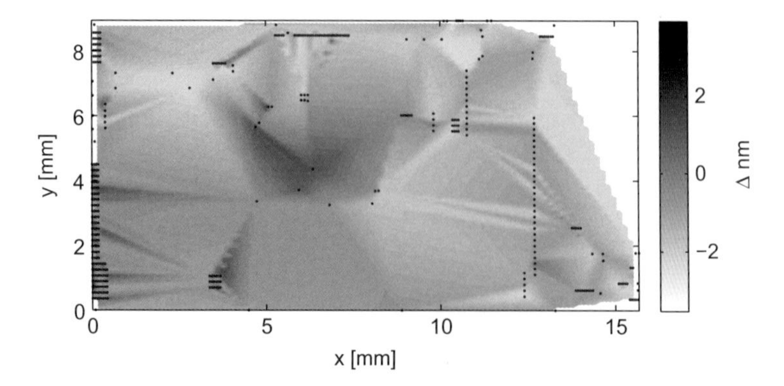

Figure 11.12 Fibre grating coupler central wavelength deviation contours (in nm) versus physical position on the chip. The "0" contour corresponds to the mean wavelength.

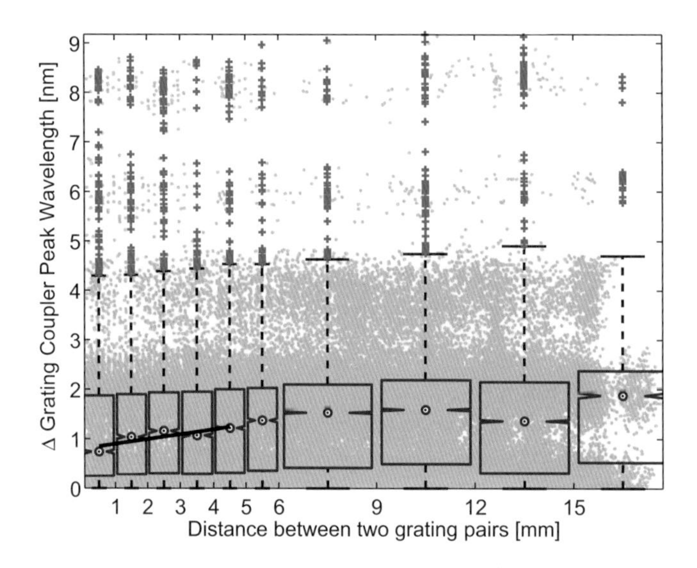

Figure 11.13 Fibre grating coupler central wavelength, related to distance between devices across the chip. Similar to Figure 11.9.

results also provide layout guidelines, namely that devices that need to be wavelength matched should be placed closely together. For example, in a transmitter consisting of ring modulators and ring resonator filters, as in Section 13.1, the ring resonators should be placed as close as possible so that their wavelengths are as closely matched as possible to minimize tuning energy consumption.

The parameters for the variability of resonators and grating couplers are also useful for process monitoring and process optimization. They are also useful for choosing which chips to use for experiments, and for binning chips before packaging based on expected performance.

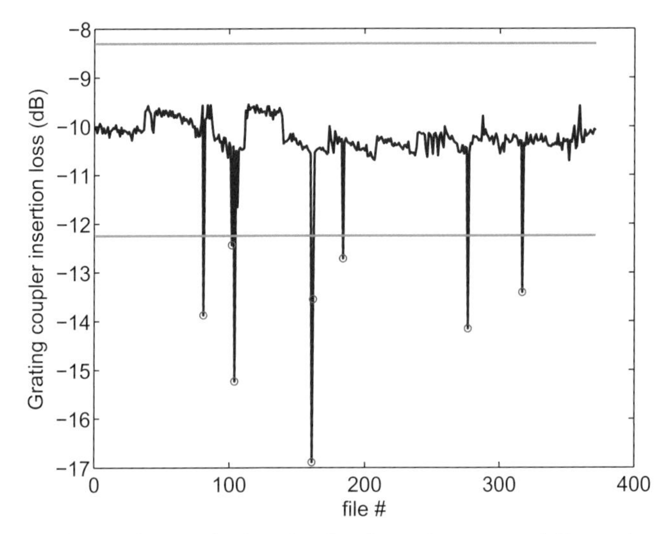

(a) Fibre grating coupler insertion for all couplers measured. Data points illustrate typical variation in loss, and indicate the number of outliers outside of the 3-σ lines.

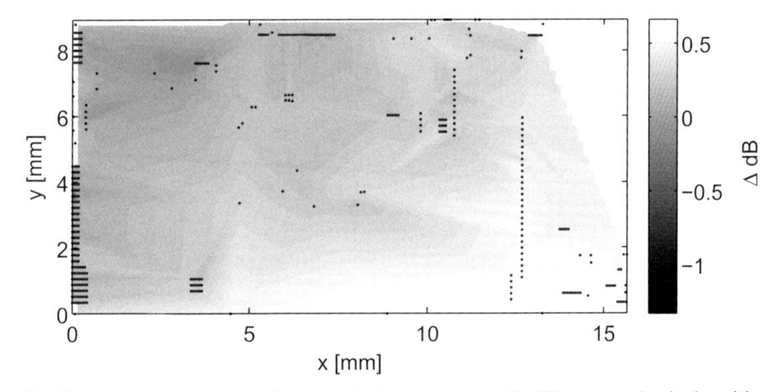

(b) Fibre grating coupler insertion loss deviation contours (in dB) versus physical position on the chip.

Figure 11.14 Grating coupler wavelength and insertion loss deviation contour maps.

11.2 Problems

11.1 Consider a fabrication process where the width and thickness of a 500 nm × 220 nm strip waveguide are both expected to vary by up to 1 nm. A wavelength division multiplexed system consisting of series-coupled ring resonators is to be implemented for a sensor application, where it is necessary to be able to distinguish between each ring. What is the minimum free spectral range and maximum ring radius necessary for resonators operating near 1550 nm?

11.2 Consider a coarse wavelength division multiplexing (CWDM) communication system utilizing on-chip lasers where the wavelength is defined by a Bragg grating.

Assuming maximum variations of up to 10 nm in each fabrication parameter, what is the anticipated wavelength variation? Identify suitable parameters for the CWDM system and necessary optical multiplexers, specifically, the channel spacing and bandwidth.

References

[1] Michael L. Cooper, Greeshma Gupta, William M. Green, *et al.* "235-Ring coupled-resonator optical waveguides". *Conf. Lasers and Electro-Optics* (2010) (cit. on p. 368).

[2] W. A. Zortman, D. C. Trotter, and M. R. Watts. "Silicon photonics manufacturing". *Optics Express* **18**.23 (2010), pp. 23 598–23 607 (cit. on p. 368).

[3] A. V. Krishnamoorthy, Xuezhe Zheng, Guoliang Li, *et al.* "Exploiting CMOS manufacturing to reduce tuning requirements for resonant optical devices". *IEEE Photonics Journal* **3**.3 (2011), pp. 567–579. DOI: 10.1109/JPHOT.2011.2140367 (cit. on p. 368).

[4] Shankar Kumar Selvaraja, Wim Bogaerts, Pieter Dumon, Dries Van Thourhout, and Roel Baets. "Subnanometer linewidth uniformity in silicon nanophotonic waveguide devices using CMOS fabrication technology". *IEEE Journal of Selected Topics in Quantum Electronics* **16**.1 (2010), pp. 316–324 (cit. on p. 368).

[5] Xu Wang, Wei Shi, Han Yun, *et al.* "Narrow-band waveguide Bragg gratings on SOI wafers with CMOS-compatible fabrication process". *Optics Express* **20**.14 (2012), pp. 15 547–15 558. DOI: 10.1364/OE.20.015547 (cit. on p. 368).

[6] L. Chrostowski, X. Wang, J. Flueckiger, *et al.* "Impact of fabrication non-uniformity on chip-scale silicon photonic integrated circuits". *OSA Optical Fiber Communication Conference* (2014), Th2A–37 (cit. on pp. 372, 373, 374, 375, 376, 377).

[7] Yun Wang, Jonas Flueckiger, Charlie Lin, and Lukas Chrostowski. "Universal grating coupler design". *Proc. SPIE* 8915 (2013), 89150Y. DOI: 10. 1117/12.2042185 (cit. on p. 372).

[8] N. Rouger, L. Chrostowski, and R. Vafaei. "Temperature effects on silicon-on-insulator (SOI) racetrack resonators: a coupled analytic and 2-D finite difference approach". *Journal of Lightwave Technology* **28**.9 (2010), pp. 1380–1391. DOI: 10.1109/JLT.2010.2041528 (cit. on p. 372).

[9] Tsung-Yang Liow, JunFeng Song, Xiaoguang Tu, *et al.* "Silicon optical interconnect device technologies for 40 Gb/s and beyond". *IEEE JSTQE* **19**.2 (2013), p. 8200312. DOI: 10.1109/JSTQE.2012.2218580 (cit. on p. 376).

[10] R. Boeck, W. Shi, L. Chrostowski, and N. A. F. Jaeger. "FSR-eliminated Vernier racetrack resonators using grating-assisted couplers". *IEEE Photonics Journal* **5**.5 (2013), p. 2202511. DOI: 10.1109/JPHOT.2013.2280342 (cit. on p. 377).

12 Testing and packaging

In this chapter, we discuss techniques for testing and packaging silicon photonic chips. We present an automated probe station, with the software made available open-source. Testing needs to be taken into account at the design stage, to ensure the testability of the designs.

12.1 Electrical and optical interfacing

The primary consideration in testing, and packaging, silicon photonics chips is the method of connecting to the chips. This involves both optical and electrical connections. In this section, we describe the common methods of providing input–output (IO) connectivity.

12.1.1 Optical interfaces

There are two common approaches for optical interfacing to a silicon photonic chip, as shown in Figure 1.5: (a) vertical coupling using grating couplers, or (b–c) edge coupling. These approaches are used for testing, including automated tests (Section 12.2), and for packaging. Two other techniques for experimentation may also be of interest, namely free-space coupling and fibre tapers.

In this section, we describe the on-chip structures required to couple light in and out, and describe what is used to bring the light to the chip. There are several approaches to probe optical structures on-chip, as illustrated in Figure 12.1.

Grating couplers

Grating couplers, described in Section 5.2, are used to couple light in–out of a chip at a near-normal incidence angle. Grating couplers have advantages including wafer and die scale testing and easy automated alignment with reduced alignment sensitivity.

The disadvantages include a reduced optical bandwidth, typically higher insertion loss (below 1 dB is possible [1, 2]), a challenge with polarization, and packaging challenges. One-dimensional grating couplers couple only a single polarization. Polarization diversity can be implemented by using polarization splitting grating couplers [3, 4], albeit with higher insertion loss. The challenges with packaging are that: (1) they require near-normal optical incidence, hence result in a non-planar package, and (2) access to

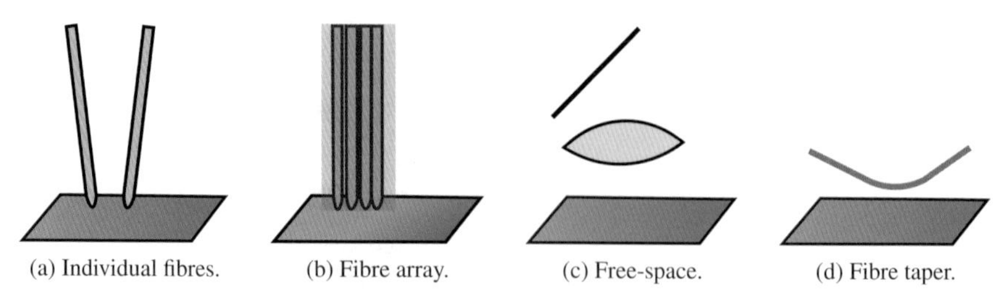

Figure 12.1 Methods of optical probing via the surface of the chip.

the surface of the chip is necessary, thereby making flip-chip bonding to electronic chips more difficult. These challenges have been successfully addressed by the packaging strategies of Luxtera, TeraXion, Tyndall [5], PLC Connections, and others, described in Section 12.1.1.

Edge couplers

Edge couplers, described in Section 5.3, are used to couple light in from the side of the chip. Edge coupling is the most common approach for packaging photonic components, such as lasers, semiconductor optical amplifiers, and detectors, hence there is significant industrial expertise. The advantages include a wide optical bandwidth, low insertion loss (lower than 0.5 dB [6, 7]), and the ability to couple both TE and TM polarizations. Coupling both polarizations enables a polarization diversity approach if used with a polarization splitter rotator [8, 9].

The challenges include alignment difficulty due to the alignment sensitivity (see Figure 5.25). The fabrication is more complicated, namely: (1) tapers, mode converters, lenses, or lensed fibres (Figure 12.2) are required to match the beam from the highly confined SOI waveguides to the optical fibre, and (2) cleaving, polishing, or etching of the facets, is required, and anti-reflection coatings may be desirable.

Individual fibres

In this approach, individual fibres are positioned above the chip for grating couplers, and on the side for edge couplers. The fibre positions are independently controlled using micro-steppers and/or piezo-actuator micro-positioners, allowing freedom in the layout approach; this, for example, allows one to test a 1-N device, such as a WDM components, such as an arrayed waveguide grating with a single input and numerous (e.g. 20) outputs. In this case, only the output fibre needs to be positioned to test separate channels. The fibres can be cleaved or lensed fibres (Figure 12.2); they can be single-mode, polarization maintaining, or even polarizing fibres. An automated optical probe system using grating couplers is described in detail in References [10, 11]. The disadvantages of this approach are that it does not scale to numerous optical fibres (e.g. even four fibres is difficult to construct); the efficiency in the compactness of the layout may not be as high as using a fibre array, mainly due to the physical limitations to how close the fibre grating couplers can be positioned; and there are challenges with instability and vibrations in the fibres.

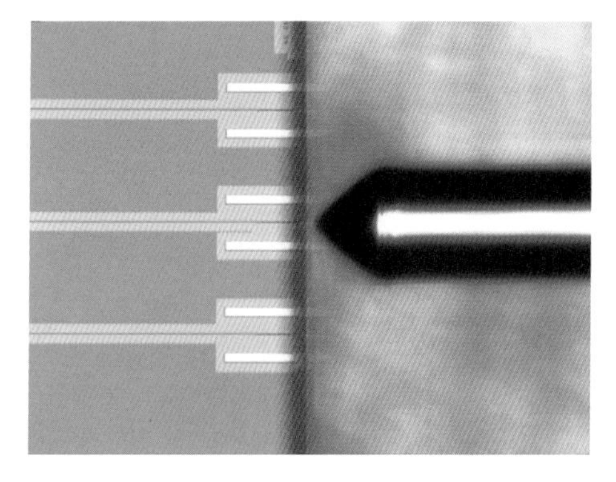

Figure 12.2 Lensed fibre for edge coupling to silicon photonics chips. Courtesy of PLC Connections [12].

Individual fibres can be used for packaging to silicon photonics chips [5]. Fibres can be glued to grating couplers, either perpendicular to a chip, or by polishing the fibre to introduce a reflection, thereby allowing the fibre to be mounted nearly horizontally, suitable for compact packaging [13]. Similarly, for edge coupling, they can be aligned and mounted in a carrier; they can be glued to the chip, an index matching fluid used, or left with an air gap.

Spot-size converter

For edge coupling, the mode field diameter and numerical aperture of the silicon photonic interface typically do not match conventional single-mode fibres. This requires a mechanism to reduce the mode size from the optical fibre down to the chip. This is commonly done by using free-space lenses in packaging for InP lasers, modulators, etc. The lens can also be on the tip of the fibre, and is thus named a lensed fibre. High numerical aperture fibres, that are cleaved, can also be used (Section 5.3.1), but this shifts the problem to the interface between the high-NA fibre and the single-mode fibre (if single-mode fibre is necessary).

Spot-size converters adiabatically convert the mode from the large mode in the single-mode fibre, down to the small mode of the silicon photonic edge coupler. Spot-size conversion can be implemented in a tapered optical fibre, either as an individual fibre, or as numerous optical fibres interfacing to a 1D array of edge couplers [14]. Spot-size conversion can also be implemented using a planar lightwave circuit (PLC) chip [12], as shown in Figure 12.3.

An improved approach demonstrated by IBM evanescently couples to the silicon waveguides, rather than coupling to their edges. It uses a polymer chip with mode-matching waveguides to transition from a fibre mode to evanescently couple to silicon waveguides [15]. The configuration is similar to an edge coupler spot-size converter except that the polymer chip is placed on top of the silicon chip instead of on the side. The structure offers improved coupling efficiency and a reduced alignment tolerance.

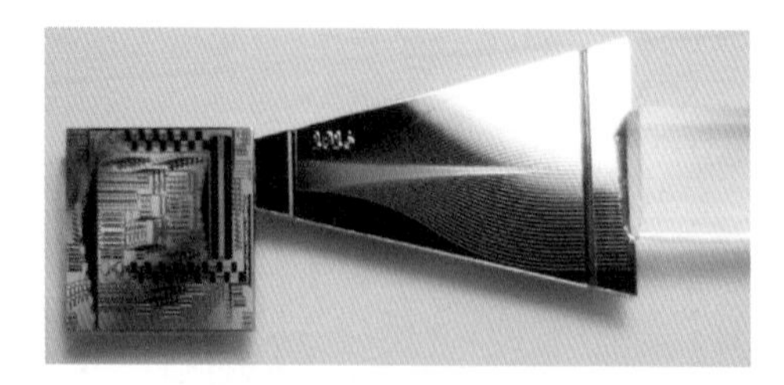

Figure 12.3 Planar lightwave circuit (PLC) fan-in assembly for edge coupling to silicon photonics chips. Courtesy of PLC Connections [12].

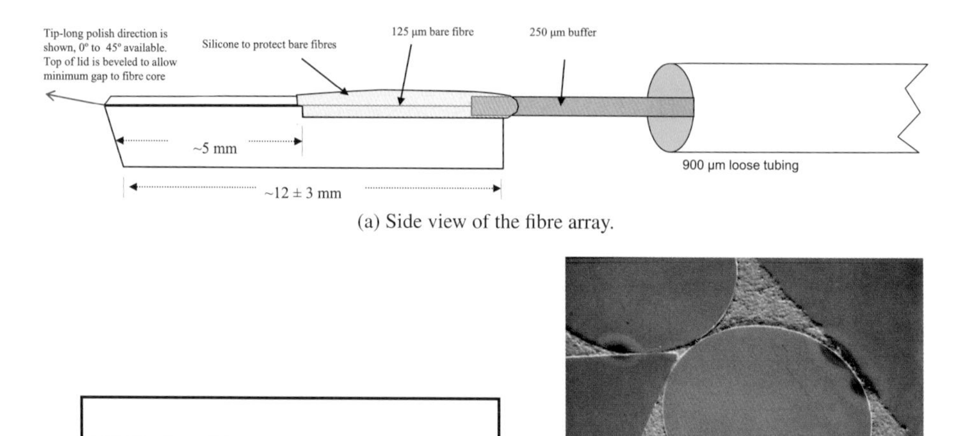

(a) Side view of the fibre array.

(b) Fibre array V-groove, showing polarization maintaining fibres with slow axis in the vertical direction.

(c) Fibre array V-groove, SEM image.

Figure 12.4 Illustration of a fibre array. Courtesy of PLC Connections [12].

Fibre array

Optical fibres can be placed in an array by using a V-groove assembly. Typical array sizes are four or eight, but 64 and beyond are presently commercially available. These arrays are available from manufacturers including PLC Connections (Figure 12.4), OZ Optics, and Corning.

These arrays can be used for automated testing or for packaging. For testing, the chip is placed on an automated micro-positioning stage while the fibre array is fixed during the measurement (or vice versa). With this approach, all the fibres are simultaneously aligned, which speeds up the alignment process. The array is a relative large object (millimeter-scale) hence is not prone to vibration problems resulting in very stable and repeatable measurements; this is particularly important for bit error rate testing (e.g. measuring down to a BER of 1E-12) where long-term stability is required. Using a fibre array, however, constrains the layout to have the optical inputs/outputs in a fixed pattern,

(a) Vertical mounting.

(b) Horizontal mounting using reflector.

Figure 12.5 Fibre array aligned to grating couplers, and glued to the surface of a silicon photonics chip. Courtesy of PLC Connections [12].

though this constraint can simplify the layout task. A fibre array is also less fragile than a single fibre, hence less prone to breaking or damage. In the event of damage, the fibre arrays can be easily re-polished. This approach lends itself well to automated wafer-scale probing, as used by Luxtera [4] and the OpSIS foundry service [16]. The OpSIS setup is shown in Figure 12.8. This approach can also be used for chip-scale testing, in detail in Reference [11]. Section 12.2 describes the implementation and considerations for an automated probe stations using these fibre arrays.

Fibre arrays can be glued to the silicon photonic chips and used for packaging [5, 12, 17]. As shown in Figure 12.5, the fibre arrays can be attached to the grating couplers either vertically or horizontally.

Free-space coupling

There are silicon photonics applications for which optical coupling needs to be done at a distance, which can be accomplished by free-space optics [18]. Example requirements include having the sample placed in a vacuum for cryogenic applications, having the sample completely immersed in a solvent, the need for high-speed scanning between numerous devices such as biosensors as used in the Genalyte system [19], or the performance benefit of measuring numerous outputs simultaneously using an infrared camera whereby one input is multiplexed to numerous devices and outputs (e.g. 128 channels in [20]).

Automated measurement of chips can be implemented using free-space optics. One approach is to use mirrors mounted on galvanometers to scan the incident and reflected light from the chip [19]. Another approach is to capture an image of the light leaving the chip, for highly multiplexed applications [20], although the use of a camera limits the frequency of operation to the scan rate of the imaging device.

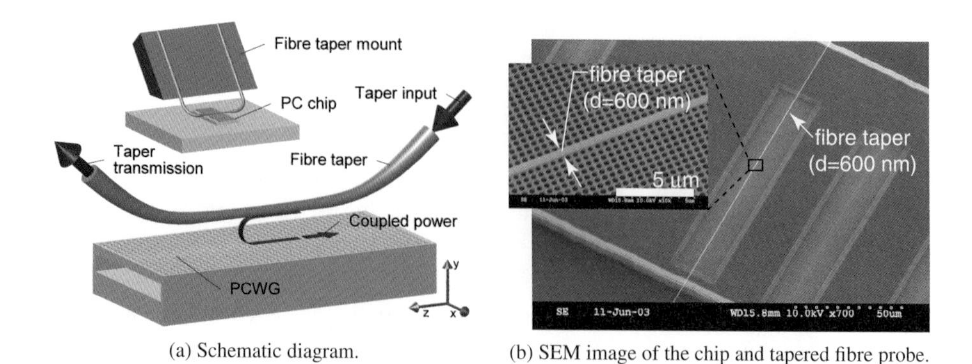

(a) Schematic diagram. (b) SEM image of the chip and tapered fibre probe.

Figure 12.6 Fibre taper probing technique [21]. Courtesy of Paul Barclay.

Fibre taper coupling

Another technique for coupling light from an optical fibre to a silicon photonic chip is by use of an optical fibre taper [21, 22]. The taper is manufactured by heating the fibre with a flame and simultaneously stretching it, until the diameter of the fibre is reduced from 125 μm to about 1 μm. For this small size, the glass fibre becomes the core of the waveguide, while the surrounding air is the cladding. The evanescent tail of the mode can be brought into proximity of a device under test, for example a ring, disk or photonic crystal resonator, as shown in Figure 12.6.

This technique can be used to test numerous devices on a wafer scale, and offers the benefit of high density [23]. However, it requires the silicon photonic device to have air as the cladding, whereas typical silicon photonic foundry processes cover the silicon with an oxide cladding. Since selective oxide removal is available via the IME foundry, this approach can be used to measure light at specific test points. This is attractive for diagnosis since it allows the designer to insert numerous test points without incurring significant excess optical loss in the circuit, when not in use. This is in contrast to grating couplers, which would require a tap (e.g. 10% directional coupler) to permanently remove part of the light from the circuit.

12.1.2 Electrical interfaces

Electrical connections can be temporary for testing (e.g. probes), or permanent for system integration (e.g. wire-bonds).

Bond pads

Connections to electrical circuits are made using bond pads, which are metallized areas on the surface of the chip. They are typically squares, drawn on the last metal layer, with an opening in the upper-most passivation layer. The size of the pad can range from 25 to 100 μm. The pad size must be chosen based on the technique used to connect to it. For probing, the pad size and pitch on-chip must match the probe. For flip-chip bonding, it is necessary to use the same pad size throughout the chip. The pads are often made of aluminum. Bond pads also have a capacitance, though typically small (e.g. 30 fF), which should be taken into account during circuit modelling.

Probing

There are several probe options for electrically contacting chips. Of primary consideration are (a) the frequency requirement, e.g. low frequency versus high frequency; (b) the number of contacts; and (c) the choice of metal tips to ensure low contact resistance, e.g. a nickel alloy or tungsten for aluminum pads (tungsten is strong; beryllium-copper has lower resistivity and suitable for high currents) [24]. The probes are held by a micro-positioner, positioned over the chip, and lowered. When contact is made, the surface oxide often requires the probe to be "skated" by over travelling in the vertical direction. This slightly scratches the pad and breaks through the surface oxide.

Needle probes

For low-frequency probing (<MHz) individual devices with few contacts, e.g. two contacts, individual micro-positioners with needle probes can be used. These are available from numerous vendors [24, 25]. These are convenient as they allow arbitrary size (50 μm in size is convenient) and location of bond pads. The needles can be manipulated so that they do not physically interfere with other test assemblies (e.g. optical inputs), hence offer excellent flexibility.

Multi-contact probes

For testing silicon photonic circuits, testers will often require numerous DC connections, including biasing voltages or currents for phase-shifters in Mach–Zehnder modulators, thermal tuning for ring modulators, low-frequency photocurrent measurements, etc. When the number of connections becomes too numerous for effective testing with individual probes, a multi-contact probe can simplify the footprint required and simplify the alignment. The number of pins and pitch can be customized (e.g. 12 pins on a 150 μm pitch is convenient for rapid manual testing). Such probes are available from several vendors [26, 27]. These probes are large and require consideration for physical test constraints. These probes can also be constructed to offer a combination of RF and DC probes.

RF probing

Making high-speed connections requires careful consideration. Signals are usually launched into an on-chip microwave transmission line, and the impedance of the test equipment, probe, and on-chip device should be well matched to minimize reflections and loss. For single small devices, such as ring modulators or detectors, contact can be made using a ground–signal (GS) probe. Ground–signal–ground (GSG) probes are used for microwave co-planar waveguides such as those used by travelling wave modulators (see Figure 12.7).

Probe manufacturers offer electrical probe design rules for the selected probes, which should be taken into account in the design of the DC and RF electrical probe pads. For example, the Cascade Infinity probes require a minimum probe pad size of

Figure 12.7 Microscope image of a GSG RF probe used for characterizing silicon photonic devices. Nine devices are visible in the image, where the three probe pads of one of the devices are clearly visible after the device was probed and slightly scratched to make contact.

25 μm × 35 μm (best case for manual probing); however, the recommended minimum pad size for automated or semi-automated probing is 50 μm × 50 μm [28]. The probe tips can be spaced in a typical range of 50–150 μm pitch. Ensure that sufficient distance is provided between the RF probe pads and the optical IOs, keeping in mind the size of the optical fibre opto-mechanics and RF probes.

Wire bonding

Wire bonding is the most common method of permanently connecting a chip to the external world or system. The chip is typically placed on a package, and interconnections are made between the two. Wire-bonding machines are available from numerous vendors [29] and can be manual or automated.

There are several types of metal wires used, including gold, copper, and aluminum. Two common types of bonds are wedge bonds and ball bonds. Wire bonders also have design rules, such as bond pad size and pitch. The bond pads are placed on the outside edges of the chips to minimize the lengths of the wire bonds, since wire bonds have a length-dependent inductance. Typically a 25 μm wire has a 1 nH inductance per mm of length. Wire bonds can be modelled in electromagnetic numerical software packages, e.g. Agilent ADS [30]. Wire bonds typically introduce a low-pass filter characteristic to the system and reduce the system bandwidth.

Flip-chip bonding

Flip-chip bonding, as the name implies, is a technique whereby the photonic or electronic integrated circuit is flipped and bonded onto another chip, wafer, or package. The connections between the two components are made using an array of bond pads, and solder bumps, which are deposited on the bond pads; the components are aligned, heated, and compressed to form the bond; finally an under-fill material is inserted for mechanical and environmental stability.

Flip-chip bonding has a big advantage in reduced parasitic capacitance, compared to other packaging techniques such as wire bonding. Specifically, the inductance is significantly reduced (0.18 nH versus 1.6 nH for the wire bond in [31]), negligible resistance is virtually eliminated (1 mΩ in [31]) and some capacitance is introduced between adjacent balls (50 fF in [31], smaller than the bump pad capacitance of 177 fF). Thus, flip-chip bonding is very attractive for high-frequency applications, when the number of connections is very large (e.g. 1000) and finally, flip-chip bonding reduces the package size. There are challenges with flip-chip bonding, namely the cost, yield, and challenges in test and repair.

There are manufacturing design rules associated with different flip-chip bond technologies. Typically, the bond pads should be matched in size (and position), with the same size used throughout the design. The layout of the two components can be verified and simulated [32], for example, to ensure that the bond pads match with the desired connectivity. There are rules for minimum and maximum die size, maximum number of bumps, and minimum pad size (e.g. 50 μm).

The interested reader is referred to textbooks on packaging for further information, e.g. [33, 34], and recent literature on silicon photonics packaging, e.g. [5, 13, 35].

12.2 Automated optical probe stations

Automated measurements are critical in several applications.

- Automated wafer-scale testing, prior to dicing and packaging, is critical to improve the yield in large-scale manufacturing. Devices can be tested and only the chips that meet the specifications are chosen for subsequent packaging steps [4].
- Automated testing is also critical for device and component research, and library development, particularly if a large range of parameter variations is considered [4].
- Automated measurements enable the testing of large quantities of devices, from numerous wafers and die, to assess the manufacturing process and stability, device yield, and so on [4].
- Automated optical alignment and probing is useful for system experiments, particularly if the experiments are conducted prior to packaging.

In the following, we describe the hardware and software that can be used in an automated probe station using a fibre array, with additional details and source code posted online [36]. Several probe stations were built using a variety of hardware platforms, including vendors such as ThorLabs and PI miCos, as shown in the photograph in Figure 12.8 and in the CAD design illustrations in Figures 12.9 and 12.10.

This probe station design provides important benefits: (1) it provides automated optical alignment to achieve high-speed testing (over 2000 devices per day); (2) it enables active silicon photonic and electronics experimentation; and (3) it was designed to be the

Figure 12.8 Wafer-level automated probe station.

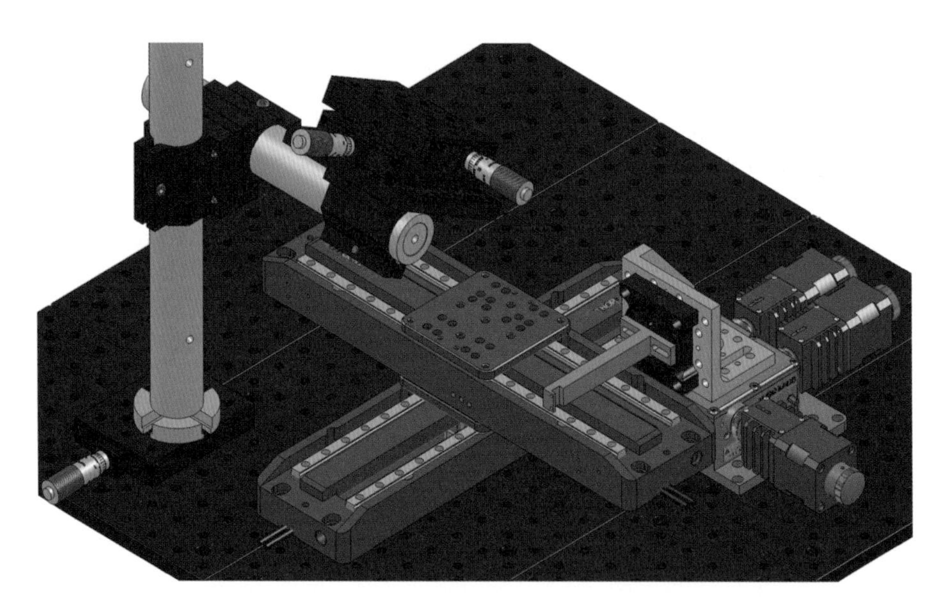

Figure 12.9 CAD diagram of the automated optical probe station mechanics. Top/left: optical microscope mount and micro-positioners; bottom/right: custom optical fibre array holder, with piezo-nano-positioner and stepper motors (ThorLabs); centre: sample stage, micro-steppers (ThorLabs); not shown: custom chip mount with temperature control and vacuum chuck, microscope, micro-positioner and piezo-controllers, computer, monitor, cables.

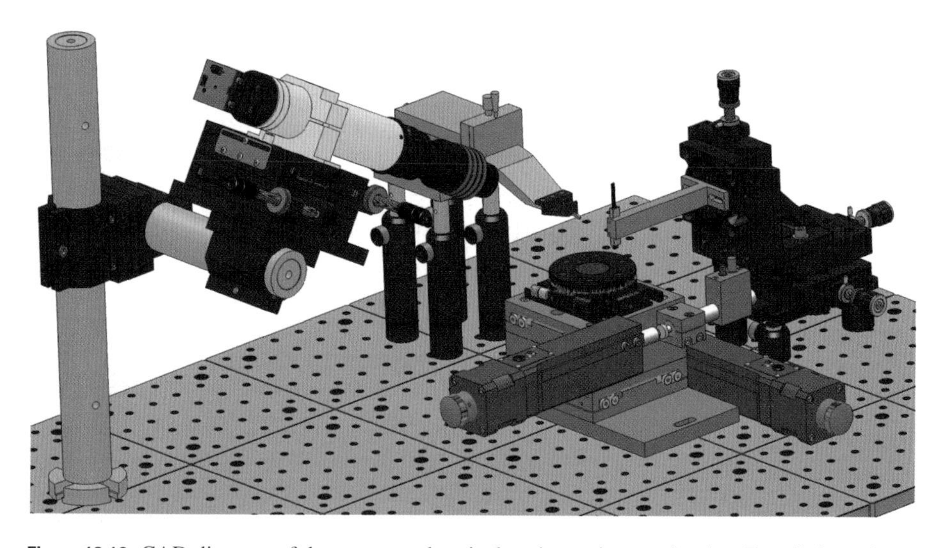

Figure 12.10 CAD diagram of the automated optical probe station mechanics. Front/left: optical microscope, assembly, and micro-positioning; back/left: RF Probe (GGB Industries) and holder (Signatone); back/right: optical fibre array (PLC Connections), custom fibre array holder, and manual positioner (ThorLabs), shown at an angle of 20°; front/centre: sample stage, micro-steppers and piezo actuators (ThorLabs); not shown: custom chip mount with temperature control and vacuum chuck, micro-positioner and piezo-controllers, computer, monitor, cables.

lowest-cost solution without compromising the performance (insertion loss, stability, repeatability).

The probe station consists of a sample stage controlled via micro-stepper motors to align the silicon photonic sample to an optical fibre array (containing optical inputs and outputs). The system will optically characterize all devices on the sample – the coordinates of the grating couplers are extracted from the mask layout file. Then, the user can electrically probe chosen devices for electrical measurements. The system can connect to existing equipment, including temperature controllers, tuneable lasers and detectors to measure the optical response of devices, optical vector network analyzers, and high-speed electrical test equipment such as Bit Error Rate testers (BERT), high-speed Vector Network Analyzers (VNA), etc.

12.2.1 Parts

Details for the components chosen for the chip-scale automated setup are as follows.

Sample stage

(1) Automated positioning using XY stepper motors (Figure 12.11 – x, y): these parts position the silicon photonic chip relative to the optical fibre array. A 100 mm travel is chosen to easily accommodate the largest die size typically available in silicon photonics foundries (25 × 32 mm). Larger travel stages can be chosen for wafer-scale testing. The stages need a resolution and repeatability of better than 1 μm, in order to accurately align the optical fibres to the silicon chips.

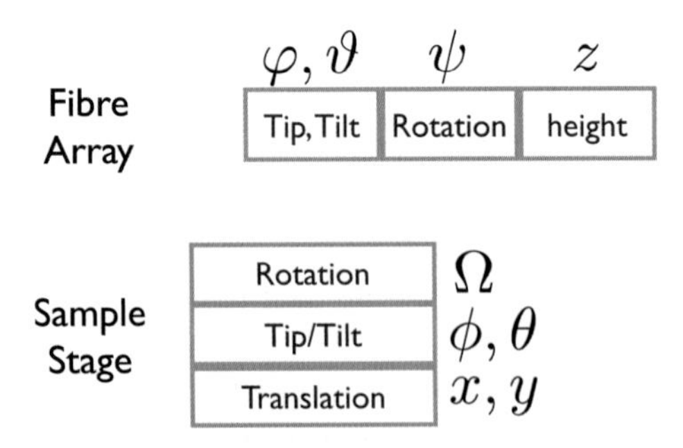

Figure 12.11 Degrees of freedom in the automated test setup, with five degrees for the sample stage, and four for the fibre array.

Hence piezo-electric stages with nanometer-scale precision are not required, and stepper-motor-based stages are sufficient. The stepper controllers utilize optical encoders to improve upon the accuracy.

(2) Motion control: the stepper motors are controlled by micro-controller, connected to the computer using a USB or RS-232 interface.

(3) High-precision rotation mount: this is used to precisely rotate the silicon photonic sample relative to the fibre array (Figure 12.11 – Ω).

(4) Tip/tilt stage: a 2-axis stage is used to ensure that the chip is flat relative to the axes of the XY stage (Figure 12.11 – ϕ, θ). This is used to ensure that the distance between the chip and the fibre array is constant across the full range of motion.

(5) Custom chip mount: a vacuum sample holder with a thermoelectric module with a thermistor for temperature control.

Fibre array probe

(1) Optical fibres are used to couple the light in and out of the silicon chips. There is a variety of options in configuring the arrays. A common standard for the fibre array pitch is 127 μm (250 and 81 μm are also possible). Given the polarization dependence of silicon photonics, generally testing requires polarization controllers or use of polarization-maintaining components. Thus, it is beneficial to use polarization-maintaining fibres such as PM 1550 PANDA. Optical connectors such as FC/APC are used for reduced reflections. The fibre array is polished at an angle determined by the grating couplers; 8° is commonly available, but custom polish angles are available and should be chosen based on the fabrication process.

(2) Tip/tilt stage: the fibre array is mounted on a two-axis stage that is used to ensure that the fibre array is parallel relative to the silicon photonic chip, in two degrees of freedom (Figure 12.11 – φ, ϑ).

(3) An aluminum block for holding the fibre array, custom machined.

(4) This is attached to a Z-axis translation stage, which is used to control the distance between the fibre array and the chip (Figure 12.11 – z). This can be a manual translation stage (e.g. 25 mm travel), or a controlled stage. The distance between the chip and the array is typically adjusted to be in the range of 5 to 50 μm.

(5) A rotation mount is used to change the angle of the fibre relative to the grating coupler (Figure 12.11 – ψ). This is used to ensure that the fibre array surface is parallel with the silicon chip to minimize the gap between the two. Another useful functionality is that, because grating couplers are wavelength sensitive, the spectral response can be tuned by changing the angle, e.g. from 10° to 40°. If this functionality is desired, the fibre array should be polished at an angle greater than the optimal polish angle, to allow for a plus/minus control of the central wavelength. This is particularly useful for experimentation with grating coupler designs, or if the fabrication process is not stable.

Electrical probes

(1) A four-axis micro-positioner is used to electrically probe the devices. This is either bolted down to the system, or, more conveniently, magnetically attached. Three axes are used for positioning (XYZ). For electrical probes with numerous electrical pins, adjusting the rotation (theta) is necessary to ensure that all the pins touch on the sample at the same time.

(2) Different probes can be used in the system, depending on the frequency response required and configuration. This can include DC, RF (e.g. 10, 40, 67 GHz), or combination probes, in configurations including GS, GSG, etc. The probe can be operated manually, or the vertical motion can be automated for automated probing.

Electrostatic discharge (ESD) protection for electronic and electro-optic components should also be included.

Microscopes

A microscope is required for imaging the fibre array and silicon photonic chip (Figure 12.12), to find the optical grating couplers and the electrical pads. A microscope with a long working distance is beneficial to reduce the physical interference with the other components (electrical and optical probes). A field of view ranging from 0.5 to 2 mm is desired, and focusing capability is necessary. A camera is attached to the microscope and the video will be displayed on the computer monitor during alignment. Translation stages are used to position the camera.

It is beneficial to have two cameras in the system. One camera is used for near-horizontal imaging to view the gap between the fibre array and the chip, and to find the correct grating couplers. A second camera is near vertical to image the electrical connections.

12.2.2 Software

This section describes the requirements for a typical automated software setup. It provides the methodology that was used in implementing the open-source automated setup software [36] and how to conduct experiments.

The instrument control software is written using MATLAB, using the Instrument Control Toolbox. The software:

Figure 12.12 Image captured using the microscope. An eight-fibre array is visible at the top, with a silicon photonic chip and grating couplers being probed. The grating couplers are aligned with the fibres in the array.

- communicates with the opto-mechanical setup and the test equipment;
- configures the experiments (wavelength span, number of points, integration time, temperature, etc.);
- displays the video from the microscope;
- *coarse search*: performs a 2D mapping of the fibre-coupled light intensity by translating the sample stage (Figure 12.11 – x, y). This generates a "heat map", and is used to identify the approximate location of grating couplers and optical alignment structures. The mapping is commonly on a distance ranging from 100 to 500 µm. Alternatively the scan is performed until a threshold optical power is reached, indicating that a device is found;
- performs a *fine align* to optimize the fibre coupling efficiency;
- registers the chip, to correlate the mask layout (GDS) to the motor positions. This is done by computing transformation matrix and estimation of error;
- reads an input file, consisting of the fibre grating coupler locations and device names (extracted from the GDS layout files). It automatically positions the chip to a user-selected device. A fine align is performed.
- performs swept measurements of the optical spectrum using a laser and detector, and records the output power versus wavelength;
- performs electrical measurements using additional equipment.

12.2.3 Operation

The first step to optical measurement is to align the chip and fibre array to one another. This is accomplished in two parts: (1) the chip is aligned to "absolute" axes, defined by the directions of the three translation stages (Figure 12.11 – x, y, z); and (2) the fibre array is aligned separately to the same axes. The alignment is done in two parts so that when one part is changed (e.g. new wafer, or new fibre array), only one of the alignment procedures needs to be performed. The procedure below uses manual adjustments (Figure 12.11 – $x, y, z, \phi, \psi, \theta, \Omega, \varphi, \vartheta$), as well as a computer-controlled coarse search and fine align, using the sample stage (Figure 12.11 – x, y).

Loading and aligning a chip/wafer

The procedure for loading and aligning a sample is as follows.

(1) Load the sample on the stage, coarse align by eye or with stage guides. Turn on the vacuum to hold the sample.
(2) Bring the fibre array in proximity to the chip (Figure 12.11 – z), and in the field of view of the microscope. Focus the microscope on the surface of the chip.
(3) Align the chip to an "absolute" flatness and rotation.
   (i) Translate the position of the chip (Figure 12.11 – x, y) so that the fibre array is positioned over one corner of the chip, either to an optical alignment marker such as a grating coupler, or to the edge of the chip.
   (ii) Estimate and note the distance between the chip and the fibre array, in the z-direction.
   (iii) Translate the position of the chip (Figure 12.11 – x, y) along one direction, x, reaching the other corner, again reaching an alignment structure.
   (iv) Estimate and note the distance between the chip and the fibre array, in the z-direction. Note the distance between the fibre array and the alignment structure, in the y-direction.
   (v) Adjust the sample tip/tilt stage (Figure 12.11 – ϕ or θ as appropriate), to reduce the mismatch between the two height estimates.
   (vi) Adjust the sample rotation stage (Figure 12.11 – Ω), to reduce the mismatch in the y-direction.
   (vii) Repeat the above until there is no mismatch in the z- and y-directions, when translating the sample stage in the x-direction to opposite ends of the chip.
   (viii) Perform the same procedure as above, except for y-translation, while observing the mismatch in the x-direction.

Aligning the fibre array

When replacing the fibre array, or for the first time, the array needs to be aligned and parallel to the wafer.

(1) Bring the fibre array in proximity to the chip (Figure 12.11 – z), and in the field of view of the microscope. Focus the microscope on edge of the fibre array.
(2) Estimate the distance of the fibre array to the chip, on both sides of the array. This is most easily accomplished when the microscope is placed horizontally, i.e. parallel to the sample, and perpendicular to the fibre array.
(3) Adjust the fibre tip/tilt stage (Figure 12.11 – φ, ϑ) until the array is parallel to the chip.
(4) Use the computer-controlled coarse align to locate a grating coupler alignment/test structure. Then fine align to maximize the optical power.

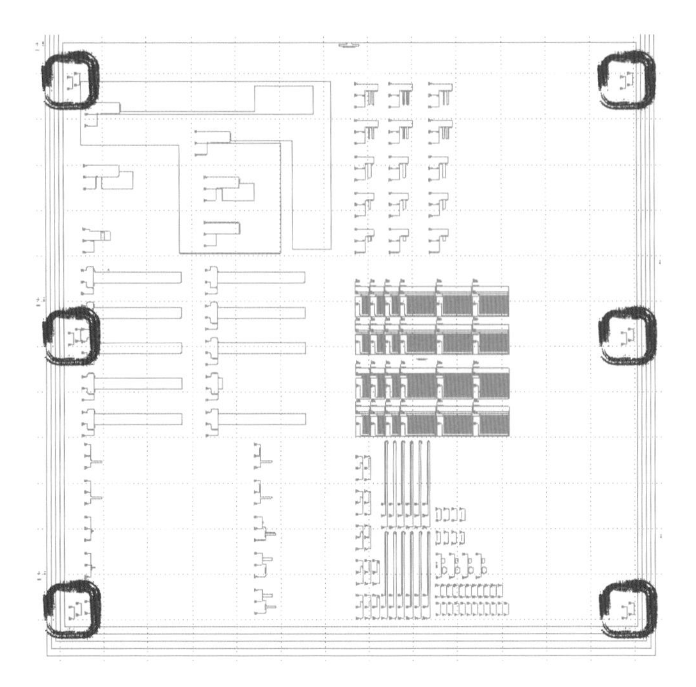

Figure 12.13 Fibre grating coupler alignment structures (see Figure 12.14), placed on the corners and edges of a chip, for automated measurement registration.

(5) Adjust the fibre tip/tilt stage (Figure 12.11 – φ, ϑ), the fibre array distance to the chip (Figure 12.11 – z), and the fibre array rotation (Figure 12.11 – ψ), to maximize the optical power. For each adjustment made, use the fine align to maximize the optical power.

Chip registration

For automated measurements, the system needs to be able to align to all known devices. A script in the GDS mask layout viewer (kLayout) exports the locations of the fibre grating couplers (inputs only) from the GDS design file, with the labels as shown in Figure 12.14. The user needs to physically locate the alignment structures on the chip, Figure 12.14, perform a fine align, and register the alignment structures' GDS coordinates. Typically this is done on the corners of the chip, as illustrated in the sample layout in Figure 12.13. This generates a matrix, T, as shown in Figure 12.15, which is subsequently used for the automated measurements.

Automated device testing

Based on the list of test devices, the system will automatically align the chip to the fibre array for each devices, one by one. Measurements will then be performed such as optical spectrum acquisition, using a tuneable laser and detector. Temperature sweeps can also be performed.

For electrical tests, this can be done manually, whereby the optical system aligns the device and the user manually probes. Alternatively, the automated system will first

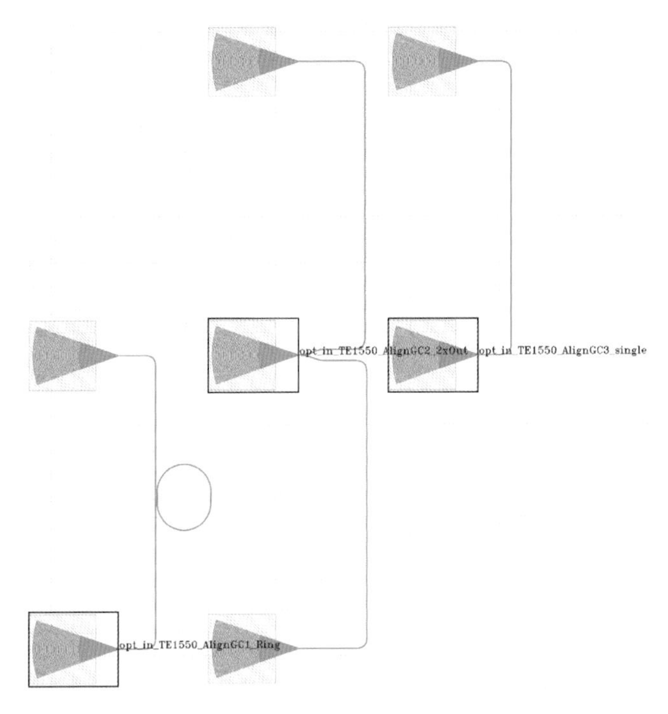

Figure 12.14 Fibre grating coupler alignment structure, for automated measurement registration. Alignment structure can include alignment for different polarizations and wavelengths, and can also include basic characterization – in this example, a ring resonator. The optical inputs are highlighted, and labelled with a text field to be read by the automated software.

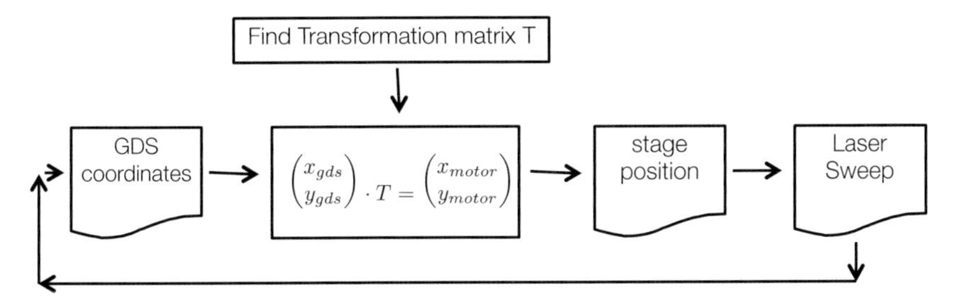

Figure 12.15 Registration of chip/wafer to the fibre array, for automated measurements.

perform optical alignment, then the electrical probe descends, makes a contact, and performs automated measurements.

12.2.4 Optical test equipment

Numerous lasers are available for testing silicon photonic devices and systems. To measure resonators, it is necessary to have a laser with a low linewidth and good stability, so that the resonator linewidth can be well resolved. For measurements where a high extinction ratio is desired, e.g. greater than 20 dB, laser noise also becomes a limiting factor. As shown in Figure 12.16, a laser with a high output power typically

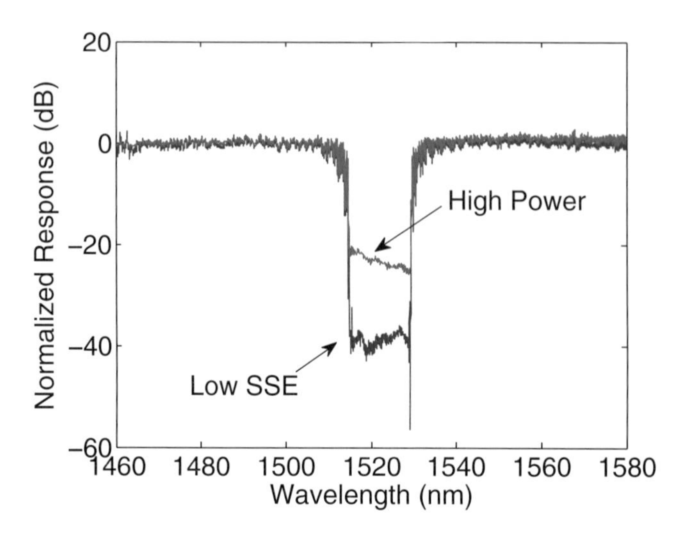

Figure 12.16 Impact of the tuneable laser source spontaneous emission (or relative intensity noise specifications) on the measurement accuracy.

has a high laser-relative intensity noise and spontaneous emission. This light, which is outside the linewidth of the laser, will be transmitted through the device and detected. Thus it leads to a noise floor which limits the maximum extinction ratio that can be measured. Tuneable filters can be used to reduce this noise, or lasers with a lower source spontaneous emission specification can be used as in Figure 12.16, where an extinction ratio of 40 dB was measured for a waveguide Bragg grating. For high-speed swept measurements, synchronization between the laser and the detector is also desirable to allow for a simple implementation for data acquisition.

Swept wavelength measurements are available in instruments known as optical vector network analyzers. These instruments are convenient in that they offer full device characterization including amplitude and phase, for both transmission and reflection of the device under test.

Optical amplifiers are also useful for both device and system characterization. Optical amplifiers introduce noise, hence optical filters may also be desirable to filter the out-of-band noise.

Testing electro-optic devices requires additional equipment such as low-noise DC electrical supplies. High-speed measurements are performed by using an electrical vector network analyzer, signal and pattern generators, oscilloscopes, and bit error rate testers.

12.3 Design for test

Design for test (DFT) encompasses design approaches that add features to simplify or to enable the testing of a design. In CMOS electronics, chip designers can include methods whereby digital chips are tested for functionality, typically by applying test vectors. Another approach is to include a probe point within a circuit, e.g. to measure a voltage at a specific node, or to change the circuit operation. For leading-edge semiconductor processes, DFT is concerned with diagnosing the fabrication process performance and yield, as well as for debugging. In the case of failures, the debugging allows for failure analysis.

In silicon photonics, and photonics integrated circuits, several considerations are important and are described in the following.

Test methodology

Each design needs to have a clear plan for how it will be tested. If it is to be used in a product, a packaging needs to be in place.

Test structures

Test structures are included in a layout for local metrology for material properties (thicknesses, composition, roughness), tests for individual components (coupling coefficients in ring resonator gaps, waveguide losses, waveguide bend losses, waveguide group index), sub-circuits (Mach–Zehnder). Some of these structures may be included by the foundry as part of their metrology.

Examples of test structures that may be used in all fabrication runs include the following.

- Critical dimension (CD) lithography test structures: e.g. isolated lines, line pairs; wafer-level uniformity data. These are measured using an SEM.
- Waveguide loss: strip and rib waveguides with varied width; waveguide with doping including pn junction; bend loss.
- Electrical resistance: doped silicon; metal interconnects; contact vias.
- Optical fibre grating coupler insertion loss.
- Directional couplers: coupling coefficients.

Some test structures are used to develop library components, and may not need to be critically monitored on every fabrication run.

- Optical fibre edge couplers.
- Thermal heater efficiency: N++/N/N++ waveguide; N++/N/N++ next to a waveguide; metal above a waveguide.
- Splitters: Y-branch, MMI.
- Waveguide crossing: insertion loss and cross-talk.
- Ring/disk resonators with doping; e.g. to find critical coupling for modulation and filtering.
- Inter-waveguide cross-talk.
- Bragg gratings; lithography effects.
- Lithography prediction test structures.
- Photonic crystals.
- Wavelength division multiplexers: Echelle grating; arrayed waveguide grating; ring resonators.
- Ring/disk modulator, with thermal tuning.

- Microwave transmission lines (GS, GSG), with and without pn junction.
- Detectors.

For testing sub-circuits such as Mach–Zehnder modulators, as shown in Figure 1.7a and in Reference [37], it is useful to include an optical path length mismatch between the two arms to help characterize the waveguide properties and simplify testing. In this example, a 100 μm path length mismatch gives rise to approximately a 6 nm free-spectral range. This intentional imbalance simplifies the task of determining the modulator's tuning coefficients, for both the thermal tuner and the carrier depletion phase shifter, by observing the shift in the fringes in the optical spectrum. It also can be used to determine the waveguide group index.

Trimming

Owing to fabrication non-uniformity and silicon's temperature sensitivity, it is often necessary to include phase adjustments in phase-sensitive circuits. This includes Mach–Zehnder modulations, ring modulators, and filters, etc. Trimming can be implemented on-chip via thermal tuning or carrier injection (see Section 6.5). It can also be done as a post process, e.g. by selective annealing of deep-level defects created by an ion implantation step [38] which can provide up to 0.02 change in index of refraction.

12.3.1 Optical power budgets

Experiments with silicon photonic circuits require an understanding of the available optical power, how much power will be transmitted through the system, and the sensitivity of the detection system.

The power budget needs to include all sources of optical loss. This includes the optical interface losses, the experimentally measured waveguide losses, insertion losses of on-chip devices, etc. An example power budget analysis is as follows.

0 dBm (1 mW)	laser output
−6 dB	input grating coupler
−10 dB	circuit loss
−6 dB	output grating coupler
= −22 dBm	fibre coupled power

It is important to understand how much power is necessary for the application. For digital communications application, this is known as the sensitivity of the receiver, e.g. −20 dBm for a 10 Gb/s receiver. For device characterization, particularly at low frequency, detectors are much more sensitive and it is possible to measure signals as small as −80 dBm. Optical amplifiers can improve the power budget, but consider their availability, gain, spectral response, and noise contributions. For example, a typical erbium doped fibre amplifier (EDFA) can provide 20–30 dB of gain at wavelengths near 1530 nm.

12.3.2 Layout considerations

When designing the layout for silicon photonic chip, the space limitations imposed by the electrical and optical probes, micro-positioners, wire bonds, etc., need to be taken into account. It is convenient to measure the components (probes, fibre array), and draw them on the mask layout, in order to ensure that no overlap in parts is accidentally obtained. Several example test arrangements are illustrated in Figure 12.17. When testing requires only a single electrical probe, such as when characterizing individual high-speed components, the layout can be quite compact. As shown in Figure 12.17b, the minimum spacing between the grating couplers and the bond pads is determined

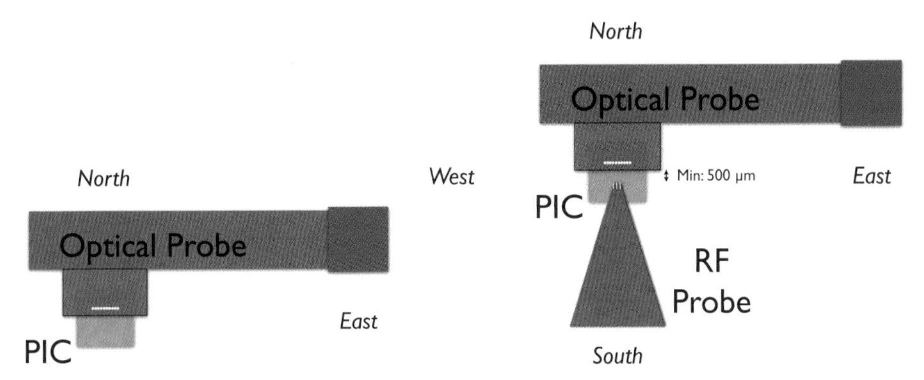

(a) Optical tests using a fibre array on the North, with its micro-positioner mounted on the East.

(b) Optical and RF tests using a fibre array on the North, and RF probe on the South.

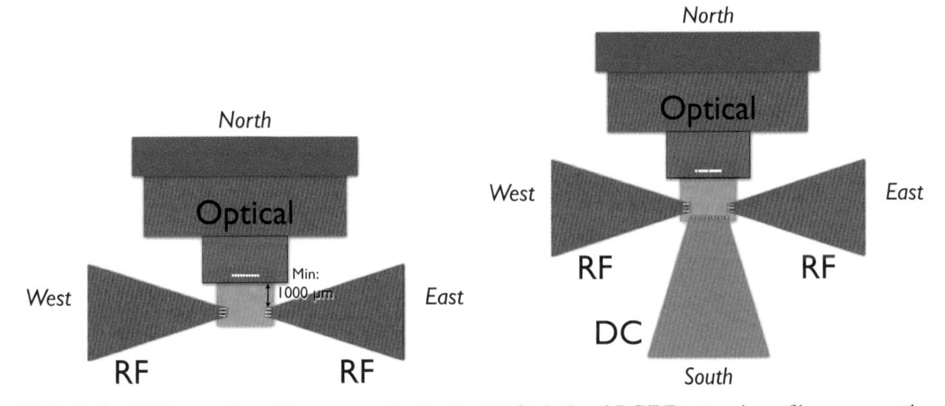

(c) Optical and RF tests using a fibre array on the North, with its micro-positioner mounted on the North. Two RF probes on the East and West.

(d) Optical and DC/RF tests using a fibre array on the North. Two RF probes on the East and West. One DC multi-contact probe on the South.

Figure 12.17 Configurations for testing photonic chips, including optical and electrical interface geometric considerations. PIC: photonic integrated circuit; East, West, North, South: common methods of describing the location of the probes relative to the chip. Illustration is drawn to scale, where the PIC is 2.5 × 2.5 mm; the fibre array is 3.75 × 2.0 mm and contains 10 fibres on a 127 μm pitch; the electrical RF probe is a GSG probe with a 150 μm pitch; the electrical DC probe is a multi-contact probe with 12 contacts on a 150 μm pitch.

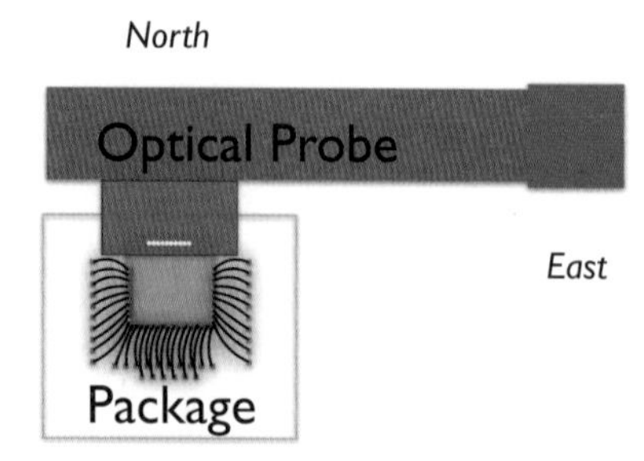

Figure 12.18 Optical interfacing to a packaged chip using a fibre array. Electrical connections wire-bonded to a package.

by the size of the fibre array. For the array in Figure 12.4, the suggested minimum space is 500 μm. When two RF probes are required and need to face each other, such as for testing high-speed travelling wave modulators with a GSG probe for the drive signal and a second GSG probe for the termination, a different configuration can be considered. As shown in Figure 12.17c, the fibre array and micro-positioner are placed on the North side, whereas the two RF probes and micro-positioners are placed on the East and West sides. In this case, the distance between the grating couplers and the RF probe pads needs to be a minimum of 1000 μm. A DC multi-contact probe can also be added to this configuration on the South side, as shown in Figure 12.17d.

As shown in Figure 12.18, the silicon photonic chip can be wire-bonded to a carrier, and optically probed using the fibre array. In the case, the wire bonding needs to be carefully done to leave room for the fibre array.

The above design rules should be evaluated and modified for the specific test situation of the designer.

12.3.3 Design review and checklist

Prior to sending a design for manufacturing, it is standard practice to perform a review of the design. A checklist is useful to consider all the main points, including manufacturability and testability.

An example checklist for an active Silicon Photonics Design Review is on the following page.

Manufacturability

- ☐ Is the design DRC (error) clean (include explanation and DRC-exclusions for the ones that will not be fixed)?.
- ☐ Has the design been tiled?
- ☐ Will the structure work for the known process variations (include design with varying parameters)?
- ☐ Is the layout considered “safe” from a manufacturability perspective, namely does the layout avoid using a lot of minimum feature sizes?
- ☐ Are there fallback designs or subsets of the system?

Mask layout

- ☐ Are all waveguides connected? Are all waveguide paths converted to waveguides with appropriate radius? Is the layout space-efficient?
- ☐ Are the waveguides and optical structures designed in the layout such that they will have low optical loss (e.g. are other structures far enough away)?
- ☐ Are all cell names correct in the .gds file?

Post-processing

- ☐ Does the layout include any necessary alignment markers for subsequent lithography?
- ☐ Are sufficient space allowances included for the release step for suspended structures?

Optical interface

- ☐ For edge-coupling waveguides, are the waveguides tapered?
- ☐ Is the probing approach defined (edge coupling or vertical coupling)?
- ☐ Fibre array or single fibre? Is the fibre array or single fibre pitch ok?
- ☐ Are the optical IOs interfering with each other and/or with electrical IOs?
- ☐ Would you be able to visualize the alignment with the microscope (will the fibre or probe block your view or shade parts you need to see for the alignment)?

Electrical interface

- ☐ Which IOs are RF, which ones are DC? Are their impedance, trace length, etc., ok?
- ☐ Are the IOs being probed only, wirebonded only, both? Are the pads pitch/area ok?

Testing

- ☐ Is the chip testable as a standalone? Does it need to be wirebonded to another device to be able to be measured (e.g. TIA)? Can all signals be probed? (Have structure that can be probed.)
- ☐ If DC pads are wirebond and RF pads are probed – can the probe possibly land on bond wires?
- ☐ Are there optical test structures included that can be measured by the automated probe station?

Packaging

- ☐ Will it be packaged? Is the IO spacing requirement for packaging met?
- ☐ Does the design follow the available packaging rules (e.g. offsets from edge of the chip, spacing, etc.)?

Design Review Acceptance Date:
Designer:
Signature:

References

[1] A. Mekis, S. Abdalla, D. Foltz, *et al.* "A CMOS photonics platform for high-speed optical interconnects". *Photonics Conference (IPC)*. IEEE. 2012, pp. 356–357 (cit. on p. 381).

[2] Wissem Sfar Zaoui, Andreas Kunze, Wolfgang Vogel, *et al.* "Bridging the gap between optical fibers and silicon photonic integrated circuits". *Optics Express* **22**.2 (2014), pp. 1277–1286. DOI: 10.1364/OE.22.001277 (cit. on p. 381).

[3] Dirk Taillaert, Harold Chong, Peter I. Borel, *et al.* "A compact two-dimensional grating coupler used as a polarization splitter". *IEEE Photonics Technology Letters* **15**.9 (2003), pp. 1249–1251 (cit. on p. 381).

[4] A. Mekis, S. Gloeckner, G. Masini, *et al.* "A grating-coupler-enabled CMOS photonics platform". *IEEE Journal of Selected Topics in Quantum Electronics* **17**.3 (2011), pp. 597–608. DOI: 10.1109/JSTQE.2010.2086049 (cit. on pp. 381, 385, 389).

[5] Bradley W. Snyder and Peter A. O'Brien. "Developments in packaging and integration for silicon photonics". *Proc. SPIE*. Vol. **8614**. 2013, pp. 86140D–86140D–9. DOI: 10.1117/12.2012735 (cit. on pp. 382, 383, 385, 389).

[6] Na Fang, Zhifeng Yang, Aimin Wu, *et al.* "Three-dimensional tapered spot-size converter based on (111) silicon-on-insulator". *IEEE Photonics Technology Letters* **21**.12 (2009), pp. 820–822 (cit. on p. 382).

[7] Minhao Pu, Liu Liu, Haiyan Ou, Kresten Yvind, and Jorn M. Hvam. "Ultra-low-loss inverted taper coupler for silicon-on-insulator ridge waveguide". *Optics Communications* **283**.19 (2010), pp. 3678–3682 (cit. on p. 382).

[8] M. R. Watts, H. A. Haus, and E. P. Ippen. "Integrated mode-evolution-based polarization splitter". *Optics Letters* **30**.9 (2005), pp. 967–969 (cit. on p. 382).

[9] Daoxin Dai and John E. Bowers. "Novel concept for ultracompact polarization splitter-rotator based on silicon nanowires". *Optics Express* **19**.11 (2011), pp. 10 940–10 949 (cit. on p. 382).

[10] Han Yun. "Design and characterization of a dumbbell micro-ring resonator reflector". MA thesis. 2013. URL: https://circle.ubc.ca/handle/2429/44535 (cit. on p. 382).

[11] Charlie Lin. "Photonic device design flow: from mask layout to device measurement". MA thesis. 2012. URL: https://circle.ubc.ca/handle/2429/43510 (cit. on pp. 382, 385).

[12] *PLC Connections PLCC – Silicon Photonics*. [Accessed 2014/04/14]. URL: http://www.plcconnections.com/silicon.html (cit. on pp. 383, 384, 385).

[13] Y. Painchaud, M. Poulin, F. Pelletier, *et al.* "Silicon-based products and solutions". *Proc. SPIE*. 2014 (cit. on pp. 383, 389).

[14] F. E. Doany, B. G. Lee, *et al. IEEE Journal of Lightwave Technology* **29**.475 (2011) (cit. on p. 383).

[15] T. Barwicz, Y. Taira, "Low-Cost Interfacing of Fibers to Nanophotonic Waveguides: Design for Fabrication and Assembly Tolerances", *IEEE Photonics Journal*, Vol. **6**, no. 4, 6600818, 2014, DOI: 10.1109/JPHOT.2014.2331251 (cit. on p. 383).

[16] Tom Baehr-Jones, Ran Ding, Ali Ayazi, *et al.* "A 25 Gb/s silicon photonics platform". *arXiv:1203.0767v1* (2012) (cit. on p. 385).

[17] Amit Khanna, Youssef Drissi, Pieter Dumon, *et al.* "ePIX-fab: the silicon photonics platform". *SPIE Microtechnologies*. International Society for Optics and Photonics. 2013, 87670H (cit. on p. 385).

[18] Ellen Schelew, Georg W. Rieger, and Jeff F. Young. "Characterization of integrated planar photonic crystal circuits fabricated by a CMOS foundry". *Journal of Lightwave Technology* **31**.2 (2013), pp. 239–248 (cit. on p. 385).

[19] Muzammil Iqbal, Martin A Gleeson, Bradley Spaugh, *et al.* "Label-free biosensor arrays based on silicon ring resonators and high-speed optical scanning instrumentation". *IEEE Journal of Selected Topics in Quantum Electronics* **16**.3 (2010), pp. 654–661 (cit. on p. 385).

[20] S. Janz, D.-X. Xu, M. Vachon, *et al.* "Photonic wire biosensor microarray chip and instrumentation with application to serotyping of *Escherichia coli* isolates". *Optics Express* **21**.4 (2013), pp. 4623–4637 (cit. on p. 385).

[21] Paul E. Barclay, Kartik Srinivasan, Matthew Borselli, and Oskar Painter. "Probing the dispersive and spatial properties of photonic crystal waveguides via highly efficient coupling from fiber tapers". *Applied Physics Letters* **85**.1 (2004), pp. 4–6 (cit. on p. 386).

[22] J. C. Knight, G. Cheung, F. Jacques, and T. A. Birks. "Phase-matched excitation of whispering-gallery-mode resonances by a fiber taper". *Optics Letters* **22**.15 (1997), pp. 1129–1131 (cit. on p. 386).

[23] C. P. Michael, M. Borselli, T. J. Johnson, C. Chrystal, and O. Painter. "An optical fiber-taper probe for wafer-scale microphotonic device characterization". *arXiv preprint physics/0702079* (2007) (cit. on p. 386).

[24] *Probe Tips – Signatone.* [Accessed 2014/04/14]. URL: http://www.signatone.com/products/tips_holders/ (cit. on p. 387).

[25] *S-725 Micropositioner – Signatone.* [Accessed 2014/04/14]. URL: http://www.signatone.com/products/micropositioners/s725.asp (cit. on p. 387).

[26] *Welcome to Cascade Microtech: Cascade Microtech Inc.* [Accessed 2014/04/14]. URL: http://www.cmicro.com (cit. on p. 387).

[27] *Technoprobe – Advanced Wafer Probe Cards.* [Accessed 2014/04/14]. URL: http://www.technoprobe.com (cit. on p. 387).

[28] Cascade Microtech. *Application Note: Mechanical Layout Rules for Infinity Probes.* [Accessed 2014/04/14]. URL: http://www.cmicro.com/file/mechanical-layout-rules-for-infinity-probes (cit. on p. 388).

[29] *West-Bond, Inc., Home Page – Wire Bonders, Wire Bonding.* [Accessed 2014/04/14]. URL: http://www.westbond.com/ (cit. on p. 388).

[30] *Advanced Design System (ADS) – Agilent.* [Accessed 2014/04/14]. URL: http://www.home.agilent.com/en/pc-1297113/advanced-design-system-ads (cit. on p. 388).

[31] Fu-Yi Han, Kung-Chung Lu, Tzyy-Sheng Horng, *et al.* "Packaging effects on the figure of merit of a CMOS cascode low-noise amplifier: flip-chip versus wire-bond". *Microwave Symposium Digest, 2009. MTT'09. IEEE MTT-S International.* IEEE. 2009, pp. 601–604 (cit. on p. 389).

[32] *3D-IC Design and Test Solutions – Mentor Graphics.* [Accessed 2014/04/14]. URL: http://www.mentor.com/solutions/3d-ic-design (cit. on p. 389).

[33] Richard K. Ulrich and William D. Brown. *Advanced Electronic Packaging.* Wiley-Interscience/IEEE, 2006 (cit. on p. 389).

[34] Andrea Chen and Randy Lo. *Semiconductor Packaging: Materials Interaction and Reliability.* CRC Press, 2012 (cit. on p. 389).

[35] Peter De Dobbelaere, Ali Ayazi, Yuemeng Chi, *et al.* "Packaging of silicon photonics systems". *Optical Fiber Communication Conference.* Optical Society of America. 2014, W3I–2 (cit. on p. 389).

[36] *Automated Probe Station – siepic.ubc.ca.* [Accessed 2014/04/14]. URL: http://siepic.ubc.ca/probestation (cit. on pp. 389, 393).

[37] T. Baehr-Jones, R. Ding, Y. Liu, *et al.* "Ultralow drive voltage silicon traveling-wave modulator". *Optics Express* **20**.11 (2012), pp. 12 014–12 020 (cit. on p. 400).

[38] J. J. Ackert, J. K. Doylend, D. F. Logan, *et al.* "Defect-mediated resonance shift of silicon-on-insulator racetrack resonators". *Optics Express* **19**.13 (2011), pp. 11 969–11 976. DOI: 10.1364/OE.19.011969 (cit. on p. 400).

13 Silicon photonic system example

Photonic integrated circuit made of silicon photonics can be used to construct a variety of system applications, as described in Section 1.3. In this chapter, one example of a system design is presented – for optical communication systems.

13.1 Wavelength division multiplexed transmitter[1]

Wavelength division multiplexed (WDM) systems are used extensively for long-distance optical communications, whereby very high aggregated data rates can be transmitted on individual optical fibres (Terabits/s). They have been recently considered for building ultra-high data rate optical networks and optical interconnects. Given that interconnects are considered the main bottleneck in the next-generation computing systems [2], micro-ring WDM systems are attractive for optical interconnects owing to small footprint, small capacitance and low power consumption. Much progress has been made in the past decade in the design of micro-ring-based modulators, filters, switches and lasers, etc. [2–5].

This section describes the implementation of two types of compact wavelength division multiplexed (WDM) transmitters based on micro-ring modulators and micro-ring-based add-filters (multiplexers) [1]. The demonstrated systems are (1) a 320 Gb/s eight-channel transmitter using a conventional common-bus architecture and (2) a 160 Gb/s four-channel transmitter using a "Mod-Mux" architecture. We discuss and compare the two designs and highlight their complementary merits. Both designs exhibit 32 fJ/bit modulation power efficiency and occupy less than 0.04 mm^2 chip area excluding the driving pads.

13.1.1 Ring-based WDM transmitter architectures

The commonly used ring-based WDM transmitter architecture is shown in Figure 13.1a, where a series of ring modulators share one bus waveguide, referred to as the "common-bus" architecture. This configuration does not require each ring modulator to be

[1] A version of this section has been published [1]: Yang Liu, Ran Ding, Qi Li, Zhe Xuan, Yunchu Li, Yisu Yang, Andy Eu-Jin Lim, Patrick Guo-Qiang Lo, Keren Bergman, Tom Baehr-Jones, Michael Hochberg. "Ultra-compact 320 Gb/s and 160 Gb/s WDM transmitters based on silicon microrings," *OSA Optical Fiber Communication Conference*, p. Th4G–6, 2014. Reprinted by permission.

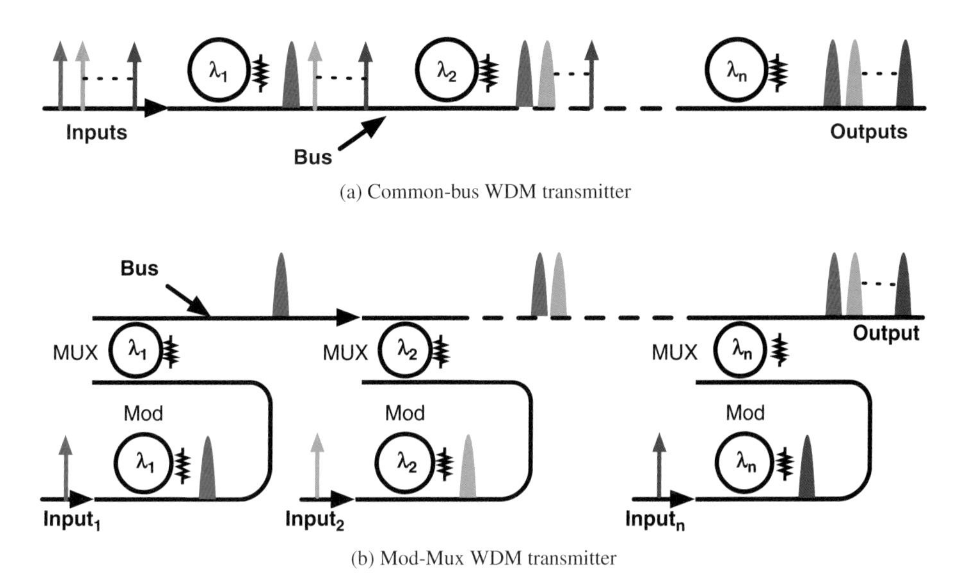

Figure 13.1 Schematic diagram two-ring modulator transmitter architectures [1].

associated with a specific wavelength in the WDM system; instead it offers the flexibility of assigning rings to the closest wavelength so as to minimize the overall tuning power [6]. However, a comb laser or pre-multiplexed laser source is required at the common input. Another issue is that cross-talk and cross-modulation may be introduced since the light in the bus waveguide passes through multiple ring modulators [7]. Finally, automated thermal stabilization is particularly challenging in the common-bus design, owing to the fact that multiple wavelengths are always present at the bus waveguide and interact with each ring modulator; yet the monitoring photo detector is naturally insensitive to wavelength.

An alternative approach, "Mod-Mux," overcomes the challenges of the common-bus design. As shown in Figure 13.1b, the laser for each channel is first fed into a ring modulator (Mod) and then the modulated light is multiplexed onto the bus waveguide by a ring add-filter (Mux). This architecture offers several key advantages: (1) avoiding cross-modulation, due to the fact each laser only passes through one ring modulator; (2) Mod-Mux enables simpler thermal stabilization schemes compared to the common-bus architecture since each Mod-Mux branch operates with only one laser wavelength; monitor detectors can be included for closed-loop control for each channel in the system [8]; (3) since the modulator and the multiplexer are separate components, they can each be optimized separately. This design partitioning allows the ring modulator to have the best tuneability with the maximum allowable quality factor, and the filter to have the optimal bandwidth with low loss and low cross-talk.

A major difference between the two approaches is in the delivery of the laser. The common-bus approach requires all the laser lines to be already multiplexed on a single waveguide [9]. This leads to an implementation with the smallest number

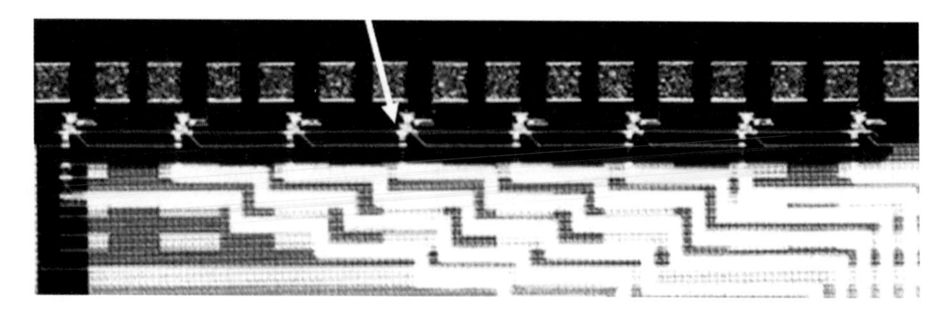

Figure 13.2 Photograph of fabricated eight-ring transmitter based on traditional common-bus design [1].

of components: one laser, and N modulators. While it places the challenge on the laser design, only a single wavelength stabilization is required for the single laser. In contrast, the Mod-Mux approach uses N lasers, which each may require wavelength stabilization. It also requires N modulators, and N optical filters. However, the laser design is more simple and the system is compatible with previous implementations of silicon photonic integrated lasers (see Chapter 8). One challenge of the approach is that the lasers, modulators, and filters need to be wavelength matched.

Both designs were fabricated via an OpSIS-IME multi-project-wafer run [10]. Grating couplers are used to couple the laser light onto and off the chip. A fibre array is attached to the silicon chip, similar to the illustration in Figure 12.5, making the optical coupling more stable during the test.

13.1.2 Common-bus WDM transmitter

The common-bus design is implemented with eight channels, as shown in Figure 13.2. The ring modulators are designed using a rib waveguide with a 500 nm width, a 90 nm slab height and a 7.5 μm radius resulting a free spectral range (FSR) of 12.8 nm. Approximately 75% of the waveguide was doped with pn-junction for high-speed modulation and the tuneability of the ring modulator is 28 pm/V near 0 V bias. The combination of low capacitance (~25 fF) and intentionally lowered quality factor Q (5000) enabled an ultra-high electro-optic bandwidth of 35 GHz. The quality factor was reduced by using a double-bus architecture. An integrated heater with 620 Ω resistance is formed by doping 15% of the ring with n-type dopants. The thermal tuneability was measured to be 150 pm/mW. The circumferences of the rings are designed to be slightly different so that the spacing between the resonance peaks of two adjacent rings is 1.6 nm, i.e. one-eighth of the FSR, in order to achieve cyclic operation with minimum tuning power.

The overall on-chip insertion loss is 12 dB, including 0.9 dB from each ring modulator, and 5 dB due to unexpected high loss from the 2 mm long routing waveguide covered by metal interconnects. In the test, we first tuned the eight rings so that their resonance peaks were evenly distributed with 1.6 nm channel spacing (Figure 13.3). The overall tuning power was 17 mW. To operate each channel, a tuneable CW laser was aligned to each of the resonance peaks. The modulated light was boosted by an EDFA,

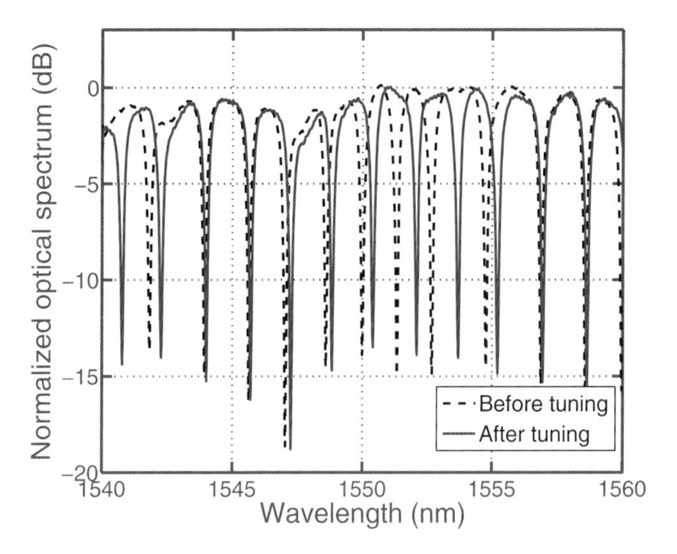

Figure 13.3 Optical spectra for the common-bus eight-ring transmitter, before and after thermal tuning [1].

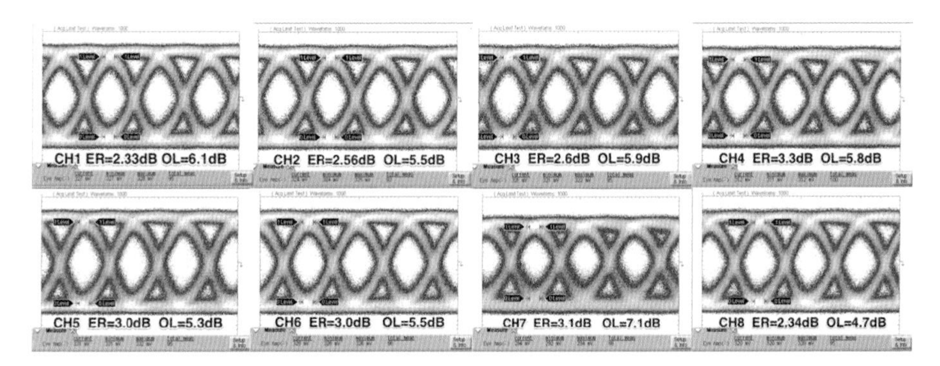

Figure 13.4 40 Gbits/s eye diagrams for the common-bus eight-channel transmitter [1]. ER: extinction ratio. OL: on-state loss.

detected by a u2t DPRV2022A receiver module and then connected to an Agilent digital communication analyzer (DCA). A Centellax TG1P4A pattern generator followed by a Centellax modulator driver amplifier, a 6 dB RF attenuator and a 40 GHz bias tee produced a 40 Gbits/s $2^{31} - 1$ PRBS signal with 2.27 V peak-to-peak centred at 0.8 V (reverse biasing the pn-junction) to drive the ring modulator. In order to avoid RF reflections, we used a 50 Ω terminated probe to contact the rings. Owing to the AC coupling feature of the receiver, the extinction ratio of the eye diagram could not be measured directly by the DCA. Instead, we measured the average optical power of the modulated light, the optical power when the laser wavelength is set to off resonance, and the amplitude of the eye diagrams. The extinction ratio of the optical eye and the on-state loss were calculated separately. Clear eye opening with ~3 dB extinction ratio and ~6 dB on-state loss were observed on all eight channels. The power consumption of each channel is estimated to be 32 fJ/bit (Figure 13.4).

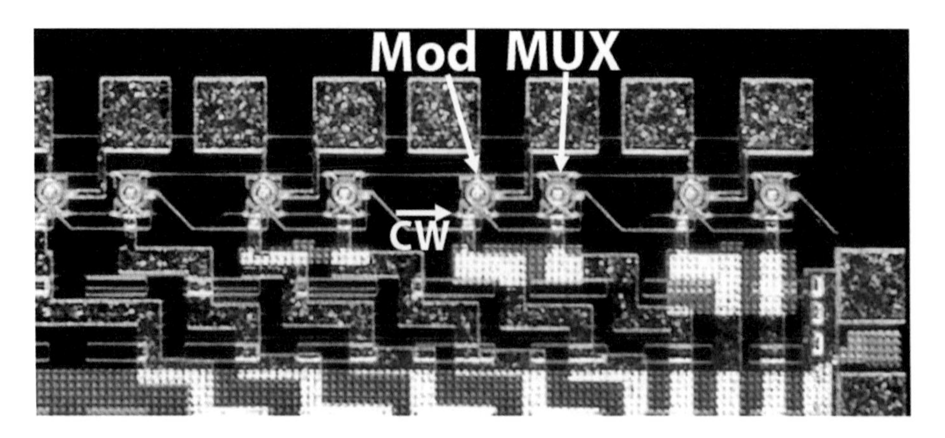

Figure 13.5 Photograph of fabricated four-ring transmitter based on Mod-Mux architecture [1].

13.1.3 Mod-Mux WDM transmitter

In the Mod-Mux design, each ring modulator is followed by a ring add-filter [11] multiplexing the ring modulator output to the bus waveguide. As shown in Figure 13.5, a four-channel prototype based on this principle was implemented. This can be scaled up to higher channel counts. The ring modulator in this transmitter is similar to those described in Section 13.1.2, where the FSR is the same, but the channel spacing is now 3.2 nm. The filters are sized to align with the modulators. Thermal tuning of the filters is achieved through a 200 Ω resistor formed by n-type doped rib waveguides (see Figure 6.18) covering 60% of the ring circumference. The thermal tuneability was on average 250 pm/mW, which corresponds to 51.2 mW/FSR (see Section 6.5.2). The add-filter achieved less than 1 dB insertion loss, a 0.8 nm (100 GHz) 3 dB optical bandwidth, and better than −20 dB cross-talk when the channels are correctly spaced.

The test setup closely resembles that in Section 13.1.2. During testing, the optical power at the bus output was monitored, when sending a tuneable CW laser into each input as illustrated in Figure 13.1b. The filters were tuned to achieve the target channel spacing; then the modulators were tuned to approximately align with the filter. The before and after tuning spectra for each channel are shown in Figure 13.6. For optimum performance, the system needs to be tuned as follows: (1) the laser wavelength should be at the peak of the optical filter to minimize loss and optical filtering of the data stream, and (2) the modulator resonance should be slightly detuned from the laser wavelength, as it is in a single-ring modulator, to generate desired extinction ratio (see Figure 6.12). This leads to the dip near the peak of the ring filter transmission spectrum, as seen in Figure 13.7.

Transmission experiments were conducted at 40 Gbits/s. The eye diagrams are shown in Figure 13.8, with the four channels operating at an estimated 32 fJ/bit modulation power efficiency. We observed ~2 dB ER and ~4 dB on-state loss with clear eye opening. We note that fine-tuning the alignment of the modulator resonance and laser wavelength was necessary in the measurement due to the drift of modulator resonance caused by RF power dissipation on the ring modulators.

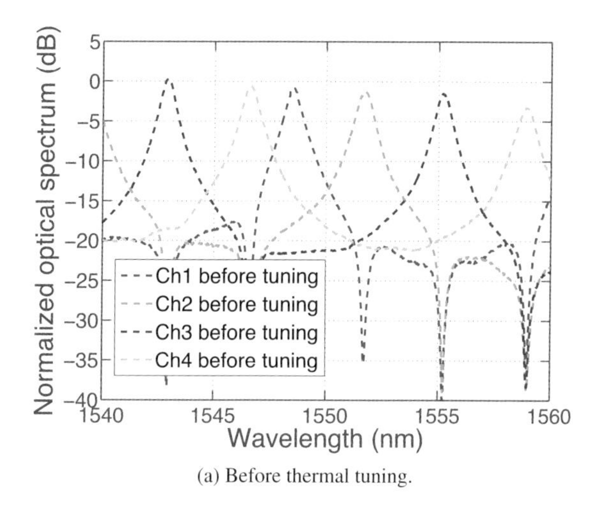

(a) Before thermal tuning.

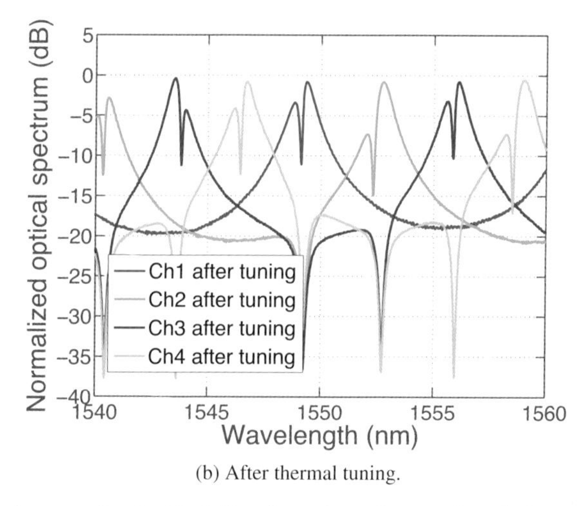

(b) After thermal tuning.

Figure 13.6 Optical spectra for the Mod-Mux four-channel transmitter, before and after thermal tuning [1].

13.1.4 Conclusion

This chapter described the implementation of high aggregate data rate WDM transmitters. Two architectures were considered, and both designs achieved 40 Gbits/s/channel data rates with 32 fJ/bit and therefore are suitable for energy-efficient high-data-rate applications. The architectures feature complementary characteristics: the conventional common-bus design allows flexible wavelength assignment, uses a minimal number of components, and can be particularly advantageous in minimizing overall thermal tuning power. The Mod-Mux architecture avoids cross-modulation entirely and does not require a comb source, therefore it is an attractive approach to achieve highly dense WDM and allow convenient integration with single-wavelength lasers.

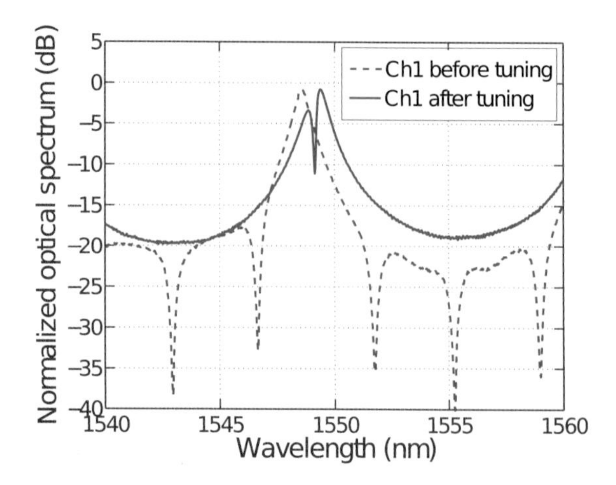

Figure 13.7 Optical spectra for one of the channels in the Mod-Mux transmitter, before and after thermal tuning [1].

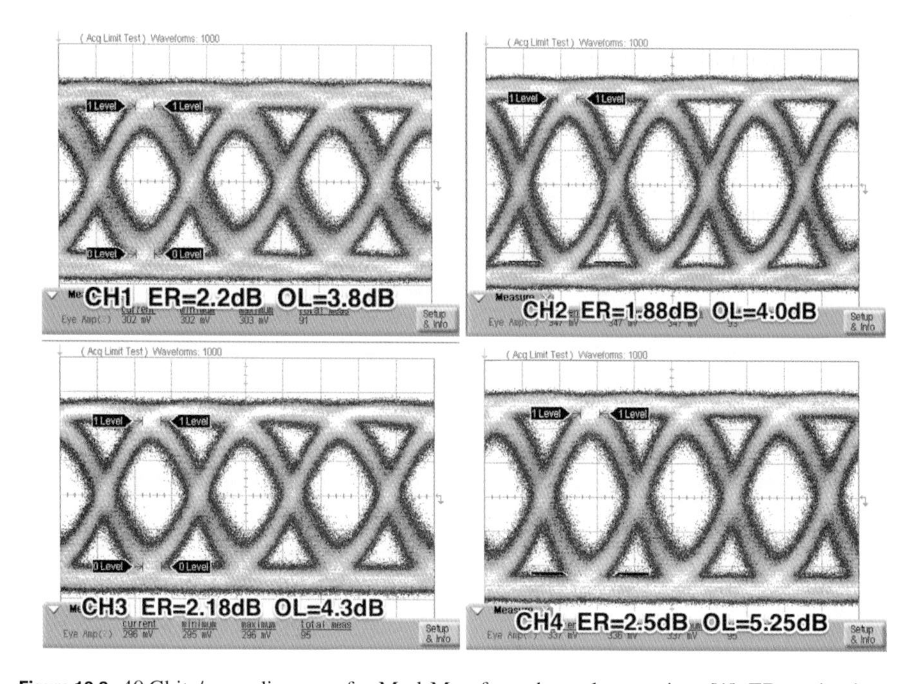

Figure 13.8 40 Gbits/s eye diagrams for Mod-Mux four-channel transmitter [1]. ER: extinction ratio. OL: on-state loss.

References

[1] Yang Liu, Ran Ding, Qi Li, *et al.* "Ultra-compact 320 Gb/s and 160 Gb/s WDM transmitters based on silicon microrings". *OSA Optical Fiber Communication Conference*. 2014, Th4G–6 (cit. on pp. 406, 407, 408, 409, 410, 411, 412).

[2] Guoliang Li, Ashok V. Krishnamoorthy, Ivan Shubin, *et al.* "Ring resonator modulators in silicon for interchip photonic links". *IEEE Journal of Selected Topics in Quantum Electronics* **19**.6 (2013), p. 3401819 (cit. on p. 406).

[3] Chao Li, Linjie Zhou, and Andrew W. Poon. "Silicon microring carrier-injection-based modulators/switches with tunable extinction ratios and OR-logic switching by using waveguide cross-coupling". *Optics Express* **15**.8 (2007), pp. 5069–5076 (cit. on p. 406).

[4] Chen Chen, Paul O. Leisher, Daniel M. Kuchta, and Kent D. Choquette. "High-speed modulation of index-guided implant-confined vertical-cavity surface-emitting lasers". *IEEE Journal of Selected Topics in Quantum Electronics* **15**.3 (2009), pp. 673–678 (cit. on p. 406).

[5] S. Akiyama, T. Kurahashi, T. Baba, *et al.* "1-V pp 10-Gb/s operation of slow-light silicon Mach-Zehnder modulator in wavelength range of 1 nm". *Group IV Photonics (GFP).* IEEE. 2010, pp. 45–47 (cit. on p. 406).

[6] Ivan Shubin, Guoliang Li, Xuezhe Zheng, *et al.* "Integration, processing and performance of low power thermally tunable CMOS-SOI WDM resonators". *Optical and Quantum Electronics* 44.12-13 (2012), pp. 589–604 (cit. on p. 407).

[7] Xuezhe Zheng, Eric Chang, Ivan Shubin, *et al.* "A 33-mW 100Gbps CMOS silicon photonic WDM transmitter using off-chip laser sources". *National Fiber Optic Engineers Conference.* Optical Society of America. 2013, PDP5C–9 (cit. on p. 407).

[8] Kishore Padmaraju, Dylan F. Logan, Xiaoliang Zhu, *et al.* "Integrated thermal stabilization of a microring modulator". *Optics Express* **21**.12 (2013), pp. 14 342–14 350 (cit. on p. 407).

[9] Andrew P. Knights, Edgar Huante-Ceron, Jason Ackert, *et al.* "Comb-laser driven WDM for short reach silicon photonic based optical interconnection". *Group IV Photonics (GFP).* IEEE. 2012, pp. 210–212 (cit. on p. 407).

[10] Tom Baehr-Jones, Ran Ding, Ali Ayazi, *et al.* "A 25 Gb/s Silicon Photonics Platform". *arXiv:1203.0767v1*(2012) (cit. on p. 408).

[11] Xuezhe Zheng, Ivan Shubin, Guoliang Li, *et al.* "A tunable 1x4 silicon CMOS photonic wavelength multiplexer/demultiplexer for dense optical interconnects". *Optics Express* **18**.5 (2010), pp. 5151–5160 (cit. on p. 410).

Index